RECENT ADVANCES IN THE UNDERSTANDING OF SOLAR FLARES

Recent Advances in the Understanding of Solar Flares

Proceedings of the U.S.-Japan Seminar held at Komaba, Tokyo, 5-8 October 1982

Edited by

S. R. KANE

University of California at Berkeley, Berkeley, California, U.S.A.

Y. UCHIDA and K. TANAKA

Tokyo Astronomical Observatory, University of Tokyo, Mitaka, Tokyo, Japan

and

H. S. HUDSON

Center for Astrophysics and Space Science, University of California at San Diego, La Jolla, California, U.S.A.

Reprinted from

Solar Physics, Vol. 86, Nos. 1 and 2

Springer-Science+Business Media, B.V.

ISBN 978-94-009-7230-8 ISBN 978-94-009-7228-5 (eBook)
DOI 10.1007/978-94-009-7228-5

Originally published by D. Reidel Publishing Company, Dordrecht, Holland in 1983
Softcover reprint of the hardcover 1st edition 1983

TABLE OF CONTENTS

(Recent Advances in the Understanding of Solar Flares)

IV. ENERGY TRANSPORT, CHROMOSPHERIC HEATING AND EVAPORATION

(Chairmen: T. Hirayama and H. Zirin)

V. EMISSION PROCESSES AND SOURCE STRUCTURE

(Chairmen: T. Takakura and M. R. Kundu)

VI. HIGH ENERGY PHOTONS, NUCLEAR PROCESSES, AND PARTICLE ACCELERATION

(Chairmen: R. Ramaty and S. R. Kane)

VII. OPEN DISCUSSION OF CONTROVERSIAL POINTS

(Chairmen: Y. Uchida and C. de Jager)

1 F. Moriyama
2 K. Nishi
3 H. Kurokawa
4 Y. Suematsu
5 T. Kosugi
6 M. Kanno
7 W. Unno
8 K. Ichimoto
9 Y. Nikaido
10 H. Ogawa
11 S. Tsuneta
12 Cao Tien-Jun
13 G. Brueckner
14 K. Shibazaki
15 M. R. Kundu
16 J. C. Brown
17 Y. Funakoshi
18 T. Takakura
19 T. Nishikawa
20 H. S. Hudson
21 T. Sakurai
22 K. Shibata
23 G. Emslie
24 J. Kubota
25 E. R. Priest
26 M. Yoshimori
27 K. Hurley
28 S. S. Degaonkar
29 R. Ramaty
30 M. Nishio
31 T. Watanabe
32 A. Duijveman
33 J. Sakai
34 G. M. Simnett
35 K. Akita
36 Z. Suemoto
37 T. Bai
38 S. Hayakawa
39 H. Zirin
40 O. Kaburaki
41 E. L. Chupp
42 C. de Jager
43 R. Pallavicini
44 A. L. Kiplinger
45 Y. Uchida
46 K. Kawabata
47 Y. Yagi
48 G. Doschek
49 S. Yanagida
50 H. Nakajima
51 K. J. Frost
52 G. A. Dulk
53 L. W. Acton
54 K. Kai
55 T. Hirayama
56 I. Suzuki
57 N. Shibuya
58 S. Enome
59 A. Gabriel
60 Z. Svestka
61 L. Orwig
62 K. Widing
63 E. Antonucci
64 R. Berton
65 E. Hiei
66 E. Tandberg Hanssen
67 J.-C. Henoux
68 S. R. Kane
69 H. Yoshimura
70 H. Washimi
71 K. Tanaka

FOREWORD

These Proceedings result from the papers and discussions of the U.S.-Japan Seminar "Recent Advances in the Understanding of Solar Flares" held in Tokyo October 5-8, 1982. The meeting was sponsored jointly by the Japan Society for the Promotion of Science and the U.S. National Science Foundation.

The principal aim of the meeting was to obtain the most up-to-date physical picture of solar flares by bringing together results from the recent observations by the Solar Maximum Mission (SMM) and HINOTORI satellites, together with other satellite data and ground-based data at both optical and radio wavelengths. These data cover the recent maximum of the solar activity cycle. The SMM and HINOTORI introduced a new dimension in flare observations by carrying out the first hard X-ray imaging observations, and the organizers especially felt that an intense discussion of the significance of these results in the context of flare theories would be important.

Starting with an introductory presentation of the characteristics of the instruments on board the satellites, the sessions of the first day and the beginning of the second dealt mainly with energy transport in flares and with the formation process of the hot plasma which is created in the corona by flares. These discussions emphasized the use of spectral line diagnostics and the method of differential emission measures compared with numerical simulations. Sessions on the second and third days dealt with the behavior of high-energy particles through observations by microwave interferometry and the imaging observations of hard X-ray sources by SMM and HINOTORI. The latter in particular led to a keen anticipation of new knowledge of when, where, and how the high-energy electrons are created and dump their energy into the flare plasma. Supplementary information on hard X-ray source locations has also come from stereoscopic viewing from two additional satellites, ISEE-3 and PVO. The session of the latter half of the third day was devoted to discussions of the processes of particle acceleration to higher energies, using the remarkable gamma-ray observations from SMM and HINOTORI.

Some of the highlights of the discussion are related to the discovery of hard X-ray sources in the high corona, which was not expected from the generally-accepted picture in which energetic electrons bombard the chromosphere; the occurrence of a very wide spectrum of highly-ionized iron ions already existing before the start of the impulsive hard X-ray emission; the discovery of iron K-alpha radiation coinciding with the hard X-ray emission; the observations by the VLA that the high-frequency source is very small and does not coincide with the low-frequency source; and so on. These and other important observations provide clues to the behavior of the accelerated electrons. It is also

Solar Physics **86** (1983) ix–x.

remarkable that separate microwave sources may accompany the hard X-ray and gamma-ray bursts that denote electron and ion acceleration, respectively.

The concluding session of the meeting was an open discussion centering on controversial loose ends in the preceding days' activities. This proved to be very useful in helping to establish the proper perspective for understanding some of the problems met in the main part of the meeting. The reproduction of this open discussion is based upon tape recordings plus heavy editing. The editors hope that this did not alter essential parts of the discussion substantially. Discussions after individual papers were transcribed from written question-and-answer forms since unfortunately no tape recordings were available. It is the editors' regret that these forms were not always scrupulously filled out, especially for some of the most interesting papers for which the discussions were very active.

Finally, the editors wish to acknowledge the Japan Society for the Promotion of Science and the U.S. National Science Foundation for their support. This was the third U.S.-Japan Seminar in Solar Physics under auspices. The editors thank Prof. M. Oda for giving the initial impetus for the Seminar, and they thank the local organizing committee for doing an outstanding job of handling all of the arrangements. We are also thankful to Mrs. H. Suzuki for her assistance in editing these Proceedings, and to Marge Currie, Space Sciences Laboratory, Berkeley, for her assistance in the organization of the Seminar.

S.R. Kane, Y. Uchida, K. Tanaka, and H.S. Hudson

LIST OF PARTICIPANTS

ACTON, L.W.	Lockheed Palo Alto Research Lab., USA
AKITA, K.	Tokyo Astronomical Obs., Univ. of Tokyo, Japan
ANTONUCCI, E.	Universita di Torino, Italy
BAI, T.	Institute for Plasma Research, Stanford Univ., USA
BERTON, R.	Tokyo Astronomical Obs., Univ. of Tokyo, Japan
BROWN, J.C.	Univ. of Glasgow, UK
BRUECKNER, G.	Naval Research Lab., USA
CAO, TIEN-JUN	Purple Mountain Obs., China
CHUPP, E.L.	Univ. of New Hampshire, USA
DEGAONKAR, S.S.	Physical Research Lab., India
DE JAGER, C.	Space Research Lab., Holland
DOSCHEK, G.	Naval Research Lab., USA
DUIJVEMAN, A.	Space Research Lab., Holland
DULK, G.A.	Univ. of Colorado, USA
EMSLIE, A.G.	Univ. of Alabama in Huntsville, USA
ENOME, S.	Research Institute of Atmospherics, Nagoya Univ., Japan
FROST, K.J.	NASA Goddard Space Flight Center, USA
FUNAKOSHI, Y.	Kwasan-Hida Obs., Univ. of Kyoto, Japan
GABRIEL, A.	Rutherford Appleton Lab., UK
HAYAKAWA, S.	Nagoya Univ., Japan
HENOUX, J.-C.	Observatoire de Paris-Meudon, France
HIEI, E.	Tokyo Astronomical Obs., Univ. of Tokyo, Japan
HIRASIMA, H.	Rikkyo Univ., Japan
HIRAYAMA, T.	Tokyo Astronomical Obs., Univ. of Tokyo, Japan
HUDSON, H.S.	Univ. of California, San Diego, USA
HURLEY, K.	Centre d'Etude Spatiale des Rayonnements, CNRS, France
ICHIMOTO, K.	Univ. of Kyoto, Japan
KABURAKI, O.	Tohoku Univ., Japan
KAI, K.	Tokyo Astronomical Obs., Univ. of Tokyo, Japan
KANE, S.R.	Space Science Lab., Univ. of California, Berkeley, USA
KANNO, M.	Kwasan-Hida Obs., Univ. of Kyoto, Japan
KAWABATA, K.	Nagoya Univ., Japan
KAWAGUCHI, I.	Kyoto Univ., Japan
KIPLINGER, A.L.	NASA Goddard Space Flight Center, USA
KONDO, I.	Institute of Cosmic Ray Research, Univ. of Tokyo, Japan
KOSUGI, T.	Tokyo Astronomical Obs., Univ. of Tokyo, Japan
KUBOTA, J.	Osaka Univ. of Economics, Japan
KUNDU, M.R.	Univ. of Maryland, USA
KUROKAWA, H.	Kwasan-Hida, Obs., Univ. of Kyoto, Japan
MAKISHIMA, K.	Institute of Space and Astronautical Science, Japan
MAKITA, M.	Tokyo Astronomical Obs., Univ. of Tokyo, Japan
MATSUOKA, M.	Institute of Space and Astronautical Science, Japan
MORIYAMA, F.	Tokyo Astronomical Obs., Univ. of Tokyo, Japan

Solar Physics **86** (1983) xi–xii.

NAKAJIMA, H.	Tokyo Astronomical Obs., Univ. of Tokyo, Japan
NIKAIDO, Y.	Univ. of Kyoto, Japan
NISHI, K.	Tokyo Astronomical Obs., Univ. of Tokyo, Japan
NITTA, N.	Univ. of Tokyo, Japan
NISHIKAWA, T.	Univ. of Kyoto, Japan
NISHIMURA, J.	Institute of Space and Astronautical Science, Japan
NISHIO, M.	Research Institute of Atmospherics, Nagoya Univ., Japan
ODA, M.	Institute of Space and Astronautical Science, Japan
OGAWA, H.	Nagoya Univ., Japan
OGAWARA, Y.	Institute of Space and Astronautical Science, Japan
OKUDAIRA, K.	Rikkyo Univ., Japan
OHASHI, T.	Institute of Space and Astronautical Science, Japan
OHKI, K.	Tokyo Astronomical Obs., Univ. of Tokyo, Japan
ORWIG, L.	NASA Goddard Space Flight Center, USA
PALLAVICINI, R.	Osservatorio Astrofisico di Arcetri, Italy
PRIEST, E.R.	Univ. of St. Andrews, UK
RAMATY, R.	NASA Goddard Space Flight Center, USA
SAKAI, J.	Toyama Univ., Japan
SAKURAI, T.	Univ. of Tokyo, Japan
SHIBASAKI, K.	Research Institute of Atmospherics, Nagoya Univ., Japan
SHIBATA, K.	Aichi Univ. of Education, Japan
SIMNETT, G.M.	Univ. of Birmingham, UK
SUEMATSU, Y.	Univ. of Kyoto, Japan
SUEMOTO, Z.	Univ. of Tokyo, Japan
SUZUKI, I.	Nagoya Univ., Japan
SVESTKA, Z.	Space Research Lab., Holland and Univ. of California, San Diego, USA
TAKAKURA, T.	Univ. of Tokyo, Japan
TANAKA, K.	Tokyo Astronomical Obs., Univ. of Tokyo, Japan
TANAKA, Y.	Institute of Space and Astronautical Science, Japan
TANDBERG-HANSSEN, E.	NASA Marshall Space Flight Center, USA
TSUNETA, S.	Univ. of Tokyo, Japan
UCHIDA, Y.	Tokyo Astronomical Obs., Univ. of Tokyo, Japan
UNNO, W.	Univ. of Tokyo, Japan
WAKI, I.	Institute of Space and Astronautical Science, Japan
WASHIMI, H.	Institute of Atmospherics, Nagoya Univ., Japan
WATANABE, T	Tokyo Astronomical Obs., Univ. of Tokyo, Japan
WIDING, K.	Naval Research Lab., USA
YAGI, Y.	Institute of Space and Astronautical Science, Japan
YANAGIDA, S.	Ibaraki Univ., Japan
YOSHIMORI, M.	Rikkyo Univ., Japan
YOSHIMURA, H.	Univ. of Tokyo, Japan
ZIRIN, H.	California Institute of Technology, USA

I. SPACECRAFT WITH MAJOR SOLAR INSTRUMENTS

Chairman: M. Oda

INTRODUCTION TO HINOTORI

Y. Tanaka
Institute of Space and Astronautical Science,
Meguro-ku, Tokyo 153, Japan

ABSTRACT. The HINOTORI mission is dedicated to solar flare observations mainly in the X-ray energy band. The instruments on board conduct (1) hard X-ray imaging of solar flares by means of rotating modulation collimators, (2) wide energy band spectroscopy in the range from 2 keV to several MeV as well as (3) high resolution crystal spectroscopy of iron emission lines. Several hundred flares have been recorded and analyses are under way. A description of HINOTORI and the characteristics of the instruments are presented.

This paper presents an outline of the HINOTORI mission. HINOTORI is a satellite dedicated to solar flare observations during the solar maximum period. This mission is unique in the sense that virtually all solar physicists in Japan participated in various phases to a certain degree, from the mission definition through the post-launch data analysis.

In the endeavour to maximize the scientific return from this spacecraft of limited capabilities, we decided to concentrate the observations into the X-ray energy band, and to take the design philosophy of studying the solar flares on three axes; (1) the space axis, (2) the energy axis and (3) the time axis.

For the space axis, we attempt to image solar flares in the hard X-ray band to reveal the site of primary energy release. This is implemented by the rotating modulation collimator (RMX) technique. As for the energy axis, energy spectra of solar flares with coarse-to-moderate energy resolution over a very wide range, ranging from 2 keV to several MeV, are obtained with three different detectors. A high resolution Bragg crystal spectrometer is also included providing a powerful tool for the diagnostics of high-temperature plasmas produced in solar flares. All of these spectra are taken with a time resolution of a few seconds. For the high time resolution studies, data in bins ranging from 100 m sec to less than 10 m sec are available for several broad energy bands.

HINOTORI is a 188 kg satellite which is spin-stabilized at about 5 rpm. It was launched in February 1981. Detailed information

Solar Physics **86** (1983) 3–6. 0038–0938/83/0861–0003 $ 00.60.

on the spacecraft, its instruments, and the data hundling, have been given in the HINOTORI Symposium Proceedings (ISAS, 1982). Table I lists key parameters of the scientific instruments for solar flare observations on board HINOTORI. All solar experiments are co-aligned with the spacecraft spin (Z) axis and their fields of view cover the entire solar disk. The Z=axis is pointed about 1° off the center of the solar disk. This offset pointing enables us to scan the required wavelength band for the Bragg spectrometer (SOX) and also to scan the flare region with the rotating modulation collimators (SXT) by using the spacecraft rotation instead of employing a driving mechanism.

Since only one tracking station (Kagoshima Space Center) is available, the real-time coverage is quite limited. HINOTORI, therefore had to rely on an on-board data recorder to store the flare observation data. We therefore developed an autonomous flare detection and storage logic. Whenever a sharp rise of the counting rates in X-ray counters (HXM, FLM) is detected, the logic checks whether the event is accompanied by a similar counting rate increase in the particle detector (PXM), as in the case of crossings of the South Atlantic geomagnetic anomaly. If not, the logic sets a flare flag which then switches the data format from the low bit-rate (1 kbps) Quiet Mode to the high bit-rate (8 kbps) Flare Mode and at the same time changes the data recorder speed accordingly. Flare data lasting about 20 minutes are stored until the play-back occurs on command. In addition, an on-board external memory keeps on storing high-bit-rate data for the previous $\sim$ 30 seconds at any instance. Therefore, once a flare flag is set, these pre-flag data are transferred to the data recorder so that the flare-rise data are saved. Another trick in the flare detection logic is that we can choose two optional threshold levels (small flare and big flare) which are substantially different. Big flares have the priority for storage; even if the data from a small flare were already stored, the data from a big flare, whenever it occurs, overwrite the previous data.

This autonomous flare detection logic has been proven to be successful. In 16 months of the HINOTORI operation, we had recorded 675 flares as of July 1982, in which 31 X-class events are included. This score shows that about 60% of the X-class flares were recorded, which corresponds to nearly 100% detection for the spacecraft day time. The largest flare that HINOTORI recorded is an X-12 flare occurred on June 6, 1982.

Now, I shall briefly explain the individual experiments on board. The soft X-ray monitor (FLM), the hard X-ray monitor (HXM) and the solar gamma ray detector (SGR) altogether cover a wide energy range from 2 keV to $\sim$ 7 MeV. The FLM is a gas scintillation proportional counter which is a new type of X-ray detector, having twice the energy resolution of conventional proportional counters. It is the first time that this new type of counter has been flown on board a satellite. HXM and SGR are NaI(Tl) and CsI(Tl) scintillation counters, respectively. The FLM spectrum of a flare typically shows distinct peaks corresponding to emission lines from various elements. It is worth mentioning that the three different detectors have overlapping energy ranges to permit

TABLE I. Science Instruments of 'HINOTORI' for Solar Flare Observations

	DETECTOR	ENERGY RANGE	RESOLUTION
SXT	113 cm^2 NaI (x2)	17 - 40 keV (10 - 27 keV before Aug. 1)	∿ 6 sec 30 arc sec (FWHM)
SXA	Fine Solar Aspect Sensor (RMC)		5 arc sec
HXM	57 cm^2 NaI	17 - 40 keV 40 - 340 keV	7.8 msec (HXM-1) 125 msec (HXM-2-7)
SGR	62 cm^2 CsI	0.21 - 6.7 MeV (17.5 keV x 32, 41.8 x 48, 81.3 x 48)	128 ch/2 sec.
FLM	0.5 cm^2 Xe SPC	2 - 12 keV Counts in L/H bands	128 ch/4 sec. 125 msec.
SOX	SiO_2 NaI Sci. SiO_2 NaI Sci.	1.72 - 1.99 Å 1.83 - 1.89 Å	.00068 Å .000042 Å
PXM	2.2 cm^2 str	100 - 800 keV particles (Radiation Veto Counter)	125 msec/ch.

intercomparisons and to verify the consistency of their data. An example of the comparison between HXM and SGR appears in the paper by Nitta et al. in these proceedings.

The soft X-ray crystal spectrometer (SOX) includes coarse and fine spectrometers to measure the individual emission lines of iron. The coarse spectrometer covers a wide wavelength range which includes lines from FeXIV through FeXXVI, Kα and Kβ. The fine spectrometer performs a high-dispersion analysis of the lines up to FeXXV. Tanaka et al. (1982) have presented some initial observations from this unique instrument.

One of the unique features of HINOTORI is the hard X-ray flare imaging (SXT) by means of the rotating modulation collimator technique. The energy range is 17-40 keV. The instrument consists of two sets of a bigrid modulation collimator. As the spacecraft rotates around the spin axis, the modulation collimator scans the flare from continuously changing directions. One can reconstruct a flare image from a set of modulation patterns of the counting rate obtained for 180° rotation (∿ 6 sec.). Accurate position determination for a flare is aided by a fine solar aspect sensor (SXA) which is good to 5 arcsec or better. The X-ray image resolution is determined by the sharpness of the triangular response of the modulation collimator transmission fruction. This function has been determined in-orbit and is triangular to an accuracy of ∿ 7 arc sec.

In concluding this paper, I should like to mention that a government level agreement between NASA and the Space Activities Commission of Japan has been made for the collaborative solar studies through the SMM and HINOTORI. As a matter of fact, the present Japan/US Seminar is fully backed up by the spirit of this bilateral collaboration program. With the unprecedented richness of the observational data produced by these successful missions, we should be able to make a big step forward for the understanding of the solar flare phenomenon.

References

ISAS: 1982, Proceedings of the HINOTORI Symposium (ISAS, Tokyo).

Tanaka, K., Watanabe, T., Nishi, K., and Akita, K.: 1982, Astrophys. J. (Lett.) 254, L59.

THE SOLAR MAXIMUM MISSION

K. J. Frost
NASA Goddard Space Flight Center, Greenbelt, Maryland, U.S.A.

EDITORS' NOTE. A full description of the Solar Maximum Mission appears in Solar Physics, Vol. 65, pp. 5-109. Individual papers in this issue also give in detail the parameters of the individual experiments carried by the spacecraft. A paper describing the Solar Maximum Mission by Frost will therefore not appear in these proceedings.

Solar Physics **86** (1983) 7. 0038–0938/83/0861–0007 $ 00.15.

SOLAR INSTRUMENTS ON THE P78-1 SPACECRAFT

G. A. Doschek
E. O. Hulburt Center for Space Research, Naval Research Laboratory, Washington, DC 20375, U.S.A.

ABSTRACT. The solar experiment package on the P78-1 spacecraft is described. The satellite was launched on 24 February 1979 by the U.S. Air Force and contains high resolution Bragg crystal spectrometers, hard X-ray proportional counters, and a white light coronagraph. The high resolution spectrometers were built by the Aerospace Corporation and the Naval Research Laboratory. The hard X-ray spectrometers were built by the Aerospace Corporation and the white light coronagraph was supplied by the Naval Research Laboratory. Most of these instruments are still returning data.

1. INTRODUCTION

The P78-1 spacecraft was launched by the U.S. Air Force on 24 February 1979, as part of the U.S. Department of Defense Space Test Program (STP). The satellite itself was built with assistance from NASA, which provided the stabilization and solar pointing control systems and other flight spare components from NASA's Orbiting Solar Observatory, OSO-7. The spacecraft lies in a nearby circular and polar orbit about 500 km above the Earth. The orbital plane is inclined 97° to the equator and precesses such that the satellite remains in an approximately noon-midnight meridian as the Earth circles the Sun. The orbital period is 97 minutes. The spacecraft has a solar pointed section that houses X-ray crystal spectrometers built by the Naval Research Laboratory (NRL) and the Aerospace Corporation, a white light coronagraph built by NRL, and hard X-ray proportional counters built by the Aerospace Corp. Most of these instruments have operated from March 1979 up to the present time.

In this paper summaries of the solar experiment package and instrumental characteristics are given, and some of the most important results are briefly described. A more detailed summary of the results from the X-ray crystal spectrometers is given in a paper by Doschek in this volume.

Solar Physics **86** (1983) 9–17. 0038–0938/83/0861–0009 $ 01.35.

2. INSTRUMENTATION

A summary of the P78-1 solar instrumentation is given in Table I. The individual experiments are discussed in more detail below.

TABLE I

Solar Experiments on the P78-1 Spacecraft

Experiment	Organization and Principal Investigator	Spectral Range	Spectral Resolution	Spatial Resolution
SOLFLEX	Naval Research Laboratory (R.W. Kreplin, P.I.)	1.82 - 1.97 Å	2.4×10^{-4} Å	full Sun
		2.98 - 3.07 Å	4.1×10^{-4} Å	"
		3.14 - 3.24 Å	4.1×10^{-4} Å	"
		8.26 - 8.53 Å	1.0×10^{-3} Å	"
SOLEX	Aerospace Corporation (P.B. Landecker, P.I.)	3 - 10 Å	10^{-3} Å at 8.2 Å	60" or 20"
		7 - 25 Å	2×10^{-2} Å at 16 Å	"
MONEX	Aerospace Corporation (P.B. Landecker, P.I.)	1 - 18 keV (LEM)	six channels	full Sun
		18 - 230 keV (HEM)	six channels	"
SOLWIND	Naval Research Laboratory (D.J. Michels, P.I.)	4000 - 7000 Å	broadband	1.25'
MAGMAP	Naval Research Laboratory (R.W. Kreplin, P.I.)	8 - 9 Å	broadband	60"
		8.5 - 9.5 Å	"	"

Four uncollimated crystal spectrometers were designed by NRL to record the X-ray spectra of flares below 8 Å over narrow selected wavelength ranges. The acronym for this experiment is SOLFLEX (= solar flare X-rays). The choices of crystals, wavelength ranges, and time resolution were based on previous analyses of X-ray spectra obtained by NRL and Goddard Space Flight Center spectrometers flown on OSO spacecraft and spectrometers flown by Mandelstam's group at the Lebedev Physical Institute on Intercosmos spacecraft. The final parameters chosen resulted in very high resolution spectrometers. They enable the diagnostically important lines to be clearly identified and resolved from nearby lines, and they enable dynamics of flares to be studied from line profiles and relative wavelength shifts.

The wavelength ranges of the four spectrometers are: 1.82-1.97 Å, 2.98-3.07 Å, 3.14-3.24 Å, and 8.26-8.53 Å. These wavelength ranges were chosen to cover, respectively, the Fe XXV lines and innershell lines of lower iron ionization stages including Kα Fe II; the Ly α Ca XX lines and associated satellites; the Ca XIX lines and innershell lines of Ca XVIII and Ca XVII; and finally, the Mg XII Lyα lines and $2\ell - 4\ell'$ transitions in Fe XXIII and Fe XXIV.

In order to resolve the lines and measure line profiles, Ge crystals (2d = 4.00 Å) were chosen for the three short wavelength spectrometers and an ADP crystal (2d = 10.64 Å) was used for the 8 Å instrument. Different wavelengths are scanned by stepping the crystals through Bragg angles of about 20" using a combination stepper motor and harmonic drive. The crystals are mounted on a common shaft, offset from each other in angle in order to cover their respective wavelength ranges, and are continuously driven back and forth over these ranges by the harmonic drive assembly. The wavelength ranges consist of 450 steps (1 step = 20"), and for most of the spectra obtained the scan rate was 8 steps s^{-1}, or equivalently, it takes 56s to generate a spectrum.

The diffracted X-rays are detected by Ar and Xe filled proportional counters with a CO_2 quench. Because the wavelength ranges are so narrow, it is not necessary to move the detectors during a scan. Additional details concerning the instruments are given in Feldman, Doschek, and Kreplin (1980).

The Aerospace Corporation solar X-ray experiment is a combination spectrometer and spectroheliometer. Because it is not restricted to flares, the acronym is SOLEX (= solar X-rays). The spectral region 3 - 25 Å covered by the SOLEX spectrometers contains strong spectral lines of O, Ne, Mg, Aℓ, Si, S, Ar, and Fe. The strongest of these lines are the 1s-2p transitions in H-like and He-like ions and 2p-3d transitions in iron ions.

The SOLEX A channel consists of a 20" FWHM multigrid collimator, a flat RAP (2d = 26.12 Å) Bragg crystal or an ADP (2d = 10.64 Å) crystal, and a channel electron multiplier array detector (CEMA). The wavelength range of the RAP crystal is 7.8 to 25 Å; the range for the ADP crystal is 3 to 10 Å. The SOLEX B channel consists of a 60" FWHM multigrid collimator, a flat RAP or ADP crystal, and a CEMA detector. The crystals for each channel are mounted back-to-back on a single shaft that can be rotated 180°. In this manner either crystal can be positioned behind the collimator. The crystal drive stepper motor rotates the shaft in 30.2" steps at a rate of either 31.25 or 62.5 times a second.

When SOLEX is operated as a spectrometer, it is pointed at the solar target and the crystals are programmed to scan back and forth between wavelengths specified by ground command. When the instrument is operated as a spectroheliometer, the crystals are set at wavelengths selected by ground command and the satellite pointed section generates either large raster (45' x 45') patterns that cover the entire solar disk or small raster (5' x 5') patterns that cover selected active regions. Large and small rasters are generated in 491.52 s and 61.44 s, respectively. Additional details concerning the SOLEX spectrometers are given by Landecker, McKenzie, and Rugge (1979).

Examples of flare spectra recorded by the NRL and Aerospace Corp. spectrometers are shown in Figures 1 and 2. The Ca lines are due to 2p→1s type transitions, primarily in H-like, He-like, and Li-like ions. The profiles of the lines in Figure 1 are intrinsic, and note the fine structure splitting of the Ca XX Lyα lines. (An example of 2p→ 1s Fe line spectra is given in the other paper by Doschek in this volume.) The iron lines between 8 and 17 Å (Figure 2) are due mainly to 3d→2p and 3s → 2p transitions. Weaker emission due to 4d→2p and 4s→2p transitions is also present. Each ionization stage tends to produce a number of lines bunched closely together in wavelength. Some of the strongest features produced by the ion stages from Fe XVII-Fe XXIV are marked. In addition to Fe lines, the 8-23 Å region contains the He-like O VII lines (2p→1s), the O VIII Lyα lines, and similar transitions for Ne, Na, and Mg.

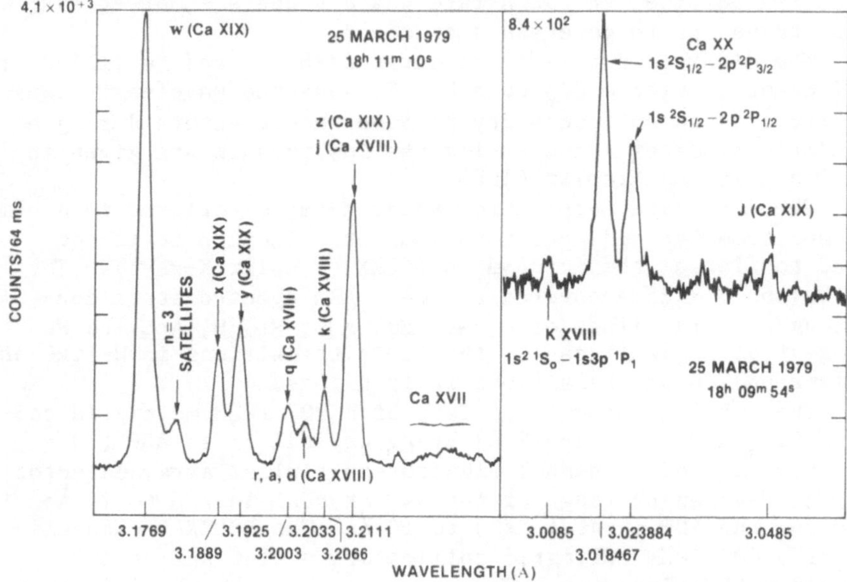

Figure 1 - Ca spectra recorded by two NRL SOLFLEX spectrometers on P78-1. The profiles are intrinsic and reflect turbulent motions in the flare plasma.

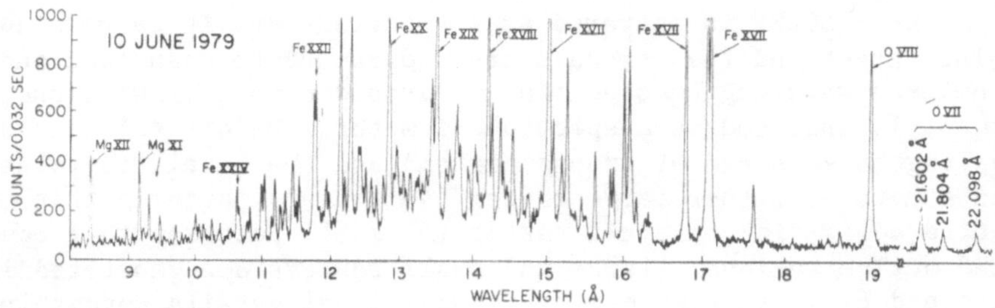

Figure 2 - Spectra beteen 8 and 23 Å recorded by an Aerospace Corp. SOLEX spectrometer on P78-1. Transitions in Fe ions produce most of the lines in this spectral region.

Many results from these crystal spectrometer experiments have already appeared in the literature. Electron temperatures of 20-25 x 10^6 K were measured in many flares, electron densities > 10^{12} cm^{-3} were found for the flare plasma near 2 x 10^6 K, and turbulent velocities up to 300 km s^{-1} were measured during the rise phase of flares. Upflowing hot plasma was also detected during the rise phase (see Doschek, this volume, for more details, and papers in the bibliography below).

In addition to high resolution spectrometers, the Aerospace Corp. provided uncollimated proportional counters which record low resolution spectra from flares in both hard and soft X-rays. The experiment is called MONEX (= X-ray monitor). There are two proportional counters: The LEM Low Energy Monitor MONEX A module operates in the 1-18 keV range, which is divided into six channels. Data are recorded every 1.024 s and the six channel sum is recorded at the faster rate of once every 0.032 s. The HEM High Energy Monitor MONEX B module operates in the 18-230 keV range, which is also divided into six channels. The time resolution is the same as for MONEX A. The six channel sum is also recorded every 0.032 s. More details concerning MONEX are given in Kane, Landecker and McKenzie (1982). A sample of the HEM data is shown in Figure 3.

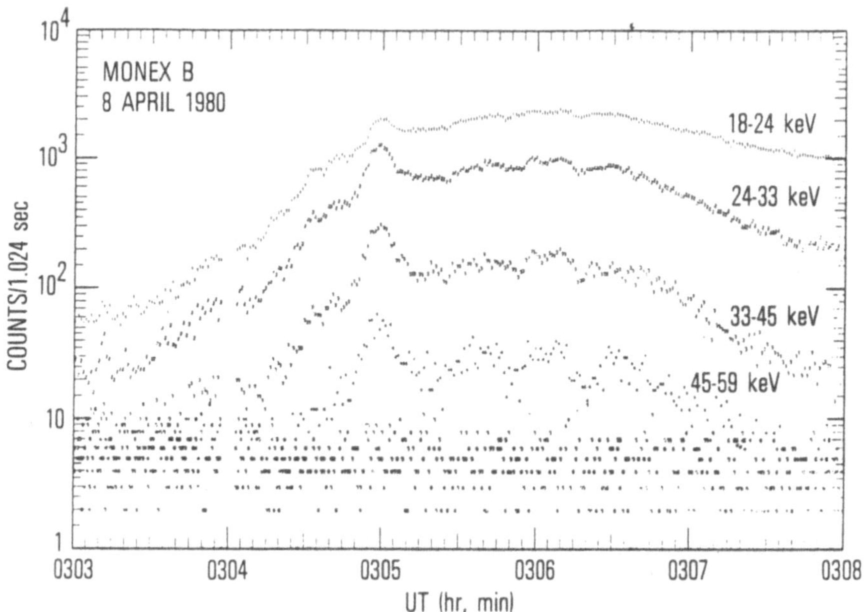

Figure 3 - A hard X-ray flare burst recorded by the Aerospace Corp. MONEX HEM spectrometer on P78-1. The time resolution is 1.024 s.

The P78-1 spacecraft also has a modified Lyot white light coronagraph, provided by NRL. The experiment is called SOLWIND (= solar wind). The Sun is occulted by three circular disks mounted in tandem, and supported in front of the instrument. The field of view is 2.6 10.0 solar radii and the spatial resolution is 1.25'. The spectral passband extends nominally from 4000 to 7000 Å. The 50% limits for the transmission function occur at 4100 and 6350 Å, but the transmission is still 42% at the Hα wavelength (6563 Å). Polarizers are provided for circular regions of the image near 5 and 8 solar radii. Images are focussed, filtered, and are recorded by an SEC Vidicon. The instrument is optically similar to the OSO-7 instrument described by Koomen et al. (1975). When the spacecraft is providing solar observations, full-field coronal images are obtained at 10 minute intervals (and sometimes at 5 minute intervals) during the approximately 1 hour sunlit portion of each orbit.

An image of a large coronal transient is shown in Figure 4. The solar disk is dubbed into the figure. In order to enhance the contrast between the background, slowly varying corona, and rapidly varying transient activity, a pre-transient image of the corona is subtracted from the images showing the coronal transients. The background image used is the image recorded as close in time as possible to the transient event. The subtraction process results in a uniformly gray background in regions of the corona where no brightness changes have occurred in the time interval between the background and transient observations. The region of the image occupied by the transient appears bright and the contrast between the transient image and the flat background can be made quite high. In a few areas of the image, the coronal brightness may have decreased after the background observation, which results in very dark regions in some of the images.

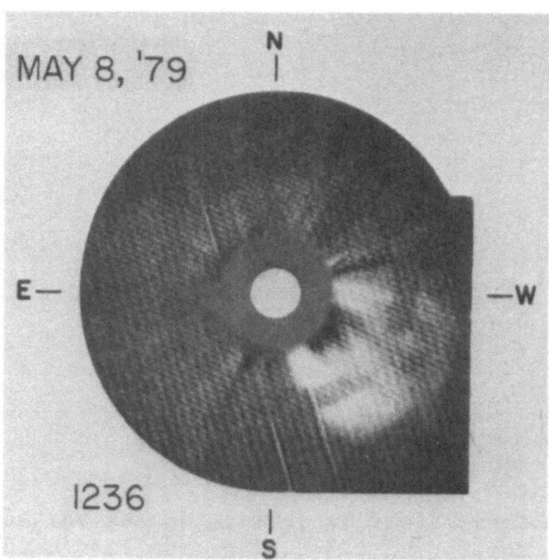

Figure 4 - A large coronal transient in progress, recorded by the NRL SOLWIND white light coronagraph on P78-1. Note the presence of the polarizing ring which dims the image within an annular arc approximately 5 solar radii from the Sun.

The transient image in Figure 4 appears to have either a loop-like or arch shape, or could instead be a three dimensional shell-like structure, with emission also occurring inside the shell. The walls of the shell would be quite thick, since no obvious "limb brightening" can be seen. The annular dark region extending across the transient image is caused by the inner polarizing ring, and demonstrates that the temperature of the transient plasma is on the order of coronal temperatures ($\simeq 10^6$ K).

A number of interesting results have emerged from the coronagraph experiment so far. These are (see additional references):

(a) About 1-2 transients per day have occurred since launch of P78-1; typical ejection speeds and the average mass are 150-900 km s^{-1} and 8×10^{14} g, respectively.

(b) The locations of events occur over a wide range of latitudes. One event occurred within 10° of the north pole.

(c) Several very large events have been observed where both coronal gas ($\simeq 10^6$ K) and cold $H\alpha$ emitting gas ($\approx 10^4$ K) were ejected. In one case (24 May 1979) the cold gas could be seen out to the edge of the coronagraph field of view (10 solar radii).

(d) One transient appeared as an expanding halo-type structure around the Sun, thus indicating that it was directed toward the Earth. The shape of this transient shows that at least some transients are three dimensional structures, rather than loops.

(e) Forty to fifty coronal transients have been found to be associated with interplanetary shocks detected by the Helios spacecraft. All interplanetary shocks observed by Helios when SOLWIND was operating are associated with coronal transients.

(f) Three comets have been discovered in the coronagraph images. These comets are apparently members of the Kruetz group of sungrazing comets, and unlike all other comets known to date, these comets either impacted the Sun, or disintegrated in the solar atmosphere, or exhausted their supply of volatiles during perhelion passage. The dates of perihelion are: 30 August 1979, 27 January 1981, and 20 July 1981.

Finally, the MAGMAP (= Magnesium Mapping) experiment listed in Table 1 consists of two proportional counters that view the Sun through the SOLEX 60" collimator. Each proportional counter is filtered to be primarily sensitive to either Mg XII ($\simeq 8.4$ Å) or Mg XI ($\simeq 9.2$ Å) radiation. It operates in the large raster mode and indicates which active regions contain the hottest pre-flare plasma. Part of this work was supported by a grant from NASA.

REFERENCES

Feldman, U., Doschek, G.A., and Kreplin, R.W.: 1980, Astrophys. J. 238, 365.

Kane, S.R., Landecker, P.B., and McKenzie, D.L.: 1982, Aerospace Corp. Space Sciences Laboratory Report No. SSL-82(2940-01)-5.

Koomen, M.J., Detwiler, C.R., Brueckner, G.E., Cooper, H.W., and Tousey, R.: 1975, Appl. Opt. 14, 743.

Landecker, P.B., McKenzie, D.L., and Rugge, H.R.: 1979, Proc. S.P.I.E. 184, 285.

Additional References

G.A Doschek, R.W. Kreplin, and U. Feldman - "High Resolution Solar Flare X-ray Spectra", Astrophys. J. Letters, 233, L157 (1979).

D.L. McKenzie, P.B. Landecker, R.M. Broussard, H.R. Rugge, R.M. Young, U. Feldman, and G.A. Doschek - "Solar Flare X-ray Spectra Between 7.8 and 23 Å, Astrophys. J., 241, 409 (1980).

G.A. Doschek, U. Feldman, R.W. Kreplin, and L. Cohen - "High Resolution Xray Spectra of Solar Flares III. General Properties of X1 - X5 Type Flares", Astrophys. J., 239, 725 (1980).

U. Feldman, G.A. Doschek, R.W. Kreplin, and J.T. Mariska - "High Resolution X-ray Spectra of Solar Flares IV. General Spectral Properties of M-Type Flares", Astrophys. J., 241, 1175 (1980).

D.L. McKenzie and P.B. Landecker - "Analysis of a Series of Solar Flare X-ray Spectra", Astrophys. J., 248, 1117 (1981).

G.A. Doschek, U. Feldman, P.B. Landecker, and D.L. McKenzie - "High Resolution Solar Flare X-ray Spectra: The Temporal Behavior of Electron Density, Temperature, and Emission Measure for Two Class M Flares", Astrophys. J., 249, 372 (1981).

D.L. McKenzie and P.B. Landecker - "The Solar Coronal X-ray Spectrum from 15.4 to 23.0 Angstroms: Lines from Highly Ionized Calcium and Chromium and Their Usefulness as Plasma Diagnostics", Astrophys. J., 254, 309 (1982).

D.L. McKenzie and P.B. Landecker - "X-ray Lines of Helium-Like Oxygen and Neon in the Solar Corona", Astrophys. J., 259, 372 (1982).

S.R. Kane, K. Kai, T. Kosugi, S. Enome, P.B. Landecker, and D.L. McKenzie "Acceleration and Confinement of Energetic Particles in the 7 June 1980 Solar Flare", Astrophys. J., in press (1983).

N.R. Sheeley, Jr., D.J. Michels, R.A. Howard, and M.J. Koomen - "Initial Observations with the Solwind Coronagraph", Astrophys. J. Letters, 237, L99 (1980).

A.I. Poland, R.A. Howard, M.J. Koomen, D.J. Michels, and N.R. Sheeley, Jr. "Coronal Transients Near Sunspot Maximum", Solar Physics, 69, 169 (1981).

D.J. Michels, N.R. Sheeley, Jr., R.A. Howard, and M.J. Koomen - "Observations of a Comet on Collision Course with the Sun", Science, 215, 1097 (1982).

N.R. Sheeley, Jr., R.A. Howard, M.J. Koomen, D.J. Michels, K.L. Harvey, and J.W. Harvey - "Observations of Coronal Structure During Sunspot Maximum", Spa. Sci. Rev., in press (1982).

R.A. Howard, D.J. Michels, N.R. Sheeley, Jr., and M.J. Koomen - "The Observation of a Coronal Transient Directed at Earth", Astrophys. J. Letters, 263, L101 (1982).

N.R. Sheeley, Jr., R.A. Howard, M.J. Koomen, and D.J. Michels - "Coronagraphic Observations of Two New Sungrazing Comets", Nature, 300, 239 (1982).

DISCUSSION

BROWN: Has there been any improvement in the original estimate of the mass of the first comet discovered by P78-1, which I believe was about 10^{15}gm?

DOSCHEK: Not as far as I am aware.

EMSLIE: You mentioned that the emission from the coronal transients appears to be diffuse and extended. Does this favor a model in which the transient corresponds to an actual outward motion of flare material, as opposed to the normal picture that the coronal material is compressed ahead of an outward moving shock wave?

DOSCHEK: Many transients do not result from flares, so I assume that by 'flare material' you also mean prominence ejecta. I am not an expert in this field, but I think that actual outward motion of either flare plasma or heated prominence material must be involved in order to account for the amount and distribution of mass observed in these 'filled in' bubble type objects.

KANNO: What is the date of your comet observations?

DOSCHEK: The comet observations were made by Michels, Howard, Koomen, and Sheeley at NRL. The first comet discovered collided with the Sun on August 30, 1979. The other two comets came to their demise on January 27, 1981 and July 20, 1981.

KANE: I understand that the comet impacted on the front side of the Sun. Did you see any soft X-ray or other enhancements associated with the impact ?

DOSCHEK: To my knowledge, no other enhancements, in X-rays or in other radiation, have yet been found.

II. GENERAL CHARACTERISTICS OF FLARES

Chairman: Z. Suemoto

DEVELOPMENT OF FLARE MORPHOLOGY IN X-RAYS, AND THE FLARE SCENARIO

C. de Jager
Kyoto University, Japan*

ABSTRACT

We define the *impulsive phase* of a flare as its first phase, characterized by: X-ray bursts of short (seconds to tens of seconds) duration, a patchy X-ray morphology, and injection of energy. It lasts some five to ten minutes. The *gradual* or *diffuse* phase starts virtually at the same time as the impulsive one and is characterized by a gradually varying X-ray flux from a larger, diffuse, area situated higher than the sources of the impulsive X-ray bursts. The diffuse cloud is initially (during the first five minutes) hotter by a few million degrees than the sources of the impulsive phase bursts and is assumed to be caused by convective motions with upward velocities of a few hundred km s^{-1}. It contains about the same number of energetic electrons as the impulsive burst patches contained initially. It cools gradually down by radiative and conductive losses, a process that may last for about an hour.

1. INTRODUCTION

In this paper we describe the development of the X-ray emitting component of solar flares, primarily on the basis of images in the 3.5 - 30 keV range obtained by the Hard X-ray Imaging Spectrometer (HXIS; Van Beek *et al.*, 1980) aboard NASA's Solar Maximum Mission. On the basis of two well-studied cases of average flares we will show that it makes sense to redefine the notion 'impulsive phase of a flare' and to describe it as a period of about 5 to 10 minutes duration during which: (a) energy is fed into the flare; (b) the X-ray emission is localized in a few discrete areas, while (c) the radiation is emitted in a number of bursts of appr. 10^1 to 10^2 s duration. The energy of the impulsive bursts photons is not primarily relevant for this definition and it is

---

* Guest Scholar at the Faculty of Sciences, Kyoto University; on leave from the Astronomical Institute at Utrecht.

Solar Physics **86** (1983) 21–32. 0038–0938/83/0861–0021 $ 01.80.

not necessarily high. The 'gradual' or 'diffuse phase' is caused by hot gas that moves upward from the heated impulsive burst areas, and that thus forms an extended diffuse cloud which increases slowly in size. Projected *on* the disk it coincides mainly in position with the impulsive phase patches but it is larger and amorphous. When seen at the *limb* it appears to be localized higher than the impulsive phase kernel and it appears as a 'convective plume'. The gradual phase thus starts practically simultaneously with the impulsive phase but it lasts much longer because after the period of energy injection the convective cloud still exists, and it normally even expands further. At the same time it will loose energy by radiation and conduction and eventually (after some 10 to 100 minutes, depending on the case) it will have faded away.

2. A FLARE PROJECTED ON THE DISK

Figure 1 (Boelee and De Jager, 1983) shows the flare of 12 November, 1980, 02:50 UT in the second energy channel (5.5 - 8.0 keV) of HXIS at two times during its development: at 02:50 UT near the end of the impulsive phase, and seven minutes later, in the gradual phase. It illustrates that the X-ray flare is patchy and discrete in the impulsive phase but amorphous or diffuse in the gradual phase. This is a feature common to many of the flares observed by HXIS (cf. Duijveman *et al.*, 1982).

As a next step we want to show the difference between the X-ray light curves for the areas representative for the impulsive phase emission and those representative for the emission in the gradual (diffuse) phase. We then encounter the difficulty that these two areas overlap to a large extent because we see the flare projected on the

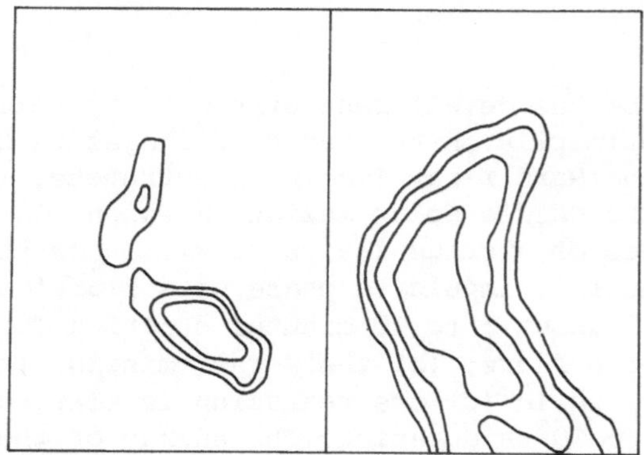

Figure 1. The flare of 12 November, 1980 seen with HXIS in the second energy channel (5.5 - 8.0 keV) at 02:50:05 and 02:57:00 UT respectively. Note the patchy aspect in the early phase and the diffuse character later-on (from Boelee and De Jager, 1983). The length of the ordinate corresponds to 8 x 10^4 km.

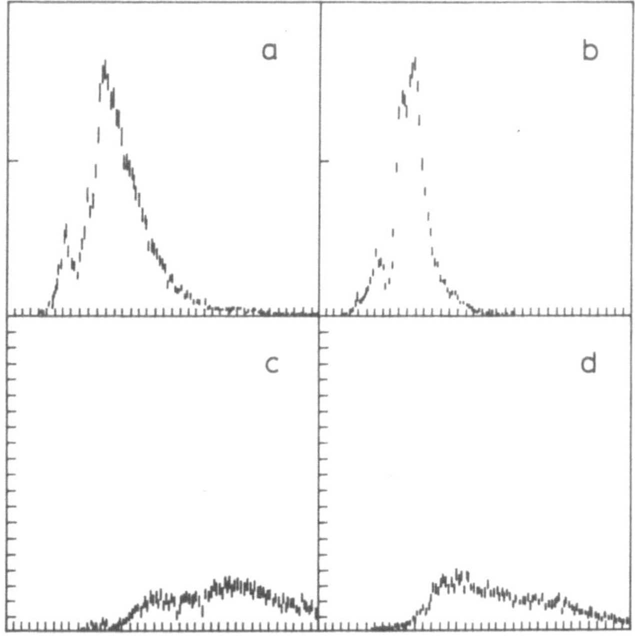

Figure 2. X-ray lightcurves (5.5 - 8.0 keV) of flare pixels mainly responsible for the impulsive phase emission (2a, 2b) or for the gradual phase emission (2c, 2d) (flare of 12 November, 1980; 02:50 UT). The abcissa gives the time (dashes denote minutes), the ordinate the X-ray counting rate in an arbitrary scale (from Boelee and De Jager, 1983).

disk. But this difficulty may partly be overcome because the diffuse image is larger than the impulsive one and thus one has to look for lightcurves for the outermost pixels of the flaring area, to have information characteristic for the diffuse phase.

A few such lightcurves are shown in Figure 2 and these demonstrate the fundamental difference between the light curves in the two areas: the 'impulsive phase area' has lightcurves showing a number of discrete bursts superimposed over a gradual curve and the 'gradual phase area' shows the characteristic gradual development without the bursts. It is natural then to go one step further: since the two kinds of areas overlap, the lightcurve of the 'impulsive phase area' also contains a contribution from the 'gradual phase area'. It is easily ascertained, from a comparison of the two kinds of X-ray light curves, that substraction of the gradual phase component from the impulsive phase component would just let the bursts remain, which therefore allows us to *conclude* that the *characteristic emission of the discrete impulsive phase patches is just a series of bursts*. Note that this conclusion is based on observations in a low energy channel of HXIS (5.5 - 8.0 keV). So far, the impulsive phase of a flare has always been identified with burst emission in hard X-rays. It is true that the hard X-ray bursts of a flare are more strictly related to the very first phase of a flare and are clearer to identify than the soft ones, because the gradual component is absent in radiation of high enough energies (say > 20 keV); furthermore, the so-called 'X-ray flare footpoints' have been discovered (Hoyng *et al.*, 1981) by the study of the areas of emission of the hard X-ray bursts. Nevertheless, we want to stress that the energy is not the most essential parameter: also in lower energies the impulsive phase is characterized by patchy areas (the 'footpoints') emitting discrete bursts of X-radiation. These arguments support the new definition of the 'impulsive phase of a flare', given in Section 1; which is more general than the classical definition and, we think, more appropriate.

3. ENERGY INPUT, AND THE DURATION OF THE IMPULSIVE PHASE

From the radiation of the various HXIS pixels in several energy channels the temperature T and emission measure $Y = \int n_e^2\,dV$ can be determined for each pixel. Assuming that Y can be approximated as $n_e^2 V$, and with the usual assumptions on the thickness of the flaring volume (mostly assuming spherical or cylindric symmetry) it is possible to derive n_e for each pixel, and hence the total energy content of the flare:

$$E = \int n_e kT\,dV \approx N_e kT \qquad (1)$$

as a function of time.

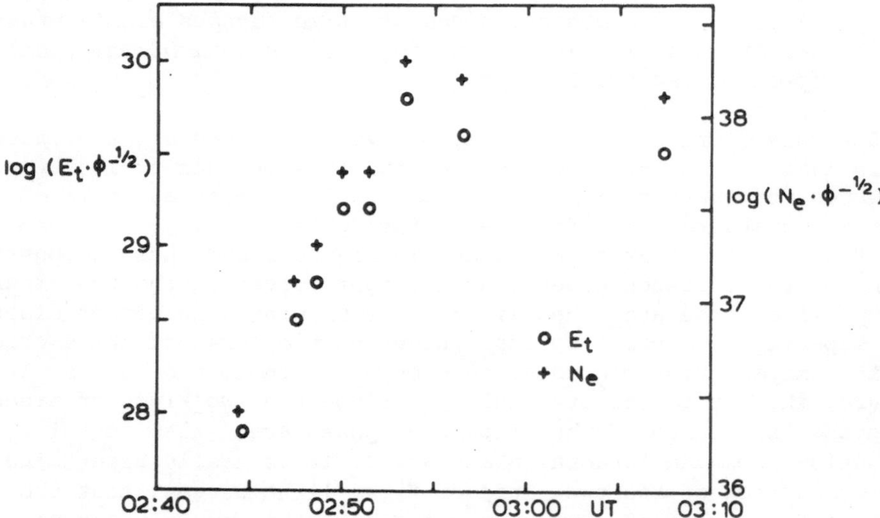

Figure 3. Energy input into the flare of 12 November, 1980. The diagram shows the total content of energetic electrons $E \cdot \phi^{1/2}$, *where* ϕ *is the filling factor of the flare (left ordinate), and the total number of energetic electrons* $N_e . \phi^{-\frac{1}{2}}$ *(right). During the impulsive phase both quantities increased steadily. This is another characteristic of the impulsive phase.*

There are indications (De Jager *et al.*, 1983) that flares are not homogeneous and for the flare of April 30, 1980 the 'filling factor' ϕ is $\approx 10^{-2}$. In that case the integral in Eq. (1) is equal to $E \cdot \phi^{1/2}$, which would make the real value of E about 0.1 times the value computed for a homogeneously filled volume, if we would assume the same filling factor as the value derived for the flare of April 30, 1980.

Figure 3 gives the variation of $E \cdot \phi^{-\frac{1}{2}}$ as a function of time and shows that from about 02:44 UT till 02:52 UT there was a constant net exponential energy input, while also $N_e = n_e V$ increased. From this diagram one would furthermore conclude, extrapolating, that the impulsive phase started at about 02:43 UT, and in any case before 02:45 UT, while it lasted till about 02:52 UT. During that period of seven to ten minutes there was an exponential net energy input with burst-associated variations. After 02:52 UT energy losses dominated over the input. We conclude that the impulsive phase of this flare was characterized by a *virtually constant net energy input during a period of about seven to ten minutes.*

4. A FLARE SEEN AT THE LIMB

When seen at the limb we are in a better position to spatially disentangle the two components because gas moving upward can be seen separated from the heated footpoint areas. The flare of 30 April, 1980, 20:30 UT (the 'Queens' Flare') was such a case (De Jager *et al.*, 1983).

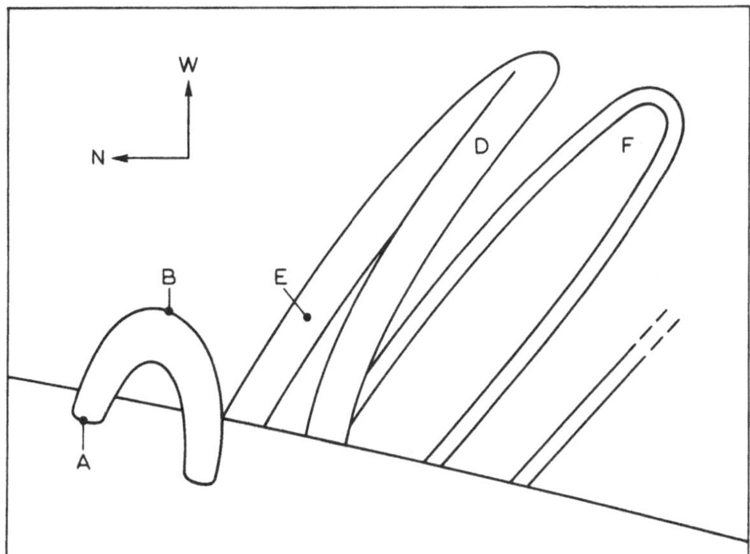

Figure 4. Preflare fluxtube configuration of the Queens' Flare derived from C IV observations with the UVSP instrument aboard SMM. The length of the ordinate corresponds to about 4 x 10^4 km.

The preflare fluxtube configuration is shown in Figure 4: it consists of a small loop and one or more larger ones. At about 20:15 UT, some five minutes before the onset of the flare was observed in Hα and X-rays, C IV emitting material was seen to rise near the small loop, initially with ≈ 40 km s^{-1}, and after some five minutes with ≈ 100 km s^{-1}. At 20:15 UT a first very faint X-ray emission was observed at the level of the top of the small loop (as could be inferred from HXIS count integration over 30 s time intervals) followed by gradually increasing X-ray emission in one of the legs of the small loop. Hα emitting material started to rise at 20:20 UT. When that gas had reached a height about equal to that of the small loop (at 20:23 UT) a short radioburst was observed, followed by a temperature and X-ray flux increase of the plasma in one of the legs of the small loop.

A similar radioburst, again associated with heating and X-ray flux increase was observed some three minutes later. Practically immediately at the start of the flare the area covered by the larger loop started to radiate in X-rays. During the first ten minutes the lower kernel (identified with one of the legs of the small loop) was still the brightest part of the flare, but after 20:30 UT it had faded away, leaving an extended tongue (Figure 5) that gradually became less bright, losing energy by conduction and radiation and becoming invisible for HXIS after ≈ 90 minutes.

This course of events leads to the following interpretation. The two loops that are mainly involved in the flare process are both identified with current circuits. The change of the electric configuration caused by the emerging plasma (and field) makes the two flux tubes to interact, causing conductive or inductive coupling. A first weak coupling occurred in the top of the small loop at 20:15 UT, at the very start of the flare, followed by continuous coupling (reconnection of field lines) and impulsive coupling at 20:23 and 20:27 UT.

The lower parts of the legs of the small loop were filled with energetic electrons, and practically at the same time the flow of electrons to the loop started, presumably by convection, along paths determined by the electromagnetic configuration.

Hence, we are here meeting a case in which X-ray observations show that there is evidence for the very first flare activity in the *top* of a loop structure. This confirms evidence from VLA observations that the first flare instability can take place in the top of a loop (Hoyng *et al.*, 1982). X-ray observations in general do not show first emission in the top of a loop, rather in the footpoints, but in the case of the Queens' Flare (30 April, 1980) the integration of counts over 30 seconds made it possible to detect a faint primary emission in the top of the small loop.

In addition, the Queen's Flare shows clearly that the longer lasting source of gradual emission (the 'tongue') extends higher than the emission source in the kernel which we identify with the source of the impulsive heating. This proves that the gradual component of a flare is indeed due to convectively rising hot material.

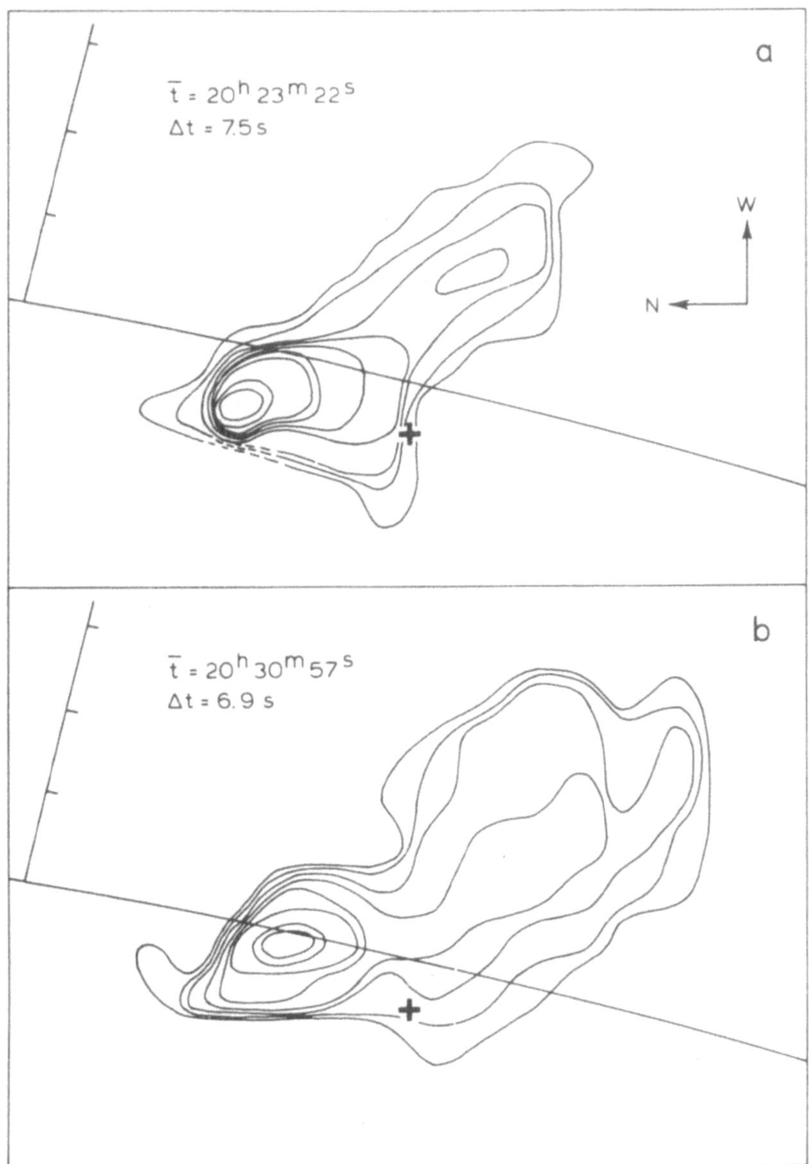

Figure 5. The X-ray flux of the Queens's Flare observed by HXIS in the energy bands 3.5 – 8.0 keV during the phase of maximum brightness of the kernel (20:23 UT) and virtually at the end of the impulsive phase (20:31 UT). The curved line is the Sun's limb. The dashès along the height ordinate are at 10^4 km distances. Note that the main emission in the kernel shifted between these observations from the Northern to the Southern leg of the small loop. Note also the continuously increasing size of the convective plume.

5. THE EVIDENCE FOR CONVECTIVE MOTIONS

In this section we summarize the various evidences that the gradual phase is a manifestation of convection (a term that should be preferred above the usual 'evaporation', which is physically wrong).

a. Direct observations, as in the two cases described here, particularly the limb flare, show already that immediately after the footpoints heating gas ascends. The pre-existing magnetic structure of the flaring area defines the shape that the convected gas cloud will assume. There is evidence for some expansion, which indicates that the gas pressure of the convected gas can exceed the confining magnetic pressure.

b. Observations made with the X-ray Polychrometer aboard SMM (Antonucci *et al.*, 1982) show that *upward motions* with velocities of a few hundred km s^{-1} occur in the impulsive phase of a flare. These velocities are consistent with the height reached by the convected cloud ($\approx 2 \times 10^4$ km s^{-1}) and the time apparently needed for its formation ($\approx$ a minute).

TABLE I. Temperature differences between the diffuse (convected) cloud and the impulsive (footpoint) areas for the two flares discussed in this paper. The temperature differences are in 10^6 K; $t = t - t_o$ where t is the time of observation and $t_o = 20{:}20{:}00$ UT for the Queens' Flare and 02:43:00 UT for the flare of 12 November, 1980. The ΔT values have an estimated uncertainty of ± 1 MK.

t (min)	ΔT (Queens' Flare)	ΔT (12 Nov. Flare)
3	5.5 MK	
4	5.5	
5	5	
5.5		4.7
6	3.5	
7	3	2.5
8	3.5	
8.5		0.3
9	2	
10	1.5	
10.5		- 0.4
13.5		+ 0.4

c. Another observation that may be related to convective motions is the fact that the *diffuse cloud is initially hotter* by a few million K than the impulsive flare patches (Table I). The proposed explanation (Boelee and De Jager, 1983) is that the impulsive areas (the 'footpoints') consist of elements of various temperatures (such an assumption is not incompatible with the observed small - 10^{-2} - filling factor of the Queens' Flare). The measured temperature of the footpoints is the average of the various temperatures. The hottest gas elements ascend earlier and quicker than those with lower temperatures. After a period of about 8 minutes (cf. Table 1) the whole area has acquired the same temperature.

d. There is some *theoretical support* for the above statement. To that end we use Siedentopf's expression for the convective velocity (cf. Unsöld, 1955, p. 223):

$$v^2 = g\ell\Delta T/T \quad , \tag{2}$$

where g is the acceleration of gravity and ℓ the main distance travelled by a convective element. We assume

$$\ell = \alpha H = \alpha RT/\mu g \quad , \tag{3}$$

where H is the atmospheric scale height, and α is a dimensionless constant, assumed to be of order unity. From Eqs. (2) and (3) we derive

$$\Delta T = v^2\mu/\alpha R \quad ;$$

hence

$$v\alpha^{-1/2} = 10^4\Delta T^{1/2} \tag{4}$$

for a completely ionized cosmical plasma.
Hence, for $\Delta T = 4 \times 10^6$ K one would find $v\alpha^{-1/2} = 200$ km s^{-1}, a value that is compatible with the XRP observations aboard SMM and with the other evidence shown above. We *conclude* that the identification of the gradual phase with the phase of fierce upward convective motions seems sound.

6. THE FLARE SCENARIO

We propose the following course of events:

a. Assuming the circuit model of a flare (Alfvén and Carlqvist, 1968; Alfvén, 1981; Spicer, 1982) as shown in Figure 6, we assume the flare to start by *circuit coupling* causing high energy particles to originate at the location of coupling.

b. These cause X-ray Bremsstrahlung to originate in the densest part of the fluxtube (circuit), hence in the lowest, near-chromospheric part.

c. Apparently the circuit coupling takes place in several discrete bursts, each of 10^1 to 10^2 s duration, during a period of 5 to 10 min; this is the impulsive phase, during which period there is quasi-continuous energy input into the flare.

d. Immediately after the first burst heated chromospheric material starts to rise with velocities of the order 10^2 km s^{-1}, thus forming an extended diffuse convective plume; this is the gradual or diffuse phase of a flare.

e. The convective plume cools gradually down, by radiative and conductive losses. This may take a time of the order of an hour.

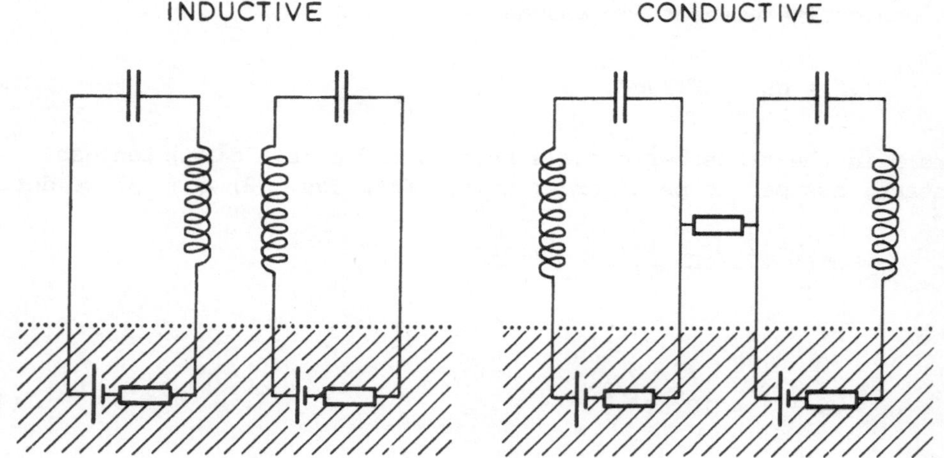

Figure 6. A flare originating by circuit coupling. The coupling can be inductive (left) or conductive. The shaded area is the resistive domain, where the specific electrical resistance is high (the photosphere); in the chromosphere-corona part the resistance is low: the conductive domain.

ACKNOWLEDGEMENT

This paper was written when the author held a position of Guest Scholar at the University of Kyoto. He expresses his thanks to the Japan Society for the Promotion of Science for enabling his stay in Japan.

REFERENCES

Alfvén, H.: 1981, Cosmic Plasma, Reidel, Dordrecht.

Alfvén, H., and P. Carlqvist: 1967, Solar Phys. 1, 220.

Antonucci, E., A.H. Gabriel, L.W. Acton, J.L. Culhane, J.G. Doyle, J.W. Leibacher, M.E. Machado, L.E. Orwig and C.G. Rapley: 1982, Solar Phys. 78, 107.

Boelee, A., and C. de Jager: 1983, Solar Phys., submitted.

De Jager, C., M.E. Machado, A. Schadee, K.T. Strong, Z. Svestka, B.E. Woodgate, and W. van Tend: 1983, Solar Phys. 84, 205.

Duijveman, A., P. Hoyng and M.E. Machado: 1982, Solar Phys. 81, 137.

Hoyng, P., A. Duijveman, M.E. Machado, D.M. Rust, Z. Svestka, A. Boelee, C. de Jager, K.J. Frost, H. Lafleur, G.M. Simnett, H.F. van Beek and B.E. Woodgate: 1981, Astrophys. J. (Letters), 246, L155.

Hoyng, P., K.A. Marsh, H. Zirin and B.R. Dennis: 1983, Astrophys. J. 268, 865.

Spicer, D.S.: 1982, Space Sci. Rev. 31, 351.

Unsöld, A.: 1955, Physik der Sternatmosphären, Springer, Berlin.

Van Beek, H.F., P. Hoyng, H. Lafleur and G.M. Simnett: 1980, Solar Phys. 65, 39.

DISCUSSION

SIMNETT: You suggest that the trigger, or ignition, of the April 30 flare took place at the top of the loop that contained the main part of the flare. Given that, in a typical flare, the soft X-ray brightness is greatest from the top of the flare loop and, following the energy release, hot plasma rises to the top of the loop. Would it not be feasible that the preflare soft X-ray emission you ascribe to the ignition phase of the flare could be from accumulated hot gas convected to the looptop following small, preflare energy released at or near photospheric level? I assume that the X-ray emission at photospheric level from such preflare activity is below the HXIS sensitivity level. The advantage of this suggestion is that all primary energy releases occur at the same place, rather than in two separate places.

DE JAGER: There is no objection, in principle, to your suggestion but the observations show that during the main part of the April 30 flare, the top of the small loop is in no case the brightest part of the flaring structure. Material seems rather to be convected higher up into the area occupied by the larger loop structure. Therefore I am still inclined to believe that the first (preflare) brightening in the top of the small loop is (related to) the first sign of the primary acceleration.

ZIRIN: There is abundant evidence that the density is highest at the loop tops, and this anti-hydrostatic behavior is not compatible with a convective, or evaporative model.

DE JAGER: The situation described by you refers to preflare conditions. But these change drastically when the lower part of the loop is heated to the (measured) values of 20 to 25 x 10^6 K as compared to the preflare value of only a few times 10^6 K. Then the Archimedean force driving the convection will be large. Actually, the derived electron densities in the footpoints of the two flares described by me were significantly higher than those in the diffuse areas (which may be identified with the tops of the loop structures). This shows that the electron density behaviour during the flares differs from the preflare situation described by you.

UCHIDA: If I saw it correctly, structures with different temperatures in the site of the April 30 event seem to be co-spatial with each other. This suggests the existence of finer substructures of different temperatures in one and the same source, at least in this event.

DE JAGER: No, there are so far no indications for the occurrence of different temperatures in the same area of the April 30 flare. But we did observe during the first ten minutes that the tongue was a few million degrees hotter than the kernel, while we also found that the filling factor of the flare is $\sim 10^{-2}$, which thus shows that the X-ray source is highly inhomogeneous.

MAGNETIC THEORIES OF SOLAR FLARES

E. R. Priest
Applied Mathematics Department, The University,
St Andrews, Scotland

ABSTRACT. The basic processes of evaporation and magnetic reconnection have recently been developed in much greater detail. They may be important in the two main types of flaring event, namely simple-loop flares and two-ribbon flares. The first type could be produced by kink instability, thermal nonequilibrium, or emerging flux. The second type is thought to be the result of an eruptive MHD instability that is either spontaneous or triggered from outside. After the eruption the magnetic field lines that have been blown open reconnect back down in a way that has now been simulated numerically.

1. INTRODUCTION

There seem to be two distinct types of flare produced by different physical mechanisms, namely *simple-loop flares* and *two-ribbon flares* (e.g. Pallavicini et al., 1977; Heyvaerts et al., 1977; Priest, 1981). In the one case the magnetic field is static and contains the flare, whereas in the other case the whole magnetic structure goes unstable and blows open. It is this physical classification which is important, rather than any botanical subdivision. There may be exceptions, or it may be shown to be too simplistic in future, but at present it appears to be a helpful subdivision for the theoretical models and it is important to try and seek an overall pattern in the jungle of observations. I recommend that, in future, observers make it clear, if possible, which type of flare they are describing. For instance, it is not sufficient just to refer to a "flare loop", since the physical processes in compact and 'post'-flare loops may well be quite different.

The properties of the two categories of flare (essentially, small flares and large flares) are as follows. Most subflares and flares are of simple-loop type. They consist of a single loop, typically 10 000 km high, with a temperature of 2×10^7K and a density of 10^{17} - 10^{18} m^{-3}. The loop brightens in soft X-rays and remains unchanged in position and shape. By comparison, most major events are two-ribbon flares. They take the form of an arcade of hot loops which rises at 20 km s^{-1} or more in the initial stages and at only 0.5 km s^{-1}

Solar Physics **86** (1983) 33–45. 0038–0938/83/0861–0033 $ 01.95.

later on, reaching an altitude of 100 000 km. The summit density and temperature are about $10^{17}m^{-3}$ and 2×10^7K at first, falling to $10^{16}m^{-3}$ and 5×10^6K after a few hours. This type of flare occurs near a dark Hα filament which, during the preflare phase, erupts slowly and enhances the soft X-ray emission. Then the eruption suddenly becomes violent and Hα knots are produced. Two Hα ribbons form and separate, being joined by the rising arcade of 'post'-flare loops. In soft X-rays, simple-loop flares have small volumes, large pressures and short time-scales, whereas two-ribbon flares display the opposite features (Pallavicini et al., 1977). Simple-loop events show at most a single spike ($\lesssim$ 1 min) in hard X-rays and the energy may be released only at the impulsive phase. Two-ribbon events may sometimes have multiple hard X-ray spikes ($\gtrsim$ 5 min) and the energy release continues through the main phase. They may also produce a coronal transient and are often associated with spot motions, emerging flux and reconnections (Zirin, 1974; Rust, 1976). Two-ribbon flares may be subdivided into: ones that are slow, long-lived and thermal due to the eruption of a quiescent filament (with a weak magnetic field and often no Hα emission); and ones that are violent, fast and nonthermal due to a plage filament eruption (with a magnetic field that is strong and complex).

2. BASIC THEORY

The fundamental theories of loop flows and magnetic reconnection have recently made great strides. They are possibly important for both types of flare and have entered a stage of sophisticated numerical study.

2.1 Loop flows

The thermal response of plasma in a loop to a sudden release of heat has been studied by many authors (e.g. Antiochos and Sturrock, 1976; Somov et al., 1977; Nagai, 1980; Craig, 1981; Wu et al., 1981; Mariska et al., 1982; Doschek et al., 1982; McClymont and Canfield, 1982). The basic approach is to solve the one-dimensional single-(or two-)fluid equations for flow along a rigid flux tube, namely

$$\frac{D\rho}{Dt} + \rho\,\frac{\partial v}{\partial s} = 0\ ; \qquad \rho\,\frac{Dv}{Dt} = -\,\frac{\partial p}{\partial s} + \rho g \cos\theta;$$

$$p = R\rho T; \quad \frac{\rho}{\gamma-1}\,\frac{D}{Dt}\left(\frac{p}{\rho}\right) = \frac{\partial}{\partial s}\left(\kappa_{\parallel}\,\frac{\partial T}{\partial s}\right) - p\,\frac{\partial v}{\partial s} - \mathcal{R} + H.$$

Switching on a heat source causes a conduction front and a shock wave to propagate along the loop down to the chromosphere, which is then heated and expands upwards (i.e. *evaporation*). Eventually, the hot plasma cools and *drains* back down, the whole process having oscillations superimposed on it at the sound travel time (L/c_s). If the heat input is increased in magnitude by a factor 10 it tends to double the maximum temperature, while smaller loops tend to have lower summit temperatures and higher summit densities (Nagai, 1980). Also, a heat source near the base rather than the summit has the effect of raising the temperature and pressure there and driving an upflow immediately, rather than after a downflow (i.e. expansion) from the summit.

Difficulties in performing realistic simulations include an adequate treatment of the very narrow transition region and of radiative transfer in the chromosphere. Plasma motions in a flare can be exceedingly complex and of different types. Emerging flux and an erupting filament move up, while plasma drains down the legs. Surges may be ejected from the flare site. Evaporation may be driven by either fast particles or thermal conduction. Thus, when a red or a blue shift is reported, it is important to know what type of motion is being observed.

2.2 Magnetic reconnection

The importance of magnetic reconnection for solar flares has been reviewed by, for instance, Priest (1976, 1981, 1982), Spicer and Brown (1981), Spicer (1981) and Van Hoven (1976, 1981). Reconnection may occur spontaneously by resistive instability in a current sheet or a more generally sheared field (Furth et al., 1963). The linear growth time is typically

$$\tau \sim (\tau_d \, \tau_A)^{1/2},$$

in terms of the diffusion time ($\tau_d = d^2/\eta$) and the Alfvén travel time ($\tau_A = d/v_A$) across a shear length d. It is typically days or weeks in the solar corona. The corresponding instability in a flux tube is known as the resistive kink (Coppi et al., 1976), which grows somewhat faster with

$$\tau \sim (\tau_d \, \tau_A{}^2)^{1/3}.$$

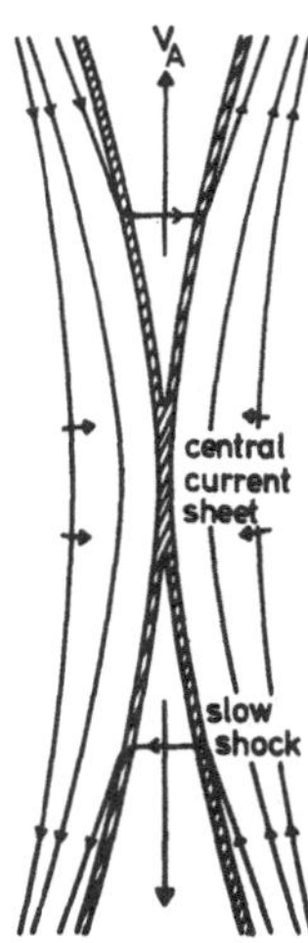

Figure 1. Petschek's mechanism, in which plasma is heated and accelerated at the slow shocks.

Reconnection may instead be driven during the dynamic formation of a current sheet either due to an ideal magnetic instability (or nonequilibrium) or due to the motion of separate flux systems that are forced together. Whether the reconnection is *spontaneous* or *driven*, its nonlinear development depends critically on the assumed boundary conditions. In many cases it can approach a nonlinear steady state known as Petschek's mechanism (Petschek, 1964; Soward and Priest, 1977, 1982), with a maximum inflow speed of between 0.1 v_A and 0.01 v_A.

We have now entered a period of detailed numerical study of time-dependent reconnection. Ugai and Tsuda (1977, 1979) have considered a magnetic field that is initially antiparallel and at rest. The boundary conditions are free, so that plasma can flow into the numerical box easily. Reconnection is driven by a local resistivity enhancement by a factor 100 so as to give an effective magnetic Reynolds number (R_m) of 25. After 20 Alfvén travel-times, the flow evolves to a steady-state configuration, whose first quadrant is shown in Figure 2a. The reconnection rate agrees with the maximum allowable value from Petschek theory and there is a slow shock wave present. Sato and Hayashi (1979) have found similar results when the reconnection is driven by an imposed inflow. The current density structure for all four quadrants (Figure 2b) shows clearly the four slow MHD shocks emanating from the central reconnection region.

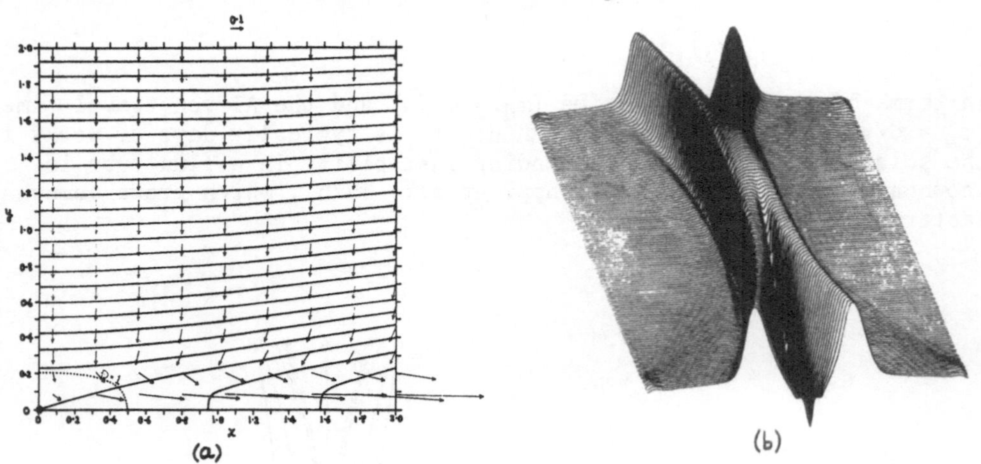

Figure 2. The numerical experiments of (a) Ugai and Tsuda and (b) Sato and Hayashi.

3. SIMPLE-LOOP FLARE

3.1 Kink instability

The main problem for understanding the simple-loop flare is to discover how the heat is generated. One possibility is the kink instability, as first analysed in the flare context by Raadu (1972) and later stressed by Spicer (1977). Consider a loop of length 2L with dominant field components $B_\theta(r)$, $B_z(r)$ that vary with distance (r) from the magnetic axis (Figure 3a), so that the amount by which a field line is twisted in going from one end to the other is

$$\Phi = \frac{2LB_\theta}{rB_z} .$$

An important effect for a solar coronal loop is line tying of the loop footpoints in the dense photosphere, which keeps the loop stable until the twist exceeds a critical value (Φ_{crit}). Raadu (1972), Giachetti *et al.* (1977), Hood and Priest (1979), Einaudi and Van Hoven (1981) found bounds on Φ_{crit}, and later Hood and Priest (1981a) determined the threshold (Φ_{crit}) exactly as follows.

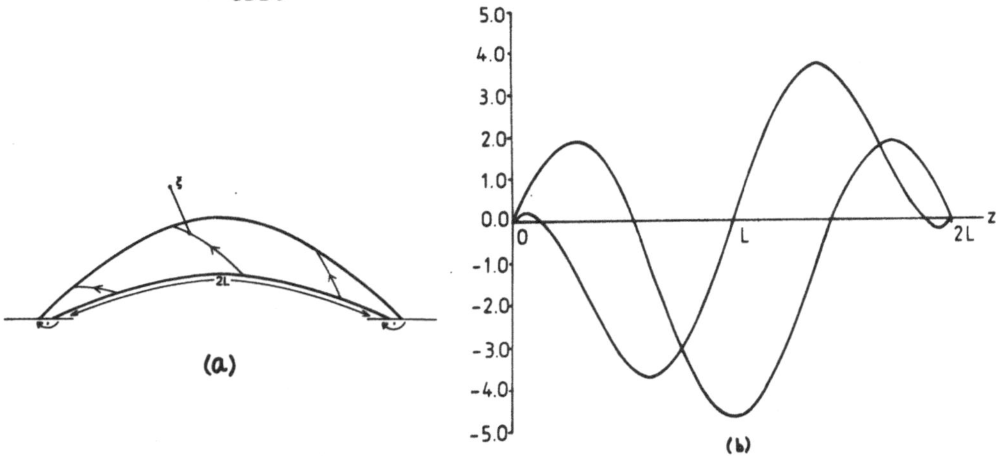

Figure 3. (a) A twisted loop. (b) The unstable perturbation $\xi_r(z)$.

A line-tied perturbation of the form

$$\underset{\sim}{\xi} = \underset{\sim}{\xi}(r,z)e^{i(\theta + \omega t)}$$

is considered, where ξ_1 vanishes at the ends ($z = \pm L$) of the loop and at large distances ($r \to \infty$). The linearised equation of motion is

$$\rho_0 \frac{\partial^2 \underset{\sim}{\xi}}{\partial t^2} = \underset{\sim}{j}_1 \times \underset{\sim}{B}_0 + \underset{\sim}{j}_0 \times \underset{\sim}{B}_1,$$

where $\underset{\sim}{j} = \nabla \times B/\mu$ and $B_1 = \nabla \times (\underset{\sim}{\xi} \times B_0)$. It reduces to a pair of partial differential equations for $\xi_r(r,z)$ and $\xi_\theta(r,z)$, which are solved numerically to give $\Phi_{crit} \approx 2.5\pi$ for a force-free field of uniform twist. The real and imaginary parts of the radial component (ξ_r) of the eigenfunction that first gives kink instability are shown in Figure 3b.

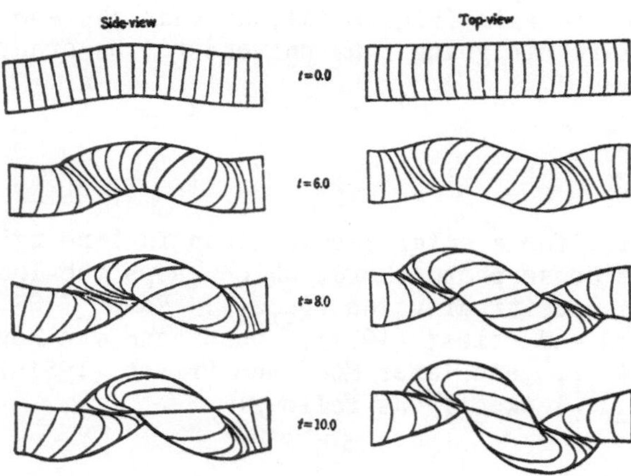

Figure 4. Kinking of a flux tube (Sakurai, 1976).

The present analysis needs to be refined in several ways. The effects of loop curvature, a non-uniform cross-section, shear stabilization and internal modes should be included (e.g. Priest, 1982). The effect of the resistivity in possibly lowering the threshold is important too. Recently, Mok and Van Hoven (1982) have found that line tying completely stabilizes the resistive kink mode and all other resistive modes except for $m = 0$. This resistive sausage is only unstable when the axial current is large enough and when there is a reversal in the axial magnetic field, which is most unlikely in practice. Thus, it may well be the case that the *ideal* kink modes are the relevant ones for solar flares. Another important task is to follow the nonlinear development in order to see whether the instability saturates or grows explosively. Sakurai (1976) has made a start by considering an infinitely long cylinder of uniform current or force-free field and simulating line-tying roughly by requiring that the axial wavenumber be π/L. His numerical solutions (Figure 4) show how the flux tube rises and twists up. Tubes with a weak twist develop strong helical kinks and rise less than those with strong twist because of the stabilizing tension force from the axial field.

3.2 Thermal nonequilibrium

Small X-ray flares may occur when the cool core of a coronal loop loses thermal equilibrium and heats up (Kahler and Kreplin, 1970; Hood and Priest, 1981b), essentially because the radiation is no longer efficient enough to balance the heat sources. This is the case when the heating becomes too large or the loop pressure becomes too small. Even though this is a basically thermal process, the resulting plasma motions may create shock waves which accelerate particles. The role of thermal instability (or nonequilibrium) in causing coronal plasma to cool down and give prominences is well established, and so it would be surprising if a similar process could not also heat plasma up. For recent numerical simulations of thermal instability in loops, the reader is referred to McClymont and Craig (1982) and McClymont and Canfield (1982).

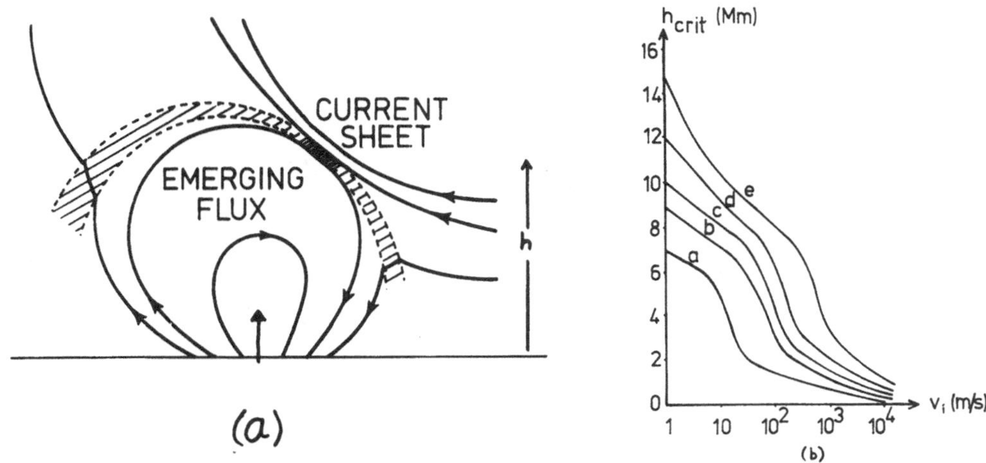

Figure 5. (a) Reconnection of newly emerging flux.
(b) Critical height for flare onset.

3.3 Emerging (or evolving) flux

Topologically separate flux systems may interact when, for instance, new magnetic flux emerges from below the photosphere (Figure 5a) or when satellite sunspots move horizontally. The interaction takes place at a current sheet on the interface between the two systems and leads to a transfer of flux from one system to the other by magnetic reconnection. As more flux emerges or the horizontal motion varies, so the height (h) of the current sheet above the photosphere varies. When this height exceeds a critical value (h_{crit}) the current density inside the sheet is so large that the plasma suddenly becomes turbulent and the flare is triggered (Canfield et al., 1974; Heyvaerts et al., 1977; Tur and Priest, 1978). The variation of h_{crit} with emergence speed (v_i) is shown in Figure 5b for several magnetic field strengths (B_i) in G: a,b,c,d,e refer to $B_i = 10^3, 10^{2.5}, 10^2, 10^{1.5}, 10$, respectively. It is calculated by solving the energy balance equation inside the reconnecting sheet and deducing the current density there. If v_i or B_i is so small that the location of the sheet does not reach h_{crit}, then the emerging flux does not produce a flare.

4. TWO-RIBBON FLARE

The overall behaviour of the magnetic field during a two-ribbon event is sketched in Figure 6. The initial configuration is a magnetic arcade containing a plage filament along which the magnetic field is directed. During the preflare phase (Figure 6a) the filament and its overlying arcade start to erupt slowly and stretch out the field lines. Then later (Figure 6b) they erupt much more rapidly, with the field lines starting to reconnect below the filament. In the main phase (Figure 6c) the reconnection continues and creates hot 'post'-flare loops and Hα ribbons.

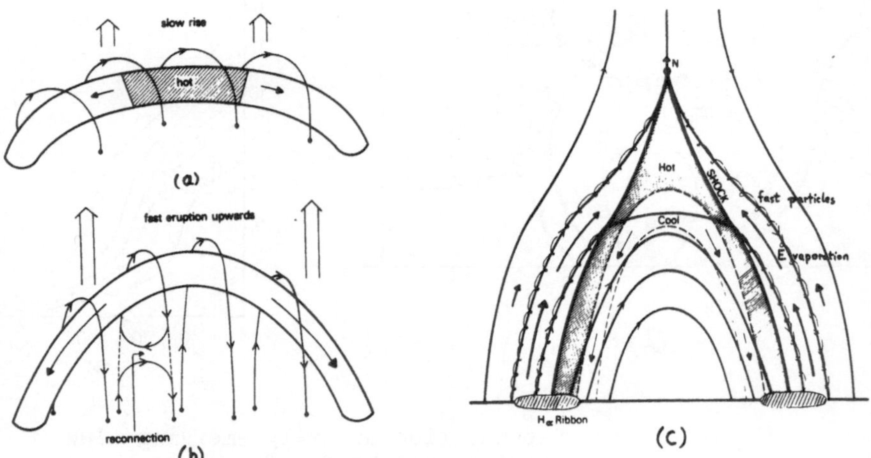

Figure 6. Overall behaviour of a two-ribbon flare.

4.1 Instability of magnetic arcade

In an attempt to find out why a coronal arcade erupts in the first place, the magnetohydrodynamic stability of model arcades has been tested, including the stabilising effect of photospheric line tying (Hood and Priest, 1980; Hood, 1983; Migliuolo and Cargill, 1983). The results show that a simple magnetic arcade with its magnetic axis below the photosphere (so that it does not support a filament) is always stable both to ideal perturbations and, according to Migliuolo and Cargill (1983), also to resistive modes. On the other hand, an arcade whose magnetic axis lies above the photosphere (and so supports a filament), becomes unstable when the height of the magnetic axis (the filament) or the twist in the overlying field are too great.

The filament eruption may therefore be caused by a spontaneous MHD instability when the filament height or field shear become too large. Alternatively, the eruptive instability may be triggered by some other agent such as: emerging flux, which pushes up the filament, tears overlying flux away and forces a large-scale reconnection by lowering the tearing mode time; thermal instability in the filament, which causes the plasma to expand and the filament to rise; a fast magnetoacoustic wave that drives tearing (Sakai and Washimi, 1982).

4.2 Reconnection and creation of 'post'-flare loops

In the main phase after the eruption, the magnetic field closes down and produces hot loops (Kopp and Pneuman, 1976; Pneuman, 1981), as shown in a section across the arcade in Figure 6c. The neutral point (N) and its trailing slow shocks rise and heat the plasma to 10^7K or more (Cargill and Priest, 1982), while the Hα ribbons at the base of the hot loops move apart. Plasma flows upwards ahead of the slow shocks and is heated and compressed at the shocks before cooling and falling back down. The upflow may be partly a reconnection-enhanced solar wind flow and partly an evaporation, driven by heat conduction and fast particles that propagate ahead of the shocks.

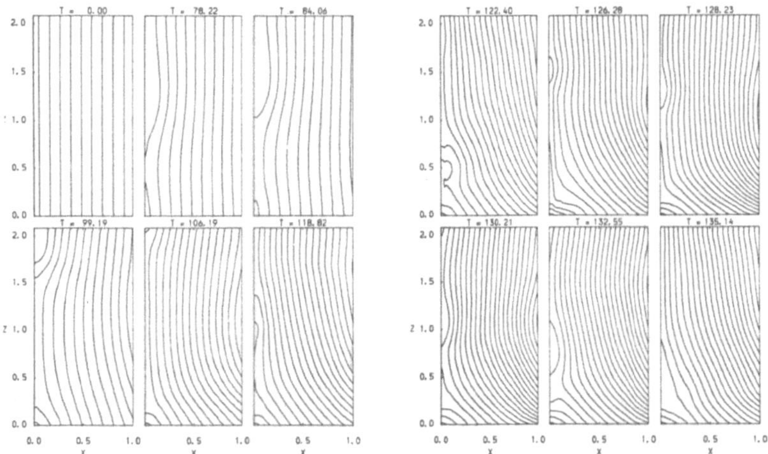

Figure 7. Line-tied reconnection (Forbes & Priest, 1983)

A numerical simulation of the reconnection process has been carried out by Forbes and Priest (1982, 1983). They begin with a vertical magnetic field that reverses at $x = 0$ and is in equilibrium. Line-tied boundary conditions are imposed at the base and free-floating ones at the top and sides. Since $x = 0$ is an axis of symmetry, only half of the magnetic configuration (i.e. $x > 0$) is shown in Figure 7, which presents some results for a magnetic Reynolds number of 150, a plasma beta of 0.1 and a current sheet width that is 0.075 of the box width. The magnetic field lines are shown at several times in units of the Alfvén travel time across the current sheet. First of all, the sheet tears near the base, and then in the nonlinear phase of Petschek-type reconnection the magnetic field closes down and the neutral point rises, while a plasmoid is ejected out of the top (first 4 frames). This process continues, but then the sheet thins (5th frame) and tears again (6th frame), creating a new pair of O- and X-type neutral points. Reconnection at the upper X dominates, and so the O is shot down rapidly (7th frame) and coalesces with the lower X. This process takes place very rapidly at Alfvénic speeds due to the coalescence instability, whereas the formation of the neutral points is on the slower tearing time-scale. Meanwhile, a new pair of neutral points is created (8th frame) and the process repeats.

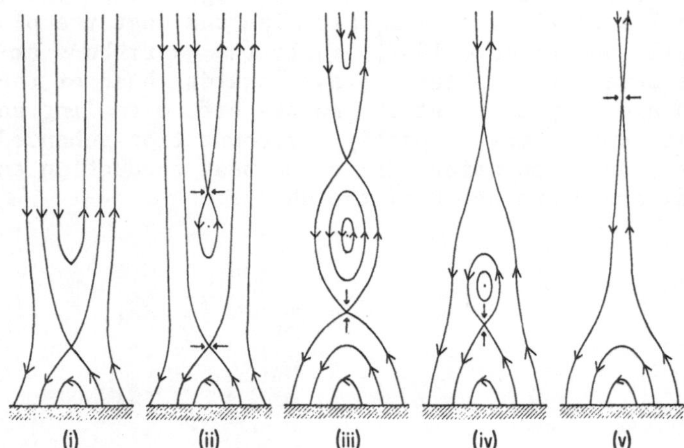

Figure 8. Creation and annihilation of neutral point pairs.

A schematic of the way neutral point pairs are created and annihilated is shown in Figure 8. The extremely rapid coalescence process is associated with strong electric fields and rapid energy release, and it should accelerate particles efficiently. The repetition of the reconnection that has been discovered in the numerical experiment may explain the double (and even multiple) flaring often found in X-rays (e.g. Strong et al., 1982)*. Another surprise in the simulation is the presence of a fast MHD shock wave below the reconnection region. It slows down the jet of plasma emitted from the region and it may provide significant particle acceleration.

5. CONCLUSION

We are at a most exciting stage of flare research as the basic processes of evaporation, MHD instability and magnetic reconnection are being developed and applied to both types of flare. A great deal of progress in our understanding of them should be forthcoming over the next few years. In particular, the details of the instability processes that initiate flares and of the dynamic energy release should be much better understood.

REFERENCES

Antiochos, S.K. and Sturrock, P.A.: 1976, Solar Phys. 49, 359.
Canfield, R.C., Priest, E.R. and Rust, D.M.: 1974, Flare-related magnetic field dynamics (ed. Nakagawa, Y. and Rust, D.M.) NCAR, Boulder.
Cargill, P.J. and Priest, E.R.: 1982, Solar Phys. 76, 357.
Coppi, B., Galvao, R., Pellat, R., Rosenbluth, M.N. and Rutherford, P.M.: 1976, Sov. J. Plasma Phys. 2, 533.

* or Simnett (1982) IAU Colloq. No.71.

Craig, I.J.D.: 1981, Chapter 5 of *Solar flare MHD* (ed. E.R. Priest), Gordon & Breach, London.
Doschek, G.A., Boris, J.P., Cheng, C.C., Mariska, J.T. and Oran, S.: 1982, Ap. J.
Einaudi, G. and Van Hoven, G.: 1981, Phys. Fluids 24, 1092.
Forbes, T.G. and Priest, E.R.: 1982, Solar Phys. 80,
Forbes, T.G. and Priest, E.R.: 1983, submitted.
Furth, H.P., Killeen, J. and Rosenbluth, M.N.: 1963, Phys. Fluids 6, 459.
Giachetti, R., Van Hoven, G. and Chiuderi, C.: 1977, Solar Phys. 55, 371.
Heyvaerts, J., Priest, E.R. and Rust, D.M. 1977, Ap. J. 216, 123.
Hood, A.W. and Priest, E.R.: 1979, Solar Phys. 64, 303.
Hood, A.W. and Priest, E.R.: 1980, Solar Phys. 66, 113.
Hood, A.W. and Priest, E.R.: 1981a, Geophys. Astroph. Fluid Dyn. 17, 297.
Hood, A.W. and Priest, E.R.: 1981b, Solar Phys. 73, 289.
Hood, A.W.: 1983, submitted.
Kahler, S.W. and Kreplin, R.W.: 1970, Solar Phys. 14, 372.
Kopp, R.A. and Pneuman, G.W.: 1976, Solar Phys. 50, 85.
Mariska, J.T., Boris, J.P., Oran. E.S , Young, T.R., Doschek, G.A.: 1982, Ap. J. 255, 783.
McClymont, A.N. and Canfield R.C.: 1982, submitted.
McClymont, A.N. and Craig, I.J.D.: 1982, submitted.
Migliuolo, S. and Cargill, P.J.: 1983, Astrophys. J.
Mok, Y. and Van Hoven, G.: 1982, Phys. Fl. 25, 636.
Nagai, F.: 1980, Solar Phys. 68, 351.
Pallavicini, R., Serio, S. and Vaiana, G.S.: 1977, Ap. J. 216, 108.
Petschek, H.E.: 1964, AAS-NASA Symp. on Solar Flares, NASA SP-50, p.425.
Pneuman, G.W.: 1981, Chapter 7 of *Solar flare MHD* (ed. E.R. Priest), Gordon & Breach, London.
Priest, E.R.: 1976, Solar Phys. 47, 41.
Priest, E.R.: 1981, Solar flare MHD, Gordon & Breach, London.
Priest, E.R.: 1982, IAU Colloq. No. 71, Activity in red dwarf stars.
Raadu, M.A.: 1972, Solar Phys. 22, 425.
Rust, D.M.: 1976, Phil. Trans. Roy. Soc. Lond. A281, 427.
Sakai, J. and Washimi, H.: 1982, Astrophys. J.
Sakurai, T.: 1976, Pub. Astron. Soc. Japan 28, 177.
Sato, T. and Hayashi, T.: 1979, Phys. Fl. 22, 1189.
Somov, B.V., Spektor, A.R. and Syrovatsky, S.I.: 1977, Iso. Acad. Sci. USSR Phys. Ser. 41, 273.
Soward, A.M. and Priest, E.R.: 1977, Phil. Trans. Roy. Soc. A284, 369.
Soward, A.M. and Priest, E.R.: 1982, J. Plasma Phys.
Spicer, D.: 1977, Solar Phys. 53, 305.
Spicer, D. and Brown, J.C.: 1981, The Sun as a star (ed. S. Jordan), p. 413.
Strong, K. *et al.*: 1982, Solar Phys.
Tur, T.J. and Priest, E.R.: 1978, Solar Phys. 58, 181.
Ugai, M. and Tsuda, T.: 1977, J. Plasma Phys. 17, 337.
Ugai, M. and Tsuda, T.: 1979, J. Plasma Phys. 21, 459; 22, 1.

Van Hoven, G.: 1976, Solar Phys. 49, 95.
Van Hoven, G.: 1981, Chapter 4 of *Solar flare MHD* (ed. E.R. Priest), Gordon & Breach, London.
Wu, S.T., Kan, L.C., Nakagawa, Y., Tandberg-Hanssen, E.: 1981, Solar Phys. 70, 137.
Zirin, H.: 1974, Vistas in Astron. 16, 1.

DISCUSSION

ACTON: Do you predict processes with time scales as short as the time scales of elementary hard X-ray flare burst', say 0.1 sec?

PRIEST: The formation of neutral point pairs occurs on the tearing mode time scale $(\tau_d \tau_A)^{1/2}$, which may be typically 10^2 - 10^3 sec. The coalescence of islands takes placed much faster, namely on the Alfven time scale τ_A, which is a factor of $R_M^{-1/2}$ shorter and therefore possibly as small as 10^{-2} - 1 s. Of course, these values depend very much on the field strength and transverse scale of the structure. What we have done is to identify MHD structures and dynamic behaviour which may lead to particle acceleration. But, having simulated these macroscopic processes, there is a need to consider the microphysics of particle acceleration within them. In particular, I feel it is most important to consider the role of shocks in such acceleration.

KANE: Could you give some specific numbers such as time constants, rate of energy release, etc?

PRIEST: The overall energy release takes place on some multiple of the tearing time, while the additional impulsive energy release at times of coalescence is on the Alfvén time. In order to calculate specific numbers you need to know the plasma parameters such as n, T, B, w. These results are brand-new and, although we expected the basic reconnection, the appearance of a fast shock and of the creation and impulsive annihilation of neutral point pairs was a great surprise. We hope to put some numbers in and make more quantitative estimates next year during the SMM workshops.

UCHIDA: In calculations by Ugai and others, the implicit non-zero constant E is equivalent to the assumption of a forced inflow from infinity towards the reconnecting point. In your case, how about this point, and what determines the location of the first and the successively created neutral points? Isn't it due to the effect of the standing fast-mode wave caused by the boundary condition?

PRIEST: The calculations of Sato and Hayashi have an imposed inflow from large distances. By comparison our numerical experiment has free boundary conditions at the sides and the initial state is in equilibrium. It goes unstable and the side boundary conditions allow it to reconnect as it wishes, driven locally by the instability rather than from outside. The first neutral point is found to form a distance above the photosphere, equal to the initial current sheet width. Subsequently, the reconnection point rises and the new neutral points occur just above the old one when it has reached such a height that the sheet has thinned enough for the secondary tear to occur. The to occur. At higher magnetic reynolds numbers (R_M) one may expect the wavelength of the fastest growing mode to enter; this scales as $R_M^{1/4}$.

EMSLIE: In your thermal instability trigger mechanism, the magnetic field plays no role. The flare then corresponds to transfer of energy from cool (10^4K) to hot (10^7 K) plasma, with no net increase in the energy content of the active region loop. This does not fit the accepted definition of a flare, namely, an event in which the energy content of the plasma increases.

PRIEST: The point of the model is to show how one cool equilibrium can cease and evolve to a new hot equilibrium. It may produce small X-ray brightenings by redistributing the energy in a loop.

BROWN: The point is that the total luminosity will not increase by this process. Indeed, in the case of plasma heated by thermal instability, the total luminosity will fall, both during the transient phase (due to fall of emission function with temperature) and the final state (due to fall in plasma density).

PRIEST: You may well be right, but the consequences of the dynamic catastrophic process following a lack of equilibrium have not yet been studied. Also, this alternative for creating simple-loop events should be regarded very much as a speculative suggestion compared with the other two much better established possibilities. If you want to work on the model to see whether it works, I would be delighted! At present, however, I am keeping an open mind.

HIEI: Is there any difference in the predicted energy for your type 1 flare (simple-loop) and in type 2 (two-ribbon flare)?

PRIEST: Simple-loop flares usually have much less total energy than two-ribbon flares since they are smaller, but sometimes a low-lying simple-loop event may have a very strong magnetic field and so a fast release rate. Both types cover a large range of size. Simple-loop flares include tiny X-ray brightenings at one end of the spectrum up to quite large events at the other. Two-ribbon flares may range from relatively small plage filament events with strong fields up to much larger quiescent filament events with weak fields and a long slow rise and fall. Thus, in order to estimate the energy and its release rate for a specific event, you need the field strength and dimensions of the loop or arcade.

III. ENERGY TRANSPORT, CHROMOSPHERIC HEATING AND EVAPORATION (Part 1)

Chairmen: Z. Svestka and A. Gabriel

SOLAR FLARE X-RAY SPECTRA FROM THE P78-1 SPACECRAFT

G. A. Doschek
E. O. Hulburt Center for Space Research, Naval Research Laboratory, Washington, DC 20375, U.S.A.

ABSTRACT. Results from the high resolution solar X-ray spectrometer experiments on the P78-1 spacecraft are discussed. These results concern physical quantities such as electron temperature and density, turbulence, mass motions, and state of ionization equilibrium, characteristic of the thermal soft X-ray emitting flare plasma, and the time behavior of these quantities during flares. In addition, a brief description of the instruments is given, the plasma diagnostics used in interpreting the spectra are summarized, and the origin of the thermal soft X-ray emitting plasma is discussed in light of the P78-1 results, earlier data, and numerical simulations of magnetic flux tubes heated to solar flare temperatures.

1. INTRODUCTION

The P78-1 spacecraft was launched by the U.S. Air Force on 24 February 1979, as part of the U.S. Department of Defense Space Test Program (STP). It has a solar pointed section that houses X-ray crystal spectrometers built by the Naval Research Laboratory (NRL) and the Aerospace Corporation. These instruments have operated from about March 1979 up to the present time. In this paper results obtained with the crystal spectrometer experiments are described.

2. THE SPECTROMETERS AND SPECTRA

Four uncollimated crystal spectrometers were designed by NRL to record the X-ray spectra of flares below 8 Å over narrow selected wavelength ranges. The acronym for this experiment is SOLFLEX (= solar flare X-rays). The wavelength ranges of the four spectrometers are: 1.82-1.97 Å, 2.98-3.07 Å, 3.14-3.24 Å, and 8.26-8.53 Å. These wavelength ranges were chosen to cover 2p → 1s type transitions in Fe II - Fe XXV, Ca XVII - Ca XX, and Mg XII. The spectra have sufficient resolution for resolving all important

Solar Physics **86** (1983) 49–58. 0038–0938/83/0861–0049 $ 01.50.

diagnostic lines, for measuring intrinsic line profiles, and for measuring relative wavelength shifts.

The Aerospace Corporation solar X-ray experiment is a combination spectrometer and spectroheliometer. Because it is not restricted to flares, the acronym is SOLEX (= solar X-rays). The instrument is spatially collimated to either 20" or 60". The spectral region 3-25 Å covered by the SOLEX spectrometers contains strong spectral lines of O, Ne, Mg, Aℓ, Si, S, Ar, and Fe. For further details of these instruments, see the other paper by Doschek in this volume.

3. PLASMA DIAGNOSTICS

With high resolution X-ray spectroscopy, physical conditions in the hot coronal flare plasma such as electron temperature and density can be determined much more accurately than is possible with broadband detectors. It is even possible to determine the degree of ionization equilibrium in the plasma, which is essential in comparing either derived differential emission measures or relative line intensities with models such as numerical simulations. Below, some of the most important plasma diagnostic techniques are described.

a) Electron Temperature

Electron temperatures are deduced from the intensity ratios of lines formed by dielectronic recombination and electron impact excitation. The theory for this technique was discussed by Gabriel and Jordan (1972) and was developed independently by Vainstein and colleagues, e.g., Vainstein and Safronova (1980). Since the earlier work, in which only transitions from the L shell, i.e., $2p \rightarrow 1s$, were considered in Li-like and He-like ions, a number of revisions or extensions of this work have been carried out (e.g., Vainstein and Safronova, 1980; Doschek, Feldman, and Cowan, 1981; Dely-Dubau et al., 1982a, b; and Phillips et al., 1983).

As an example and for those unfamiliar with the spectroscopy, one of the most useful dielectronic lines for Fe is called j and is the Fe XXIV transition, $1s^2 2p\ ^2P_{3/2} - 1s2p^2\ ^2D_{5/2}$ at 1.8660 Å. The resonance line is called w and is the Fe XXV transition, $1s^2\ ^1S_0 - 1s2p\ ^1P_1$, at 1.8500 Å. Similar lines are present in the Ca spectra. These lines are shown for a typical P78-1 Fe-line spectrum in Figure 1.

b) Electron Density

It was shown by Gabriel and Jordan (1972) that the ratio (R) of the He-like ion forbidden line, $1s^2\ ^1S_0 - 1s2s\ ^3S_1$, to the He-like ion intercombination lines, $1s^2\ ^1S_0 - 1s2p\ ^3P_{1,2}$, is sensitive to to electron density. In addition, the ratio (G) of the sum of the forbidden and intercombination lines to the resonance line,

$1s^2\ ^1S_0$–$1s2p\ ^1P_1$, is sensitive to electron temperature. For the wavelength region accessible to the SOLEX spectrometers, electron densities can be deduced from He-like ion lines of O VII, Ne IX, Mg XI, and also from lines of Ca XV, which is C-like. Since these lines are formed at temperatures $< 10^7$ K, the densities refer to rather low temperature plasma. There is no good line ratio method for obtaining densities at temperatures $> 20 \times 10^6$ K, where the Fe XXV lines are formed.

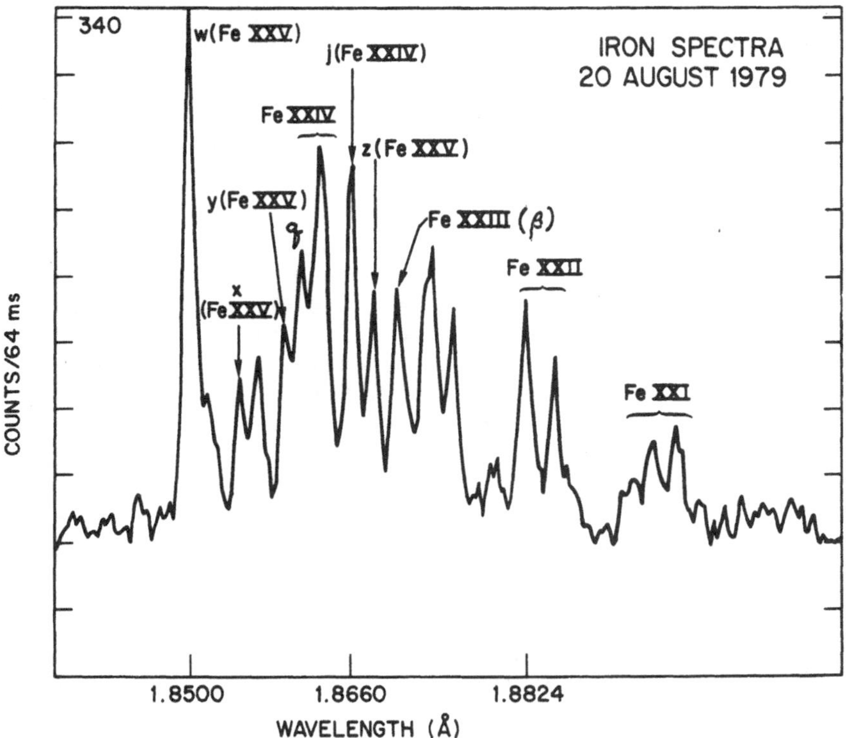

Figure 1 – A typical SOLFLEX Fe-line flare spectrum. Line q is adjacent and on the long wavelength side of line y.

c) Ionization Equilibrium

Some of the innershell X-ray lines of the type 2p→1s, in ionizaion stages below He-like, are produced by electron impact excitation rather than dielectronic recombination. Examples are line q, Fe XXIV (1.8610 Å), and line β, Fe XXIII (1.8704 Å) $1s^2 2s^2\ ^1S_0$–$1s2s^2 2p\ ^1P_1$ (see Figure 1). If the electron temperature T_e is known from j/w ratios, then the q/w or q/j ratio can be used to determine the Li-like to He-like abundance ratio at T_e. If the abundance ratio agrees with ionization equilibrium calculations for T_e, then the plasma is in ionization equilibrium, at least up to He-like ionization stages. If the abundance ratio is different than expected in ionization equilibrium at T_e, then the plasma may be either transiently ionizing or recombining.

The q/w intensity ratios are linearly proportional to the Li-like to He-like ion abundance ratios. A good set of calculations of these latter ratios is that of Jacobs et al. (1977). More recently, Doyle and Raymond (1981) and Shull and Van Steenberg (1982) have also published equilibrium abundances. A comparison of these different calculations shows that there is about a factor of 2 uncertainty in the Li-like to He-like ion abundance ratios for Ca and Fe. The uncertainty arises mainly from uncertainties in ionization and dielectronic rate coefficients, which are still not precisely known.

d) Plasma Dynamics

Since the X-ray lines covered by SOLFLEX are optically thin, asymmetric line profiles are indicative of anisotropic plasma motions at the temperatures at which the lines are formed. Also, if the electron temperature is known from j/w ratios, the ion temperature or random nonthermal motions (turbulence) in the plasma can be determined from the line widths. Finally, for X-type flares, the line widths (FWHM) of the Fe lines vary between about 1.9 and 1.3 m Å. If an average FWHM of 1.6 m Å is assumed for a line, its relative position can be measured with an accuracy of about one-third of the FWHM, or to about 0.5 m Å. This wavelength interval corresponds to a Doppler shift line-of-sight velocity of about 80 km s^{-1}.

4. RESULTS FROM THE P78-1 SPECTROMETERS

A fair number of X-type and M-type flares have been observed by the P78-1 spectrometers, including some very impulsive flares (e.g., McKenzie and Landecker, 1981; Feldman, Doschek, and Kreplin, 1982). The most important conclusions reached from application of the diagnostics discussed in Section III to the flare spectra are:

(a) The peak electron temperature reached during a flare by the bulk of the thermal soft X-ray emitting plasma is about 25 x 10^6 K, for most M and X-type flares. This result is obtained from the j/w ratio for Fe, and the temperature of 25 x 10^6 K is defined as the j/w temperature plus 4 x 10^6 K. Occasionally, the j/w temperature is higher. For the intense 20 August 1979 flare the maximum j/w temperature was about 26 x 10^6 K, and similar values have been found for flares observed by the Hinotori spacecraft. When the j/w temperature is this high, the maximum temperature in the flaring loop or loops might be as high as 35 x 10^6 K.

(b) The electron temperature rises very rapidly to nearly maximum values, and then remains nearly constant or increases by a few million degrees as the X-ray flux continues to increase by a factor of about 20. The Fe XXV resonance line (w) is present in the very first Fe spectra that are recorded during a flare. The rise from ambient electron temperature to near maximum appears to occur in a time less than $\simeq$ 10 s, if the X-ray flux in the Fe XXV line is extrapolated backward in time to zero flux.

(c) After peak flux in the Fe XXV lines is reached, the behavior of the electron temperature varies from flare to flare. In some flares the temperature falls in a few minutes by about 5×10^6 K and then remains nearly constant for long time periods of about 30 minutes. In other flares the temperature simply falls monotonically to values of about 10^7 K in a time on the order of 10 minutes. In the case of impulsive flares, the temperature can fall from 22×10^6 K to 10×10^6 K in about 1 minute. The constancy or slow decline of this temperature during the decay phase of flares indicates continuous energy input into the thermal flare plasma.

(d) The electron density of the preflare loop must be $> 3 \times 10^{10}$ cm^{-3}, and the electron density at peak X-ray flux near 20×10^6 K is $> 10^{11}$ cm^{-3}. The former result is deduced from the observation that the Fe XXV lines are present in the first spectra recorded during a flare. If the density were $< 3 \times 10^{10}$ cm^{-3}, the ionization times of iron ions above Fe XVII would be sufficiently long that transient spectra would be observed.

(e) The electron density at lower temperatures ($< 5 \times 10^6$ K) for several flares has been determined from lines of Ca XV and O VII. For two flares observed by SOLEX spectrometers, the O VII density reaches a maximum value of about 10^{12} cm^{-3} shortly before peak flux in Fe XXV lines is reached (Doschek et al., 1981). Thereafter the density falls to values of about 10^{11} cm^{-3}. Before the flares, the active region density was around 10^{10} cm^{-3}. Typical observed variations in R, the O VII forbidden to intercombination line ratio, are shown in Figure 2. In Figure 2 the smallest ratio R is $\simeq 1$, which corresponds to 10^{11} cm^{-3}. The time variations of density for the two flares discussed by Doschek et al. (1981) are suggestive of heating by the high energy electrons that also produce the hard X-ray bursts. That is, the O VII emission may arise from chromospheric heating to temperatures of $1\text{-}3 \times 10^6$ K, produced by energy deposition of high energy electrons. At the time of peak density, the volume V of O VII emitting plasma can be determined from the volume emission measure obtained from the O VII resonance line (w) and the density deduced from R. The volumes are around 10^{24} cm^3, which corresponds to a very small length, $(10^{24})^{1/3} = 10^3$ km.

(f) The flare plasma near 20×10^6 K appears to be near ionization equilibrium (IE), at least up to He-like ionization stages. As discussed in Section III, the line ratio q/w can be used to infer Li-like to He-like ion abundance ratios at known values of the electron temperature. The measured q/w ratios are similar in all flares observed. The interpretation of the ratios in terms of equilibrium or nonequilibrium conditions depends on the IE calculations adopted. If the calculations of Jacobs <u>et al</u>. are adopted, then the plasma appears to be transiently ionizing. However, this condition persists even during the decay phase, when the plasma is actually cooling. The Doyle and Raymond (1981) calculations give only a very slight ionizing condition during the rise phase, and a recombining condition or IE condition in the decay phase.

(g) The volume emission measure ($N_e^2 V$) determined from the flux in the Fe XXV resonance line increases only slightly (by a factor of 3 or less) after peak flux and electron temperature are reached in the Fe XXV line. The peak in emission measure occurs shortly after peak flux and then begins to fall.

(h) Element abundances relative to oxygen determined for Cr and Ca from lines emitted near 20 Å are 0.0036 ± 0.0018 and 0.015 ± 0.005, respectively (McKenzie and Landecker 1982). Several lines of Cr and Ca were identified for the first time in solar spectra. In all, about 60 lines of this type between 15.4 and 23 Å were detected by the SOLEX spectrometers.

(i) Both random mass motions and upflowing plasma are evident during the rise phase of many flares. The lines appear very broad in the first spectra recorded; much broader than expected at the electron temperature determined from j/w ratios. If the excess widths are interpreted as Doppler broadening due to random nonthermal motions in the plasma, then velocities of 160 km s^{-1} are typical, but velocities as high as 300 km s^{-1} have been observed. The velocities decrease monotonically as the flux in the X-ray lines continues to rise. By the time of peak flux, and during the decay phase, the velocities have become much smaller, about 80 km s^{-1} or less.

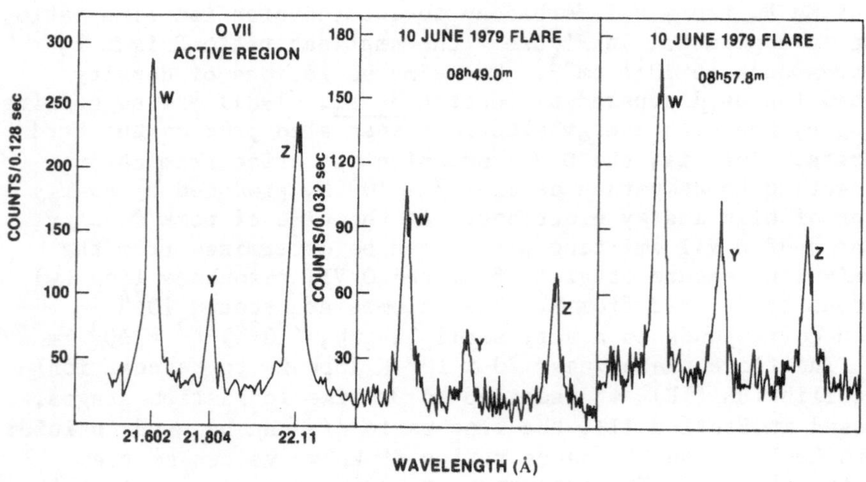

Figure 2 - SOLEX O VII spectra of an active region and a solar flare.

In addition to line broadening, blue wing asymmetries are also observed during the rise phase of most, if not all, flares. For the X-type flares, the intensity of the blue wing of the Ca XIX resonance line is about 15% of the intensity of the unshifted or stationary component, and indicates typical line-of-sight velocities of several hundred km s^{-1}. The blue wing decreases during the rise phase and is very weak or absent after peak flux in the Ca XIX line is attained and the decay phase begins. A typical rise phase profile is shown in Figure 3. For two M-type flares the blueshifted emission was much more intense relative to the unshifted emission, and in the first observed spectra was about 1/2 as intense as the unshifted emission (Feldman et al., 1980).

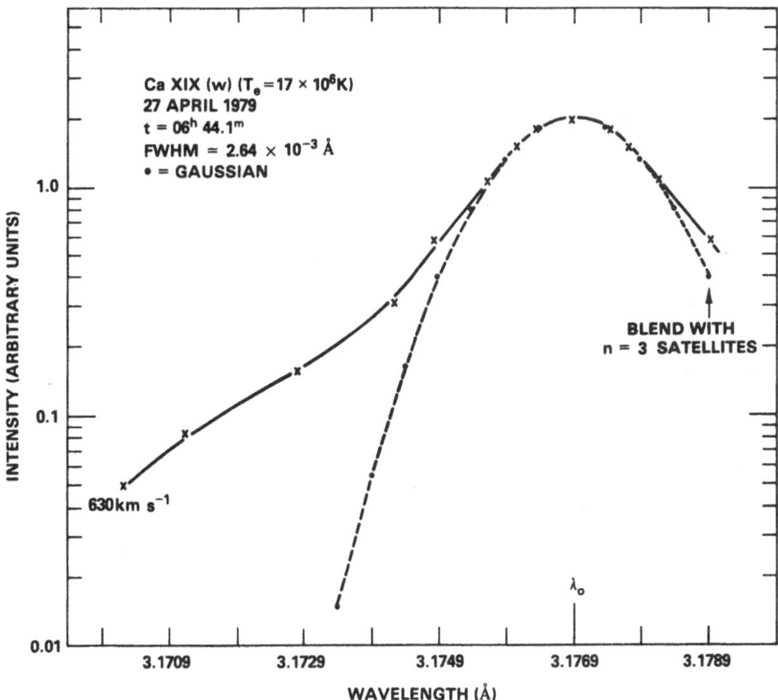

Figure 3 – A typical flare rise phase profile of the Ca XIX line. The dashed line is a Gaussian fit to the unshifted component.

5. INTERPRETATION OF RESULTS

Although energetic electrons that produce hard X-ray bursts are not adequate to account for the entire soft X-ray flare (Feldman, Cheng, and Doschek, 1982), the general idea of chromospheric heating and expansion into flux tubes has gained rather wide acceptance (e.g., Sturrock, 1973). The precise nature of the heating mechanism need not be known in order to postulate and explore the mechanism, generally known as "chromospheric evaporation". Here the process is referred to as ablation.

The basic idea behind the chromospheric ablation process is that the increase in soft X-ray flux and emission measure is produced entirely by chromospheric plasma ablating into flux tubes. The high resolution NRL XUV spectroheliograph data obtained from Skylab do not give any convincing evidence for the particular ablation process we are discussing. With the advent of the recent space missions, such as P78-1, SMM, and Hinotori, interest in directly observing the process has heightened.

NRL also had a slit spectrograph on Skylab that covered the 1100-2000 Å region with a resolution of 0.06 Å. Only one strong high temperature line ($\simeq 9 \times 10^6$ K) falls within this wavelength region, the Fe XXI forbidden line at 1354.1 Å. Although the line was observed in many flares, no net Doppler shift of the line greater than about 4 km s^{-1} was observed. However, in at least some cases the instrument was probably observing cold loops rather than the footpoints of the hot loops from which Fe XXV emission arises.

One way to approach the ablation issue is to attempt to model a flare with numerical simulations. Detailed numerical simulations were carried out by Nagai (1980, 1982), who calculated the ablation dynamics assuming several different heating functions. Similar studies were recently carried out by Cheng et al. (1983) and Pallavicini et al. (1983). The comparison with observations for the Cheng et al. (1983) simulations was presented in a companion paper by Doschek et al. (1983). The dynamical results found by all of the above authors are essentially the same.

In the paper by Doschek et al. (1983), line intensities, emission rates within the loop, and line profiles were calculated for the Fe XXV and Ca XIX X-ray lines, and the Fe XXI forbidden line at 1354.1 Å. The results are: (a) the blueshifted asymmetries on the wings of the Ca XIX and Fe XXV lines are larger than observed in P78-1 data, and, (b) the calculated Fe XXI line was blueshifted by very large amounts ($\simeq$0.4-0.7 Å). Such wavelength shifts are not observed, as mentioned.

Thus, at first sight numerical simulations do not predict spectroscopic parameters that entirely agree with observations. The blueshifts and asymmetries that are observed are less than predicted by the adopted model. However, one possible explanation for at least the discrepancy regarding the intensities of the blueshifted components of the X-ray lines is that the X-ray observations pertain to times after the temperature has reached about 20 x 10^6 K, i.e., after the virtually explosive phase during which the loop is filled with ablating plasma. The spectrometers may lack the sensitivity to observe events during this explosive phase. In this case there may be no discrepancy, since the calculated intensities of the blueshifted spectral components relative to the stationary components decrease considerably at high temperatures after the loop peak temperature reaches about 20 x 10^6 K and is maintained at this value. The problem with the Doppler shift of the Fe XXI line still appears to remain, however. An attempt should be made to simulate an actual flare for which upflows are observed.

The observed blueshifted asymmetries clearly show that some ablation is occurring. It is possible that this ablation is not entirely confined to the hot flare loops, and is therefore not entirely responsible for the observed high densities. That is, a scaled-down version of the ablation idea may operate. High densities in soft X-ray emitting loops might be produced by heating an initially high density loop, or by plasma compression, as discussed by Feldman, Cheng, and Doschek (1982).

Although the results from the P78-1 spacecraft and similar instruments have considerably increased our understanding of solar flares, much work yet remains to be done, both experimentally and theoretically. On the theoretical side, it appears necessary to consider multidimensional models and problems such as flux tube emergence.

Part of this work was supported by a grant from NASA. The author acknowledges useful and stimulating discussions concerning this paper with Drs. C.-C. Cheng, K.P. Dere, U. Feldman, J.T. Mariska, and K.G. Widing.

REFERENCES

Bely-Dubau, F., Dubau, J., Faucher, P., and Gabriel, A.H.: 1982a, Mon. Not. R. astr. Soc. 198, 239.

Bely-Dubau, F., Dubau, J., Faucher, P., Gabriel, A.H., Loulergue, M., Steenman-Clark, L., Volonte, S., Antonucci, E., and Rapley, C.G.: 1982b, Mon. Not. R. astr. Soc. 201, 1155.

Cheng, C.-C., Oran, E.S., Doschek, G.A., Boris, J.P., and Mariska, J.T.: 1983, Astrophys. J., in press.

Doschek, G.A., Feldman, U., and Cowan, R.D.: 1981, Astrophys. J. 245, 315.

Doschek, G.A., Feldman, U., Landecker, P.B., and McKenzie, D.L.: 1981, Astrophys. J. 249, 372.

Doschek, G.A., Cheng, C.-C., Oran, E.S., Boris, J.P., and Mariska, J.T.: 1983, Astrophys. J., in press.

Doyle, J.G., and Raymond, J.C.: 1981, Mon. Not. R. astr. Soc. 196, 907.

Feldman, U., Doschek, G.A., Kreplin, R.W., and Mariska, J.T.: 1980, Astrophys. J. 241, 1175.

Feldman, U., Cheng, C.-C., and Doschek, G.A.: 1982, Astrophys. J. 255, 320.

Feldman, U., Doschek, G.A., and Kreplin, R.W.: 1982, Astrophys. J. 260, 885.

Gabriel, A.H., and Jordan, C.: 1972, in Case Studies in Atomic Collision Physics, Vol. 2, ed. E.W. McDaniel and M.C. McDowell (Amsterdam: North-Holland).

Jacobs, V.L., Davis, J., Kepple, P.C., and Blaha, M.: 1977, Astrophys. J. 211, 605.

McKenzie, D.L., and Landecker, P.B.: 1981, Astrophys. J. 248, 1117.

McKenzie, D.L., and Landecker, P.B.: 1982, Astrophys. J. 254, 309.

Nagai, F.: 1980, Solar Phys. 68, 351.

Nagai, F.: 1982, preprint.

Pallavicini, R., Peres, G., Serio, S., Vaiana, G., Acton, L., Leibacher, J., and Rosner, R.: 1983, Astrophys. J., in press.

Phillips, K.J.H., Lemen, J.R., Cowan, R.D., Doschek, G.A., and Leibacher, J.W.: 1983, Astrophys. J., in press.

Shull, J.M., and Van Steenberg, M.: 1982, Astrophys. J. Suppl. 48, 95.

Sturrock, P.A.: 1973, Ramaty and Stone (ed.), Symp. High Energy Phenomena on the Sun, Goddard Space Flight Center, X-693-73-193, p. 3.

Vainstein, L.A., and Safronova, U.I.: 1980, Atomic Data and Nuclear Data Tables 25, 311.

DISCUSSION

PRIEST: (1) How do you know that there must be continued heating for your 20 min event, given that the conduction may be turbulent? (2) In your numerical simulation of loop flow, how do you treat the transition region and the chromosphere?

DOSCHEK: (1) Even if conduction is turbulent, cooling by radiation at the observed densities should produce faster temperature decreases than observed. (2) The transition region is not resolved in our simulation, but this does not appear to affect the gas dynamical results in a significant way. The chromosphere is idealized, i. e., a standard model such as the VAL is not used. The chromosphere is coupled dynamically to the transition region and corona and serves as a source of mass and momentum during variable heat input. Our results are about the same as those obtained by Nagai, who does use a more detailed chromospheric model. The gross dynamical effects resulting from flare heating fortunately appear to be independent of chromospheric details.

HUDSON: You have distinguished between two types of flares from the behavior of the temperature decay during the gradual phase. Is there any evidence for different behavior of these flares during the impulsive phase?

DOSCHEK: No, I do not believe so. Flares with long decay times can have either long or rather impulsive rise times. The rise times are shorter than the decay times.

NON-THERMAL AND NON-EQUILIBRIUM EFFECTS IN SOFT X-RAY FLARE SPECTRA

A. H. Gabriel[1], E. Antonucci[2], and L. Steenman-Clark[3]

[1] Rutherford Appleton Laboratory, Chilton, Didcot, Oxon, U.K.
[2] University of Turin, Italy
[3] Nice Observatory, Nice, France

ABSTRACT

Processes leading to the excitation of soft X-ray line spectra are discussed in relation to their thermal or non-thermal nature. Through analysis of calcium spectra from the XRP experiment on SMM, it is shown that the ionization balance during the gradual phase of flares is effectively in the steady-state. A search of suitable complex flares with multiple impulsive features has shown indications of soft X-ray line intensity anomalies, consistent with the presence of a non-thermal electron component.

1. INTRODUCTION

It is customary to refer to two stages of a flare as impulsive and gradual. The impulsive phase is assumed to follow closely from the primary energy release. It is characterised by a rapid pulse of intensity and is normally assumed to involve many non-thermal characteristics, eg. fast electron streams and high rate of change of plasma energy content. The gradual phase is characterised by a slower intensity variation and is often referred to as the thermal phase, indicating an assumption regarding the form of the plasma energy. The relationship between these two phases is clearly crucial in understanding the flare mechanism, and it is therefore of great interest to determine the extent to which the observed plasma is truly in a thermal or a quasi steady-state condition.

It is now well known that the impulsive phase is evident in the intensity profile of ultraviolet and hard X-ray spectra, but not usually seen in the soft X-ray region. This is in part due to the structure of the solar atmosphere which causes the downward flow of flare energy to be absorbed in the transition region and upper chromosphere regions, emitting line spectra in the ultraviolet. However the paradox is also explained by the fact that the soft X-ray and UV regions are dominated by spectral lines, whereas the hard

Solar Physics **86** (1983) 59–66. 0038–0938/83/0861–0059 $ 01.20.

X-ray region is basically continuum. The continuum emission responds to the instantaneous electron energy distribution, whether thermal or non-thermal, whereas the production of lines at short wavelengths must normally await the development of higher stages of ionization. The soft X-ray line spectra (or integrated spectra, since the lines dominate) are thus indicative of the gradual phase of the flare. Exceptions to this are the soft X-ray continuum, which is difficult to measure on the SMM, and the Kα type X-ray lines, which are dealt with elsewhere (Parmar et al., 1983).

Having established that the soft X-ray lines generally follow the gradual phase, we can still consider two aspects of their possible departure from thermal equilibrium. The first concerns the possibility of the ionization balance being different from the steady-state value due to the rapid time development of the plasma. The second is concerned with the possibility that the relative spectral intensities could reveal excitation by an electron distribution that departs from a simple Maxwellian.

2. IONIZATION BALANCE

The technique used is based upon line intensities in the complex of $n = 2 \rightarrow 1$ transitions in the He-like ion and associated dielectronic satellite lines. Figure 1 shows such a spectrum for calcium, recorded by the X-Ray Polychromator (XRP) experiment on the Solar Maximum Mission. The instrumentation has been described by Acton et al.; (1980). It has long been realised (eg. Bhalla, Gabriel and Presnyakov, 1975) that the ratio of some pairs of lines (k/w) is sensitive only to electron temperature T_e, whereas others (q/w) are sensitive to the ratio of Li-like to He-like ions. It is thus possible from a single spectrum to compare the ion ratio to that predicted in the steady-state from ionization balance theory. In this way one can determine the extent of departure from steady-state ionization. It is important however to consider the accuracy of the atomic theory involved. The calculated intensity of the dielectronic satellite lines (Bely-Dubau et al., 1982) is probably good to ~ 5%. The intensities of the directly excited lines q and w, given in the same reference, is perhaps good to ~ 10%. However, the theoretical ionization balance (see eg. Jacobs et al., 1980) is the result of many processes difficult to evaluate, and is probably not better than a factor of 2. Thus, although we expect T_e and the measured Li-like/He-like ion balance to be good to 10%, the conclusions on departures from steady-state ionization are more difficult to formulate.

Starting with Bhalla et al. (1975) many investigators have determined that the plasmas observed usually show a high value of Li-like/He-like ion ratio, and might therefore be considered as in a transient ionizing state. This has been found in both solar active regions and in flares. Since the predicted time-constant for

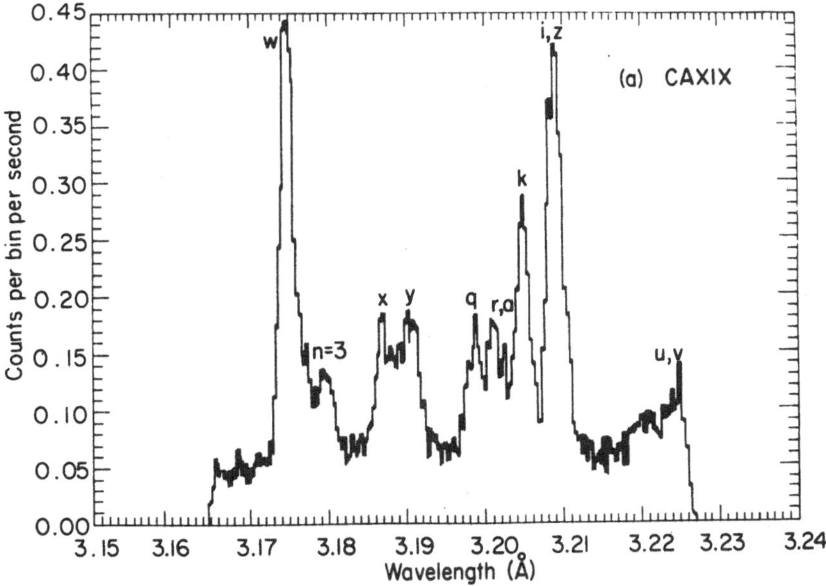

Figure 1. Solar flare calcium spectrum recorded by the Bent Crystal Spectrometer of XRP.

steady-state ionization is much shorter than the duration of these sources, their transient nature is difficult to understand. Koshelev and Kononov (1982) proposed that the flares are really composed of a large number of successive elementary flares of short 3
duration, and that they are thus really in a transient state. An alternative explanation has been proposed (Korneev et al., 1981) in which the problem is attributed to the presence of supra-thermal electrons produced by the large temperature gradients found in these plasmas. Doschek and Feldman (1981) try to explain the discrepancies by an argon line blended with the calcium q line. However they concluded that this was insufficient to account for the effect. We propose that it is unnecessary to invoke complex mechanisms and that the explanation lies in predictable errors in the ionization balance theories.

Using data from the XRP, calcium spectra were analysed taking 368 different spectra, recorded at times throughout the development of some 18 flares and active regions. These data points are shown in Figure 2, plotting measured ion ratios against measured electron temperatures. The fact that these points lie on a single curve, and that the rising portion of the flares (shown as crosses) and the

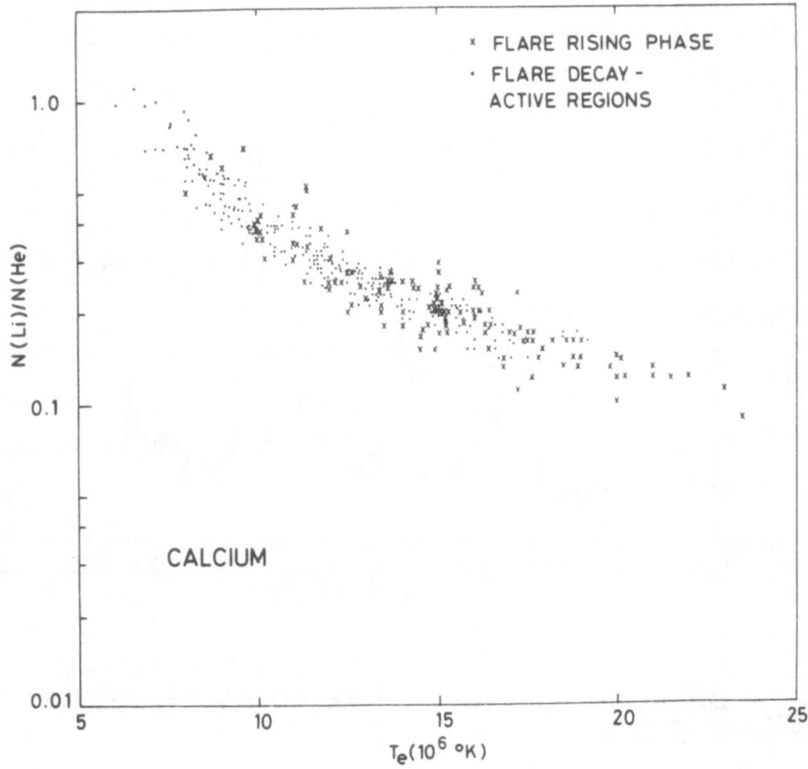

Figure 2. Ionization ratios derived from solar flare and active region observations.

decaying portion (shown as dots) define the same curve, leads us to conclude that the flares are in steady-state ionization. This analysis effectively assumes the plasma to be isothermal. However, a careful study by modelling a number of non-isothermal plasmas has shown that the isothermal assumption does not introduce any significant systematic errors in this case (Antonucci et al., 1982).

In principle one can use the measured curve of Figure 2 in order to correct the theoretical ionization balance. However this requires an evaluation of the intensity of the argon line blended with the calcium q line, and depends on the poorly known abundance ratio of argon to calcium. Work is continuing on this problem.

3. NON-THERMAL ELECTRONS

As indicated earlier, soft X-ray lines are not normally enhanced during the impulsive phase of flares, since the primary electrons pass easily through the low density of the corona without being stopped. The soft X-ray intensity builds up, when the density of plasma in the coronal loops increases by means of chromospheric evaporation. However, in complex multiple events, one can expect the primary acceleration to occur repeatedly, after the coronal density is enhanced. In such events one might look for the effects of non-thermal excitation on the soft X-ray line ratios. In one such well studied event on November 6, 1980 (Svestka et al., 1982) we have searched for the signature of these electrons.

It can readily be shown that the only lines which are effectively excited by high energy electrons (> 10 keV) are the w and q lines, since these are normal optically allowed excitations. The triplet excitations for x, y and z proceed only through electron exchange, and therefore fall off rapidly with increasing energy. The dielectronic capture lines such as k are not excited at all by these electrons. We therefore search for impulsive spikes in which the ratio of w to the blend z+j is abnormally high. One such event was found at time 11:43:19 UT lasting < 6 sec. The time profile is shown in Figure 3, while Figure 4 shows the spectrum just before and during the event. For such short integration times, the photon noise on the spectra becomes relatively large. However, the best

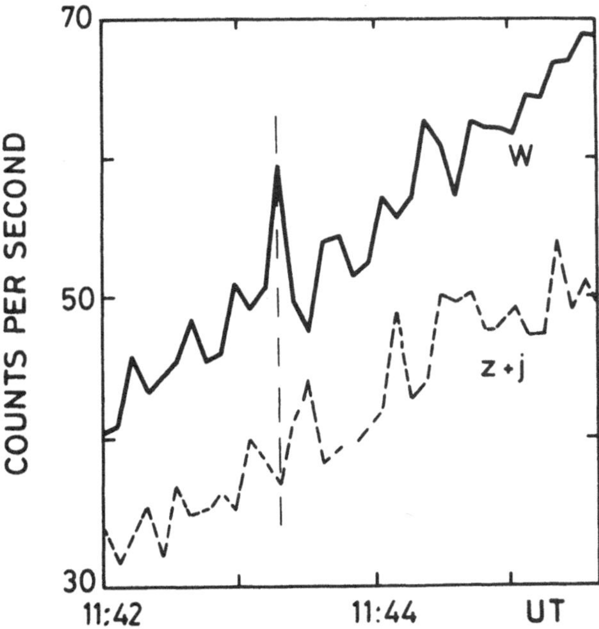

Figure 3. Time development of calcium lines w and j+z.

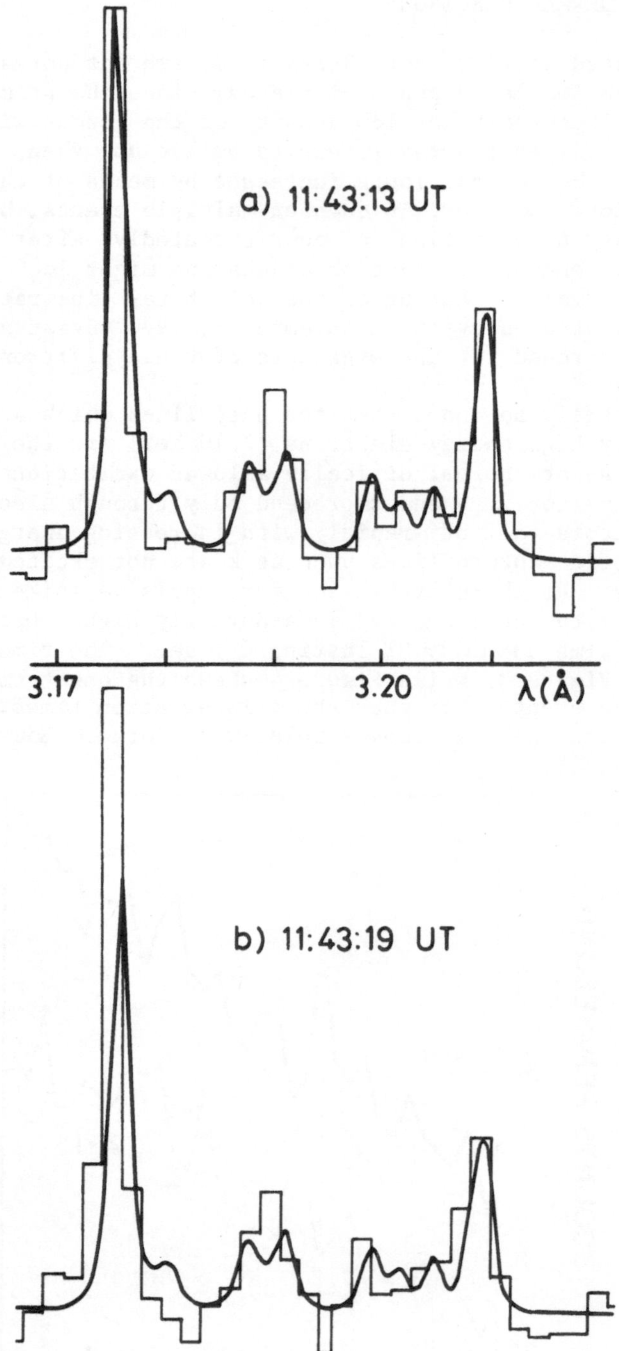

Figure 4. BCS spectra of calcium during 6 sec intervals on November 6 1980.

fit theoretical spectra to the pre-event observation at 11:43:13 UT can be seen superimposed on the 11:43:19 UT spectrum, and clearly shows that the w line has been greatly enhanced relative to the others. There is also some indication of an enhanced q line as predicted.

In view of the poor statistics, one hopes to find further examples of this phenomenon before being confident of the effect. Analysis of Figure 4b indicates that the ratio of non-thermal to thermal electrons is about unity if it is assumed that the event duration is 6 sec, the time resolution of this particular data set. Such a pulse of non-thermal electrons could produce a measurable increase in the ionization of the plasma for a short time (~ 10 sec) following it. The apparent increase in intensity of the line z soon after 11:43:19 UT might be due to the formation of this line by recombination from the H-like ion, following such ionization.

4. CONCLUSIONS

We find that the gradual phase of solar flares as observed in soft X-ray lines is normally in overall steady-state ionization balance. In complex flares with repeated impulsive bursts, there are indications of short bursts of non-thermal electrons producing selective excitation of certain lines.

5. ACKNOWLEDGEMENTS

The X-Ray Polychromator experiment is a joint project between Lockheed Palo Alto Research Laboratory, Mullard Space Science Laboratory and Rutherford Appleton Laboratory. Part of the work described has benefitted from grants from NATO and from CNR.

REFERENCES

Acton, L.W., Culhane, J.L., Gabriel, A.H., Bentley, R.D., Bowles, J.A., Firth, J.G., Finch, M.L., Gilbraith, C.W., Guttridge, P., Hayes, R.W., Joki, E.G., Jones, B.B., Kent, B.J., Leibacher, J.W., Nobles, R.A., Patrick, T.J., Phillips, K.J.H., Rapley, C.G., Sheather, P.H., Sherman, J.C., Stark, J.P., Springer, L.A., Turner, R.F. and Wolfson, C.J.: 1980, Solar Physics **65**, 53.

Antonucci, E., Gabriel, A.H., Doyle, G., Dubau, J., Faucher, P., Jordan, C. and Veck, N.J.: 1983, to be submitted.

Bely-Dubau, F., Dubau, J., Faucher, P., Gabriel, A.H., Loulergue, M., Steenman-Clark, L., Volonté, S., Antonucci, E. and Rapley, C.G.: 1982, Mon. Not. R. Astron. S. **201**, 1155.

Bhalla, C.P., Gabriel, A.H. and Presnyakov, L.P.: 1975, Mon. Not. R. Astron. S. **172**, 359.

Doschek, G.A. and Feldman, U.: 1981, Astrophys.J. **257**, 792.

Jacobs, V.L., Davis, J., Roberson, J.E., Blaha, M., Cain, J. and Davis, M.: 1980, Astrophys. J. 239, 1119.
Korneev, V.V., Mandelstam, S.L., Opariu, S.N., Urnov, A.M. and Zitnik, I.A.: 1981, paper presented at COSPAR meeting.
Koshelev, K.N. and Kononov, E. Ya.; 1982, Solar Physics 77, 177.
Parmar, A.N., Wolfson, C.J., Culhane, J.L., Phillips, K.J.H., Acton, L.W., Dennis, B.R. and Rapley, C.G.: 1983, submitted to Astrophys. J.
Svestka, Z., Dennis, B.R., Pick, M., Raoult, A., Rapley, C.G., Stewart, R.T. and Woodgate, B.E.: 1982, Solar Physics 80, 143.

DISCUSSION

DOSCHEK: Do you see any enhancement of the Fe XXV line as well as for the Ca XIX line?

GABRIEL: This was a very cool event and the iron spectrum was consequently too weak for such an analysis.

OBSERVATION OF CHROMOSPHERIC EVAPORATION DURING THE SOLAR MAXIMUM MISSION

E. Antonucci[1] and B. R. Dennis[2]

[1] Universita di Torino, Italy
[2] NASA Goddard Space Flight Center, Greenbelt, Md., U.S.A.

ABSTRACT. A sample of flares detected in 1980 with the Bent Crystal Spectrometer and the Hard X-Ray Burst Spectrometer on the Solar Maximum Mission satellite has been analysed to study the upward motions of part of the soft X-ray emitting plasma. These motions are inferred from the presence of secondary blue-shifted lines in the Ca XIX and Fe XXV spectral regions during the impulsive phase of disk flares. Limb flares do not show such blue-shifted lines indicating that the direction of the plasma motion is mainly radial and outward. The temporal association of these upward motions with the rise of the thermal phase and with the impulsive hard X-ray burst, as well as considerations of the plasma energetics, favour the interpretation of this phenomenon in terms of chromospheric evaporation. The two measureable parameters of the evaporating plasma, emission measure and velocity, depend on parameters related to the energy deposition and to the thermal phase. The evaporation velocity is found to be correlated with the spectral index of the hard X-ray flux and with the rise time of the thermal emission measure of the coronal plasma. The emission measure of the rising plasma is found to be correlated with the total energy deposited by the fast electrons in the chromosphere by collisions during the impulsive phase and with the maximum emission measure of the coronal plasma.

1. Introduction

The Soft X-Ray Polychromator observations during the Solar Maximum Mission (SMM) have shown a systematic presence of upward moving, high temperature plasma in the corona during the impulsive phase of many flares (Antonucci *et al.*, 1982; Antonucci, 1982). Part of the plasma at a temperature exceeding 10^7 K rises in the solar atmosphere at a velocity of the order of 400 km/sec or less at flare onset. This plasma flow presumably results from the chromospheric evaporation process, which was proposed to explain the appearance of large amounts of hot plasma in the corona during flares (Neupert, 1968; Hudson, 1973; Sturrock, 1972). In fact,

Solar Physics **86** (1983) 67–77. 0038–0938/83/0861–0067 $ 01.65.

evidence for rising hot plasma is observed from the beginning of the impulsive phase until the peak of the thermal phase. The rate of increase of the fractional emission measure of the thermal coronal plasma is found to be proportional to the velocity of the upward motion. For a few flares whose characteristics can be more accurately determined, the energy and mass transferred by the upward moving plasma are sufficient to account for all of the thermal coronal plasma at the peak of the gradual phase on the assumption of magnetic confinement of the rising plasma. The observations suggest that chromospheric evaporation is a common occurrence in X and M class flares.

The soft X-ray data for this analysis were obtained with the Bent Crystal Spectrometer (BCS) on SMM in the spectral regions containing atomic lines of Ca XIX (3.165 - 3.266 Å) and Fe XXV (1.843 - 1.896 Å) in the period from March to November 1980. The BCS integrates the soft X-ray emission over the whole flaring region, but it was the first spectrometer to measure X-ray spectra simultaneously in various wavelength regions with a time resolution of 1 to 6 seconds (Acton _et_ al., 1980). Hence, it was the first instrument capable of studying the rapidly changing dynamic conditions of the soft X-ray plasma during the impulsive phase of flares. For this paper, we have considered the 36 largest flares detected by BCS, all of which were class M or X. Only results from the calcium spectral region will be reported although similar results are obtained by analysing other high temperature emission lines.

The hard X-ray data were obtained with the Hard X-Ray Burst Spectrometer (HXRBS, Orwig _et al._, 1980) also on SMM. Power-law fits to the 15 channel spectra measured every 0.128s were obtained for all 36 flares studied. The hard X-rays are assumed to be produced in thick-target interactions as the fast electrons deposit their energy in the chromosphere. This hypothesis is supported by correlations found between the hard X-ray spectral index and the upward velocity of the evaporating plasma, and between the total energy deposited by the electrons and the maximum emission measure of the hot coronal plasma.

2. Role of Mass Flows in the Development of the Thermal Phase of Flares

A typical spectrum obtained with the BCS in the Ca XIX region during the impulsive phase of a disk flare is shown in Figure 1. This is an observation at $03^h05^m04^s$ UT of a class M4 flare on April 8, 1980. The data are averaged over 27 seconds. This spectrum shows a secondary component which is blue-shifted and reduced in intensity with respect to the principal component. The best fit to the data, shown as a continuous line in the upper part of the figure, is obtained by superposing two syn-

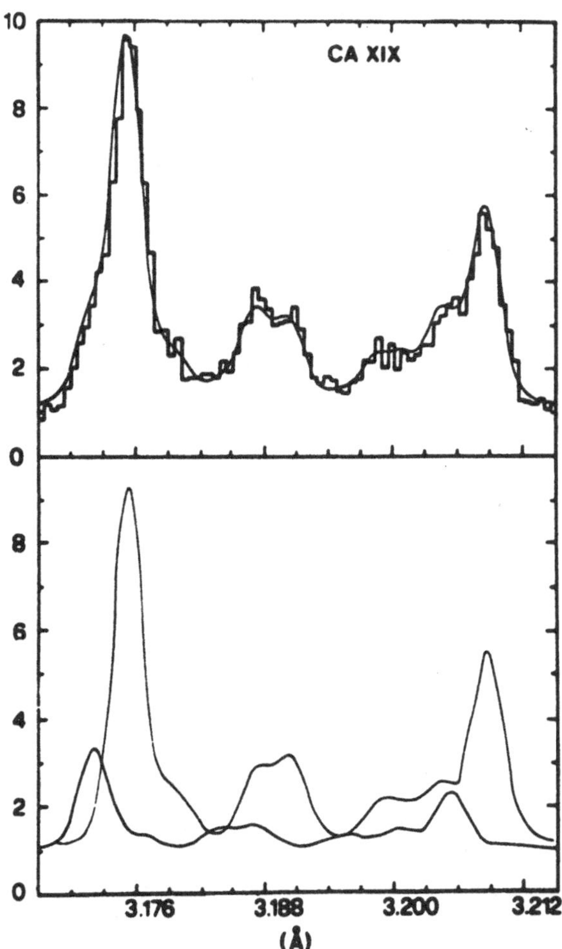

Figure 1. Ca XIX spectrum during the April 8, 1980 flare at $03^h05^m04^s$ UT. The intensity is expressed in counts/sec and has been averaged over a period of 27 seconds.

thesised spectra computed for the same set of physical parameters. These two spectra are shown in the lower part of the figure. The secondary spectrum is shifted by 3mÅ towards shorter wavelengths and its intensity is a factor of 0.17 lower than that of the principal spectrum. The method for computing these synthesised spectra is described by Antonucci et al. (1980). Both spectra are computed for a plasma electron temperature of 1.4×10^7 K and a Doppler temperature of 5.6×10^7 K. The additional parameters determining the emission line intensities are the population ratio of the Li-like to He-like ions, taken to be 0.2, and the ratio of the H-like to the He-like ions, taken to be 0.08. This spectral analysis indicates that 17% of the plasma at 1.4×10^7K moves upward with a velocity of 310 km/sec. In addition, the high non-thermal broadening of the emission lines of the principal spectrum, which is also a characteristic of the early phase of flares (Doschek *et al.*, 1980; Antonucci *et al.*, 1982), is likely to be due to turbulent motions in the coronal plasma at 130 km/sec.

The belief that the plasma motions are systematically upward is supported by the absence of blue-shifted secondary spectra in all of the flares occurring at longitudes exceeding 60°, while blue-shifted spectra are detected in 90% of all disk flares. As a consequence of this difference between limb flares and disk flares, the upward velocity derived from the observed blue-shift must be corrected for the longitude of the flare site.

According to the above results, the soft X-ray plasma at the flare onset consists of two basic components, a dynamic component that later disappears, and a stationary component that is present throughout the event. The dynamic component, we believe, is due to chromospheric evaporation resulting from the energy deposition during the impulsive phase. The stationary component is formed at coronal heights by the accumulation of the plasma flowing into the corona from the chromosphere. Moreover, the increase of the soft X-ray emission of the stationary coronal plasma should continue as long as chromospheric evaporation continues. Therefore, the emission measure and the thermal energy of the stationary coronal plasma should continue to increase as long as blue-shifted X-ray spectra are observed.

The temporal evolution of the emission from the dynamic and stationary plasma components for the May 21, 1980 class X1 flare is shown in Figure 2. The velocity of the rising plasma decreased from 380 km/sec at $20^h55^m20^s$ UT, the time of the main hard X-ray burst, to 120 km/sec at $21^h06^m48^s$ UT. Any upward motion with a velocity below this limiting value is not detectable since, for such a case, the shifted and principal spectra would not be resolved. The emission measure of the stationary plasma continues to increase as long as plasma motions are detected, and begins to decrease when the upward motion disappears, as expected from the plasma accumulation hypothesis. Emission measures are computed on the assumption that the calcium atoms contributing to the emission are at the temperature derived for each time interval from the spectral analysis discussed above (Antonucci *et al.* 1980). Only in 25% of the events considered is the emission measure of the coronal plasma still increasing after plasma flows are no longer observed. However, in those cases the increase stops within two minutes after the disappearance of the plasma flow. An analysis involving higher temperature emission lines than calcium XIX would tend to improve the temporal association between the plasma flow and the increase in the amount of coronal plasma. In fact, we are making the simplifying assumption of isothermal plasmas even though it is known that the intensity of such higher temperature lines as Fe XXV and Fe XXVI peak earlier than the intensity of the Ca XIX line.

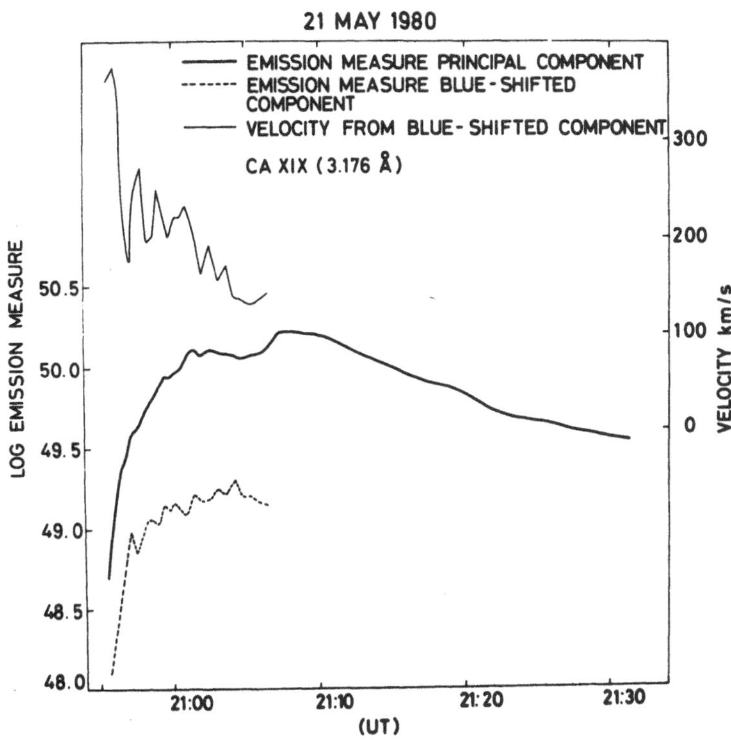

Figure 2. Time evolution of the velocity and emission measure of the dynamic component compared to that of the emission measure of the stationary plasma component during the May 21, 1980 flare.

For some of the flares analysed, the Hard X-Ray Imaging Spectrometer on SMM has resolved separate sources of hard X-ray emission in the energy range 16-30 keV (Hoyng et al., 1981). It is suggested that these sources are at the footpoints of magnetic loops where the electron energy is deposited, and that this is where chromospheric evaporation occurs. For these flares, the magnetic configuration can be inferred by assuming that the hard X-ray sources are at the footpoints of semicircular loops.

The fact that different footpoints may be activated successively during flares is not crucial to our treatment since the BCS observes the integrated emission of the whole flaring region. Once the magnetic configuration is inferred, the mass and energy input to the corona due to evaporation can be derived on the hypothesis of magnetic confinement and continuous mass flow through the footpoints during the impulsive phase. However, since the electron temperature, T_e', and the density, n_e', of the upward moving plasma are unknown, this coronal input cannot be derived directly. Note that the electron temperature has been considered to be the same for both the principal and the secondary spectrum in the analysis of the data shown in Figure 1, but it is, in fact, accurately determined only for the principal component. The emission measure, EMB, and the velocity, V, of the dynamic plasma are, however, determined directly from the spectral analysis. We have inferred values for T_e' and n_e' from the measured quantities by requiring that the input rates of mass and energy, integrated over the time that chromospheric evaporation is taking place, account for the increase in the number of coronal electrons, ΔN_e, and in their thermal energy, ΔE_{th} (Antonucci et al., 1982). Since the time of evaporation coincides with the rising part of the gradual phase, ΔN_e and ΔE_{th} correspond to the total mass and thermal energy increase in the corona during the flare.

In Table I the peak values of electron density, $n_{e,max}$, and temperature, $T_{e,max}$, of the coronal plasma compared with the average values of the density, $\overline{n_e'}$, and temperature, $\overline{T_e'}$, derived for the evaporating mass. The observed coronal plasma in a flare can be easily explained for reasonable physical conditions of the evaporating material. Therefore, we suggest that the accumulation process following chromospheric evaporation is the main mechanism that builds up the thermal phase of flares.

TABLE I

Date of Flare (1980)	Start Time (UT)	$n_{e,max}$ (10^{11} cm^{-3})	$\overline{n_e'}$ (10^{10} cm^{-3})	$T_{e,max}$ (10^6K)	$\overline{T_e'}$ (10^6K)
8 April	0303	3.1	3.6	15.5	20.0
10 April	0917	3.3	7.5	16.0	15.0
9 May	0711	6.9	11.0	18.0	17.0
21 May	2055	2.7	3.6	16.6	20.0
5 November	2233	2.0	9.9	20.4	16.0

We have searched for correlations between the two measurable parameters of the plasma flow, V and EMB, and parameters of the coronal plasma. We find that V is associated with R, the average rate of increase of the fractional emission measure of the coronal plasma. This parameter, R, is defined by the relation, $R = 100 (E_{max}-E_{pre})/(E_{max}\Delta t)$ % min^{-1}, where E_{max} is the maximum emission measure recorded during the flare, E_{pre} is the emission measure existing prior to the flare, and Δt is the time taken to reach E_{max}. The maximum value of V, V_{max}, is plotted versus R in Figure 3, where the positive correlation can be seen. The correlation coefficient is 0.8. The peak emission measure of the upward moving plasma, EMB_{max}, is also positively correlated with EM_{max} with a correlation coefficient of 0.7.

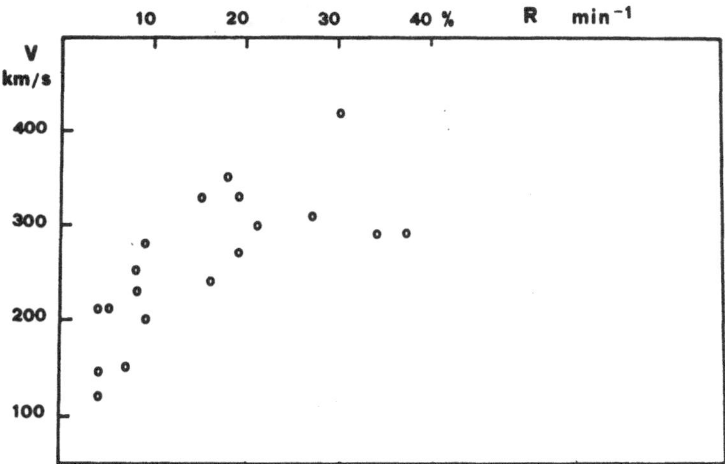

Figure 3. Evaporation velocity versus rate of increase of the emission measure of the coronal plasma.

3. Association of Primary Energy Release and Plasma Motions.

The secondary, blue-shifted lines in the soft X-ray emission appear in coincidence with the start of the hard X-ray emission. The hard X-ray power-law spectral index, γ, and the photon flux at 50 keV have been computed throughout each flare from the HXRBS data. The energy deposited in the chromosphere by the primary non-thermal electrons that are assumed to produce the observed hard X-rays is still quite low - a few times 10^{27} erg s^{-1} - at the time of the first observation of chromospheric evaporation. However, energies of this order are sufficient to sustain the process at this stage. This can be deduced by comparing the energy deposition rate and the rate of energy transferred into the corona by the evaporating plasma at the onset of the flares listed in Table I, for which the evolution of the energy transfer is known. For these flares, the total energy of the non-thermal electrons (> 25keV) assumed to produce the hard X-rays in thick-target interactions appears to be sufficient to power the thermal flare. Qualitatively, for all flares, chromospheric evaporation lasts throughout the time of the impulsive hard X-ray burst.

The velocity of the evaporating plasma is not correlated with the energy deposited in the chromosphere assuming the thick-target model. It is, however, correlated with the spectral index of the hard X-ray flux. Figure 4 shows that the highest value of the evaporation velocity during a flare is correlated with the lowest value of the power-law index of the hard X-ray spectrum with a correlation coefficient of -0.6. This result, together with the correlation found between V_{max} and R, implies that R should also be correlated with the hardness of the X-ray spectrum. This correlation between R and γ can be see! in Figure 5 and the correlation coefficient is -0.7. Limb flares have been included in Figure 5 since observations of blue-shifted lines are not required to determine a value for R.

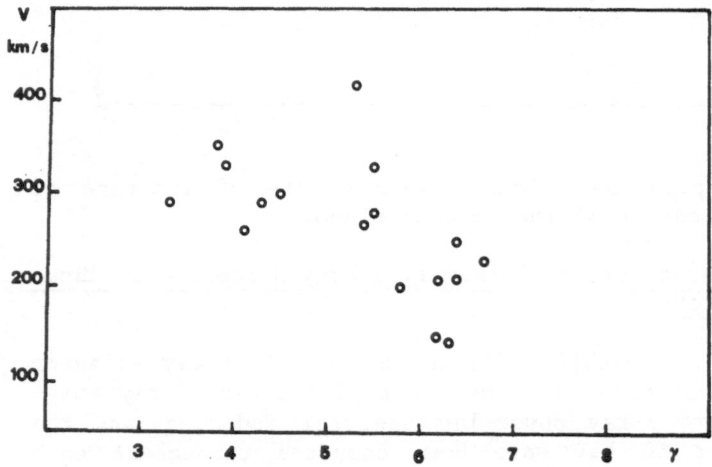

Figure 4. Evaporation velocity versus the lowest, hard X-ray spectral index, γ, measured during the flare.

The total energy deposited in the chromosphere by fast electrons (>25 keV) assuming a thick target model appears to be related to the peak emission measure of the evaporating plasma, but the correlation coefficient is only 0.4. This correlation together with the dependence found between the emission measure of the dynamic and stationary components implies a positive correlation between the total energy deposited by the electrons in the chromosphere and the peak emission measure of the thermal coronal plasma. This correlation has been tested, including also limb flares, and the coefficient is found to be 0.7.

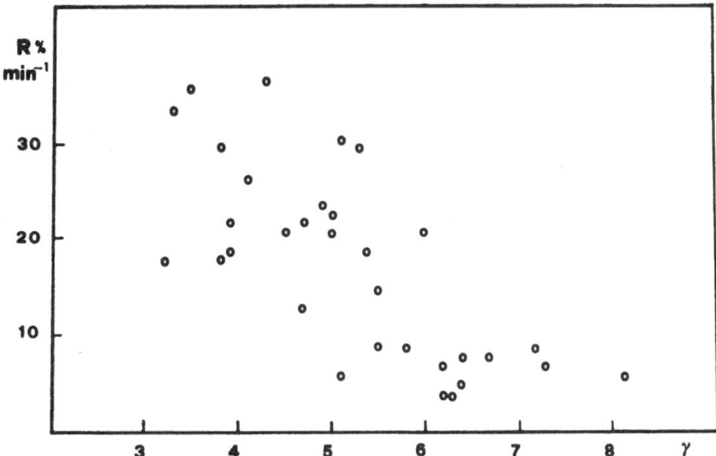

Figure 5. Rate of increase of the fractional emission measure of the coronal plasma versus the lowest hard X-ray spectral index, γ, measured during the flare.

4. Conclusion

We conclude that upward motions of the soft X-ray plasma are temporally associated with the build up of the thermal phase of flares and with the period of energy deposition as indicated by the hard X-ray emission. Moreover, the hardness of the hard X-ray spectrum, the evaporation velocity, and rate of increase of the gradual phase are correlated. The total electron energy deposited in the chromosphere, the peak emission measure of the evaporating plasma, and the peak emission measure of the thermal coronal plasma also appear to be correlated.

Acknowledgments

The Soft X-Ray Polychromator Experiment is a collaboration of three institutions: the Lockheed Palo Alto Research Laboratories, the Mullard Space Science Laboratory and the Rutherford Appleton Laboratory. We are indebted to Dr. A.H. Gabriel for his interest and encouragement in this work, to Shelby Kennard of the Computer Science Corporation for the computation of the hard X-ray spectra, and to D.P. Mathur of the Rutherford Laboratory for his help in the development of the software for the XRP data analysis.

References

Acton, L.W. et al.:1980, Solar Phys. 65, 53.
Antonucci, E. et al.:1982, Solar Phys. 78, 107.
Antonucci, E.:1982, Memorie Soc. Astr. Ital. 53, 495.
Doschek, G.A., Feldman, U., Kreplin, R.W., Cohen L.: 1980, Astrophys.J. 239, 725.
Hoyng, et al.:1981, Astrophys.J. Letters 246, L155.
Hudson, H.S.:1973, Symp. High Energy Phenomena on the Sun, Goddard Space Flight Center X-693-73-193, p. 207.
Neupert, W.M.:1968, Astrophys.J. Letters 153, L59.
Orwig, L.E., Frost, K.J., and Dennis, B.R.:1980, Solar Phys. 65, 25.
Sturrock, P.A.:1973, Symp. High Energy Phenomena on the Sun, Goddard Space Flight Center X-693-73-193, p. 3.

DISCUSSION

HIEI: Evaporation may occur not only in one flare loop, but in several loops, successively. Could you find such an evidence in the SMM soft X-ray data?

ANTONUCCI: The Bent Crystal Spectrometer detects the soft X-ray emission integrated over the flare plasma, therefore single loops cannot be spatially resolved. It is also difficult to extract this information from the time structure of the profile of the evaporation velocity during the impulsive phase.

EMSLIE: If the data to which you refer are indeed to be explained by chromospheric evaporation, then the instantaneous rate of increase of the emission measure of the stationary component should be proportional to the product of the emission measure in the upward moving component times its velocity. The data you present do not appear to satisfy this requirement, even though the event-integrated meission measure in the moving component does correspond with the emission measure in the stationary component, for reasonable densities of this moving material.

ANTONUCCI: The data presented do satisfy this requirement. The instantaneous rates of increase of the thermal energy and mass observed in the stationary component are indeed consistent with the computed energy and mass input rates, due to the moving component. For instance, at 2100 UT of the 21st of May flare, the increase in thermal energy and mass observed in a 20 interval is 3×10^{29} ergs and 2.4×10^{37} electrons. For the same interval, the computed energy input would be 2.6×10^{29} ergs and the mass input 3.6×10^{37} electrons.

PRIEST: If the heating were at the top of a loop you would expect a downflow in the initial stages. Is the lack of an observed redshift due to a lack of sensitivity to that particular plasma or is it because of heating at footpoints?

ANTONUCCI: Redshifts at the onset of flares were not observed during the SMM period. In some events, there is a significant soft X-ray emission in pre-flare conditions, but no redshift is observed; however non-thermal broadenings of spectral lines can precede the appearance of blueshifts. Hence, it seems reasonable to think in terms of heating at the footpoints.

PALLAVICINI: It is true that the conduction front moves very fast from the loop top to the footpoints. However, in addition to that, there is also a pressure wave which moves downwards much more slowly. This pressure wave may produce a redshifted component. The problem is that the flare loop is initially almost empty, before evaporation starts, so that the emission measure corresponding to the redshifted component is very small and you may not see it with the Bent Crystal Spectrometer.

BRÜECKNER: Have you (or anybody) considered the charge imbalance and resulting electric field caused by an electron beam?

BROWN: The problem of charge imbalance in electron beams was solved 10 years ago by recognition that a neutralizing return current is established in the background plasma, just as in the laboratory beam experiments. This point has beem made in some tens of papers already.

TRANSPORT AND CONTAINMENT OF PLASMA, PARTICLES AND ENERGY WITHIN FLARES

L. W. Acton, W. A. Brown, M. E. C. Bruner,
B. M. Haisch, and K. T. Strong
Lockheed Palo Alto Research Laboratory

ABSTRACT. Results from the analysis of flares observed by the Solar Maximum Mission (SMM) and a recent rocket experiment are discussed. We find evidence for primary energy release in the corona through the interaction of magnetic structures, particle and plasma transport into more than a single magnetic structure at the time of a flare and a complex and changing magnetic topology during the course of a flare. The rocket data are examined for constraints on flare cooling, within the context of simple loop models. These results form a basis for comments on the limitations of simple loop models for flares.

1. OVERVIEW

Observations of the solar corona have shown that magnetic loops are a common structural form in flares. The most obvious examples of this are the X-ray and EUV images obtained from ATM (Moore et al., 1980 and references therein). At present, practically all theoretical models of flares make the assumption that the magnetic field acts solely as a rigid, impermeable, symmetric container for the plasma, with inflow and outflow of material restricted to the open-ended footpoints rooted in the chromosphere. The dynamics of the flare process are restricted to the responses of the confined plasma to *ad hoc*, time dependent heating functions introduced at some point, most often the apex, within the loop. It is the purpose of this paper to discuss solar contexts in which the common assumption of an isolated plasma in a low beta loop of uniform cross section is inadequate. Our comments will be focussed on four flares observed by the Solar Maximum Mission (SMM) and one recently observed by a rocket-borne instrument.

Solar Physics **86** (1983) 79–90. 0038–0938/83/0861–0079 $ 01.20.

2. SELECTED SMM FLARE OBSERVATIONS

We first discuss several flares that illustrate the transport of flare plasma and particles into dramatically different magnetic structures. This type of phenomenon had already been recognized from ground-based observations (see, for example, Tang and Moore, 1982); however, the SMM has provided especially useful new observations of this kind and we will draw our examples from this work.

A. Flare of 24 June 1980

Excellent SMM and ground-based observations from Sacramento Peak Observatory reveal interesting plasma transport for the flare at 1522 UT on 24 June 1980. Figure 1 illustrates the structure of the flare with respect to the magnetic neutral line. The primary (most intense) flare took place in an asymmetrical arcade of loops that spanned the neutral line of an isolated pole and extended toward the west. These loops ranged from 10" to 60" in length with the near footpoints (+ polarity) of the arcade clustered within the isolated pole and the remote footpoints (- polarity) distributed over the western part of the region. The shortest of these loops, with approximately NE-SW orientation, brightened in soft X-rays at the same time as the 28-54 keV hard X-ray burst. The hard X-rays took about one minute to reach maximum; the soft X-rays about one minute longer.

Both H-alpha and X-ray data reveal a separate, apparently single, loop with its near footpoint close to the arcade footpoints (- polarity). Its remote footpoint was approximately 70" east on the

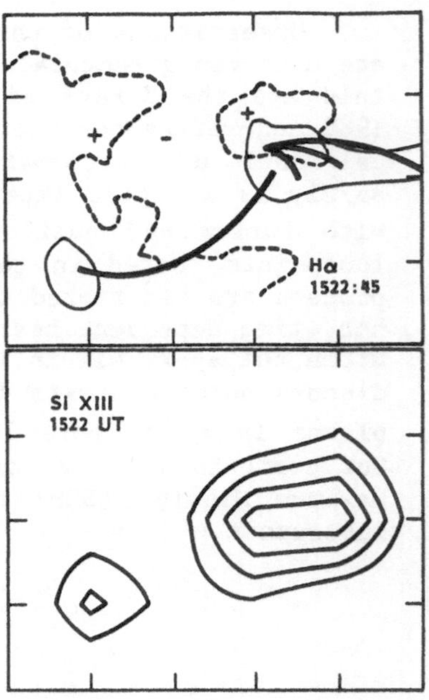

Figure 1. Comparison of Si XIII (10^7K) and Hα contour diagrams taken during the rise of the hard X-ray burst of a flare on 24 June 1980. Tic marks are 20" apart, which is the resolution of the X-ray image; north is up, west to the right. Magnetic neutral lines are indicated by broken lines in the upper panel. The arcade of loops, shown schematically, originates in the (+ polarity) region within the isolated pole and extends to the west; the single loop originates nearby but on the other side of the neutral line (- polarity) surrounding the isolated pole and terminates across the neutral line to the east. We infer that energetic particles and/or hot plasma were introduced into the single loop and into the shortest of the arcade loops by a single energy release process in the (- polarity) region.

other (+ polarity) side of the primary magnetic neutral line of the active region. As recorded by the Hard X-Ray Imaging Spectrometer (HXIS) and the Flat Crystal Spectrometer (FCS), this remote (eastern) footpoint of the single loop brightened in synchronism with the hard X-ray burst. The longer loops of the arcade to the west took several minutes to reach their maximum X-ray extent and intensity. Their remote (western) footpoints were not prominent in X-rays. Subsequently, both loop systems showed evidence, in soft X-rays, for post-maximum heating with no associated hard X-ray signature.

From these observations we conclude that an energy release process, probably induced by loop interaction at the time of the hard X-ray burst, introduced energetic particles and/or hot plasma into two separate loop systems; namely a process in the negative polarity region where the near footpoint of the single loop and the negative polarity footpoint of the shortest of the arcade loops are in close proximity. The rapid response of the remote (eastern) footpoint of the single loop stretching to the east required particle or wavefront velocities on the order of 1000 km/s. Plasma moved into both loop systems from the chromosphere as evidenced by unreversed H-alpha line profiles at the footpoints (Acton et al., 1982; T.A. Gunkler, private communication).

B. Flare of 5 July 1980

The detailed X-ray and chromospheric morphology of a flare on 5 July 1980 has been studied by Harrison et al. (1982). They find evidence, from the location of the hard X-ray burst, for primary energy release near one end of a magnetic structure the footpoints of which were 14000 km apart. This caused the loop to fill with flare plasma that was hottest at the top of the loop. A second hard X-ray burst appeared throughout the loop, followed by further filling and expansion of the soft X-ray source. Finally, from both the chromospheric and X-ray data they find evidence for "breaking out" of the hot flare plasma into a larger magnetic structure which subsequently evolved into a typical two-ribbon flare. Topological evolution (i.e. expansion, reconnection and reorientation) seemed to proceed throughout the event, again indicating processes that may be fundamentally different than energizing of plasma within the confines of a single, rigid, isolated loop.

C. Flare of 31 August 1980

Strong et al. (1982) have used optical, radio and SMM data to study a double (i.e. two in succession, 3 minutes apart) impulsive flare that occurred around 1250 UT on 31 August 1980. They conclude that the apparent spectral differences between the two flares was primarily the result of initial conditions; that is, the second flare took place in the already hot, dense, ionized plasma left over from the first flare. It follows that the hard X-ray producing electrons were produced with access to the same, or at least a magnetically connected, volume in both cases. Furthermore, the acceleration site must have been in the corona because in the second flare the soft X-ray plasma responded immediately to the energy injection that produced each electron burst. Also, the fast electrons must have had direct access into yet larger loops within which the observed U-type radio bursts were produced. Finally, there was evidence for expansion of the primary X-ray loop(s) and, thus, topological evolution of the flare volume.

For this double flare the spectrum and timing of the hard X-ray and microwave U-bursts provide convincing evidence that electrons from the primary acceleration location had simultaneous access to both the short loops containing the hot, dense X-ray plasma and higher, less dense, loops which were about ten times longer. Different loops were involved at different times and the radio spectra indicate that the field strengths within these structures must have differed by a factor of 4. These observations lend support to flare energy release models in which particle acceleration takes place due to magnetic reconnection at a neutral sheet between two interacting loop systems, in this case a large and a small one (Heyvaerts _et al._, 1977, Hood and Priest 1981).

D. Flare of 30 April 1980

Lastly, from a thorough analysis of the SMM, microwave and H-alpha data of the limb flare of 2023 UT on 30 April 1980 (discussed in several papers in _Astrophys. J._ _244_, Part 2, 15 March 1981) de Jager _et al._ (1982) find evidence for major plasma transfer from a small flare kernel into a much larger magnetic structure. They conclude that the large "tongue", as they call it, comprised filamentary strands with diameter to length ratios on the order of 1:100. In many aspects the morphology of this flare is similar to that described by Harrison _et al._ (1982), for the 5 July 1980 event, including the filling of the tongue which seems analogous to their plasma "breakout". In both cases there is the need to transfer large amounts of plasma from a compact to a much more extended magnetic structure shortly following the hard X-ray burst that signals the primary energizing of the flare plasma.

3. ROCKET OBSERVATIONS OF A FLARE ON 13 JULY 1982

The SMM flares we have just discussed indicate the complex magnetic topology and plasma transport in flares, and by implication, the deficiency of simple, single-loop, flare models to address some important flare processes. In this section we shall investigate the assumptions of constant loop cross-section and/or static atmosphere in a flare for which a simple loop model might be expected to apply. In particular, we consider constraints on flare cooling, within the context of some simple loop models, implied by our recent rocket observations.

A. Observations

We have preliminary results from a flare jointly observed by rocket-borne and ground-based instruments on 13 July 1982. The data were acquired by the following instruments (Brown et al., 1979; Bruner et al., 1980):

1) A 5 meter, grazing incidence, Rowland circle X-ray spectrograph operating in the 8-97 Å range with photographic recording; this instrument also employed X-ray and broadband EUV (1550 Å) photoelectric central image detectors for the purpose of locating the brightest X-ray feature prior to the spectrograph exposures.

2) A photographic and video H-alpha system for pointing information.

We were fortunate to have a C6 class soft X-ray flare begin shortly before launch time and were able to change our targeting program to observe it. The flare was of the non-impulsive filament activation type. Its beginning, maximum and ending times on 13 July, as reported by NOAA, were 1627, 1633 and 1640 UT. The GOES 1-8 Å light curve establishes that our observations were taken at the time

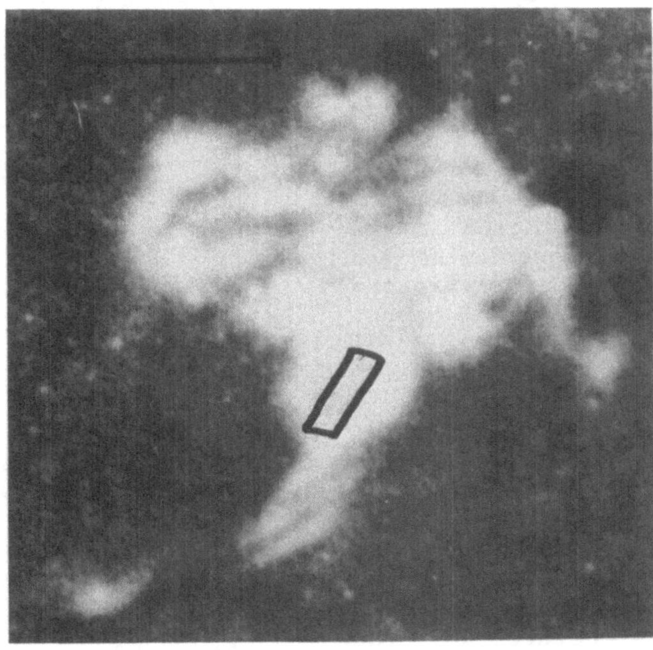

Figure 2. Area sampled by X-ray spectrograph during the exposure at 1635:33 - 1638:00 UT. Hα Cal Tech flare patrol image taken at 1635:16 UT. North is up and the line segment in upper LH corner is 60" long.

of soft X-ray maximum. Figure 2 illustrates the approximate projection of the spectrograph slit on the flare.

The coolest and hottest ions observed were Mg IX ($T \approx 7 \times 10^5$K) and Fe XIX ($T \approx 6.6 \times 10^6$K). Other observed ions include C V-VI, O VII-VIII, N VI-VII, Ne VIII-X, Mg VII-X, Si X-XII, S XIII, A XIV, Ca XIII-XV and Fe XIII-XIX. Some 298 individual spectral lines have been counted in the spectrum, exclusive of multiple orders. The rocket spectra thus provide important information not available from Hinotori or SMM. In particular, the temperature interval between $T \sim 1\text{-}3 \times 10^6$K is covered by several lines. The following preliminary line list, Table I, has been chosen for emission measure diagnostic analysis. Cases where multiplet lines have been summed to derive the given intensity are indicated.

The shape of the differential emission measure, DEM(T), derived from the rocket results is shown in Figure 3. The plasma emissivities used in our DEM(T) analysis have been taken from Mewe and Gronenschild (1981). Their results were normalized to the following logarithmic solar abundances (log H=12.00): C=8.76, N=8.08, O=8.45, Ne=7.65, Mg=7.65, Si=7.67, S=7.20, Fe=7.65. The DEM curve of Figure 3 has been computed according to the procedure of Sylwester et al. (1980). For comparison, the individual points are plotted based on a simple Pottasch (1963) type analysis in which the line is assumed to be formed over a temperature range corresponding to the full width at half maximum of the peak of its emission function and at an emissivity of 0.8 of its peak value. The points are plotted at T_{max} for the line. The scatter of the points reflect photometry and calibration error and possible discrepancies in the atomic collision strengths and abundances. The instrument calibration is based upon laboratory measurements.

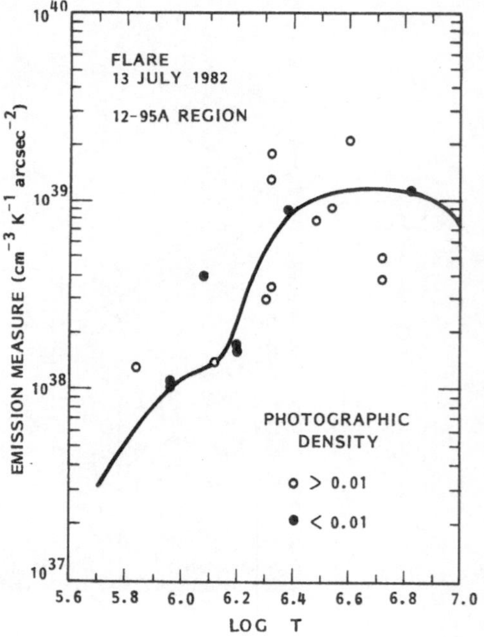

Figure 3. Emission measure distribution derived from the rocket spectra.

B. Loop Model Analysis

We estimate that the flare consisted of an ensemble of flaring loops with length scales on the order of 40" - 60", inferred from the separation of the intense kernels and the magnetic structure of the region. We thus assume for the sake of analysis that there were n identical loops, each of length, $\ell = 5\text{-}7 \times 10^9$ cm (i.e. 40" - 60" x π/2), and characterized by a single DEM(T). We note that DEM(T), defined physically as

$$DEM(T) = N_e^2(T)\ dV(T)/dT \ , \qquad (1)$$

has been normalized (per square arc-sec). There is no way of knowing what fraction of the loop ensemble contributed to our spectrum. However, we may use the observed GOES 1-8Å X-ray flux to establish an absolute calibration and find that the correction factor is approximately 2 x 10^3.

TABLE I

Ion	Wavelength (Å)	T_{max} (10^6K)	Intensity*	Comment
Ne IX	13.447	4.0	50.0	Multiplet
Fe XIX	13.5	6.6	92.0	
Fe XVIII	14.203	5.2	104.0	
Fe XVII	15.013	3.5	260.0	
Fe XVIII	16.072	5.2	100.0	
O VIII	18.969	3.0	160.0	
O VII	21.602	2.1	70.0	
N VII	24.781	2.1	23.0	
S XIII	32.410	2.4	9.0	
C VI	33.736	1.3	31.0	
C V	40.268	0.9	6.8	
Si XI	43.763	1.6	4.6	
Si XII	44.165	2.0	15.2	Multiplet
Fe XV	52.911	1.7	2.8	
Fe XVI	54.728	2.1	14.3	Multiplet
Mg IX	62.751	0.6	1.8	
Fe XIV	71.377	1.6	1.4	
Mg VIII	74.858	0.9	3.4	
Ne VIII	88.092	0.7	2.0	

* Photons $cm^{-2}\ s^{-1}$ $arcsec^{-2}$

The most significant finding from our observationally derived DEM(T) is that in the upper transition region interval, $10^{6.1} \leq T \leq 10^{6.4}$, we determine that DEM(T) ~ T^{α}, where $\alpha \simeq 10/3$. This is in sharp contrast to non-flaring quiet and active region observations of $\alpha \lesssim 0.5$, but is in agreement with compact flare observations indicating $\alpha > 2.0$ as summarized in the discussion by Antiochos (1980; note also that $q(T) = T \cdot DEM(T)$ and hence $\delta = \alpha+1$ in that discussion).

For a loop geometry, DEM(T) defined as in Equation (1) corresponds to

$$DEM(T) = A(T)\ N_e^{\ 2}(T)\ |\ dT/ds\ |^{\ -1}\ , \qquad (2)$$

where A(T) is the cross-sectional area of the loop at temperature T. Since the conductive flux is proportional to the temperature gradient,

$$F_c \simeq 10^{-6}\ T^{5/2}\ dT/ds\ , \qquad (3)$$

taking the pressure $P(T) \simeq 2N_e kT$ we rewrite DEM(T) as

$$DEM(T) \simeq 2 \times 10^{-6}\ \frac{A(T)\ P^2(T)\ T^{1/2}}{4k^2}\ |F_c|^{\ -1}, \qquad (4)$$

where the factor of 2 arises from the two loop segments.

Using this expression we can immediately rule out purely conductive cooling of a constant cross section loop, since over this narrow temperature range we expect that P ~ constant and F_c ~ constant as discussed by Antiochos and Sturrock (1976), and therefore DEM(T) ~ $T^{1/2}$.

Allowing for the possibility of chromospheric evaporation in a constant cross section loop as a cooling process, Antiochos and Sturrock (1978) again find that P ~ constant, but now F_c ~ T, and hence DEM(T) ~ $T^{-1/2}$.

Turning to radiative-dominated flare cooling models, Antiochos (1980) has considered two quasi-steady-state situations in which radiative losses predominate; he finds for both of these situations that DEM(T) ~ T^{β} where β is the power of the radiative loss coefficient, $\Lambda(T) \simeq T^{\beta}$, and this is variable and lies in the range $0 < \beta \leq 2$ for $T > 10^5$K (Rosner et al., 1978).

One final quasi-steady state analytical model for a cooling flare loop has been proposed by Antiochos and Sturrock (1982). In this model rapid radiative cooling at the base of a constant cross section loop creates strong pressure gradients; these gradients are assumed to generate near-supersonic downflows, so that energy loss by mass motion dominates over both conduction and radiation. We have examined these models, which by virtue of the assumption that the flow is in a pre-specified steady state are represented by a non-linear but numerically integrable differential equation, and find that DEM(T) ~ T^{α}, where α can indeed take on values as large or larger than that required to explain our observed DEM(T).

The problem is that neither observations nor detailed hydrodynamic models of flaring loops that are now becoming available (compare, for example, Nagai, 1980; Wu et al., 1981; Cheng et al., 1983; Pallavicini et al., 1983) show evidence for near-supersonic downflows: quite to the contrary, the bulk of the flows observed and predicted are upflows (but see Doschek et al. (1982) for examples of the predicted complexity of the velocity fields in flare models).

The most plausible way of bringing about a steep temperature dependence for DEM(T) in any simple model is by assuming that the loop converges in cross sectional area in the transition region as one moves down toward the footpoints; this is evident from equation (2) both from the temperature dependence of A(T) and from the inferred inhibition of $F_c(T)$ as the area converges near the footpoints. The shape of our DEM(T) is rather similar to that presented by Withbroe (1978) for a large solar flare, except that the emission peak seems to occur at a somewhat lower temperature in our flare. As shown in Figure 5 of Withbroe's paper, a simple cooling model that allows for a modest (factor of 4-5) variation in the cross-sectional area fits such a DEM(T) curve extremely well and far better than any of the constant cross-section models.

We would like to put forward a speculative suggestion based on the foregoing result. On both observational (Ramsey et al., 1977, Spruit en Zwaan, 1981) and theoretical (Gabriel, 1976) grounds there is reason to believe that the magnetic field necks down in passing from the corona through the transition region. We suggest that the reason flare loops tend to have a steeper DEM(T) than non-flaring loops is because the energy flow associated with flaring pushes the transition region boundary well down into this converging "flux tube funnel", thus decreasing the cross-sectional area and inhibiting conduction from the corona.

4. OBSERVATIONAL CONCLUSIONS

The complexity of flares and shortcomings of the observations do not permit absolute conclusion; a variety of interpretations are tenable in most cases. In the spirit of this seminar we put forward the following opinions based on our research experience and the work discussed in Sections 2 and 3:

1) Particle acceleration and energy release occurs when magnetic structures interact; often compact structures interact with much larger ones.

2) The primary energy release region is in the corona.

3) Energetic electrons and/or energized plasma enter both (or several) interacting loop systems.

4) Whether hard X-ray emission appears at the footpoints or up in the loop depends on the density, hence upon the particle mean free path, within the loop.

5) Energy transport (via electron beaming, shock or conduction front advance) within loops is asymmetrical - usually one footpoint is much brighter.

6) Plasma "breakout" and loop "expansion" implies that the gas pressure approaches or surpasses the magnetic pressure in some cases.

7) Conductive cooling of some flare loops is inhibiited by a decrease in loop cross-section at the footpoints.

5. DISCUSSION

The simple one-dimensional loop model for the study of flare evolution has proved useful. However, in its most popular form, namely constant cross-section and total containment of plasma, it may be sufficiently unlike most real flares to be misleading. We find evidence for energetic electrons and/or thermal plasma entering into distinctly different magnetic structures essentially simultaneously. In at least some instances this is apparently happening at coronal levels. Electrons accelerated there produce radio bursts or hard X-rays at the footpoints or in the corona, depending upon the magnetic field topology and the ambient density. There seem to be cases in which pre-existing hot flare plasma "breaks out" from a compact source into a much larger volume. In at least one well studied case this appears to be related to the phase at which the flare changes to a typical two-ribbon configuration. In several cases the data are described as an "expanding" set of loops. Thus it appears that the assumptions that the magnetic pressure dominates the gas pressure ($\beta \ll 1$) and that the plasma is totally contained may often be violated.

We have compared the slope of the low temperature end of the differential emission measure distribution of one flare with the predictions of simple loop models. We find that the slope is steeper than can be accounted for without some inhibition against conductive losses. Either high downward velocities or a constriction in the loop cross section at the footpoints could explain this. Because of the magnitude of the velocities required, and because there are observational and theoretical arguments for expansion of the magnetic tubes of force in passing upwards through the Sun's atmosphere, we prefer the latter explanation; although other factors not even considered herein may be even more important.

Presently, improvements to loop modelling are concentrating on gas dynamic effects and more accuracy in the representation of the transition region and chromosphere at the footpoints. We suggest that relaxation of the full containment hypothesis and more freedom for possible field topologies at the footpoints may both be fruitful areas for future modelling improvements. Finally, this work supports flare models involving primary energy release through the interaction, at coronal heights, of compact and extended loop systems. Plasma transport in such topologies and in trapping volumes where $\beta \gtrsim 1$ need to be investigated theoretically.

ACKNOWLEDGEMENTS

The SMM X-Ray Polychromator experiment is a collaboration including Mullard Space Science Laboratory, Rutherford Appleton Laboratory and Lockheed Palo Alto Research Laboratory. Study of the 24 June 1980 flare has benefitted from discussions with T. Gunkler and R. Canfield and data from Kitt Peak and Big Bear Observatories.

Key contributions by our long term collaborators R. Speer, A. Franks, G. Schmidtke and W. Schweizer are reflected in the success of the X-ray spectrograph experiment. We are grateful to R. Mewe for a special listing of line emissivity functions. This work has been supported by the National Aeronautics and Space Administration under contracts NAS5-23758 and NAS5-25727. LWA is grateful for an NSF travel grant under the U.S.-Japan Cooperative Science Program.

REFERENCES

Acton, L.W., Canfield, R.C., Gunkler, T.A., Hudson, H.S., Kiplinger, A.L. and Leibacher, J.W.: 1982, Astrophys. J. 263, p. 409.

Antiochos, S.K.: 1980, Astrophys. J. 241, p. 385.

Antiochos, S.K. and Sturrock, P.A.: 1976, Solar Phys. 49, p. 359.

Antiochos, S.K. and Sturrock, P.A.: 1978, Astrophys. J. 220, p. 1137.

Antiochos, S.K. and Sturrock, P.A.: 1982, Astrophys. J. 254, p. 343.

Brown, W.A., Bruner, E.C. Jr., Acton, L.W., Franks, A., Stedman, M. and Speer, R.J.: 1979, Space Optics-SPIE 184, p. 278.

Bruner, E.C., Jr., Acton, L.W., Brown, W.A., Salat, S.W., Franks, A., Schmidtke, G., Schweizer, W. and Speer, R.J.: 1980, Opt. Engr. 19, p. 433.

Cheng, C.C., Orau, E.S., Doschek, G.A., Boris, J.P. and Mariska, J.T.: 1983, Astrophys. J., submitted.

de Jager, C., Machado, M.E., Schadee, A., Strong, K.T., Svestka, Z., Woodgate, B.E. and van Tend, W.: 1982, Solar Phys. to be published.

Doschek, G.A., Boris, J.P., Cheng, C.C., Mariska, J.T. and Orau, E.S.: 1982, Astrophys. J. 258, p. 373.

Gabriel, A.H.: 1976, Phil. Trans. R. Soc. Lond. A. 281, p. 339.

Harrison, R.A., Simnett, G.M., Hoyng, P., LaFleur, H. and Van Beek, H.F.: 1982, Solar Phys., to be published.

Heyvaerts, J., Priest, E.R. and Rust, D.M.: 1977, Astrophys. J. 216, p. 123.

Hood, A.W. and Priest, E.R.: 1981, Solar Phys. 73, p. 289.
Mewe, R. and Gronenschild, E.B.H.M.: 1981, Astron. Astrophys. Suppl. Ser. 45, p. 11.
Moore, R.L., et al.: 1980, "The Thermal X-Ray Flare Plasma", in Solar Flares, P. Sturrock ed. Colorardo Assoc. Univ. Pr., p. 341.
Nagai, F.: 1980, Solar Phys. 68, 351.
Pallavicini, R., Peres, G., Serio, S., Vaiana, G., Acton, L., Leibacher, J. and Rosner, R.: 1983, Astrophys. J., to be published.
Pottasch, S.: 1963, Astrophys. J. 137, p. 945.
Ramsey, H.E., Schoolman, S.A. and Title, A.M.: 1977, Astrophys. J. Letters 215, p. L41.
Rosner, R., Tucker, W.H. and Vaiana, G.S.: 1978, Astrophys. J. 220, p. 643.
Spruit, H.C. and Zwaan, C: 1981, Solar Phys. 70, p. 207.
Strong, K.T., Benz, A.O., Dennis, B.R., Leibacher, J.W., Mewe, R., Poland, A. Schrijver, J., Simnett, G., Smith, J.B. Jr. and Sylwester, J.: 1982 in preparation for submission to Solar Phys.
Sylwester, J., Schrijver, J. and Mewe, R.: 1980, Solar Phys. 67, p. 285.
Tang, F. and Moore, R.L.: 1982 Solar Phys. 77, p. 263.
Wu, S.T., Kau, L.C., Nakagaw, Y. and Tandberg-Hanssen, E.: 1981, Solar Phys. 70, p. 137.
Withbroe, G.L.: 1978, Astrophys. J. 225, p. 641.

INTERPRETATION OF THE SOFT X-RAY SPECTRA FROM HINOTORI

K. Tanaka, N. Nitta*, K. Akita*, and T. Watanabe
Tokyo Astronomical Observatory,
University of Tokyo, Mitaka, Tokyo, 181, Japan

* Department of Astronomy,
University of Tokyo, Bunkyoku, Tokyo, Japan

ABSTRACT. We present analyses of the soft X-ray iron line spectra of flares obtained from the Bragg Spectrometer on Hinotori. We first present a case of strong Kα emission at the impulsive phase of the hard X-ray burst, and assess net Kα emission due to the electron impact by eliminating the fluorescence contribution. Second we discuss on the differences in the electron temperatures and emission measures derived respectively from FeXXVI and FeXXV spectra. A pilot two-temperatures model which can explain the two spectra is presented. Finally, we compare the temporal relations between the soft X-ray and hard X-ray intensities and show two extreme classes of flares, one characterized by the efficient formation of a hot thermal plasma above 30 million degree, and the other characterized by the spiky hard X-ray component. Energetical relation of the thermal plasma to the electron beam is discussed for the two classes.

I. INTRODUCTION

The Bragg crystal spectrometer (SOX) has operated successfully since the launch of Hinotori, and has obtained the spectra of iron complex from some 675 flares. For each flare two wavelength regions, about 1.72-1.95 Å and 1.83-1.89 Å, are scanned with spectral resolutions of 2.5 mÅ and 0.15 mÅ, respectively. The time resolution (7-10s) and wavelength ranges differ slightly from flare to flare depending on the spin period of the satellite and the location of flare, respectively. Description of the method and examples of the spectra are given by Tanaka et al. (1982b). In this paper a concise description of various results obtained from the analyses of the Fe spectra will be given.

The SOX spectra include the Kα and FeXXVI regions. Kα emission in flares has generally been attributed to fluorescence from the low temperature region as irradiated by the X-rays from the hot plasma (Bai, 1979; Feldman et al., 1980; Culhane et al., 1981; Tanaka et al., 1982a). However, some Kα emission due to electron impact is expected if an

Solar Physics **86** (1983) 91–100. 0038–0938/83/0861–0091 $ 01.50.

electron beam exists. Search for such Kα emission is important to confirm the existence of electron beams. The detection of the electron-impact Kα emission is difficult due to contamination by the dominant fluorescence Kα emission. We attempt in section 2 to assess such Kα emission by separating the fluorescence contribution from the observed Kα emission.

The FeXXVI spectrum provides a tool to diagnose the hottest part of flare plasma since the emissivity of FeXXVI lines continues to increase beyond 100M(million)K. Comparisons of the observed spectra with the theoretical spectra (Tanaka et al., 1982c) have revealed significant differences between the electron temperatures derived respectively from the FeXXVI spectrum and FeXXV spectrum. In Section 3 we discuss a two-temperatures model. Tanaka et al. (1982c) have also noted the close similarity of the temporal behavior of FeXXVI intensity to that of the hard X-ray continuum in the lower energy range in some flares. A classification of flares according to this similarity is made in section 4. We will point out two categories (cf. Tanaka, 1982), one characterized by the efficient formation of a hotter plasma component above 30 MK, and the other characterized by the dominance of spiky hard X-ray component and also by a less pronounced formation of the hotter plasma component. Finally we discuss the key parameter, which may distinguish the two types of flares.

2. ELECTRON IMPACT Kα EMISSION

Figure 1 shows a sequence of low resolution spectra for an M 3.4 flare of 1981 July 28. The spectra in the early phase indicate low-temperature characteristics, as may be recognized from the smaller peak intensity of the FeXXV resonance line(w) at 1.85Å relative to FeXXIII (1.87Å). The relative intensity is reversed in 10s, suggesting rapid heating. The electron temperatures are derived from the fitting of the observed spectra to the theoretical spectra (cf. Fig.2 of Tanaka et al., 1982c). From Figure 1 it is evident that the Kα (1.94Å) intensity relative to the intensities of FeXXV, XXIV and XXIII (1.85-1.87Å) is exceptionally large in the initial phase. An intense hard X-ray burst was observed in this phase. This already may suggest electron impact Kα emission in the initial phase.

The fluorescence Kα emission is proportional to the total X-ray flux above 7.11 keV. We derived this flux as a function of time from the derived emission measure and temperature and evaluated the Kα fluorescence contribution by assuming that the observed Kα intensity after 20:11 UT was completely due to the fluorescence. The assumption may be justified since very weak hard X-ray emission was seen after 20:11 UT. The observed Kα intensity profile and evaluated fluorescence contribution are shown in Figure 2. The residual between the two (also shown in Figure 2) may be attributed to the electron impact Kα emission.

To compare this with Kα intensity predicted by electron impact theory, we applied Phillips and Neupert's (1973) formula, using the hard X-ray spectra obtained by Hinotori. The predicted Kα intensity profile (top) shows a close similarity to the time profile of the

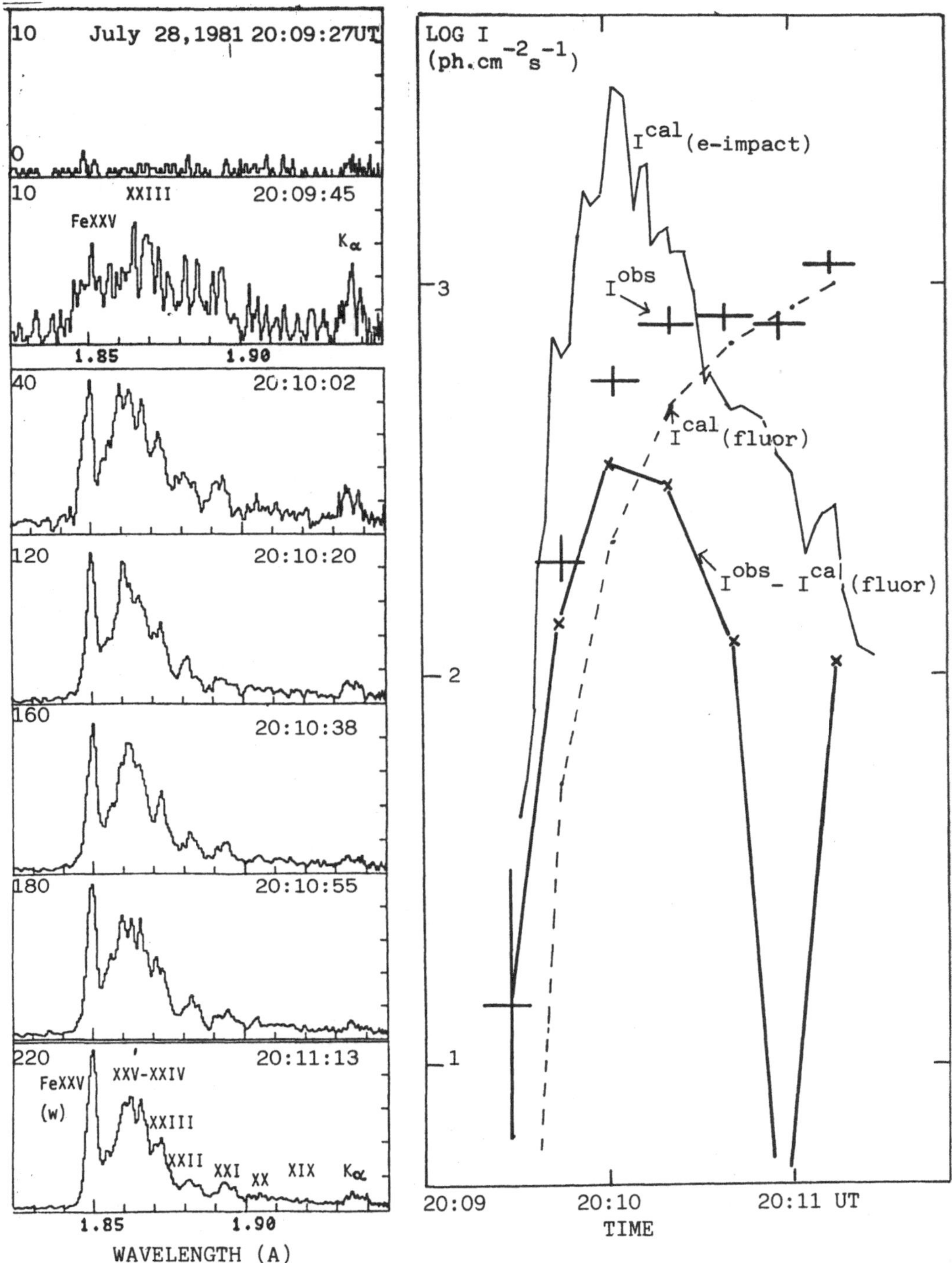

Fig.1. Time Sequence of low-resolution spectra for the 1981 July 28 flare.

Fig.2. Various Kα intensity time profiles for the 1981 July 28 flare. Error bar of one sigma level is shown for the observed intensities (I^{obs}).

observed Kα residuals, but the predicted values are about an order of magnitude larger than the observed ones. The discrepancy may partly be attributed to the thin-target assumption in Phillips and Neupert's treatment. Further comparisons with the thick-target theory recently made by Dr. Emslie (private communication) are in progress.

3. ELECTRON TEMPERATURE AND EMISSION MEASURE DIAGNOSTICS OF Fe XXVI AND Fe XXV SPECTRA

As shown in previous papers (Tanaka et al., 1982a,c), isothermal electron temperatures can be derived respectively from flux ratio of FeXXVI Lα(1.77-1.785Å) to the di-electronic recombination satellites (1.785-1.8Å) and from the intensity ratio of the FeXXV w to FeXXIV j lines. Generally the two electron temperatures, Te(FeXXVI) and Te (FeXXV), differ significantly, thus contradicting the isothermal assumption. We show in Figure 3 (lines) two sets of electron temperatures and emission measures derived for the largest flare (1982 June 6; X12 class) we observed. Because of high counts, the accuracy of these determinations should be good. The emission measures are derived from the intensities of FeXXVI Lα and FeXXV w respectively assuming ionization equilibrium of Jacobs et al.(1977). Besides the large difference between the two temperatures we find different evolutions. Towards the intensity maxima the FeXXV emission measure EM(FeXXV) increases with rather constant Te (FeXXV) equal to 25 MK, while Te(XXVI) steadily increases from 30 MK to 40 MK with a small increase of emission measure. Representative spectra at these two phases are shown in Figure 4.

The two derived sets of Te's and EM's are not self-consistent because Te and EM from FeXXVI contribute considerably to the FeXXV intensities. This contradiction may be resolved by considering multi-temperature distribution. Here I have assumed a two-temperature structure and have determined two sets of Te's and EM's which satisfy the observed intensities of the same four lines as used in the line-ratio analyses. The shaded areas in Figure 3 represent solutions which reproduced the observed intensities within 10% errors. The higher-temperature component (abbreviated as hot component hereafter) shows a Te even larger than Te(FeXXVI), while the lower-temperature component (cool component) shows Te below 20MK and EM higher than EM(FeXXV). The temperatures of the cool component seem to agree better with the electron temperature derived from Ca line ratio. (Doschek et al., 1979; Antonucci et al., 1982).

While the uniqueness of the two-temperatures structure thus derived cannot be proven until a reliable differential emission measure solution is obtained, there are some indications that the two temperature componetnts originate in different volumes and show different time behaviors. In some flares (Figure 5b) time variation of the Lα intensity which represents the hot component is found to be quite different from the intensity of FeXXV line which represent the cool component. Results of X-ray images indicate also that the FeXXVI emitting region, which is identical to the hard X-ray image in 15-40keV in some particular

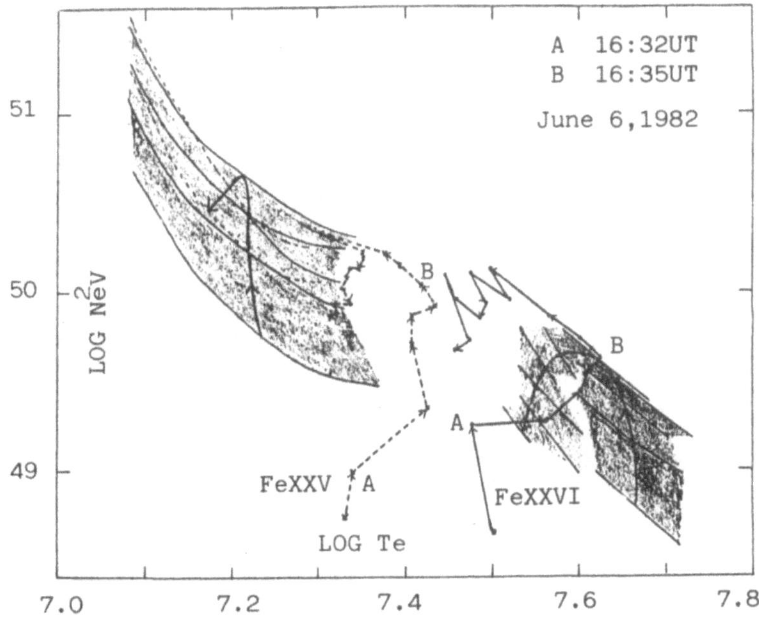

Fig.3. Evolution of temperature and emission measure for the 1982 July 6 flare (X12). Broken line: from FeXXV(j to w line ratio), Thick line: from FeXXVI, shaded area: from FeXXV and FeXXVI (self-consistent two-component model).

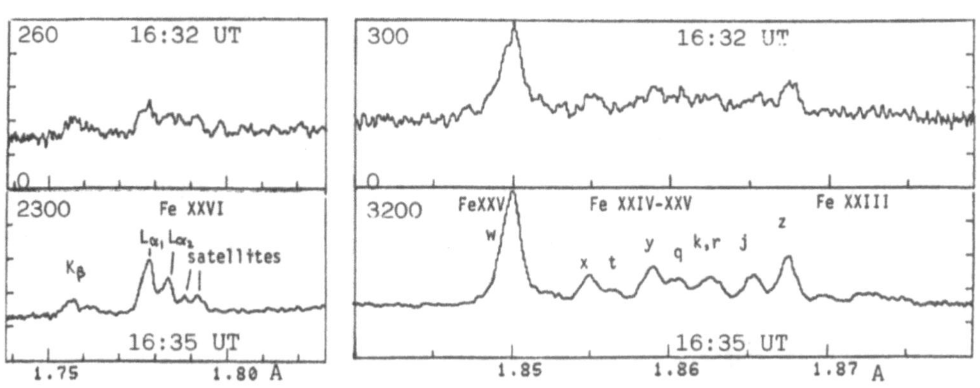

Fig. 4. Examples of FeXXVI (left) and FeXXV (right) spectra for the June 6 flare.

flares, is smaller by some factor than the FeXXV emitting region, which matches the image in the 5-10keV range.

4. RELATIONS BETWEEN HARD X-RAY AND SOFT X-RAY BURSTS

When we compare the time behavior of the hot component (30-50MK; FeXXVI line) and the cool component (20MK ; FeXXV or lower ionization lines) with hard X-ray time profiles, two extreme cases may be found. In the first class, which I call type A, the low energy hard X-ray burst below 50keV shows a very smooth time variation without spiky structure, and its time behavior is quite similar to the intensity profile of FeXXVI Lα line. The burst is short-lived but very intense. Examples are shown in Figure 5, where various plasma parameters are plotted together with various line and continuum intensities in the same logarithmic scale. In these flares the parameters derived from the FeXXVI spectrum explain the hard X-ray flux below 40keV, so the low energy smooth component of the hard X-ray burst must have a thermal origin from 30-50 MK plasma. Spike components which may represent either non-thermal beam or short-lived very hot ($\gtrsim$100MK) thermal plasma appear only in high-energy channels above 50-60keV. The hard X-ray spectrum is very soft in the type A flare.

The hard X-ray images of this kind of flare show very compact source structure (10") as shown by Tsuneta et al. (1982) for Figure 5a flare and by Ohki et al. (1982) for Figure 5b flare. Since we can safely identify the FeXXVI emitting area with the hard X-ray image in the low energy band, we obtain from EM(FeXXVI) Log Ne=11.8 and Log Ne=11.0 respectively for the flares of Figure 5a and 5b. From the hard X-ray image of 1981 July 17 flare (Figure 5b) which occurred at the limb, we find that the hot component is located in the low corona.

Another extreme case (type B flare) shows the characteristics of so-called impulsive flares (Figure 6a). While the hard X-ray time profiles show spiky structure from the low energy channel (20-30keV), Lα intensity is not similar to the hard X-ray burst except for decay phase, when the low energy hard X-ray burst becomes smoother and similar to the Lα intensity curve. When we inspect the hard X-ray spectra in the decay phase, there can be found an additional component in the low-energy side as superposed on the power law spectra which fit the higher energy range nicely. Presumably this component is thermal and is similar to the one which is prominent in type A flares. The high resolution hard X-ray spectrum observation by Lin et al. (1981) has demonstrated the apppearance of thermal component of Te=30MK in the middle phase of such a flare. The most pronounced characteristics of a type B flare would be the similarity of time profile of dI/dt (I:total iron line emission) to the hard X-ray burst profile (Figure 6a) It indicates that the production rate of cold component (Te~20MK) is closely associated with the occurrence of hard X-ray spike bursts (cf. Tanaka, Nitta and Watanabe, 1982).

Hard X-ray images of the spike components generally indicate low sources probably at footpoints (Hoyng et al., 1981; Ohki et al., this proceedings). This suggests precipitation of the electron beams to the

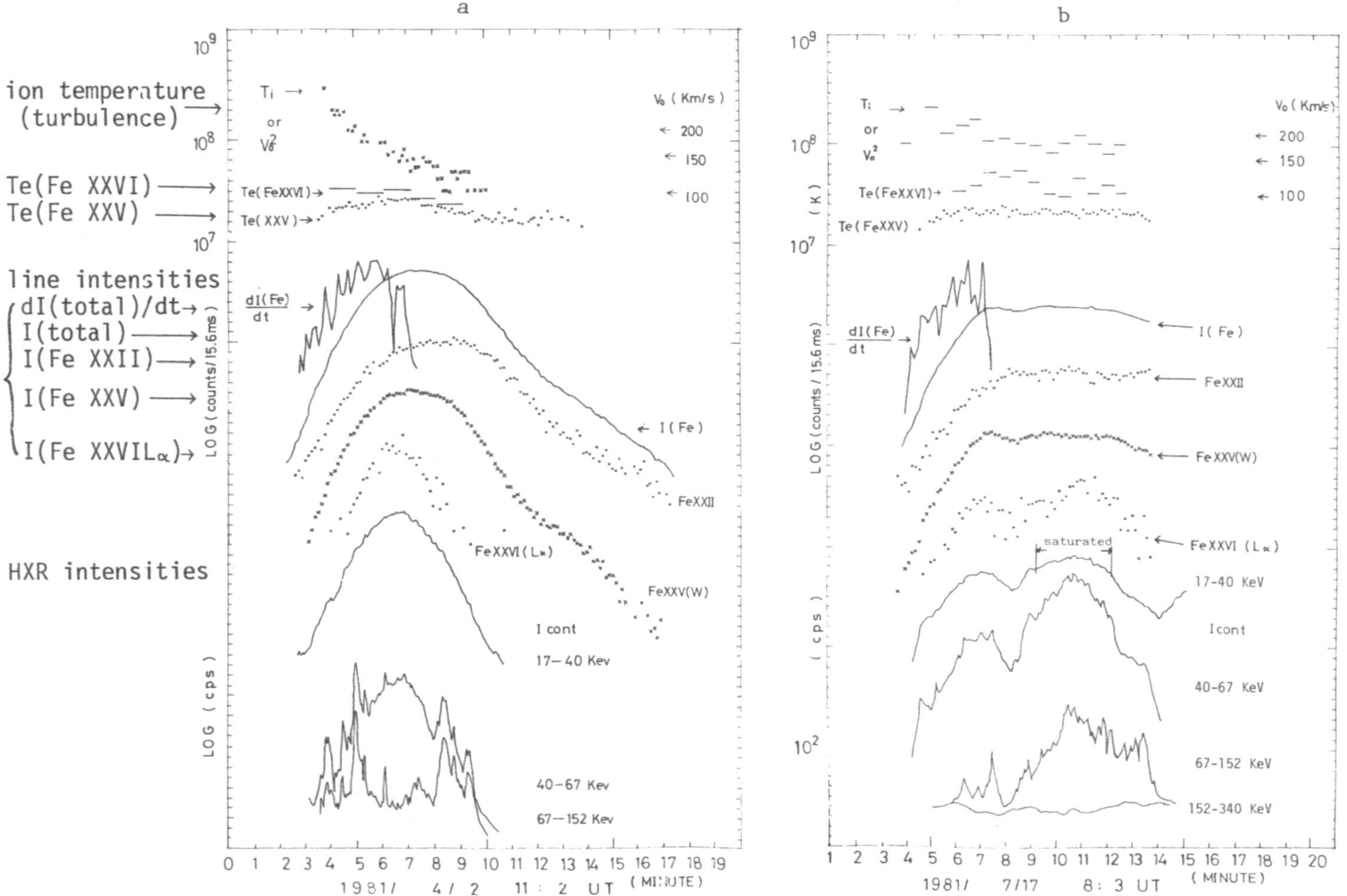

Fig.5 Time profiles of two type A flares.

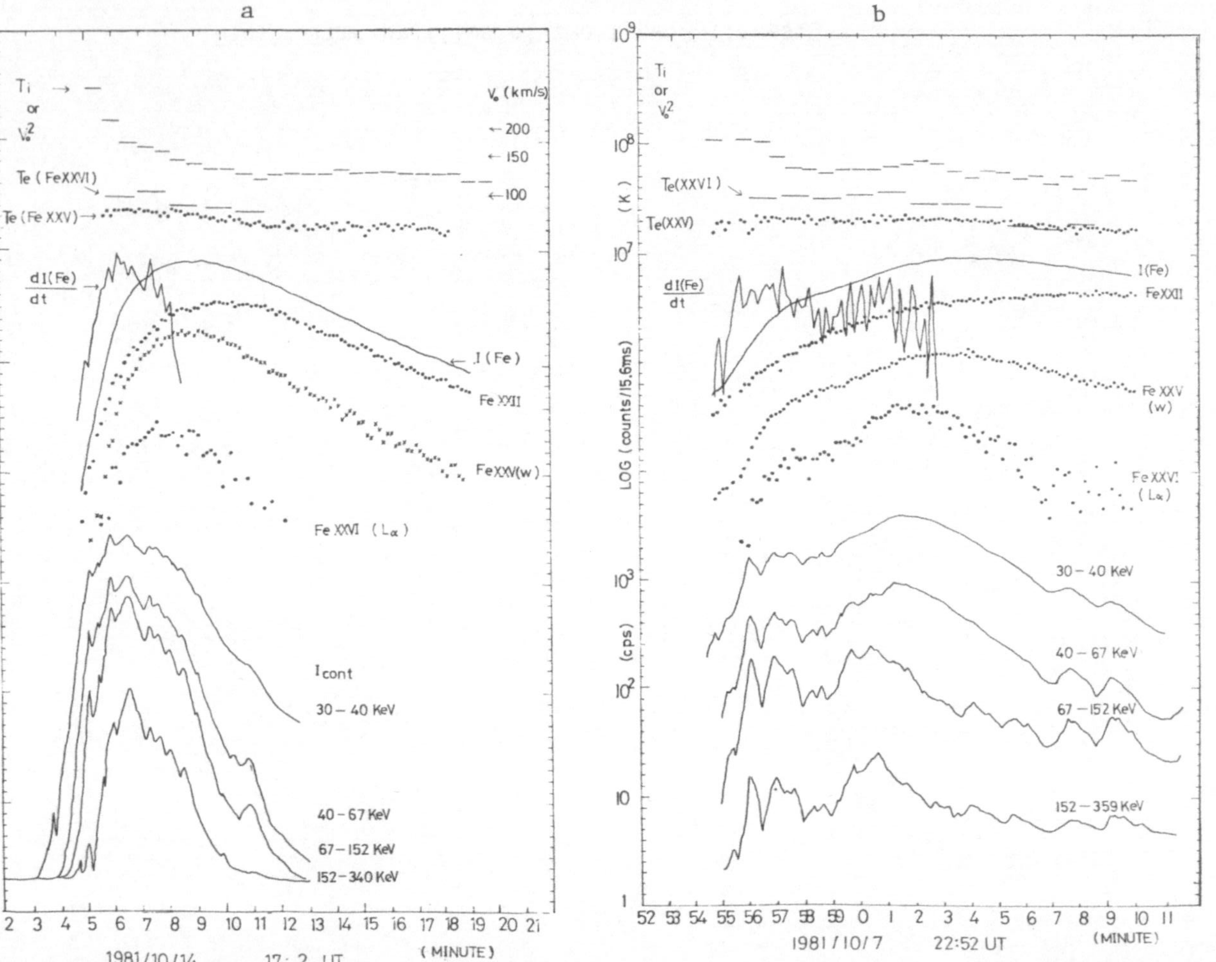

Fig.6a Time profile of type B flare.

Fig.6b Intermediate case between type A and B.

footpoints, and the above relation may suggest the formation of the 20MK plasma by the dissipation of the beam energy.

There are some cases between the extremes of type A and type B. Such flares are generally of long duration. As shown in the example of Figure 6b, the time derivative of total iron emission is proportional to profiles of the hard X-ray spike component which is seen from the low energy channel at the initial phase and only in the higher energy channels at the middle phase. From the middle phase the low energy hard X-ray burst shows gradual and smooth enhancement which is very similar to the intensity variation of FeXXVI Lα. Thus, the first half of this burst shows characteristics of type B, and the latter half of type A.

These results suggest that the hard X-ray burst generally consists of a spike component (electron beam or very hot plasma) and a hot (30-50MK), dense ($10^{11}cm^{-3}$) thermal component which also emits FeXXVI lines. The variety of flares ranging from type A to type B may be due to the relative importance of the two components. In type A flare the contribution of the hot thermal component dominates the spike component in the low-energy channel, while in type B flare hot component is produced less and contributes flux only in decay phase.

As mentioned above, a close similarity between dI/dt (cool component) and I (hard X-ray spike component) suggests the possibility that the cool thermal component is produced by the dissipation of electron beams. For the very short single spike bursts, Tanaka, Nitta and Watanabe (1982) have shown some evidences for this hypothesis. To check this hypothesis energetically for more complex flares we compared the total injected energy of the electron beam above a certain electron energy Eo with the total thermal energy of the cool component. The thermal energy is derived from the product of emission measure and temperature from FeXXV lines with the use of electron density which is assumed equal to the density of the hot thermal component.

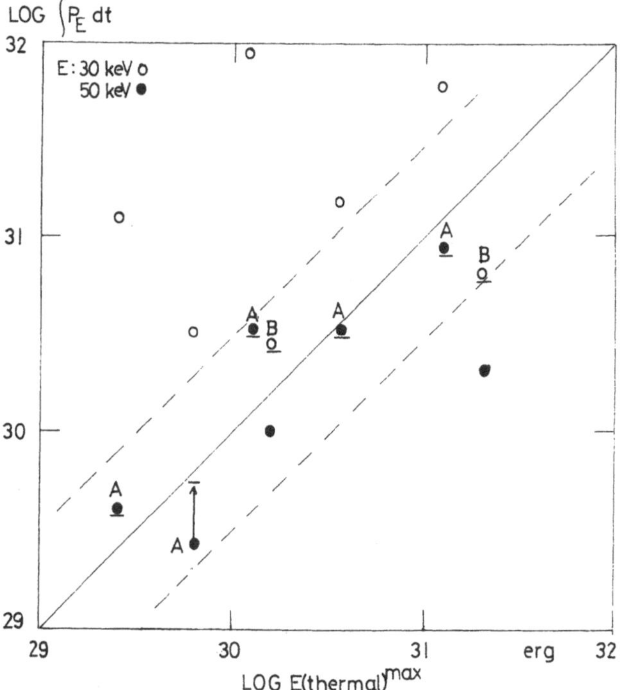

Fig.7. Total injected energy of electron beam above E(ordinate) versus maximum thermal energy of cool component, 20Mk (abscissa) for two classes of flare.

This density can be obtained from the hard X-ray image size and EM (Fe XXVI) as noted before. For the low energy threshold, E_o=30 keV and 50 keV are adopted respectively for type B and type A flares since the spike component appears above these energies approximately for each type of flare. Figure 7 shows comparisons of the two quantities for various flares, and reveals a reasonable agreement between the two. It is noted that if we use E_o=30keV for the type A burst the discrepancy becomes enormous.

Electron beams of energies 30keV and 50keV can penetrate the transition regions, respectively, of plage models (Basri et al., 1979) and of low density flare model (Machado et al., 1980). Therefore the above agreement may be explained if type A and type B flares occur respectively in high density (flare like) and medium density (plage like) corona, and the electron beams reaching the transition layer deposit their energy to produce 10-20 MK plasma from this layer. This suggests that a key parameter which distinguishes between type A and type B flares may be the ambient coronal density before the flare starts. In type A flare where the spiky hard X-ray component is missing in the low energy bands, low energy electrons below 50keV may not be produced by some effect of high density in the acceleration region, or they may be dissipated immediately in the high density corona(coronal thick target).

We acknowledge many people who have participated in the SOX experiment and the Hinotori project. We wish to thank Prof. F.Moriyama for the data reduction of 1981 July 28 flare.

REFERENCES

Antonucci, E. et al. : 1982, Solar Phys. 78, 107
Bai, T. : 1979, Solar Phys. 62, 113.
Basri, G.B. et al. : 1979, Ap.J. 230, 924.
Culhane, J.L. et al. : 1981, Ap.J. (Letters). 244, L141.
Doschek, G.A., Kreplin, R.W., and Feldman, U. : 1979, Ap.J. (Letters) 233, L157.
Feldman, U., Doschek, G.A., and Kreplin, R.W. : 1980, Ap.J. 238,365.
Hoyng, P. et al. 1981, Ap.J. (Letters) 246, L155.
Lin, R.P. Schwartz, R.A., Pelling, R.M. and Hurley, K.C. : 1981, Ap.J. (Letters) 251, L209.
Machado, M.E., Avrett, E.H., Vernazza, J.E., and Noyes, R.W. : 1980, Ap.J. 242, 336.
Ohki, K. et al. : 1982, Proc. of Hinotori Symp. Publ. by ISAS, p.102.
Phillips, K.J.H., and Neupert, W.M. : 1973, Solar Phys. 32, 209.
Tanaka, K. : 1982, IAU Col.71 "Activities in Red Dwarf Stars".
Tanaka, K., Watanabe, T., Nishi, K., and Akita, K. : 1982a, Ap.J. (Letters) 254, L59.
Tanaka, K. et al. : 1982b; Annals of Tokyo Astron. Obs. 2nd Series, Vol.18, p237.
Tanaka, K., Akita, K., Watanabe, T., and Nishi, K. : 1982c, Hinotori Symp. Publ. by ISAS, p.43.
Tanaka, K., Nitta, N. and Watanabe, T. : 1982, Hinotori Symp. Publ. by ISAS, p.20.
Tsuneta, S. et al.: 1982, Hinotori Symp. Publ. by ISAS, p.130.

POLARIZATION MEASUREMENTS USING THE BRAGG CRYSTAL SPECTROMETERS ON HINOTORI

Kyo Akita
Department of Astronomy, University of Tokyo,
Bunkyo-ku, Tokyo, 113, Japan

Katsuo Tanaka, Tetsuya Watanabe
Tokyo Astronomical Observatory, University of Tokyo,
Mitaka, Tokyo, 181, Japan

ABSTRACT. The two Bragg crystal spectrometers on board Hinotori provide a new technique for measuring linear polarization of the soft X-ray line emission from solar flares. The results of the measurements of large (X class) flares are given in this paper, and the polarization degree averaged over the soft X-ray maximum phases is estimated to be less than 4%.

1. INTRODUCTION

In investigating the energy release mechanisms of the solar flares, it is fundamentally important to know whether the energized electrons have a thermal or a nonthermal velocity distribution. A detection of polarization of the X-ray emission would be considered as direct evidence for the existence of beamed (nonthermal) electrons (Haug, 1979; Lemen et al.,1982). The X-ray polarimeter of Lemen et al.(1982) which measures the polarization degree of the continuum hard (5-30 keV) X-ray emission from solar flares is regarded as the best equipment for this purpose. However, the Bragg crystal polarimeter proposed by Novick (1975), which measures the polarization of the X-ray line emission, is sufficiently applicable especially for the polarization due to the lower ($\gtrsim$excitation threshold) energy electron beam (Shlyaptseva et al., 1981).

Krutov et al.(1981) reported anomalous values of FeXXV line intensity ratios observed by an Intercosmos satellite with a system of two Bragg spectrometers, and they attributed those anomalies to linear polarization. Due to the unique method of wavelength scanning used by the two Bragg crystal spectrometers on Hinotori (SOX) covering the wavelength range 1.7 - 1.95 Å centered on the FeXXV resonance line, we can use a new technique to detect linear polarization of line emission.

The results of the analyses of 6 large flares are given in this paper, after a brief description of the method applied here.

2. METHOD

X-rays incident upon a Bragg crystal spectrometer can be thought of as undergoing reflection by Thomson scattering, so that the reflectivity

Solar Physics **86** (1983) 101–105. 0038–0938/83/0861–0101 $ 00.75.

of the crystal depends on the polarization degree as (Novick, 1975)

$$R \propto 1 - \sin^2 2\Theta \;(1 - P\cos 2\Psi)/2, \tag{1}$$

where Θ denotes Bragg angle, and P is for the polarization degree, and Ψ is the angle of the polarization vector with the plane of incidence. The polarization degree, therefore, can be evaluated with sets of spectra of various Θ and/or Ψ.

SOX consists of a low-dispersion SOX1 ($\Theta = 16°$) and a high-dispersion SOX2 ($\theta = 52°$), which observe the same wavelength range. For wavelength scanning, it utilizes the spin of the satellite to change the bragg angle gradually as

$$\sin\Theta = \cos\delta \sin\alpha - \sin\delta \cos\alpha \cos\phi, \tag{2}$$

where α is the off set angle of the crystal, and δ is the angular distance between the flare and the spin axis, and ϕ is the phase angle of the spin. Note that Ψ changes as the satellite spins ; $\Psi = \Psi_0 + \phi$. Ψ_0 is Ψ at $\phi = 0$.

It is easily seen from Equation (1) that the reflectivity of SOX2 changes with polarization degree, while that of SOX1 does not. Hence the polarization can be found by comparing the data from the same line. There are two other estimates that use only SOX2 data. The three methods are described briefly below:

A) Ratio of SOX2 to SOX1; The ratio of the intensities of the same line of SOX1 and SOX2 is

$$I_2/ I_1 = (1 - 0.94P\cos 2\Psi_2)/(1 - 0.16P\cos 2\Psi_1) R_{12}, \tag{3}$$

where R_{12} is the SOX1/SOX2 ratio of net efficiency for non-polarized incident X-rays. This method is valid for investigating temporal variations of polarization, since the absolute R_{12} value is not known accurately. The difficulty with this method is to define the corresponding blended lines exactly in spite of the difference in dispersion between SOX1 and SOX2.

B) Relative line intensities measured by SOX2; Equation (2) shows that each line has a different polarization angle Ψ according to its wavelength, and a different polarization degree according to its transition (see, for example, Shlyaptseva et al. 1981). Therefore, the relative intensity of a certain line l normalized by, for instance, the intensity of FeXXV resonance line w (see Gabriel, 1972) will vary as

$$I_l/I_w = (1 - 0.94P_l\cos 2\Psi_l)/(1 - 0.94P_w\cos 2\Psi_w) R_{lw}, \tag{4}$$

where R_{lw} is the ratio for the non-polarized case. This method holds good as long as the line ratio is almost insensitive to temperature (e.g. x,y,z). However, temperature sensitive lines are blended with these lines to a greater or lesser degree.

C) Difference of the intensity of the same line measured in different spin phases by SOX2; In a single rotation of the satellite, the wavelength scan is done twice in opposite directions. The wavelength of a given line corresponds to spin-phase angles $\phi = \phi_+ (0<\phi_+<\pi)$, and $\phi_- = 2\pi - \phi_+$. Consequently, the polarization angle $\Psi = \Psi_0 + \phi$ changes,

$$R_{+-} = (I_+ - I_-)/(I_+ + I_-) = 0.94P\sin2\phi \sin2\Psi_0 /(1-0.94P\cos2\phi \cos2\Psi_0). \quad (5)$$

If P is constant, the maximum R_{+-} value is given when $\phi = 45°$ or 135°. Figure 1 shows the contour of R_{+-} values of the line w (ϕ= 70°).

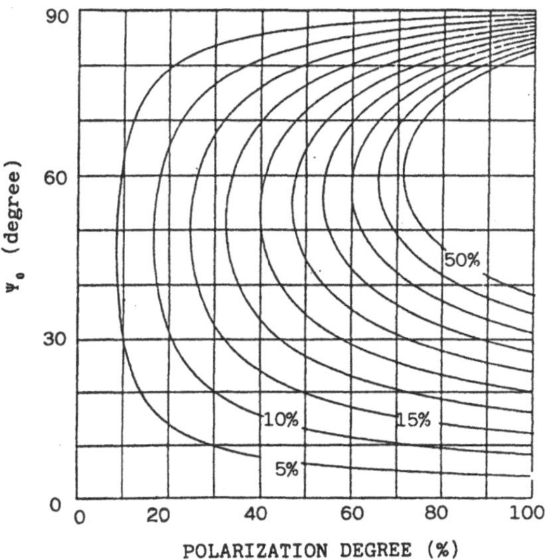

Figure 1. Contours of R_{+-} for FeXXV resonance(w)

In this paper, we have adopted method C), because only this method reflects the absolute P value, rather than the variation of P. Lines w, z, and ß (see Feldman et al., 1981) occur at adequate phase angles. Line ß, however, is too weak to give a good statistics, and line z is not supposed to be polarized at all owing to the symmetry of the upper states. Conversely, line w is the most intense line in this spectral region and is expected to have the strongest polarization, so that it was used here.

3. RESULTS AND DISCUSSION

The values of R_{+-} for six large flares are shown in Figure 2 (circles), where the temporal gaps of I_+ and I_- are corrected by interpolation.

TABLE I

Date			Onset(UT)	Class	Location	$\overline{R}_{+-}$ (%)
1981	April	2	11^h02^m	X2.2/1B	S43 W68	-0.63
	'	4	05^h01^m	X1.9/2B	S44 W85	-1.1
	October	7	22^h54^m	X3.6/1B	S19 E88	-0.45
		14	17^h04^m	X3.0/SB	S06 E86	-0.36
1982	March	30	05^h33^m	X2.8/3B	N13 W11	-1.7
	June	6	16^h30^m	X12 /1B	S10 E10	+1.5

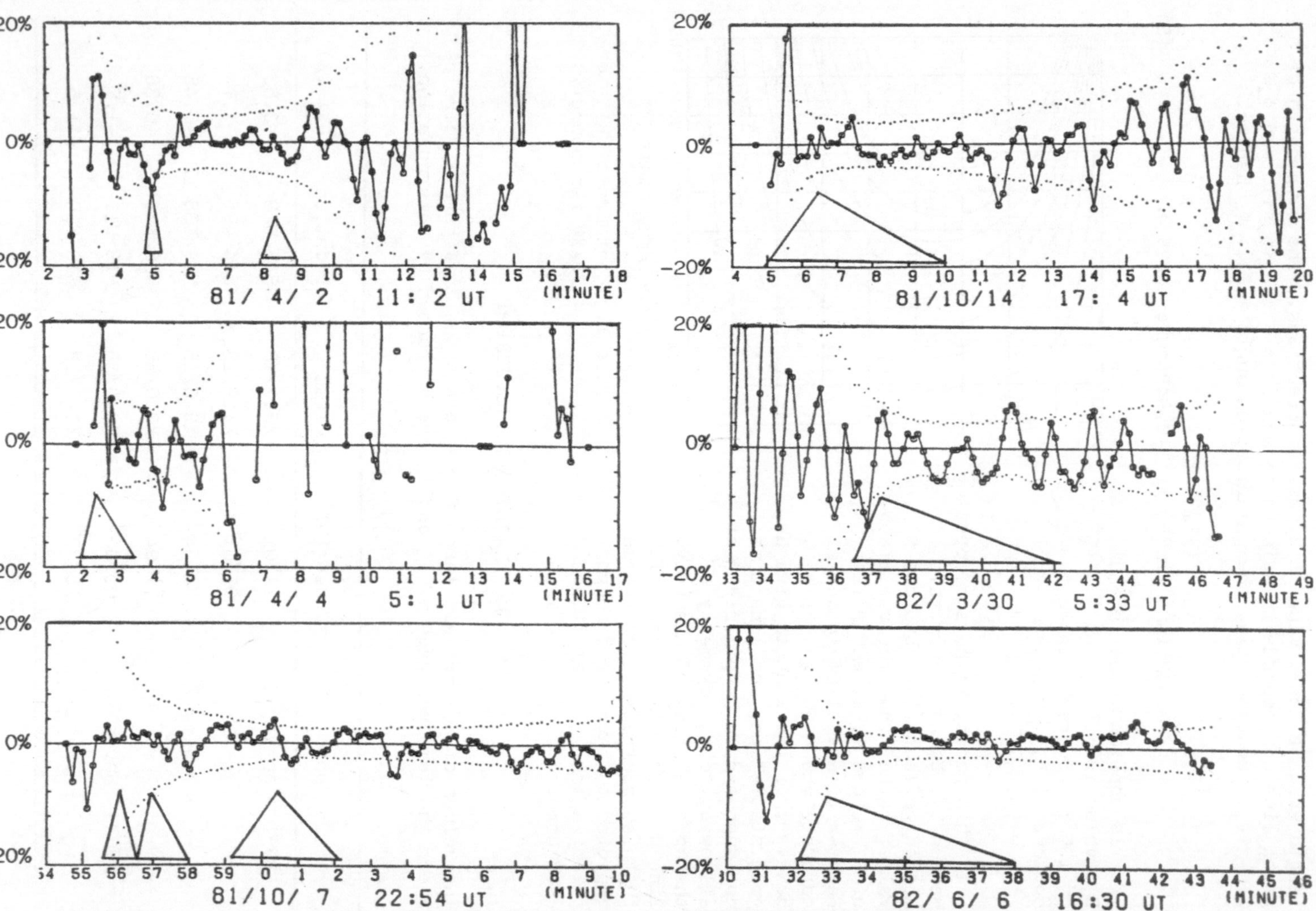

Figure 2. Temporal behavior of R+-. Triangles indicate the hard X-ray peaks.

Dots represent estimated statistical errors. Averaged R_{+-} values over the maximum phase of each flare are given in Table I.

The maximum value of the averaged R_{+-} value turned out to have a small value, 1.7%, and it implies that the average polarization degree $\overline{P}$ during the maximum flux phase of the soft X-ray flare is less than ~4% assuming that $\overline{P}$ is about the same for every flare. The resulted value is somewhat smaller than Haug's (1979) estimation for the OSO-5 flares assuming that the lower cut off energy of the nonthermal electrons is 10 KeV. This infers that the lower cut off energy of the observed flares is higher than 10 KeV or the velocity distribution depart from the beam.

As for the 1982 June 6 flare, which is the largest flare Hinotori has ever observed, the R_{+-} value of 1.5% appears to be maintained without fluctuations, implying $\overline{P} \lesssim 3\%$. There is however no denying of the possibility that these R+- values may arise from some asymmetric instrumental effects, which could still be very small.

No strong polarization as reported by Krutov et al. has been found yet. It assures the validity of our temperature analyses using relative line intensities, so far.

With regard to the temporal behaviour, R_{+-} values rarely exceed the estimated errors and seem to have no significant features related to hard X-ray bursts except for the small enhancements at 11h05m UT of the 1981 April flare (cf. Tanaka et al., 1982) and at 05h02m UT of the 1981 April 4 flare.

REFERENCES

Feldman, U., Doschek, G. A., and Kreplin, R. W.: 1980, Astrophys. J. 239,725

Gabriel, A. H.: 1972, Monthly Notices Roy. Astron. Soc., 177, 687.

Haug, E.: 1979, Solar Phys. 61, 129

Krutov, V. V., Korneev, V. V., Mandelstam, S. L., Shlyaptseva, A. S., Urnov, A. M., Vinogradov, A. V., Zhitnik, I. A.: 1981, Lebedev Physical Institute preprint 133

Lemen, J. R., Chanan, G. A., Hughes, J. P., Laser, M. R., Novick, R. Rochwarger, I. T., Sackson, M., and Tramiel, L. J.: 1982, Solar Phys. 80, 333

Novick, R.: 1975, Space Sci. Rev. 18, 389

Shlyaptseva, A. S., Urnov, A. M., and Vinogradov, A. V.: 1981, Lebedev Physical Institute preprint 193

Tanaka, K., Watanabe, T., Nishi, K., and Akita, K.: 1982, Astrophys. J. (letters) 254, L59

THERMAL EVOLUTION OF FLARE PLASMA

Tetsuya Watanabe and Katsuo Tanaka
Tokyo Astronomical Observatory, University of Tokyo,
Mitaka, Tokyo, 181, Japan

and

Kyo Akita and Nariaki Nitta
Department of Astronomy, University of Tokyo,
Bunkyo-ku, Tokyo, 113, Japan

ABSTRACT. The evolution of hot thermal plasma in solar flares is analyzed by a single-temperature model applied to continuum emission in the 5 keV < E < 13 keV spectral range. The general trend that the thermal plasma observed in soft X-rays is heated by the non-thermal electrons that emit as the hard X-ray bursts is confirmed by the observation of an electron temperature increase at the time interval of hard X-ray spikes and a quantitative comparison between thermal energy content and hard X-ray energy input. Non-thermal electrons of 10 keV < E <30 keV energy may play an important role in pre- and post-burst phases.

1. Introduction and Instrumentation

The FLare Monitor (hereafter FLM) on board Hinotori is a scintillation proportional counter, which has an energy resoltuion about twice as good as an ordinary proportional counter. It is designed for wide-band non-dispersive spectroscopy of solar X-rays in the energy range from 1.5 to 30 keV. The energy resolution of this detector is about 10% at 6 keV (see Inoue et al., 1982). The data obtained by FLM consist of two sets. One is two-channel count-rate data, and the other is 128-channel pulse-height analyzed spectral data. The count-mode data are obtained every 125 msec when the satellite is operated in the flare mode, and energy-mode data are obtained every 4 seconds. Time resolutions in satellite quiet mode operation are 500 msec and 32 sec, respectively.

With this instrument, we can obtain soft X-ray spectra of solar flares as well as contributions of quiescent active regions without using any dispersive elements. We show in this paper a simple analysis of the evolution of flare thermal plasma and make some comments on the energetics of small simple flares.

2. Method of Analysis

In order to derive temperatures and emission measures of the flaring plasma, we adopted a single temperature model and applied it to FLM energy mode data in the energy range E > 5 keV. Continuum emissions in

Solar Physics **86** (1983) 107–113. 0038–0938/83/0861–0107 $ 01.05.

this energy range mostly come from flare plasma and the contribution of quiescent active regions hardly exist. We separate the continuum and line emissions of iron and nickel, which are assumed to have Gaussian shapes due to the instrumental profile. Parameters of our model are temperature and emission measure of continuum emission for $E > 5$ keV, intensity, width, and location of the iron line at 6.7 keV, and intensity and location of the nickel feature at 7.6 keV, as well as a constant counter background. We infer the line width of Ni from the Fe line width using the fact that the instrumental width is proportional to the inverse square root of the energy.

In the fitting procedure, we have searched for the parameter set that gives the minimum χ^2 value. Quiescent background emission inferred from pre-flare data are subtracted, though contributing little in the energy range $E > 5$ keV. As for the emissivity of soft X-ray region, we have adopted the simple formula of Culhane and Acton (1974). The fitting using these rather simple assumptions suffices for an analysis of FLM energy mode data in 5 keV $< E <$ 13 keV region, and we can obtain parameters within a 90% confidence level (see Figure 1).

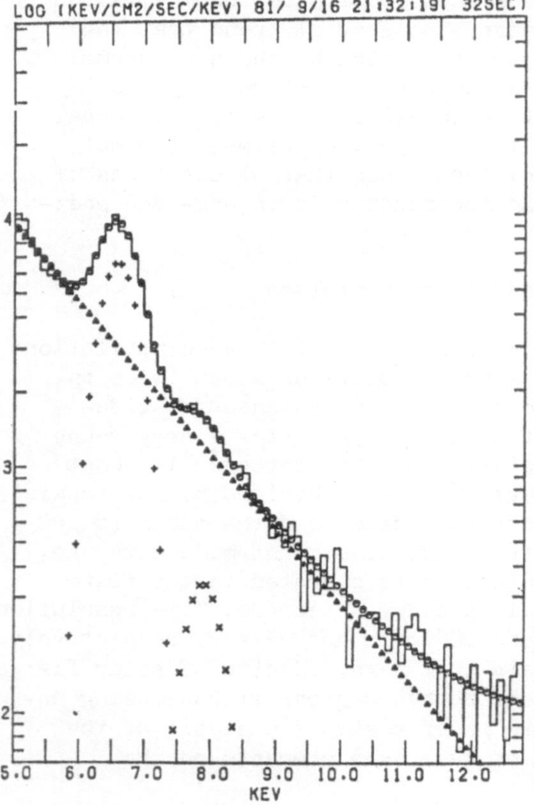

Figure 1. Energy flux obtained from FLM (histogram) and fitted flux distributions from model analysis (open circles). Plus and cross are the contributions of iron and nickel lines to the continuum emission (triangle, log T_e = 7.24, log EM = 47.19).

3. Characteristics of Thermal Evolution of Flare Plasma

3.1. T_e increase in the time interval of hard X-ray burst and counter clockwise evolution in T_e vs EM diagram

Figure 2 shows time profiles for a flare of 1981 September 16. The lower three figures are the counting rate of FLM-L (1.6-2.6 keV), FLM-H (2.6-12.7 keV), and the Hard X-ray Monitor (HXM) on Hinotori PC-1 (30-40 keV) and the upper three are the derived electron temperatures (T_e), emission measures (EM), and thermal energy content (T.E.C.), the definition of which is $4.13 \times 10^{-16}\, n_e^2 V T_e$. Electron temperatures derived from FLM increase as long as the hard X-ray burst continues. However, the emission measure reaches a maximum about three minutes later than the temperature maximum. This situation may be well

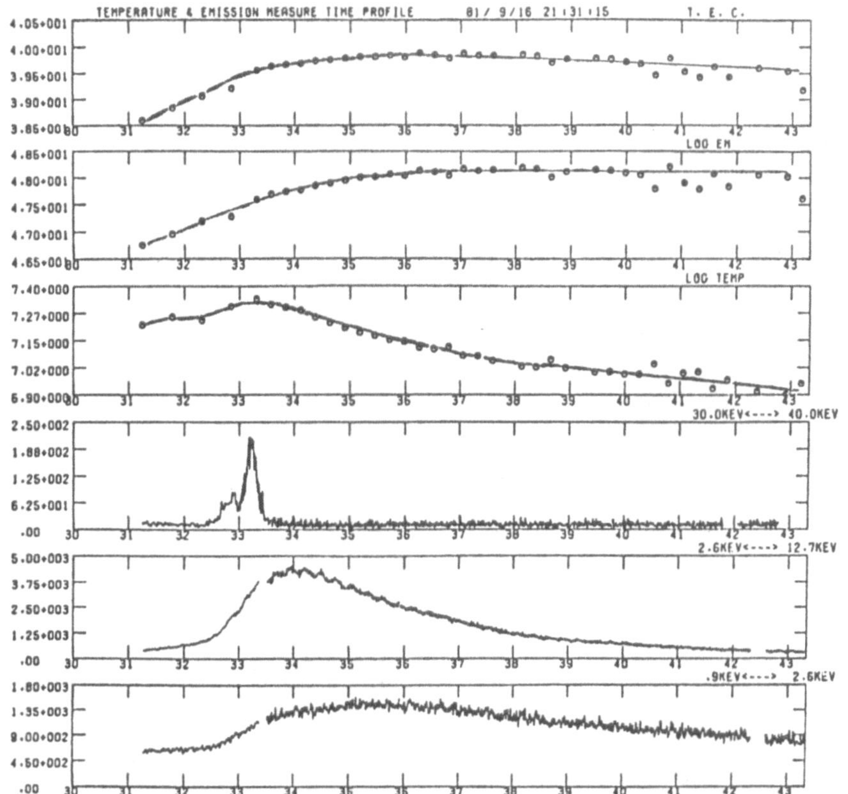

Figure 2. September 16, 1981 flare (C 5.2) time profiles of FLM-L (1.6-2.6 keV), FLM-H (2.6-12.7 keV),HXM PC-1 (30-40 keV) counting rates and of electron temperatures emission measure, and thermal energy content derived from the analysis.

understood in the T_e vs EM diagram (Figure 3), in which the evolution follows a counter-clockwise track, that is, electron temperature reaches maximum first and then the emission measure. Electron temperature increases in the interval of the hard X-ray burst and the counter-clockwise evolution of flare thermal plasma in T_e vs EM diagram are almost universal phenomena in C-class flares, which are observed by FLM and analyzed by the above procedures.

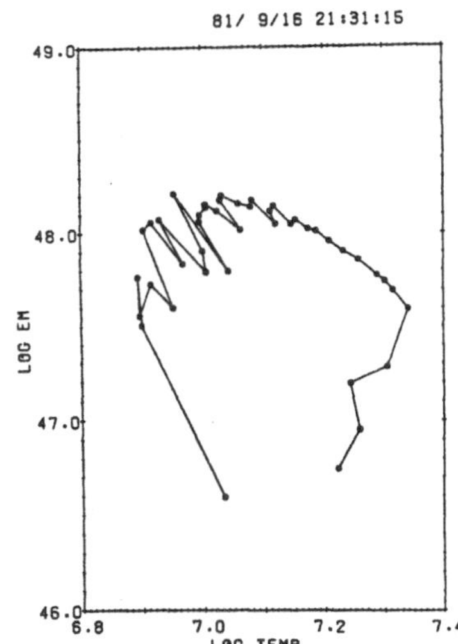

Figure 3. T_e vs EM diagram for September 16, 1981 flare.

3.2. Thermal energy content vs energy input estimated from hard X-rays

The electron temperature increase during the hard X-ray burst may suggest a scenario in which flare thermal plasma is heated by the non-thermal electrons which emit the hard X-ray burst. To support such a scheme, we have compared the thermal energy content derived from FLM to the energy power injected by electron beams derived from HXM. In order to obtain the injected power, we performed a power-law fit of the hard X-ray spectrum obtained by HXM using the formula by Hudson et al. (1978), and derived the total power of electrons in the energy range above 20 keV (P_{20}) and its time integral (Tanaka et al., 1982).

Figure 4 shows time profiles of a C 9.2 flare of 1981 November 1. In the uppermost figure showing thermal energy content, the injected power calculated from HXM is superposed. The general similarity in time profiles of thermal energy content and injected power suggests the possibility that the soft X-ray plasma was produced by electron beams and had an electron density of about $n_e \approx 10^{11.2}$ cm^{-3}.

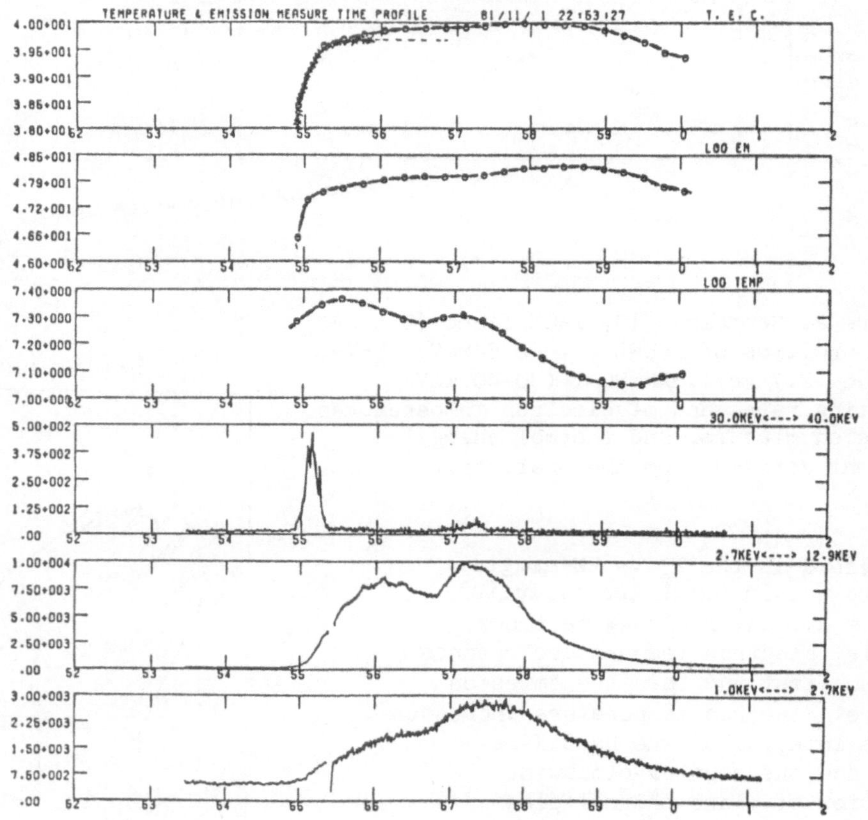

Figure 4. Same figure for November 1, 1981 flare (C 9.2) as Figure. 2. Crosses in the uppermost figures of thermal energy content is the calculated energy input from HXM with electron density of $10^{11.2}$

3.3. Pre- and post-burst heating, minimum energy for beaming electrons

The general time behavior is quantitatively consistent with the interpretation that the energy of no-thermal electrons emit the hard X-ray burst is converted (or thermalized) to the thermal energy of the soft X-ray plasma (Lin and Hudson, 1976). However, there are some flares which have small discrepancies between thermal energy content and injected power in pre- and/or post-burst phases. The electron temperature is already rather high before hard X-ray burst starts. For an explanation of this heating, there seems to be a possibility that electron beams may exist to much lower energy than 30 keV, which was the lowest energy observed by HXM. To test this idea, we have used data from after March, 1982 when FLM covered the energy range of 2 keV ~ 25 keV. We have divided this range into five keV bins, plotted in Figure 5 for a flare of 1982 May 11. This figure clearly shows that in the energy range E > 10 keV, the time profiles are different from those at lower energies and similar to the hard X-ray time profiles.

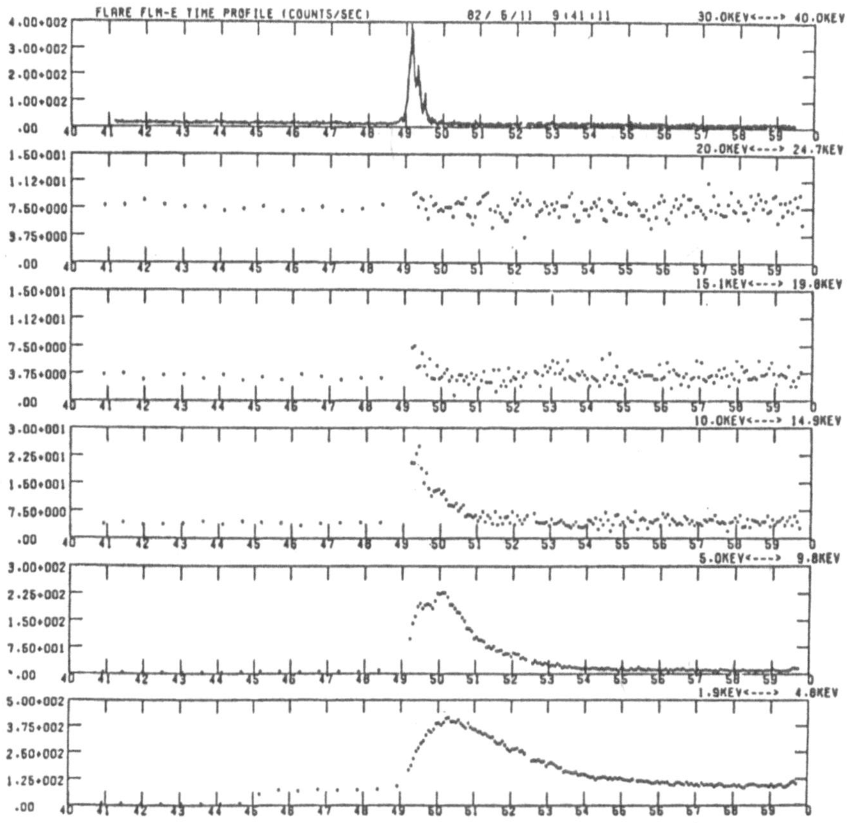

Figure 5. May 11, 1982 flare (C 1.2) time profiles of FLM energy mode data and HXM PC-1 counting rates. FLM energy mode data are divided into five energy ranges in each nearly 5 keV.

In general the FLM spectra show a strong tendency for higher-energy emission to resemble the time profile of the hard X-ray burst, even at energies low as 10 keV for some flares. These facts suggest that beam electrons may exist down to at least to 15 keV and sometimes to 10 keV, depending possibly on the ambient electron densities of the loop. Particularly in pre-burst phase, these softer electrons may have some contribution to the heating of the flare plasma.

Af for post-burst heating, we found an example of a flare of 1981 October 21, in which after the end of hard X-ray burst, emission measure continuously increased and the thermal energy content remained nearly constant for about 13 minutes. This means that after the hard X-ray burst, some additional heating must have been necessary in this time interval. We also may attribute this to electrons in the enegy range less than 30 keV.

4. Conclusion

We have analyzed FLM energy-mode data using a single temperature model and found that flare soft X-ray emissions in the energy range 5 keV $<$ E $<$ 13 keV are well fitted by an isothermal plasma. Energy input by non-thermal electrons estimated from hard X-rays is quantitatively consistent with the thermal energy content of the soft X-ray source, and the electron temperature of the thermal plasma increases only in the time interval of the hard X-ray burst. As for pre- and post-burst energy input, there seems to be a possibility that non-thermal electrons in the energy range below 30 keV play some important roles in heating thermal plasma. We should analyze spectra in these energy ranges more extensively.

References

Culhane, J. L. and Acton, L. W.: 1974, Ann. Rev. Astron. Astrophys. 12, 359.

Hudson, H. S., Canfield, R. C., and Kane, S. R.: 1978, Solar Phys. 60, 137.

Inoue, H., Koyama, K., Mae, T., Matsuoka, M., Ohashi, T., Tanaka, Y., and Waki, I.: 1982, Nuclear Instruments and Methods 196, 69.

Lin, R. P. and Hudson, H. S.: 1976, Solar Phys. 50, 153.

Tanaka, K., Nitta, N., and Watanabe, T.: 1982, Hinotori Symposium on Solar Flares, 20.

DISCUSSION

ACTON: You showed some very good isothermal fits to FLM data. Are all of your data well fitted with an isothermal model?

WATANABE: In rise phase, fitting with an isothermal is rather poor in some cases. There seems to be some contribution of high temperature plasma in high energy channels (E >10 keV). However, it is a problem

of sensitivity and one cannot clearly discriminate that from the noise.

HUDSON: Does your use of isothermal fits mean that your FLM data cannot contribute to the knowledge of soft X-ray differential emission measure? The 10 - 30 keV continuum may in principle make an important contribution to resolving the problems in the X-ray emission line spectrum.

WATANABE: At least for $5 < E < 12$ keV, which is the energy range FLM observed until this March, I think an introduction of differential emission measure is not necessary. I also think time profile changes at $E > 15$ keV indicate that the FLM spectrum in the energy range 15 - 30 keV has a key to answer such questions.

GABRIEL: It is clear that many groups are having difficulties with differential emission measure analysis, including my own. The effect of random noise on the inversion process is well-known. However, it may be that systematics are the main source of our difficulty. What would be the effect, for example, if the ionization balance were to be in error by a factor of two? This might produce a major change in the interpretation, such as, for example, a single temperature peak instead of a double peak.

CONDUCTIVE FLUX IN THE CHROMOSPHERE DERIVED FROM LINE LINEAR POLARIZATION OBSERVATION

J.-C. Henoux, G. Chambe, D. Heristchi, and M. Semel
Observatoire de Paris, Meudon, France

B. Woodgate and D. Shine
Goddard Space Flight Center, Greenbelt, Md., U.S.A.

J. Beckers
University of Arizona, Tucson, Ariz., U.S.A.

ABSTRACT. Linear polarization in two chromospheric lines (Hα and SI 1437 A) was observed in the gradual phase of solar flares. The polarized electric vector is directed towards disk center.

This polarization could be due to collisional excitation of hydrogen and SI by energetic electrons beamed in the vertical direction. Direct excitation by a highly energetic beam of electrons of order 10-100 keV. is doubtful. The heat flux in the region connecting the transition zone to the high chromosphere during the gradual phase of a flare could lead to an anisotropic excitation.

Selecting a function which represents the velocity distribution of electrons carrying heat flux, the relationship between conductive heat flux and linear line polarization has been computed. The application of the relationship between linear polarization and heat flux to the observed degree of polarization leads to the determination of the conductive heat flux in the high chromosphere. This conductive flux is of the order of magnitude of the total radiation loss in the chromosphere and below, which is also of the order of magnitude of the conductive flux in the transition zone.

INTRODUCTION

The aim of this paper is to show how the measurement of impact line polarization can provide quantitative information on the mechanisms of energy transport which take place in solar flares.

Chromospheric flares are usually interpreted as secondary phenomena assuming that energy is supplied to the chromosphere, from the flaring corona, via heat conduction, electron bombardment or X-ray irradiation. The importance of these modes of energy transfer in the energetic equilibrium of a flaring chromosphere is widely accepted through theoretical considerations. In this paper an observational test is presented which could give estimates of the relative importances of the energy transport modes.

Heat conduction, electrons bombardment, and XUV irradiation are associated with anisotropic velocity distribution functions of energetic electrons :

Solar Physics **86** (1983) 115–122. 0038–0938/83/0861–0115 $ 01.20.

- Velocity distribution of electron beams are peaked in the magnetic field direction.
- Velocity distribution of electron transporting heat flux has a tail of energetic electrons in the heat flux direction.
- Photoelectrons produced by XUV irradiation are predominantly moving in the solar horizontal plane.

Therefore the determination of the main travelling direction of energetic electrons would allow discrimination between the different modes of energy transport.

The main characteristics of the velocity distribution anisotropy could be estimated by measuring line linear polarization for atomic species collisionnally excited by these electrons :

Beams of electrons colliding with atoms could lead to the emission of linearly polarized lines. The maximum polarization is observed at 90° from the beam direction. This polarization is energy dependent and changes of sign when the energy increases. Therefore the degree of polarization gives information on the anisotropy of the colliding electrons and on the energy of these electrons.

Linear polarization observations made in two different chromospheric lines are reported in Section 1. The observed polarization is interpreted in Section 2 as impact polarization produced by electrons carrying heat flux. The conductive flux derived in Section 2 is compared in Section 3 to the conductive flux in the transition zone and to the total radiative losses in the chromosphere.

1. POLARIZATION OBSERVATIONS

Linear polarization was observed in two different chromospheric lines, SI 1437 Å and Hα, during the decay phase of two different events.

1.1. SI lines observations. We used the Ultraviolet Spectrometer and Polarimeter (UVSP) on the Solar Maximum Mission. This instrument is described by Woodgate et al. (1980) and by Calvert et al. (1979). The observational results briefly presented here are given in detail in Henoux et al. (1982).

When the instrument is used as polarimeter a magnesium fluoride waveplate is inserted behind the entrance slit of an Ebert Fastie Spectrometer and it is rotated in steps of 22.5°. The spectrometer grating serves as an analyser of polarization.

The rotation of the wave plate produced a modulation of the intensity if the incoming radiation is polarized. Fourier analysis is used to evaluate the Stokes parameters. The coefficients of the Fourier component modulated at four times the waveplate rotation frequency give the linear polarization.

Polargrams are obtained by rastering the solar image. They cover a field of view of 50" x 50" with a spatial resolution of 10" x 10". At each raster pixel the integrated line intensity was sampled at each of the 16 waveplate positions.

A flare of importance SB and class C4 was observed on July 15th 1980 from 22 48 UT to 23 07 UT. Line linear polarization was detected during the decay phase of soft X-ray emission in flaring pixels. Using a vectorial representation Figure 1 shows the degree of linear polarization and the orientation of the polarization for six flaring pixels.

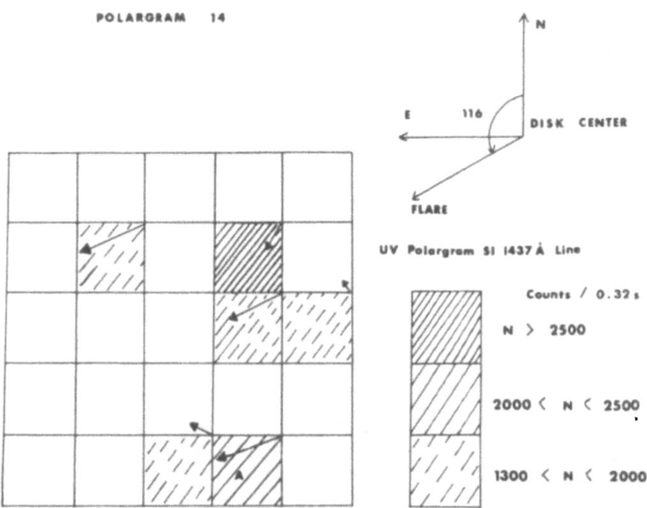

FIG. 1 - Linear polarization in the SI 1437 Å line is represented by vectors of amplitude proportional to the degree of polarization and oriented in the direction of the electric vector. The maximum degree of polarization is 25 %.

SI 1437 Å line emission is polarized at three of these six flaring points with a degree of polarization as high as 25 %. The angle between the South-North direction and the electric vector varies from 109° to 117° and is therefore very close to the angle between South-North and disk-center-to-flare direction (116°). By adding the signal from the six bright pixels we found a mean polarization of 12 % with an electric vector directed towards disk center to within 3°.

1.2. Hα line observations

We slightly modified an existing heliograph by inserting a rotating half-wave plate in front of a linear analyser in front of the Lyot Filter. This wave plate was rotated through 22,5° per step. Successive sets of 16 images were taken every minute on photographic film. A pass band of 0.75 Å centered at Hα center was used.

From each set of 16 images we derived seven determinations of the linear polarization. Four of these seven determinations are independant.

The decay phase of a 1B flare was observed on May 17, 1980. Four sets of 16 images were obtained in four minutes (from 0738 to 0741). Restricting the investigation to polarization signals of long duration (1 to 4 minutes), we integrated the signal over two and four minutes of observation, and we found polarization above Hα kernels and slightly outside these kernels. The direction of polarization was within 7° of the disk-center-to-flare direction (see Figure 2).

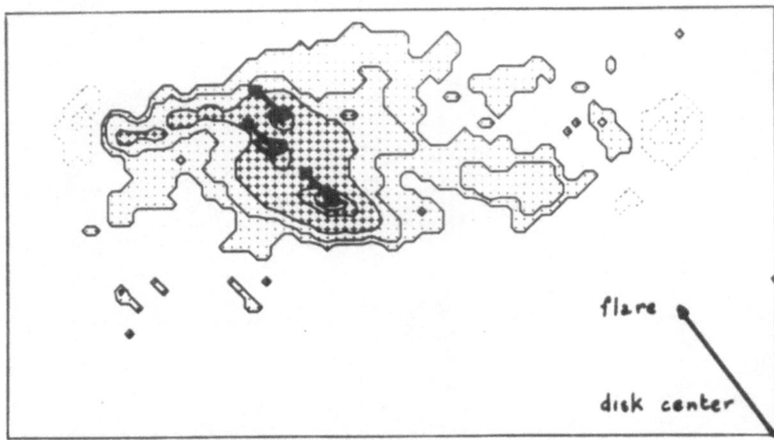

FIG. 2 - Linear polarization in the Hα line is represented by vectors of amplitude proportional to the degree of polarization and oriented in the direction of the electric vector. The maximum degree of polarization is 2 %. The arrow on the bottom right corner represents the disk-center-to-flare direction.

The direction is stable in time above the kernels during the four minutes of observation. The degree of polarization is about 2 %.

SI 1437 Å and Hα line observations lead to the same result : these two chromospheric lines are linearly polarized during the decay phase of the observed events. The polarized electric vector is directed towards disk center. Linear polarization may also be produced during the flare impulsive phase but the time resolution of the observations was not high enough to detect low duration polarization.

2. INTERPRETATION: IMPACT POLARIZATION PRODUCED BY ELECTRONS CARRYING HEAT FLUX

2.1. Impact polarization

Atoms collisionally excited by beams of electrons emit line radiation which may be linearly polarized. The theory was formulated by Percival and Seaton (1958). A review of the experimental and theoretical results on impact polarization was presented by Kleinpoppen (1969) and Heddle (1979).

The maximum polarization $P_{90}(v)$ is observed at 90° from the beam direction ($P_{90}(v) = (I_{/\!/} - I_{\perp})/(I_{/\!/} + I_{\perp})$ where $I_{/\!/}$ and $I_{\perp}$ are respectively the intensities of the vibrations parallel and perpendicular to the beam ; v is the electron velocity. This polarization is energy dependant.

The linear polarization of photons emitted at an angle β from the beam is given by $P(\beta,v) = P_{90}(v) \sin\beta/(1 - P_{90}(v) \cos\beta)$. (1)
The linear polarization P(θ) produced by collisional excitation by electrons with a specific angular velocity distributions $dn(v)/d^3v$ can

easily be computed.

We defined $P(\theta) = (I_{/} - I_{\perp})/(I_{/} + I_{\perp})$ where $I_{/}$ and $I_{\perp}$ are the intensities of the vibrations parallel and perpendicular to the plane defined by the line of sight and the solar vertical. θ is the angle between the solar vertical and the line of sight.

For any velocity distribution with cylindrical symmetry

$$P(\theta) = P(90°) \sin\theta \,/\, (1 - P(90°)\cos\theta). \qquad (2)$$

This equation is identical to Equation (1) when β and $P_{90}(v)$ are replaced by θ and $P(90°)$ where

$$P(90°) = \int_0^\infty \frac{2J_o - P_{90}(v)(J_2 - J_o)}{3 - P_{90}(v)} v\,\sigma(v)\,dv \qquad (3)$$

and $$J_n = \int_0^\pi \frac{dn(v,\alpha)}{d^3v} \cos^n\alpha \sin\alpha \, d\alpha;$$

α is the angle between an electron travel direction and the solar vertical ; $\sigma(v)$ is the line excitation cross section.

2.2. Velocity distribution associated with heat flux

A typical electron velocity distribution which has a heat flux in the z direction but no net particle current was given by Manheimer (1977) as

$$\frac{dn}{d^3v} = \frac{n_e}{(2\pi v_e^2)^{3/2}} \cdot \exp - \frac{v^2}{2v_e^2} \cdot \{1 - 0.5\frac{Q}{Q_S}[(\frac{v_z}{v_e})^3 - 3(\frac{v_z}{v_e})]\}, \qquad (4)$$

where $KT = m_e v_e^2$ and $Q_s = \frac{3}{2} n_e m_e v_e^3$ is the saturated heat flux.

Negative values of dn/d^3v obtained for large negative v_z were replaced by zero. The first three moments suffer only minor changes. Then the effective heat flux Q^* ($Q^* = \frac{1}{2}\int m_e v_z^3 \frac{dn}{d^3v} d^3v$) differs from Q.

2.3, Polarization of the SI 1437 Å and Hα lines. Computation of the polarization and comparison with observations.

P(90°) was computed using formula (3) and the modified version of Manhemier's distribution function for Hα and SI 1437 Å lines.

(1) Hα

The figure 3 shows P(90°) as a function of Q/Qs for three sets of known values of $P_{90}(v)$ and $\sigma(v)$ and for two different temperatures of formation of Hα line center ($T = 10^4$ K and $T = 2 \times 10^4$ K).

Knowing the flare solar coordinates and using formula (2) P(90°) was derived from the observed degree of polarization. P(90°) = 6 %. For such value of P(90°), the Figure 3 and the relation between Q^*/Q_s and Q/Q_s lead to the inequality $0.16 < Q^* / Q_s < 0.37$ where the two limits correspond to $T = 10^4$ and $T = 2 \times 10^4$ K.

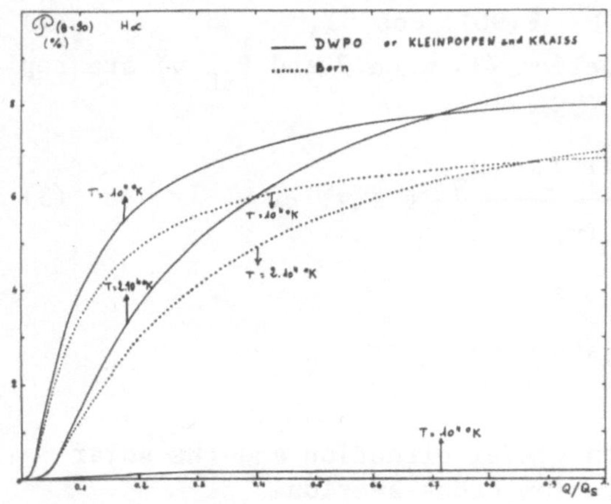

FIG. 3 - Maximum linear polarization P(90°) of the Hα line as a function of the heat flux. Calculations were made for two temperatures. Lower full line curve σ(v) and $P_{90}(v)$ from Kleinpoppen and Kraiss (1968) Upper full line curves $P_{90}(v)$ from the DWPO approximation (Syms et al. 1975). In the last two cases σ(v) is from Mahan et al. (1976).

(2) <u>SI 1437 Å</u>

The maximum polarization in the SI 1437 Å line is shown as a function of Q/Q_s in Figure 4. P(90°) was computed for two sets of values of $P_{90}(v)$ derived from the polarization of resonance transition of alkali-metal atoms and from the polarization of the hydrogen Lyα line, and for $T = 10^4$ and $T = 2 \times 10^4$ K.

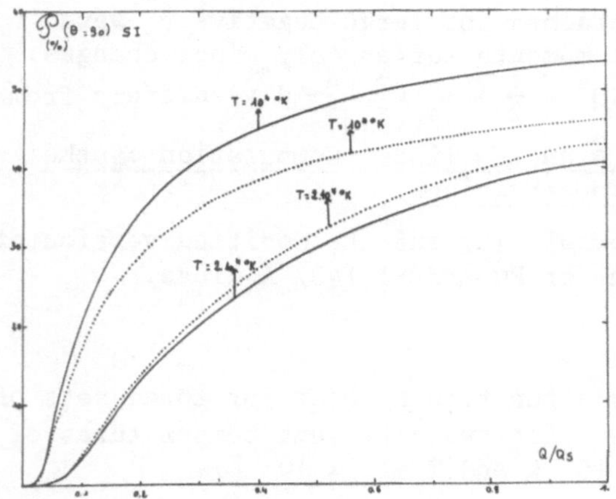

FIG. 4 - Maximum linear polarization P(90°) of the SI 1437 Å line as a function of the heat flux σ(v) and $P_{90}(v)$ were obtained from data corresponding to either alkali transitions (full line curves) or H Lyα (dotted line curve).

From Equation 2 the maximum polarization P(90) corresponding to the mean degree of polarization observed (12 %) is 31 %. Finding this value on Figure 4 leads to the following inequality $0.13 < Q^*/Q_s < T = 10^4$ K and $T = 2\ 10^4$ K and are close to the limits derived from Hα observations.

3. CONDUCTIVE FLUX IN THE HIGH CHROMOSPHERE - CONCLUSION

From the two preceeding inequalities we derived $0.16 < Q^* / Q_s < 0.33$. Assuming $n_e = 10^{12}$ the extreme values of the saturated heat flux are $Q_s = 5.6 \times 10^7$ and $Q_s = 2.1 \times 10^8$ leading to $9 \times 10^6 < Q^* < 6 \times 10^7$ ergs cm^{-2} s^{-1}.

This estimate of the heat flux can be compared to the conductive flux in the transition zone and to the radiative losses in the chromosphere. The conductive flux in the transition zone during flares was computed by Withbroe (1978) who found $F_c = 3.2\times10^8$ ergs cm^{-2} s^{-1} for a 2B flare. Machado and Emslie (1979) derived conductive flux values varying between 2×10^6 and 5×10^7 ergs cm^2 s^{-1} for 7 subflares.

If we assume that no energy is lost between the middle of the transition zone and the Lyα-Hα center formation layer then the conductive flux in the high chromosphere, leading to a reasonable agreement with our estimates.

The measured conductive flux is of the order of magnitude of the total radiation losses in the chromosphere and below : Canfield et al. (1980) gave 4×10^7ergs cm^2 s^{-1} as the radiative output of a class (-N) flare.

The measurement of linear polarization makes possible the determination of the heat flux. This method is the only one which could give the conductive flux value in the high chromosphere. Its accuracy's is still limited by the uncertainities in the atomic data and in the shape of the velocity distribution function. Nevertheless this first analysis of polarization observation clearly indicates that heat conduction plays an important role in chromospheric flare heating.

REFERENCES

Calvert J. et al. : 1979, Optical Engineering 18, 287.

Canfield R.C. et al. : 1980, Solar Flares. Edited by P.A. Sturrock, Colorado Associated University Press 451.

Heddle D.W.G. : 1979, Advances in Atom. Mol. Physics edited.

Henoux J.C., Chambe G., Semel M., Sahal S., Woodgate B., Shine D., Beckers J., Machado M. : 1982, to be published in Ap. J.

Kleinpoppen H., Kraiss E. : 1968, Phys. Rev. Lett. 20, 361.

Kleinpoppen H. : 1969, Physics of the one and two electrons atoms. Bopp F. and Kleinpoppen H. (Editors) North Holland, 612.

Machado M.E., Emslie A.G. : 1979, Ap. J. 232, 903.

Mahan A.H., Gallagher A., Smith S.J. : 1976, Phys. Rev. A 13, 156.

Percival I.C., Seaton H.J. : 1959 Phil. Trans. Roy. Soc. London Series A 251, 113.

Syms R.F., McDowell M.R.C., Morgan L.A., Myerscough V.P. : 1975, J. Phys. B. : Atom. Molec. Phys. 8, 2817.

Withbroe G.L. : 1975, Ap. J. 225, 641.

Woodgate B.E. et al. : 1980, Solar Physics 65, 73.

UPPER LIMITS ON THE TOTAL RADIANT ENERGY OF SOLAR FLARES

H. S. Hudson
Center for Astrophysics and Space Sciences, University of California, San Diego, La Jolla, CA 92093, U.S.A.

and

R. C. Willson
Jet Propulsion Laboratory, Pasadena, CA 91103, U.S.A.

ABSTRACT. We establish limits on the total radiant energy of solar flares during the period 1980 February - November, using the solar-constant monitor (ACRIM) on board the Solar Maximum Mission. Typical limits amount to 6×10^{29} erg/s for a 32-second integration time, with 5σ statistical significance, for an impulsive emission; for a gradual component, about 4×10^{32} ergs total radiant energy. The limits lie about an order of magnitude higher than the total radiant energy estimated from the various known emission components, suggesting that no heretofore unknown dominant component of flare radiation exists.

1. INTRODUCTION

A solar flare consists of an explosive release of radiation and mechanical energy near the surface of the sun, with copious acceleration of non-thermal particles. The Solar Maximum Mission made the first systematic effort to observe simultaneously all of the important radiations, and hence no accurate knowledge of the total radiant energy release previously existed. Indeed, we have no observations of some regions of the spectrum, and of others so little that we cannot estimate their contributions. Bruzek (1967) and others have described the magnitudes of the known components of the observed energy distribution. The mechanical energy release represented by the interplanetary shock wave (Hundhausen, 1972), probably contains more energy than the electromagnetic radiation; in any case the particle radiation ("solar cosmic rays") makes a negligible contribution energetically (see, for example, the review by Svestka, 1976). A direct measurement of the total excess irradiance produced by a solar flare - by bolometric techniques - will reveal any radiation that had heretofore remained hidden in an unobserved gap in the spectrum, and at the same time define the radiant part of the flare energy exactly as an aid in understanding of the flare mechanisms.

The Solar Maximum Mission includes in its instrument complement a sensitive radiometer, the Active Cavity Radiometer Irradiance Monitor or ACRIM (Willson, 1979). Although intended primarily to define the

Solar Physics **86** (1983) 123–130. 0038–0938/83/0861–0123 $ 01.20.

solar irradiance precisely as an input to climatological studies, this instrument has properties that make it useful for the study of solar variations of many causes. Sunspots in particular produce marked effects (Willson et al., 1981), resulting in a "missing flux" that represents a substantial fraction of the 6.4 x 10^{10} erg/cm^{2}s total radiant flux at the photosphere. The sunspot flux deficits cause variations in the "solar constant" at the level of tenths of a percent.

This paper examines the ACRIM data for selected important flares that occurred during its interval of good pointing, essentially 1980 February - November. We compare the upper limits thus obtained with estimated energy fluxes in Hα and soft X-rays. At present we have an incomplete survey, omitting many flares and not using the highest time resolution of the ACRIM detector.

2. THE ACTIVE CAVITY RADIOMETER

The ACRIM sensor consists of a black conical cavity in a carefully-designed thermal environment (Willson, 1979). Measurement of the absolute solar irradiance relies on a determination of the electrical power needed to maintain the cavity temperature when a shutter blocks off the solar input. The resulting absolute measurement has relatively good accuracy, but for the present purposes we only care about the relative precision. The ACRIM instrument has excellent DC stability (Willson et al., 1981) and observes solar variations on all time scales up to the Nyquist frequency of 3.815 mHz (Woodard et al., 1982).

The ACRIM, a thermal detector, has an extremely broad-band spectral response. The absorbing cone has a specular black finish with an absorptance near unity in the visible wavelengths. Although no measurements exist at other wavelengths of interest for solar flares - infrared, ultraviolet, or XUV down to a few angströms - we may assume that the absorptance remains essentially independent of wavelength from millimeter waves to soft X-rays. The ACRIM measurement thus yields a true bolometric magnitude.

ACRIM has a thermal time constant of about 1 s. It operates normally in a shutter cycle of 131.072 s duration, with one quarter of this interval used for prime measurements. In this paper we analyze only this kind of data, consisting of 32.768 s integrations spaced at 131.072 s intervals. Information also exists at the level of indivicual 1.024 s samples, but we have not yet studied these data and will describe them later. A typical orbit (∿ 96 min) contains about 27 shutter cycles. The instrument thus obtains some 864 individual measurements. An analysis of these data yields an "orbital mean" together with a statistical standard deviation; typical standard deviations lie in the range 10 - 15 parts per million of the mean irradiance.

3. FLARE SELECTION

We have examined a selection of flares culled mainly from the hard X-ray flare catalog (HXRBS team, 1982). Table I summarizes the observed parameters of these flares, of all area classifications but strongly tending towards the "B" brightness classification in Hα, presumably because of the original selection from the hard X-ray list. The list includes most of the very significant flare events during the first year of SMM observation, including the June 21 "white-light prominence" observed visually by Harvey and Duvall (Harvey, 1982). The list includes one importance 3B flare, that of 1980 October 14. None of the events produced a striking variation in the ACRIM reading.

We also surveyed for completeness a second list, also taken from the hard X-ray catalog, but for all of the 28 events with hard X-ray fluence exceeding 10^5 counts during the interval 1980 October 23 - 1980 November 15. Of these there was ACRIM coverage for 19, and in no case did a significant variation occur at the time of the flare.

4. OBSERVATIONS OF SEVERAL MAJOR FLARES

During the period under study, the Sun cooperated with only a single importance 3 flare, that of 1980 October 14. We show orbital means of the ACRIM data corresponding to this period in Figure 1.

TABLE I

Flares Examined

Date	Hα			Soft X-ray		
	Imp.	Location	Start	Max.	End	Class
5/21/80	2B	S14W15	2051	2107	2144	X1
6/04/80	SB	S14E59	0654	0655	0705	M6
6/07/80	SB	N14W70	0309	0314	0320	M7
6/21/80	1B	N17W91	0115	0120	0146	X2
7/01/80	1B	S12W38	1618	1628	1642	X2
7/05/80	1B	N28W29	2233	2246	2329	M8
7/23/80	2B	S17E13	0051	0103	0209	M8
9/04/80	1B	S09W28	2156	2220	2231	M2
10/14/80	3B	S09W07	0558	0611	0632	X3
11/06/80	2B	S12E72	0329	0352	0533	X9
11/07/80	2B	N07W11	0159	0208	0307	X2
11/12/80	1B	N10W72	0445	0452	0517	X2
11/15/80	1B	S12W53	1540	1553	1626	X1

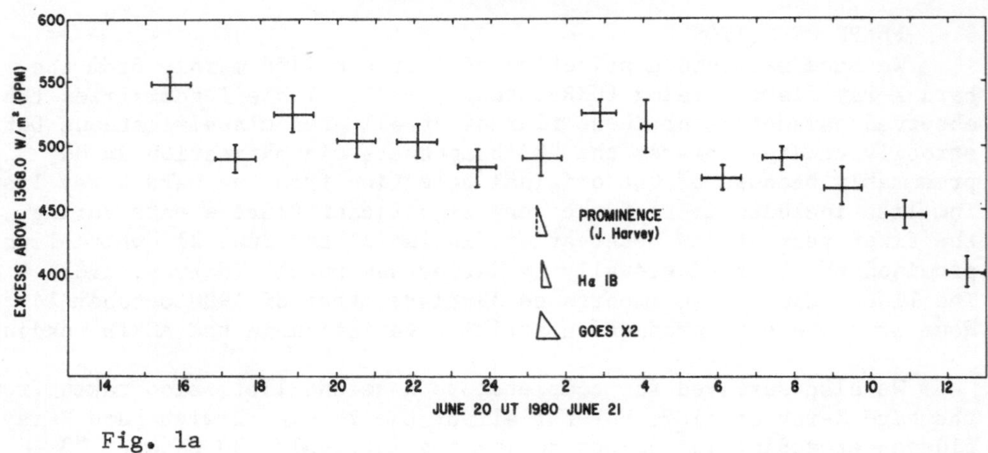

Fig. 1a

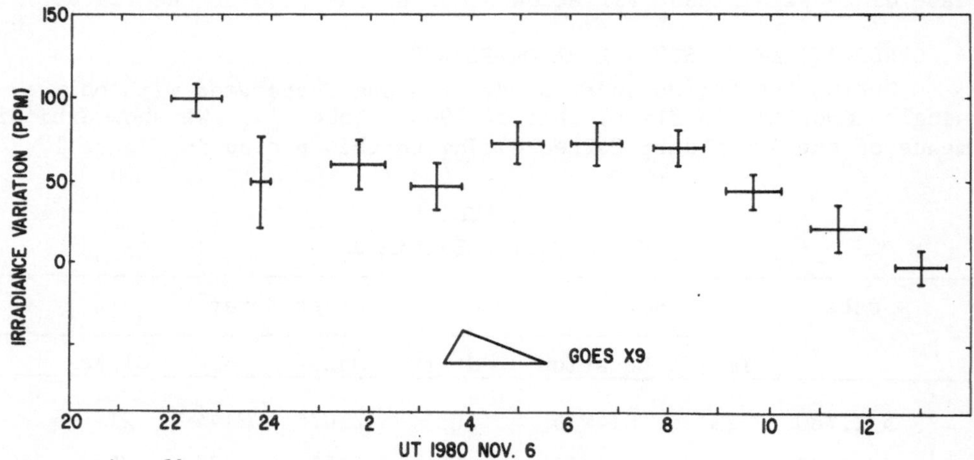

Fig. 1b

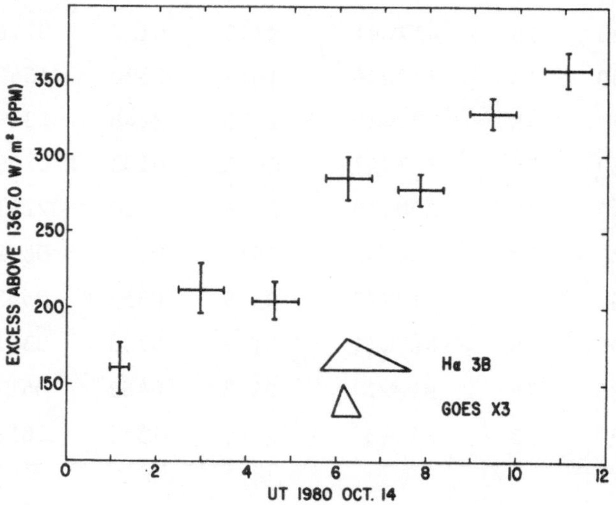

Fig. 1c

TABLE II

Typical Flare Upper Limits (5σ)

Interval (s)	Power (ergs/s)	Total Energy (ergs)
32	6×10^{29}	2×10^{31}
3700*	1.15×10^{29}	4.3×10^{32}

*Thomas (1981)

Another exceptional flare, an X9 in the GOES soft X-ray classification, occurred on 1980 November 6; we show this flare also in Figure 1 along with the limb flare of June 21 responsible for the white-light prominence and intense γ-ray emission. In none of these cases did an obvious excess irradiance occur on a time sçale comparable to an orbital period. Accordingly we will treat the data as upper limits.

For simple estimate of the upper limit on the total flare luminosity we take $\sigma = 12$ parts per million as a typical measured standard error (one orbit). Thus for a 5 σ limit we find $L_{flare} < 1.15 \times 10^{29}$ erg/s. We can obtain a limit on the total flare energy by assuming a duration of 3700 s, taken as representative of the Hα or soft X-ray gradual phase. Then $W_{flare} < 4.3 \times 10^{32}$ ergs.

Finally, we can construct a limit on shorter time scales from a single two-minute shutter cycle, corresponding roughly to the impulsive phase of a flare. The resulting 5 σ limits are $L_{impulsive} < 6 \times 10^{29}$ erg/s or $W_{impulsive} < 2 \times 10^{31}$ ergs. Figure 2 shows examples of the ACRIM data at higher time resolution; we note that ACRIM provides only 32.768 s of data out of each 131.072 s shutter cycle in its normal mode of operation, so that in any given flare the shutter-open periods may not have intercepted the impulsive emission.

5. CONCLUSIONS

The ACRIM sensitivity falls very close to that needed to detect solar flares in integrated sunlight. For example, Thomas (1981) estimates the total Hα energy as $0.31 - 1.1 \times 10^{27}$ erg/s for 2B - 3B flares. Our ACRIM limit lies about two orders of magnitude above this

← Fig. 1. Time histories of ACRIM orbital means during three major flares: (a) 1980 June 21, responsible for a white-light prominence viewed by Harvey and Duvall, as well as the most important γ-ray event of 1980; (b) 1980 October 14, the only importance 3B flare in the prime data set from SMM; and (c) the 1980 Nov. 6 event that produced the largest soft X-ray flux.

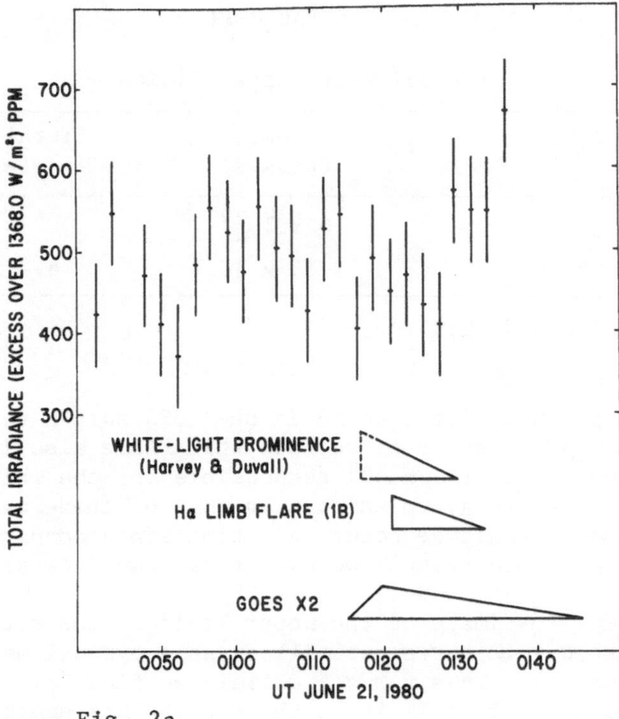

Fig. 2a

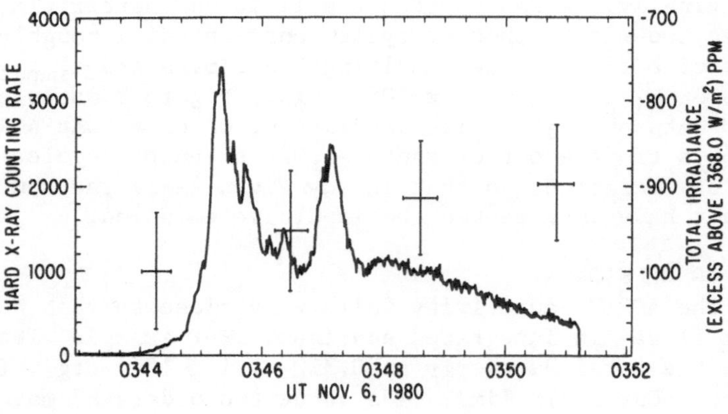

Fig. 2b

Fig. 2. Time histories for ACRIM data during two impulsive events, showing individual 32-second integrations spaced at 131-second intervals: (a) 1980 June 21, and (b) 1980 November 6. In the latter case we compare with the hard X-ray light curve for the lower-energy channel (HXRBS team, 1982), which shows the gradual phase. The impulsive energy release occurs during the increase of the X-ray counting rate.

TABLE III

Estimated Component Luminosities

Radiation	Time Scale (s)	Power (ergs/s)	Total Energy (ergs)	Source
Hα (3B)	3700	1.1×10^{27}	4.1×10^{30}	(1)
White light	600	2.8×10^{27}	6×10^{30}	(2)
EUV flash	1000	3×10^{28}	3×10^{31}	(3)
Soft X-ray Burst	2000	2.8×10^{27}	6×10^{30}	(4)

(1) Thomas (1981)
(2) Rust and Hegwer (1975)
(3) Donnelly (1971)
(4) NOAA Solar-Geophysical Data

level. The Hα line, however, should contain only some small fraction of the total flare luminosity. Unfortunately, we know little quantitatively of the detailed spectral breakdown of flare radiation. Canfield et al. (1980) find that Hα contributed about 12% of the visible line energy in a small flare observed by Skylab; however the existence of other emission components must reduce the Hα fraction of the total considerably. Canfield et al. point out, for example, that the visible continuum, usually not detected in flare patrol observations, could make a relatively large contribution. The ACRIM data set limits on the magnitude of such contributions; apparently no major predominant component of flare radiation has escaped detection.

Improved total-irradiance measurements would allow the detection of solar flares in integrated sunlight. Similarly, a panoramic detector with good photometric properties could also make such observations of stellar flares, and the successful observation of solar flares will help us to understand the common mechanisms of the two sets of phenomena. One should note that Livingston and Ye (1982) have already detected flares in integrated solar K-line data; the "K-index" increased by some 3% during an importance 3B flare. Unfortunately we cannot interpret these measurements quantitatively in terms of actual energy flux.

ACKNOWLEDGEMENTS

This work was supported by the National Science Foundation under grant ATM-81-17355 and by the National Aeronautics and Space Administration under NSG-7161. We would like to thank M. Woodard for helpful discussions.

REFERENCES

Bruzek, A: 1967, in J. Xanthakis (ed.), Solar Physics (Interscience), p. 414.

Canfield, R.C., Cheng, C.-C., Dere, K.P., Dulk, G.A., McLean, D.J., Robinson, R.D., Schmahl, E.J., and Schoolman, S.A.: 1980, in P.A. Sturrock (ed.), Solar Flares (Colorado), p. 451.

Donnelly, R.F.: 1971, Solar Physics 20, 188.

Hundhausen, A.J.: 1972, Coronal Expansion and the Solar Wind (Springer, New York).

HXRBS Team: 1982, private communication.

Livingston, W.C., and Ye, B.: 1982 Publ. Ast. Soc. Pac. (to be published).

Rust, D. and Hegwer, F.: 1975, Solar Physics 40, 141.

Svestka, Z.: 1976, Solar Flares (Reidel, Dordrecht).

Thomas, R.J.: 1981, private communication

Willson, R.C.: 1979, Applied Optics 18, 179.

Willson, R.C., Gulkis, S., Janssen, M., Hudson, H.S., and Chapman, G.A.: 1981, Science 211, 700.

Woodard, M., Hudson, H., and Willson, R.: 1982, UCSD-SP-82-32.

DISCUSSION

HIRAYAMA: When you add all the flares together, can you find any enhancement in the solar irradiance?

HUDSON: We have not yet tried to do a summed epoch analysis, but we plan to do this in the future. It would be much less ambiguous if we were able to make a direct observation of a particular flare.

IV. ENERGY TRANSPORT, CHROMOSPHERIC HEATING AND EVAPORATION (Part 2)

Chairmen: T. Hirayama and H. Zirin

ENERGETIC ELECTRONS AS AN ENERGY TRANSPORT MECHANISM IN SOLAR FLARES

A. Gordon Emslie
Department of Physics, The University of Alabama in Huntsville, Huntsville, AL 35899, U.S.A.

ABSTRACT. We review the observations and theory relating to the role of energetic electrons in the solar flare, with particular emphasis on discriminating between "thermal" and "nonthermal" origins of these electrons. We discuss diagnostics in hard X-rays, especially those relating to the recent observations of the SMM and HINOTORI satellites. We also briefly address the response of the atmosphere to energy input in the form of high energy electrons, in particular through the diagnostics of both the Fe Kα feature and optically thin transition region lines such as OV. Finally, we discuss the relative roles of electron and proton heating in γ-ray flare events.

1. INTRODUCTION

The successful launches and operation of a number of spacecraft observatories in recent years, notably the NASA *Solar Maximum Mission* (Frost, 1983), the Japanese HINOTORI satellite (Tanaka, 1983), the U. S. Air Force P78-1 satellite (Doschek 1983), and the International Sun-Earth Explorer 3 and Pioneer Venus Orbiter satellites (e.g. Kane *et al.*, 1979) have added a new wealth of data pertaining to high energy (γ-ray, hard and soft X-ray, and EUV) emissions in solar flares. This data set has the notable feature of providing observations at a variety of wavelengths *simultaneously*. This hitherto rare situation allows intercomparison of observations of the same event at different wavelengths, and so permits us to test the predictions of candidate energy transport models in regard of their observational signatures.

A solar flare can release up to 10^{32} ergs of energy over a period of some 10^3 s. This energy is believed to be released by annihilation of sheared magnetic fields near the top of a coronal loop (e.g. Spicer, 1976, 1977) and transported to the other regions of the flaring atmosphere by a combination of suprathermal charged particles (e.g. Brown, 1973; Emslie, 1978, 1980; Machado, Emslie and Brown, 1978; Hudson and Dwivedi, 1982; Emslie, 1983a), thermal conduction (e.g. Shmeleva and Syrovatskii, 1973; Antiochos and Sturrock, 1976, 1978; Machado and Emslie, 1979), hydrodynamic motions (e.g. Craig and McClymont, 1976; Nagai,

Solar Physics **86** (1983) 133–146. 0038–0938/83/0861–0133 $ 02.10.

1980), and radiative transport (e.g. Somov, 1976; Machado, 1978; Machado et al., 1980). The relative role of these processes is currently somewhat controversial, and has important consequences for the process of primary energy release itself. In particular, the role of energetic electron beams as an energy transport mechanism is currently a topic of great debate and interest. In this brief review I will consider some matters relating to energy transport by energetic electrons, and the observational signatures in hard X-rays and at other wavelengths which can help us place constraints on the viability of energetic electrons as a mechanism of energy transport in solar flares.

2. DIAGNOSTICS IN HARD X-RAYS

As a result of the intense observational effort in the last decade to study the hard X-ray emission from flares, we now have hard X-ray data from a vast number of flares, with high temporal, spectral, and spatial resolution. In addition, multispacecraft observations have given us stereoscopic views of the hard X-ray emission in some events. Most of these measurements can be used to assess the role of nonthermal electron beams in the flare process. In the remainder of this section I will address these matters. However I believe that it is important to first discuss in some detail exactly what is meant (by myself, at least) by the terms "nonthermal" and "thermal", as applied to high energy electrons in solar flares and their X-ray bremsstrahlung emission, since there appears to be a great deal of confusion in this area.

Historically, a thermal interpretation for hard X-ray emission in flares was first proposed by Chubb, Kreplin, and Friedman (1966). Kahler (1971a, b) contested such an interpretation on the grounds that an extremely hot ($\gtrsim 10^8$ K) plasma would cool extremely rapidly due to free streaming of electrons into the cooler plasma lower in the flare loop. However, Kahler did not consider the fact that such a free-streaming flux of electrons would be unstable to the generation of ion-acoustic plasma turbulence, enhancing the collision frequency and reducing the cooling rate considerably (Brown, Melrose, and Spicer, 1979). In the meantime the so-called thick target model (Brown, 1971), involving a beam of electrons of energy $E \gg kT$ interacting with a cool dense target, gained considerable acceptance in the solar physics community, fueled by the (apparently; see Section 2.1 below) nonthermal nature of the hard X-ray spectrum in many events, and by the polarization of hard X-ray emission found by Tindo and collaborators (see, e.g. , Tindo, Shuryghin, and Steffen, 1976, and references therein). However, since the process of bremsstrahlung emission is so relatively inefficient (most of the electrons' energy is lost in Coulomb collisions with ambient particles), the energy requirements of such a thick target model can be extreme (e.g. Hoyng, Brown and van Beek, 1976). This has prompted some authors (e.g. Smith, 1980) to reject such a model on energetic grounds alone, instead favoring a model in which electrons are bulk heated to temperatures in excess of 10^8 K, and so do not lose energy in collisions with ambient, cooler, electrons -- this is the so-called "thermal" model of Brown, Melrose, and Spicer

(1979), Smith and Lilliequist (1979), and Smith and Auer (1980).

Although the electrons in such a "thermal" model do lose energy by thermal conduction and convection into the cooler surroundings (Smith and Lilliequist, 1979; Smith and Auer, 1980), this energy loss can, for reasonable flare parameters, be smaller than that due to Coulomb collisional losses in a thick target scenario, thus rendering the "thermal" model more efficient, and removing the somewhat embarrassing energy requirements that the latter model imposes. Even for parameter values (e.g. low source density; Smith and Harmony, 1982) which cause this energetic advantage to be small, the flare energy is still released in a higher entropy form than in an acceleration of a relatively small fraction of the electrons in the energy release site; this appears more realizable on the basis of candidate flare energy release models (Smith, 1980).

It is thus important to realize the qualitative difference between these two models -- in the "nonthermal" model, a relatively small fraction of the electrons in the primary energy release site are accelerated to energies much greater than that of the bulk electron component, while in the "thermal" model almost all of the electrons are impulsively heated to deka-keV energies. Although the distributions of these bulk energized electrons may not have sufficient time to relax to a strict Maxwellian, I nevertheless feel that the term "thermal" is indeed appropriate for such a model.

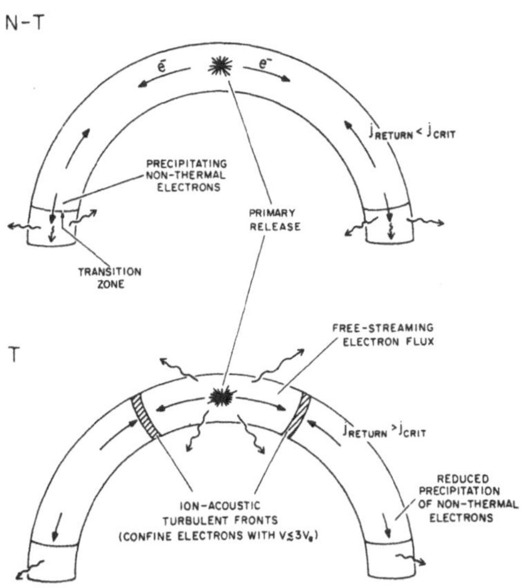

Figure 1: Schematic illustration of nonthermal (N-T) and thermal (T) models of hard X-ray production in solar flares. For details, see Emslie and Rust (1980).

In Figure 1 I show a schematic of these two hard X-ray models, nonthermal (thick target) and thermal. It is to be noted that a significant suprathermal streaming component may exist in the so-called thermal model. This component may consist of highly energetic electrons, capable of overcoming the confining thermoelectric field at the source boundary (Smith and Brown, 1980), or may be due to maser action rapidly precipitating electrons through a trap loss cone (Melrose and Dulk, 1982). Either way, such a component gives a nonthermal "impurity" to the thermal model, and means that any remarks applicable to nonthermal models may also apply, to some extent, to the thermal model as shown. It is important to realize that the terms "nonthermal" and "thermal" apply only to the hard X-ray production; other "nonthermal" processes, such as proton acceleration, may occur in a model which emits its X-rays principally "thermally". In what follows I will now discuss the hard X-ray signature of both models with regard to possible observational discrimination between the two.

2.1. Spectrum

The hard X-ray spectrum of many events is frequently observed to be power-law in character (see, e.g., Hoyng, Brown, and van Beek, 1976; Ohki *et al*., 1982), implying (Brown, 1971) a power-law character for the electron spectrum also. Although no known particle acceleration mechanism is capable of producing such power law distributions of accelerated particles, such "obviously non-thermal" spectra have been frequently cited as evidence for a nonthermal interpretation of hard X-ray emission in flares. In view of the remarks made above, it should be clear that a power-law distribution of electrons need not correspond to a suprathermal component, since a multitemperature, or non-Maxwellian "thermal" structure can also result in such a distribution of electrons (see, e.g. Brown, 1974; Emslie and Brown, 1980). Thus it is extremely important to realize that spectra, on their own, are not capable of distinguishing between the two models discussed above, so that we must resort to other observational diagnostics, as discussed below.

2.2. Polarization

Since a nonthermal model of hard X-ray production involves the directed acceleration of suprathermal particles, and since the cross-section for bremsstrahlung production is anisotropic (Gluckstern and Hull, 1953), it follows that the hard X-ray emission in such a model should be highly polarized (Brown, 1972; Langer and Petrosian, 1977; Bai and Ramaty, 1978; Leach and Petrosian, 1983). In the thermal model, a small polarization is also found, due to the asymmetry of the electron velocity distribution caused by the heat flux along the magnetic field lines (Emslie and Brown, 1980); however, this polarization is generally much less than for nonthermal models.

A review of the observation results relating to polarization of hard X-ray emission in flares can be found in Emslie and Brown (1980). These observations, principally due to Tindo and collaborators

(Tindo et al., 1970, 1972a, b; Tindo, Mandel'stam, and Shuryghin, 1973; Tindo, Shuryghin, and Steffen, 1976) show a detectable polarization at low ($\simeq$ 10 keV) energies, but with large uncertainties due to the poor statistics involved. Indeed, with the large integration time used by the INTERCOSMOS instruments, it is possible that any intrinsic polarization structure could be washed out by the time-averaging process. The results are therefore inconclusive, and, furthermore, consistent with both nonthermal and thermal models (Emslie and Brown, 1980).

On the third mission of the U. S. Space Shuttle Columbia, a hard X-ray polarimeter was flown by Novick and collaborators at Columbia University, with the intention of making solar flare measurements. However, the author is unaware of any results, positive or negative, from this experiment at the present time.

2.3. Directivity

For the same reasons as discussed in Section 2.2, one expects a large directivity of hard X-ray emission in nonthermal models, and a more nearly isotropic emission in thermal models (Emslie and Brown, 1980). Here, however, the observations are more abundant. They fall into two classes:

(a) Statistical

Datlowe et al. (1977) studied some 200 hard X-ray bursts observed by the OSO7 satellite, and concluded that, at the 95% confidence level, there was no apparent variation of burst intensity with position of the flare on the disk, consistent with isotropic emission.

(b) Multi-Spacecraft

Kane et al. (1980) report that in several flares observed simultaneously by the International Sun-Earth Explorer-3 and Pioneer Venus Orbiter spacecraft, at various relative viewing angles to the flare event, no evidence of directivity was found, again consistent with isotropic emission form the flare.

The interpretation of directivity (and polarization) observations is somewhat complicated by the photospheric backscattered component (Hénoux, 1975; Langer and Petrosian, 1977; Bai and Ramaty, 1978), which enhances the polarization and directivity for intrinsically isotropic, unpolarized sources, and diminishes these quantities for highly polarized and anisotropic sources. This, plus the possible (or, indeed, probable) inhomogeneity of the flare volume, means that although the observations are consistent with a nearly isotropic source (and so with a thermal-type scenario), a nonthermal model cannot be ruled out. The observation of significance would be to find very high directivity (factors of 2 and up) and high ($\gtrsim$ 50%; Leach and Petrosian, 1983) polarization in an event, since this would then strongly favor a nonthermal interpretation.

2.4. Spatial Location

With the advent of the SMM HXIS and HINOTORI instruments, it is now possible to form images of a solar flare in moderately hard ($\simeq$ 20 keV) X-rays. As can be deduced from Figure 1, a nonthermal model is characterized by the appearance of bright footpoints, or, if the coronal density is high enough, a fully bright loop, while the thermal model is characterized by two distinct regions of emission -- the thermal emission from the electrons confined at the top of the loop, and the (possible, see above) thick target emission from the footpoints (Emslie, 1981a). Figure 2 shows these results in the form of sketches.

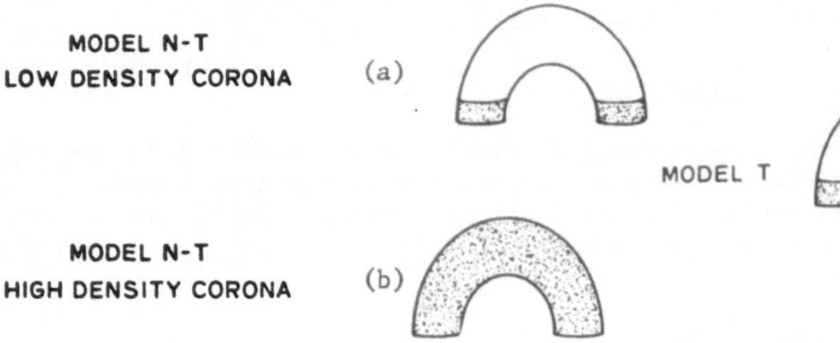

MODEL N-T
LOW DENSITY CORONA (a)

MODEL T

MODEL N-T
HIGH DENSITY CORONA (b)

Figure 2: Schematic of the appearance of flare loops in hard X-rays ($\varepsilon \simeq 20$ keV) for nonthermal (N-T) and thermal (T) models. Whether a nonthermal model has the appearance (a) or (b) depends on the coronal density and loop length; see below.

The critical density for the appearance of Figure 2(b) rather than 2(a) can be found by considering the stopping length for electrons of energy $E \gtrsim \varepsilon$ (the photon energy under observation) in a region of average density n. From the relations given in Emslie (1978), this is given by

$$n\,L \simeq 10^{17}\ [E(\mathrm{keV})]^2,$$

so that for $\varepsilon \simeq 20$ keV, we have

$$nL \gtrsim 4 \times 10^{19}\ \mathrm{cm}^{-2}$$

for the loop to appear as in Figure 2(b). For a loop of length 10^9 cm, this implies $n \gtrsim 4 \times 10^{10}\ \mathrm{cm}^{-3}$, which is a relatively large value for the coronal density at the onset of the flare. However, as the flare proceeds, energy deposited in the chromospheric material will cause it to heat, expand, and rise up the flux tube -- the process of so-called "chromospheric evaporation" (see e.g., Antiochos and Sturrock, 1978; Acton et al., 1982). Thus the hard X-ray structure in a nonthermal scenario should evolve, over a hydrodynamic timescale $L/v_{t,i}$ (where $v_{t,i}$ is the ion thermal speed $\simeq 10^4 T_i^{1/2}\mathrm{cm\ s}^{-1}$), from the situation shown in Figure 2(a) to that in Figure 2(b).

Observationally, there appears to be some uncertainty as to the geometry of the hard X-ray emission in flares, or, at least, there is evidence that many different pictures are possible. The direct imaging observations from the SMM HXIS of the 1980 May 21 event (Hoyng et al., 1981) indicate the presence of impulsive footpoint brightenings during the early stages of the flare which spread out as the flare develops until the entire loop appears bright. This is consistent with the nonthermal picture presented above. On the other hand, Ohki et al. (1982) present observations of the 1981 July 17 event using the HINOTORI SXT instrument, which indicates a single coronal source, consistent with the thermal model if the escaping nonthermal tail represents a small component of the energetic electron population. The results of Tsuneta et al. (1983) also support this observation.

Kane et al. (1979) have reported multispacecraft (ISEE-3 and PVO) observations of a behind-the-limb event on 1978 October 5. The geometry of the spacecraft lines-of-sight to the flaring region was such that while PVO could see the entire flare loop, the photosphere occulted all parts of the loop below 25 000 km as viewed by ISEE-3. The result is that the hard X-ray emission in the range 50-500 keV observed by PVO was some 600 times more intense (and also had a spectrum some 2 powers harder than that observed by ISEE-3). Brown, Hayward, and Spicer (1981) have interpreted this event in terms of a thick target nonthermal model and find both the flux and spectral observations to be consistent with such an interpretation. Brown and Hayward (1981) have also attempted to explain the Kane et al. (1979) observations in terms of a large number of very small kernels, which are inefficient at radiating thermal bremsstrahlung but whose nonthermal tails (see above) make the footpoint emission large. As pointed out by Brown and Hayward, this interpretation in fact differs little from the conventional thick target picture, and, in view of the remarks at the beginning of this section on the meaning of the terms "nonthermal" and "thermal", we can conclude that the Kane et al. (1979) observations are in fact best interpreted in terms of a nonthermal model.

Kane et al. (1982) have also reported observations of 3 flares on 1979 November 5 showing that the bulk of the hard X-ray ($\gtrsim$ 100 keV) emission in these flares comes from low ($\lesssim$ 2500 km) altitudes. One of these events lasted long enough that a variation in the spatial distribution of hard X-ray emission with time could be detected, showing a gradual increase in the high (coronal) component as the event proceeded; this component eventually saturated at around 30% of the total emission. This data is thus qualitatively consistent with the picture of chromospheric evaporation late in the event, as discussed above in reference to Figure 2.

In summary, there appears to be as yet no definitive evidence, based on spatially resolved measurements of solar flare hard X-rays, for either of the two models discussed above. Future interpretation of more such data sets does, however, appear to be a highly profitable method of attack on the problem.

2.5. Fast Time Structure

Kiplinger et al. (1983a, b) and Hurley et al. (1983a, b) have reported statistically significant fluctuations in solar hard X-rays on timescales τ of order 10 ms. When interpreted nonthermally, this implies a collisional stopping time for the electrons of this order (Emslie, 1983b); when interpreted thermally, this timescale corresponds to the conductive cooling time of the source region (see, e.g., Emslie 1981b). In a nonthermal scenario, this 10 ms value implies a mean source density (along the electron's path) of order 10^{12} cm^{-3}, which can either result from coronal densities of this order, or a very short ($\lesssim 5 \times 10^7$ cm) distance of the acceleration point above the chromosphere, so that the electrons penetrate the thin target corona and reach the high density chromosphere in the required $\simeq$ 10 ms. In a thermal model, the 10 ms value gives a kernel size of order $10^4 T^{\frac{1}{2}} \tau \simeq 10^6$ cm. It seems to this author that the latter interpretation is more likely; however, independent observations may be able to assess the viability of either of the two nonthermal alternatives. For a more detailed discussion, see Kiplinger *et al*. (1983a).

2.6. Correlation with EUV Bursts

Emslie, Brown, and Donnelly (1978) and Donnelly and Kane (1978) have shown that the relative timing of EUV and hard X-ray bursts is very close, better than $\simeq$ 1 s in the high time resolution events reported by Donnelly and Kane. This close temporal connection between these two emission implies a very fast energy transport mechanism, so fast, in fact, that energetic electrons seem to be the only feasible agent. However, the role of these electrons in the flare energy budget, or whether they form part of a "thermal" or "nonthermal" distribution, remains uncertain until more detailed studies of the chromospheric response to impulsive flare energy input can be made (see Section 4).

3. IRON Kα EMISSION IN FLARES

The Kα feature in the spectrum of iron is formed by 2p -> 1s deexcitation following the removal of an inner shell electron, either by electron input or by fluorescent irradiation (see, e.g., Phillips and Neupert, 1978; Bai, 1979). It appears (Culhane *et al*., 1981) that the fluorescent mechanism predominates at most times during the flare; however, if nonthermal beams of electrons really do exist, then there should nevertheless be a corresponding impulsive spike in the observed Kα flux corresponding to each intense hard X-ray spike. We have recently (Emslie, Phillips, and Dennis, 1983) calculated the expected Kα yield due to impact ionization of chromospheric iron atoms, with the bombarding electron flux computed from the hard X-ray flux through a thick target model. We find that the results are consistent with such a thick target model in many events, indicating the presence of a significant suprathermal component in the electron distribution in these events.

4. RESPONSE OF THE LOWER ATMOSPHERE

The energy transported into the lower atmosphere, whether it be in the form of suprathermal electrons, or thermal conduction, or whatever, will cause the chromosphere to heat rapidly to coronal temperatures (see, e.g., Somov, Syrovatskii, and Spektor, 1981). In so doing, the intensities in optically thin transition region lines (i.e., lines formed at temperatures of order 10^5K) will change. Thus simultaneous observation of transition region lines and hard X-ray bursts can provide a test of a candidate energy transport model.

On close examination of the problem, it is by no means obvious whether the intensity of a transition region line will increase or decrease in response to the energy input into the chromosphere. On one hand, the density of the material at 10^5 K will rise, enhancing the line emission measure by a factor $(n/n_o)^2$. On the other hand, however, the strong heating of the upper atmosphere (to temperatures in excess of 10^7 K) will cause a large conductive flux throughout the loop, steepening temperature gradients and reducing the thickness of the region in which a given spectral line is formed. Quantitatively, the ratio of flare to pre-flare values of the emission measure in a given line (peak formation temperature T) is

$$\frac{EM}{EM_o} = \left(\frac{n}{n_o}\right)^2 \frac{\Delta z}{\Delta z_o} = \left(\frac{n}{n_o}\right)^2 \frac{\int_T \left(\frac{dz}{dT}\right) dT}{\int_T \left(\frac{dz}{dT}\right)_o dT} \simeq \left(\frac{n}{n_o}\right)^2 \left.\frac{(dT/dz)_o}{(dT/dz)}\right|_{T\,=\,\text{const}},$$

and since T is fixed, and $\underset{\sim}{F}_c = -\kappa_o T^{5/2} \underset{\sim}{\nabla} T$,

$$\left|\frac{EM}{EM_o}\right| = \left(\frac{n}{n_o}\right)^2 \frac{F_{co}}{F_c} .$$

Finally, since

$$F_c \sim \kappa_o \frac{T_{corona}^{7/2}}{L} ,$$

where L is the half-length of the loop, we find that, since L is essentially unchanged is a result of the flare,

$$\frac{EM}{EM_o} = \left(\frac{n}{n_o}\right)^2 \left(\frac{T_{corona,\,o}}{T_{corona}}\right)^{7/2}$$

Estimates of the ratios n/n_o and $T_{corona}/T_{corona,o}$ may be found by respectively considering the energy balances between flare heating and radiative losses in the chromosphere, and between flare heating and conductive cooling in the corona (Emslie, 1982); alternatively, they may be determined observationally, resulting in good agreement with the theoretical estimates. With $(n/n_o) \simeq 10^2$ and $(T_{corona}/T_{corona,o}) \simeq 30$, we find that $(EM/EM_o) \simeq 0.1$, indicating a decrease in the line intensity compared to preflare values. Preliminary time-dependent numerical simulations of the flare heating process, using non-thermal electrons as the heating source (Emslie and McClements, 1982), confirm this conclusion. However, observations (Poland et al., 1982) show that the intensities of the λ1371 Å line of OV and of hard X-rays are positively correlated, in contradiction to the above theoretical predictions. If these preliminary conclusions are verified in detail, then the processes of energy transport in solar flares must be critically reexamined in the light of these results.

5. THE ROLE OF PROTONS IN CHROMOSPHERIC HEATING

The simultaneous measurement of γ-ray (Chupp et al., 1981) hard X-ray radiation in flares can be used to infer the ratio of number of energetic protons ($E \simeq 10$ MeV) to energetic electrons ($E \simeq 20$ keV) in the flare, and so infer the relative role each plays in heating the solar atmosphere during the flare. Emslie (1983a) has calculated the relative role of heating by non-thermal electrons, and by non-thermal protons, in the γ-ray flare of 1980 June 7. These results are shown in Figure 3. One may conclude from these results that proton heating is negligible at chromospheric levels, even in events with high γ-ray/hard X-ray flux ratio such as that of 1980 June 7, unless the hard X-ray emission is predominately thermal in origin (see Section 2). Therefore, an assessment of the character of hard X-ray emission in flares is essential to our understanding of energy transport in the lower atmosphere.

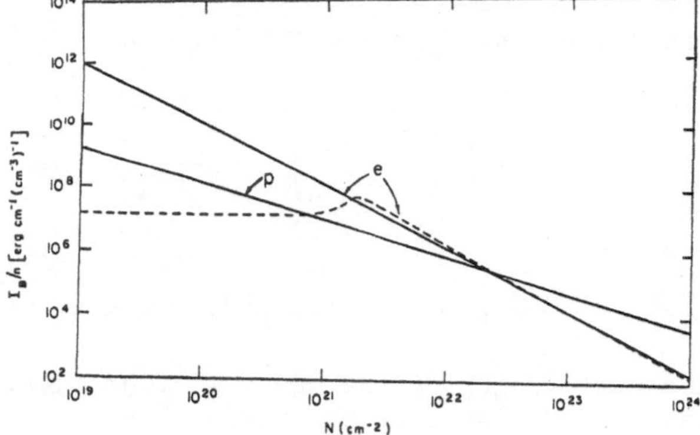

Figure 3: Energy deposited per unit vertical height per unit ambient particle as a function of column depth N by both electrons and protons in the event of June 7, 1980. The solid lines refer to thick target models for both X-ray and γ-ray productions, while the dashed line refers to the "hybrid" thermal-plus-nonthermal hard X-ray production model of Emslie and Vlahos (1980). Note the dominance of electron heating at chromospheric levels ($N \sim 10^{20}$ cm^{-2}) in a thick target interpretation of the hard X-ray emission, and the importance of proton heating in the thermal interpretation (after Emslie, 1983a).

6. CONCLUSIONS

We are currently faced with a great mass of data, much of which is directly relevant to studies of energy transport in the flaring atmosphere, and in particular the role of high energy electrons in this process. Although much theoretical work in the past has dealt with the relevance of two or more competing models for a given phenomenon, it is becoming apparent that solar flares may occur in all shades of grey, as well as in stark black and white. Whether solar flares can be understood in terms of a relatively small class of models, or whether indeed "anything is possible", remains to be determined. At present, the great quantity and detail of observations surpasses the depth of theoretical work in the relevant areas; thus the role of theoreticians in future research must be to provide as comprehensive a set of model-dependent predictions as possible, so that best use of the available data can be made, and also so the data from the present generation of instruments can shape the instrumentation and scientific research of the future.

ACKNOWLEDGEMENTS

The author would like to express his gratitude to the Japanese Society for the Promotion of Science, and to the National Science Foundation, for making his trip to Tokyo possible, enjoyable, and profitable. Much of the author's work to which this article relates was supported by NASA grant NAGW-294 and by NSF grant ATM-8200934; during the course of this work the author held a von Braun Fellowship at the University of Alabama in Huntsville. I thank Dr. S. R. Kane for numerous suggestions for improvement of this article.

REFERENCES

Acton, L.W., Canfield, R.C., Gunkler, T.A., Hudson, H.S., Kiplinger, A.L., and Leibacher, J.W.: 1982, Astrophys. J. 263, 409.
Antiochos, S.K., and Sturrock, P.A.: 1976, Solar Phys. 49, 359.
Antiochos, S.K., and Sturrock, P.A.: 1978, Astrophys. J. 220, 1137.
Bai, T.: 1979, Solar Phys. 62, 113.
Bai, T., and Ramaty, R.: 1978, Astrophys. J. 219, 705.
Brown, J.C.: 1971, Solar Phys. 18, 489.
Brown, J.C.: 1972, Solar Phys. 26, 441.
Brown, J.C.: 1973, Solar Phys. 31, 143.
Brown, J.C.: 1974, in I.A.U. Symp. 57, Coronal Disturbances (G.A. Newkirk, Jr., ed). p. 395.
Brown, J.C., and Hayward, J.: 1981, Solar Phys. 73, 121.
Brown, J.C., Hayward, J., and Spicer, D.S.: 1981, Astrophys. J. (Letters) 245, L91.
Brown, J.C., Melrose, D.B., and Spicer, D.S.: 1979, Astrophys. J. 228, 592.
Chubb, T.A., Kreplin, R.W., and Friedman, M.: 1966, J. Geophys. Res. 71, 3611.
Chupp, E.L., et al.: 1981, Astrophys., J. (Letters) 244, L171.

Craig, I.J.D., and McClymont, A.N.: 1976, Solar Phys. 50, 133.
Culhane, J.L., et al.: 1981, Astrophys. J. (Letters) 244, L141.
Datlowe, D.W., O'Dell, S.L., Peterson, L.E. and Elcan, M.J.: 1977, Astrophys. J. 212, 561.
Donnelly, R.F., and Kane, S.R.:1978, Astrophys. J. 222, 1043.
Doschek, G.A.: 1983, these proceedings.
Emslie, A.G.: 1978, Astrophys. J. 224, 241.
Emslie, A.G.: 1980, Astrophys. J. 235, 1055.
Emslie, A.G.: 1981a, Astrophys. J. 245, 711.
Emslie, A.G.: 1981b, Astrophys. J. 244, 653.
Emslie, A.G.: 1982, B.A.A.S. 14, 607.
Emslie, A.G.: 1983a, Solar Phys., in press.
Emslie, A.G.: 1983b, Astrophys. J., submitted.
Emslie, A.G., and Brown, J.C.: 1980, Astrophys. J. 237, 1015.
Emslie, A.G., Brown, J.C., and Donnelly, R.F.: 1978, Solar Phys. 57, 175.
Emslie, A.G., and McClements, K.G.: 1982, unpublished.
Emslie, A.G., Phillips, K.J.H., and Dennis, B.R.: 1983, in preparation.
Emslie, A.G., and Rust, D.M.: 1980, Solar Phys. 65, 271.
Emslie, A.G., and Vlahos, L.: 1980, Astrophys. J. 242, 359.
Frost, K.J.: 1983, these proceedings.
Gluckstern, R.L., and Hull, M.H.: 1953, Phys. Rev. 90, 1030.
Hénoux, J.C.: 1975, Solar Phys. 42, 219.
Hoyng, P., Brown, J.C. and van Beek, H.F.: 1976, Solar Phys. 48, 197.
Hoyng, P., et al.: 1981, Astrophys. J. (Letters) 246, L155.
Hudson, H.S., and Dwivedi, B.N.: 1982, Solar Phys. 76, 45.
Hurley, K., Niel, M., Talon, R., Estulin, I.V., and Dolidze, V.Ch.: 1983a, Astrophys. J. 265, in press.
Hurley, K., Niel, M., Talon, R., Estulin, I.V., and Dolidze, V.Ch.: 1983b, these proceedings.
Kahler, S.W.: 1971a, Astrophys. J. 164, 365.
Kahler, S.W.: 1971b, Astrophys. J. 168, 319.
Kane, S.R., Anderson, K.A., Evans, W.D., Klebesadel, R.W., and Laros, J.: 1979, Astrophys. J. (Letters) 233, L151.
Kane, S.R., Anderson, K.A., Evans, W.D., Klebesadel, R.W., and Laros, J.: 1980, Astrophys. J. (Letters) 239, L85.
Kane, S.R., Fenimore, E.E., Klebesadel, R.W., and Laros, J.G.: 1982, Astrophys. J. (Letters) 254, L53.
Kiplinger, A.L., Dennis, B.R., Emslie, A.G., Frost, K.J., and Orwig, L.E.: 1983a, Astrophys. J. (Letters), in press.
Kiplinger, A.L., Dennis, B.R., Emslie, A.G., Frost, K.J., and Orwig, L.E.: 1983b, these proceedings.
Langer, S.H., and Petrosian, V.: 1977, Astrophys. J. 215, 666.
Leach, J., and Petrosian, V.: 1983, Astrophys. J., submitted.
Machado, M.E.: 1978, Solar Phys. 60, 341.
Machado, M.E., Avrett, E.H., Vernazza, J.E., and Noyes, R.W.: 1980, Astrophys. J. 242, 336.
Machado, M.E., and Emslie, A.G.:1979, Astrophys. J. 232, 903.
Machado, M.E., Emslie, A.G., and Brown, J.C.: 1978, Solar Phys. 58, 363.

Melrose, D.B., and Dulk, G.A.: 1982, Astrophys. J. (Letters) **259**, L41.
Nagai, F.: 1980, Solar Phys. **68**, 351.
Ohki, K., et al.: 1982, in Hinotori Symposium on Solar Flares, Tokyo, January, p. 102.
Phillips, K.J.H., and Neupert, W.M.: 1973, Solar Phys. **32**, 209.
Poland, A.I., et al.: 1982, Solar Phys. **78**, 201.
Shmeleva, O.P., and Syrovatskii, S.I.: 1973, Solar Phys. **33**, 341.
Smith, D.F. 1980, Solar Phys. **66**, 135.
Smith, D.F., and Auer, L.H.: 1980, Astrophys. J. **238**, 1126.
Smith, D.F., and Brown, J.C.: 1980, Astrophys. J. **242**, 799.
Smith, D.F., and Harmony, D.W.: 1980, Astrophys. J. **252**, 800.
Smith, D.F., and Lilliequist, C.G.: 1978, Astrophys. J. **232**, 582.
Somov, B.V.: 1975, Solar Phys. **42**, 235.
Somov, B.V., Syrovatskii, S.I., and Spektor, A.R.: 1981, Solar Phys. **73**, 145.
Spicer, D.S.: 1976, NRL Report 8036.
Spicer, D.S.: 1977, Solar Phys. **53**, 305.
Tanaka, Y.: 1983, these proceedings.
Tindo, I.P., Ivanov, V.D., Mandel'stam, S.L., and Shuryghin, A.I.: 1970, Solar Phys. **14**, 204.
Tindo, I.P., Ivanov, V.D., Mandel'stam, S.L., and Shuryghin, A.I.: 1972a, Solar Phys. **24**, 429.
Tindo, I.P., Ivanov, V.D., Valníček, B., and Livshits, M.A.: 1972b, Solar Phys. **27**, 426.
Tindo, I.P., Mandel'stam, S.L., and Shuryghin, A.I.: 1973, Solar Phys. **32**, 469.
Tindo, I.P., Shuryghin, A.I., and Steffen, W.: 1976, Solar Phys. **46**, 219.
Tsuneta, S., et al.: 1983, these proceedings.

DISCUSSION

SVESTKA: In the impulsive phase of several flares HXIS imaged first point sources in hard X-rays and after that (some 20 - 40 seconds later) the whole loop began to be seen. Hoyng et al. have interpreted it as thick-target (non-thermal) emission changing into thermal. Isn't it possible that we actually encounter thick-target emission all the time, first in a low-density, and later in a high-density loop after evaporation?

EMSLIE: Assuming that the relevant loop lengths and time scales are such that significant filling of the loop due to evaporation can take place, I believe that this is an extremely likely possibility. Observations presented by Dr. Antonucci at this meeting show that evaporation velocities on the order of 400 km/s (the sound velocity in 10^7 K plasma) are observed for tens of seconds into the event; thus for the time scales you mention some 10^4 km of loop could be filled. This is very reasonable, justifying a thick-target interpretation throughout the entire event.

DULK: A possible way of obtaining your thick-target case (a) is to force a precipitation of trapped particles, e.g., by a maser. This mimics a beam, but does not require directive acceleration by the energy release process. (Here I take 'beam' to mean particles travelling essentially entirely in one direction, along the field liens). It is also not necessary to have nonthermal fast electrons as the main precipitating population: A hot Maxwellian ($T > 10^8$ K) can be forced to precipitate by the maser just as well.

EMSLIE: I agree. However, in the case where the precipitating electrons are simply part of a hot Maxwellian tail, one would expect a significant thermal hard X-ray component also.

ACTON: How many events with distinct $K\alpha$ bursts (excited by hard X-ray-producing electrons) do you have?

EMSLIE: A total of 7 intense hard X-ray spike events (as measured by the SMM HXRBS) had good $K\alpha$ data. Of these, 4 showed $K\alpha$ spikes above the (dominant) fluorescent intensity profile, and so were used in the analysis.

BROWN: Does your value of $L \sim 10^6$ cm refer to the thermal or non-thermal case, which are clearly different ?

EMSLIE: To the thermal case, the non-thermal L value being around 43 times greater.

BROWN: In that case, for an emission measure $n^2L^3 \sim 10^{45}\text{cm}^{-3}$, we need a density in the thermal model of $n \sim (10^{45}/10^{18})^{1/2} \sim 10^{13.5}$ cm^{-3},

EMSLIE: In our case, the emission measure is not that big.

PRIEST: How does the critical density for thick-target emission occurring from the whole of a loop instead of the footpoints vary with particle energy and loop length?

EMSLIE: An electron of energy E keV is stopped in a distance of about $L \sim 10^{17} E^2 /n$ cm, where n is the density. Since photons of energy are principally emitted by electrons of energy $\varepsilon \gtrsim E$, the critical density at $\varepsilon \sim 20$ keV is of order $4 \times 10^{19}/L$, i.e., some 10^{11}cm^{-3}.

HYDRODYNAMICS OF FLARING LOOPS: SMM OBSERVATIONS AND NUMERICAL SIMULATIONS

R. Pallavicini
Osservatorio Astrofisico di Arcetri, Firenze, Italy

G. Peres
Osservatorio Astronomico di Palermo, Italy

ABSTRACT. The hydrodynamic response of confined magnetic structures to strong heating perturbations is investigated by means of a time-dependent one-dimensional code which incorporates the energy, momentum and mass conservation equations. The entire atmospheric structure from the chromosphere to the corona is taken into account. The results of model calculations are compared with observations of flares obtained with the X-Ray Polychromator experiment on the Solar Maximum Mission.

1. INTRODUCTION

X-ray and ultraviolet observations have shown that flares consist of magnetically confined arches which extend from the chromosphere to the corona. For at least one class of flares -the compact ones- these structures remain unchanged throughout the flare evolution, while heated at temperatures of the order of 20 to 30 million degrees. The observations have also shown that the volume emission measure and density increase by one to two orders of magnitude during transient events, implying that matter in addition to energy is injected into the coronal portion of flaring loops. This mass is commonly believed to be supplied by the chromosphere through an evaporation process occurring when the dense chromospheric layers receive more energy than can be radiated away.

In this paper we use numerical simulations to investigate the gas-dynamic processes occurring in flaring loops. Models with different energy input, heating time dependence, preflare conditions and heating location are examined. We compare model predictions with X-ray observations obtained from the Solar Maximum Mission and we address the question whether the comparison can provide useful information on the processes of energy release and transfer in solar flares.

Solar Physics **86** (1983) 147–157. 0038–0938/83/0861–0147 $ 01.65.

2. THE MODEL

The hydrodynamic behaviour of flaring loops has been investigated by means of a single-fluid, one dimensional code which incorporates the full set of energy, momentum and mass conservation equations. The geometry is that of a circularly simmetric loop of semilength L and constant cross-section, which extends from the temperature minimum to the corona. The preflare state is in energy balance and hydrostatic equilibrium and is calculated as in Serio et al. (1981) for the transition region and the corona. Radiative losses and atmospheric structure for the chromosphere are derived from Vernazza, Avrett and Loeser (1980).

The preflare state is perturbed by introducing a transient flare heating term, whose amount, location, and spatial and temporal dependences can be arbitrarily varied to simulate different situations. In the models run so far, the flare heating source has been assumed to be gaussian in shape and to be located either at the loop top or in the transition region close to the loop footpoints. The non-linear, time-dependent equations are solved numerically as described in Peres et al. (1982) and Pallavicini et al.(1982).

The gasdynamic evolution predicted by our calculations is similar to that found by Nagai (1980) and Cheng et al. (1982). When energy is supplied at the top of the loop, there is an increase of temperature and pressure locally. The increase of temperature causes a conduction front to propagate rapidly towards the loop base, while the pressure increase causes a much slower downflow of matter. The transition region steepens and moves downwards. As soon as the conduction front reaches the chromosphere, part of the chromospheric material is heated to coronal temperatures and expands upwards with velocities of several hundred kilometers per second, filling the loop with high density material. When the transient flare heating ceases to be operative, the coronal temperature rapidly drops, the corona cools by conduction and radiation, and the transition region returns to its initial position. In the case when energy is supplied close to the footpoints, upward expansion of chromospheric material occurs from the very beginning, while the following evolution remains essentially the same.

3. THE OBSERVATIONS

The observations used in this paper have been obtained with the X-Ray Polychromator experiment (XRP) on the Solar Maximum Mission. The XRP consists of two separate instruments, the Flat Crystal Spectrometer (FCS) and the Bent Crystal Spectrometer (BCS). A full description of the XRP instrumentation can be found in Acton et al. (1980).

The FCS provides X-ray spectroheliograms simultaneously in seven spectral channels. The spatial resolution is 15x15 arcseconds, comparable to the linear dimensions of most compact flares. Given this moderate spatial resolution, comparison with model predictions will be limited to the integrated emission from an entire loop. If one refers to the same pixel in all FCS raster images, light curves of flares can be constructed. An example is shown in Figure 1, where normalized count rates are plotted as a function of time. The observed lines are formed over a broad range of coronal temperatures from about 2 million degrees to over 30 million degrees. Notice that the high temperature lines (e.g. Fe XXV) peak earlier and decay faster than low temperature lines (e.g. O VIII). This trend has been observed in all cases in which the temporal resolution was sufficiently high in comparison with the characteristic flare time scales.

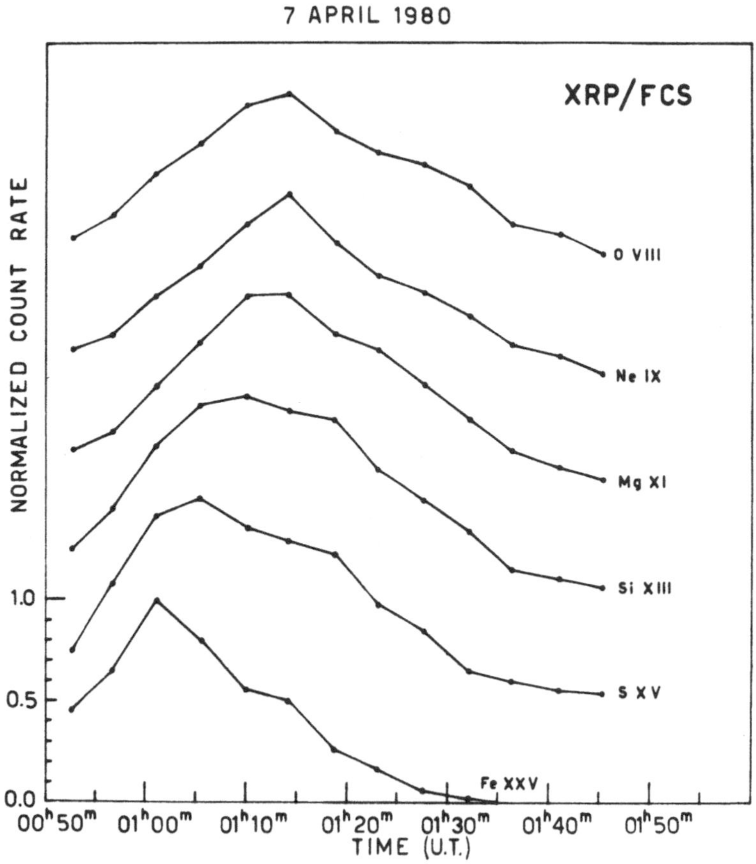

Fig.1 - Light curves in relative units for a flare observed with FCS

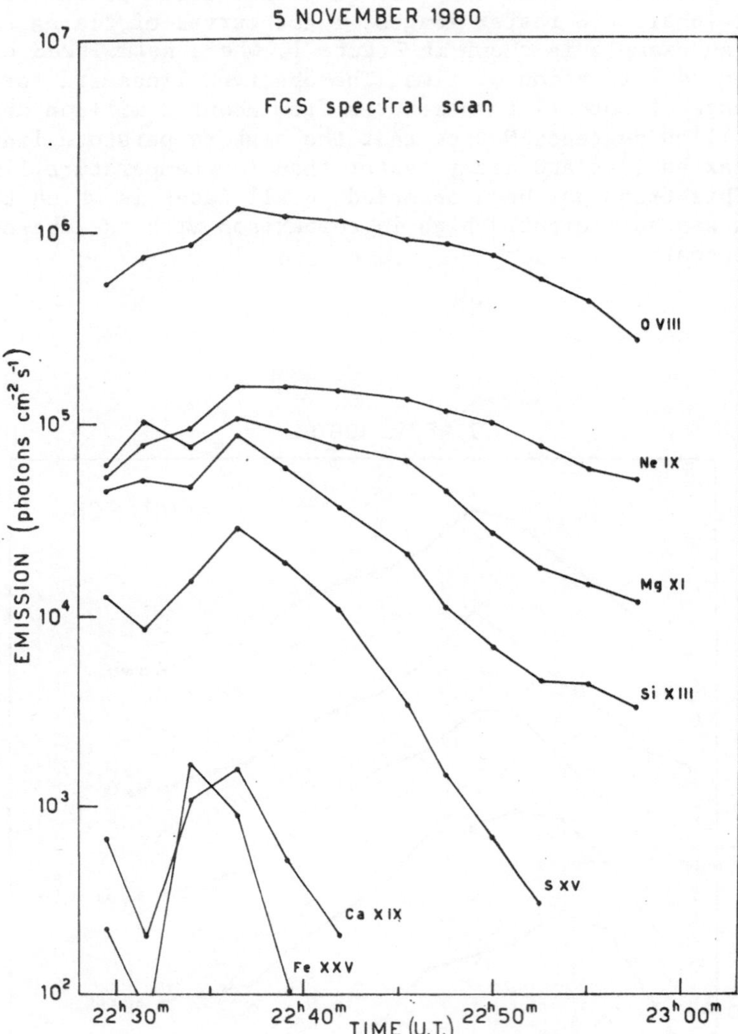

Fig.2 - Light curves in absolute flux units during a flare observed with FCS. In this case the instrument was operated in the spectral scanning mode and the observed fluxes refer to a single pixel.

For comparison with the results of model calculations, line intensities in absolute units, or at least an internally consistent relative scale, are needed. Figure 2 shows an example of FCS light curves calibrated in absolute units. In this case the instrument was operated in the spectral scanning mode, and the observed fluxes refer to a single 15x15 arcsecond pixel. While the accuracy of light curves in relative units is limited only by photon counting statistics and is of the order of 10%, light curves in absolute units may have much larger systematic errors. At present, the accuracy of the absolute fluxes is estimated to be of the order of a factor of two.

Figure 3 shows an example of flares observed with the BCS. In the Figure, fluxes in the resonance lines of Ca XIX and Fe XXV are plotted as a function of time. The BCS has a high spectral resolution which allows the application of a variety of plasma diagnostic techniques. The electron temperatures derived from Calcium and Iron lines using these diagnostic techniques are typically between 15 and 20 million degrees. In addition, line blueshifts are observed during the flare rise phase (Antonucci et al., 1982) which indicate the presence of systematic upflows with velocities of up to several hundred kilometers per second.

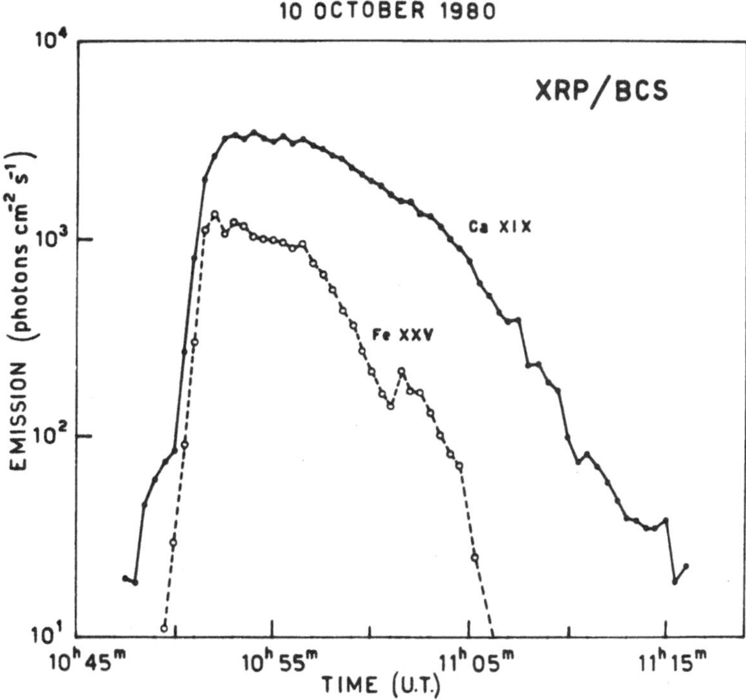

Fig.3 - Light curves in the resonance lines of Ca XIX and Fe XXV observed with BCS.

4. RESULTS

The numerical code described in section II has been used to compute spectral line intensities and velocities for comparison with XRP observations. Typical results are shown in Figures 4 and 5 for a loop of semilength $L = 2 \times 10^9$ cm and unit cross-section, in which heat is supplied around the apex for a period of 100 sec. Figure 4 shows computed light curves for spectral lines observable with FCS and BCS. The time sequence of maxima for different spectral lines and the different decay times for lines formed at different temperatures are in general agreement with the observations. Also the predicted relative values of photon fluxes are consistent with the observations, at least within the factor of two uncertainty due to XRP absolute calibration, ionization equilibrium calculations and chemical abundances. Figure 5 shows computed velocities along the loop at different times during the flare rise phase. Upward velocities of the order of a few hundred kilometers per second are predicted, in agreement with line blueshifts observed with BCS.

With the model considered above, in which a constant energy input of 10 erg cm^{-3} s^{-1} is supplied for 100 seconds, a temperature of the order of 20 million degrees is reached in the coronal portion of the loop. The density is predicted to peak a few minutes later than temperature, in agreement with observations. The time delay is due to the evaporation process which continues for a while after the heating has been switched off.

Model predictions depend on a number of free parameters to be varied to find the best agreement with specific observed cases. Figure 6 summarizes the numerical simulations run so far. We have investigated various effects such as changing energy input (by varying either heating rate or heating duration), temporal dependence of heating, preflare conditions and location of heat source (either at the loop top or close to the footpoints). The investigation of these various cases has allowed us to gain a general understanding of the hydrodynamic response of the system to heating perturbations and to clarify the importance of various parameters in affecting spectral line intensities and velocities.

5. CONCLUSIONS

The numerical simulations run so far and the comparison with XRP observations allow us to draw the following conclusions:

- There is an overall agreement between numerical simulations and observed X-ray line intensities. The agreement concerns both temporal profiles (reproduced very well) and line ratios (reproduced within a factor of two).

- The model predicts a temperature peak earlier than density peak, as observed.

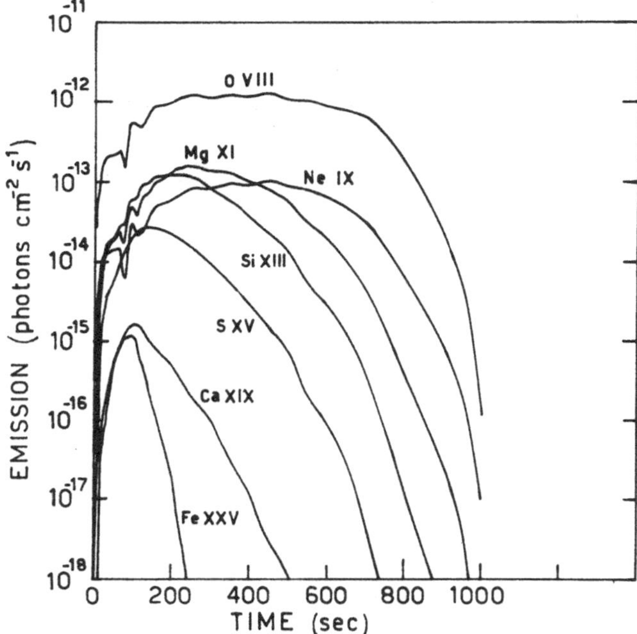

Fig.4 - Predicted line intensities as a function of time for the integrated emission of a model loop.

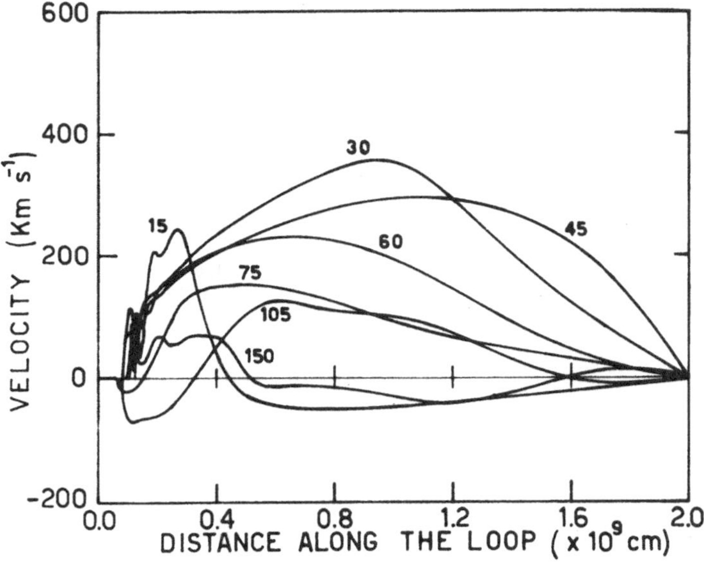

Fig.5 - Predicted velocity distributions along the loop at representative times during the flare rise phase.

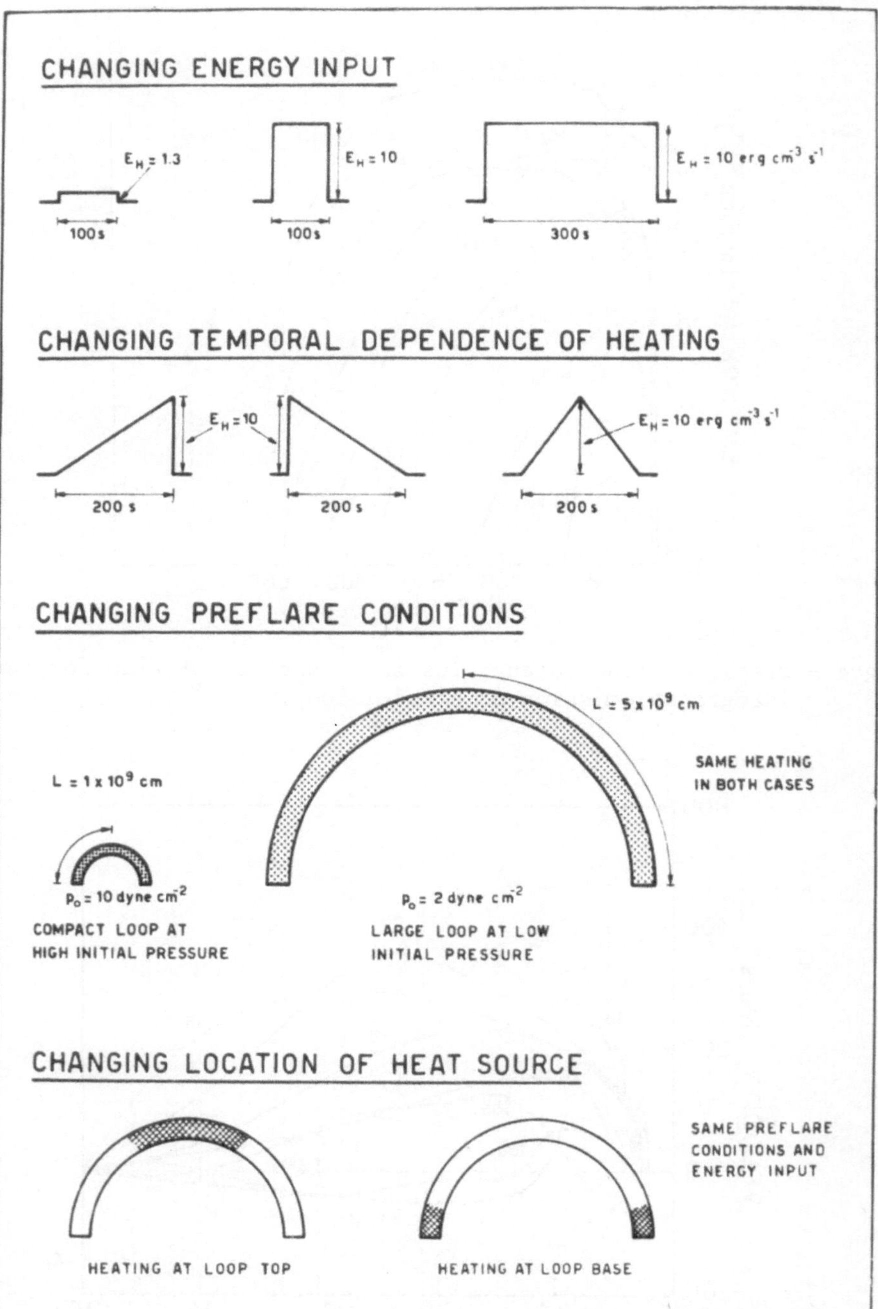

Fig.6 - Summary of numerical simulations run so far.

- The model predicts upward velocities of the same order as observed with BCS, supporting the interpretation of the observed blueshifts as due to evaporation of chromospheric material. However, the model predicts blueshifted components which are generally larger than observed. It remains to be determined whether geometrical effects can reduce the discrepancy.

- The highest temperature XRP line (Fe XXV) appears to be a useful indicator of the duration of the energy release process.

- Large differences in heating rate produce relatively small differences of peak temperature. For instance, in order to increase the peak temperature by a factor of two, we need to increase the heating rate by almost one order of magnitude. The peak temperature is also much less sensitive to heating duration than to heating rate.

- Initial preflare conditions (e.g. loop length and initial pressure) strongly affect flare evolution. In order to use the comparison of model predictions with observations to derive meaningful information on the flare heating source, one needs to find an isolated loop and to know preflare conditions with high accuracy.

- X-ray line intensities observed with XRP are not an effective indicator of the site of energy release in flares. There is no significant difference in the X-ray light curves of otherwise identical flares heated at the loop top or close to the footpoints. Even the downward motions occurring at the onset of flares heated at the top may remain largely unobservable with BCS owing to the small downward velocities and the low density in the loop in the early flare phases.

- The model predicts cooling below preflare temperature in the late decay and even, in some extreme cases, radiative instabilities leading to the formation of post-flare Hα loops. In all cases, however, the system returns eventually to preflare conditions under the stabilizing effect of the steady active region heating.

ACKNOWLEDGEMENTS

A number of people have contributed to the work presented in this paper. In particular, we thank R.Rosner, S.Serio and G.Vaiana for stimulating discussions about flare loop modeling, and L.Acton and J. Leibacher for substantial help in the analysis and interpretation of XRP data. One of the authors (R.P.) wishes to thank the Japanese colleagues and friends, and in particular Prof. Y.Uchida, for the kind invitation to attend the Tokyo Seminar. This work has been supported by the Italian National Research Council through a grant from Servizio Attività Spaziali.

REFERENCES

Acton, L.W. et al.: 1980, Solar Phys. 65, 53.
Antonucci, E., Gabriel, A.H., Acton, L.W., Culhane, J.L., Doyle, J.G., Leibacher, J.W., Machado, M.E., Orwig, L.E., and Rapley, C.G.: 1982, Solar Phys. 78, 107.
Cheng, C.C., Oran, E.S., Doschek, G.A., Boris, J.P., and Mariska, J.T.: 1982, Astrophys. J., in press.
Nagai, F.: 1980, Solar Phys. 68, 351.
Pallavicini, R., Peres, G., Serio, S., Vaiana, G., Acton, L.W., Leibacher, J.W., and Rosner, R.: 1982, Astrophys. J., in press.
Peres, G., Rosner, R., Serio, S., and Vaiana, G.: 1982, Astrophys. J. 252, 791.
Serio, S., Peres, G., Vaiana, G., Golub, L., and Rosner, R.: 1981, Astrophys. J. 243, 288.
Vernazza, J.E., Avrett, E.H., and Loeser, R.: 1980, Astrophys. J. Suppl. 45, 619.

DISCUSSION

DOSCHEK: I would just like to emphasize that, in carrying out these numerical simulations, it is important to calculate the X-ray line profiles and wavelength shifts. The numerical simulations appear to produce larger blue-shifted components than observed, but more work is needed to see if this result could perhaps be due to an observational selection effect. Also, lines formed right above the transition region, should show net wavelength shifts.

HENOUX: (1) How do you compute the radiative losses in the chromosphere? (2) Don't you think that chromospheric evaporation could depend on the amount of radiative loss?

PALLAVICINI: Radiative losses in the chromosphere have been computed empirically starting from the grid of chromospheric models of Vernazza, Avrett, and Loeser (1980, Ap. J. Suppl. 45, 619). We have looked for a general functional form of radiative losses and have found $n^2 \beta P(T)$ with the same P(T) for all cases to be appropriate. The adopted forms of P(T) can be found in Peres et. al. (1982, Ap. J. 252, 791). Although we have not carried out detailed time-dependent radiative transfer calculations. We think that this empirical approach is sufficient for our purposes. We are not interested in the computation of Hα and Ca line profiles, but only in the overall energetics and dynamics of the chromosphere in so far as they affect the coronal portion of the loop.

UCHIDA: I wonder whether it did not come to your attention that Nagai has already done an almost similar calculation and has got almost similar results.

DOSCHEK: Can you not increase the peak temperature of the loop in these numerical simulations, *i.e.*, make the temperature 30 to 40 x 10^6 K, by increasing the rate of energy input?

PALLAVICINI: We can do it provided we increase sufficiently the rate of energy input. However, in order to increase the temperature by a factor of two, we need to increase the energy imput rate by almost a factor of ten. If we keep the input rate constant, and we simply increase the total energy input by increasing the heating duration, we obtain only a negligible effect on temperature.

HUDSON: The simulations seem to do a good job of matching the observations, and they also guide one towards an understanding of the sensitivity of the models to the input parameters. When will we be confident enough to use the models actually to learn something new, for example, about pre-flare conditions?

PALLAVICINI: The simulations show that the predicted line intensities and velocities depend on a number of parameters, such as preflare conditions, heating rate, temporal and spatial dependence of heating, etc. If we want to be sure that a formal fitting of observations represents a unique solution - and hence if we want to get useful information on the flare heating source - we need additional constraints to limit the range of possibilities. This can be done if we have good observations of the geometry and preflare state of a flaring loop. We are going to use a set of XRP data which appears to satisfy this criterion (May 7, 1980 flare). I do not think, however, that we can use the comparison of model predictions with observations to derive information on the preflare state, since this requires an a priori knowledge of the spatial and temporal dependence of the flare heating source, which is just what we want to determine.

BEHAVIOR OF TRANSITION-REGION LINES DURING IMPULSIVE SOLAR FLARES

E. Tandberg-Hanssen and E. Reichmann
NASA/MSFC, AL 35812, U.S.A.

and

B. Woodgate
NASA/GSFC, MD 20771, U.S.A.

ABSTRACT. We describe briefly the Ultraviolet Spectrometer and Polarimeter on the Solar Maximum Mission and discuss data pertaining to the emissions observed in lines originating in the transition-region plasma, particularly during impulsive flares. The data pertain to lines from the following ions: SiII, CIV, OIV, SiIV, OV, and FeXXI.

1. INTRODUCTION

The concept of a "transition region" refers to the narrow region between the cool, roughly 10^4 K, chromosphere and the million, or multi-million, degree plasma of the corona. Since we no longer may consider either chromosphere or corona as spherical, let alone plane-parallel, structures, it follows that the shape of the transition region is quite complex. This is clearly brought out in the case of prominences which all may be considered as "coated" with a prominence-corona transition region, analogous to the chromosphere-corona transition region.

1.1. Transition-Region Observations

With the advent of spaceborne instrumentation increasingly more sophisticated observations of the transition region became possible, and comparisons could be made of observed data with theoretical predictions from energy-balance computations for the solar atmosphere. In particular, observations from OSO-6 and ATM on Skylab proved very successful. The reader is referred to Jordan (1977) who gives further references. Further progress was made with the observations obtained by OSO-8 instrumentation (Bruner, 1977), and the data obtained with the Ultraviolet Spectrometer and Polarimeter (UVSP) on the Solar Maximum Mission (SMM) furnished us with the latest information on the transition region (Tandberg-Hanssen, et al., 1981).

Solar Physics **86** (1983) 159–171. 0038–0938/83/0861–0159 $ 01.95.

1.2. Choice of Spectral Lines

The difficulties associated with the observation of the transition region have forced one to make some judicious choices regarding how to observe the emission from this region. In particular, if one wants information on the flaring transition region, one is forced to consider certain spectral regions. Figure 1, reproduced from Donnelly and Hall (1973), shows how the emission in flare lines maximizes for temperatures around 10^5 K; i.e., in transition-region plasmas for wavelengths longward of the Lyman continuum; i.e., 912 Å. For the UVSP experiment on the SMM, we therefore chose to observe the wavelength range available with MgF_2 coated optics; i.e., roughly > 1150 Å.

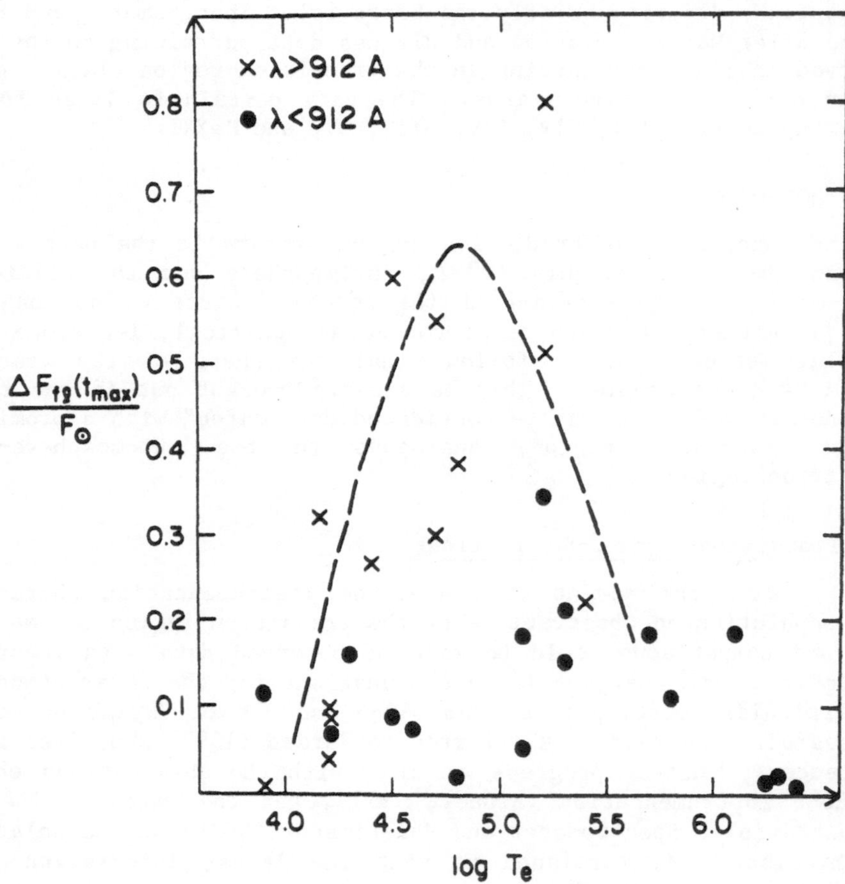

Fig. 1.

1.3. The UVSP Experiment on the SMM

The UVSP instrument consists of an aplanatic Gregorian telescope with an ellipsoidal primary and a hyperboloidal secondary. The secondary mirror can be tilted to provide rectangular rasters up to 256 x 256 arc s^2. The Ebert-Fastie spectrometer has a holographic grating, formed in photoresist, that in second order provides spectra from approximately 1150 to 1800 Å, and in first order from 1750 to 3600 Å. A polarimeter is located behind the spectrometer entrance slit, and consists of two retarders (wave plates) and a linear polarizer for one of the wave plates. The spectrometer grating acts as the analyzer (linear polarizer) for the other wave plate. For details on the UVSP experiment, see Woodgate, et al. (1980).

2. THE QUIET TRANSITION REGION

As a background for the discussion of the behavior of transition region during flares, we shall briefly consider the region during more quiet conditions.

2.1. The Transition Region Away from Sunspots

It is becoming increasingly clear that the transition region may not be considered a more or less plane-parallel layer between chromosphere and corona. Rather, the shape of the transition region is intimately linked to the notion of flux tubes making up the solar atmospheric structure. In this structure, large-scale velocity patterns are observed that show the dynamic nature of the transition zone, even in the absence of strong solar activity.

From studies of the CII (1336 Å), SiIV (1393 Å), and CIV (1548 Å) lines with the UVSP instrument, Gebbie, et al. (1981) found steady flows in the quiet transition region, consisting of persistent vertical motions with velocities ranging from 1.4 (CII) to 4.2 km/s (CIV). Similar large-scale velocity patterns have been detected by Athay, et al. (1982 a,b,c). These observations are consistent with widespread loop structures having downflow in both legs of the loops (see Figure 2).

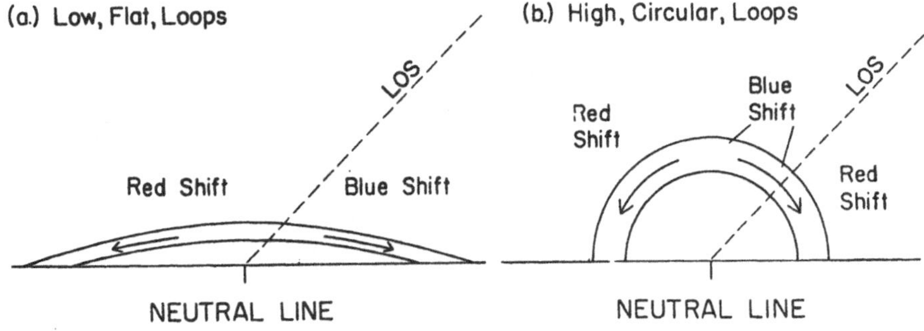

Fig. 2.

2.2. Oscillatory Motions above Sunspots

The transition region behaves quite differently above sunspots from what we have discussed so far. From time-series observations of CIV line profiles above sunspots, Gurman et al. (1982) discovered that the transition-region plasma above these spots undergoes significant oscillations in the line-of-sight velocity with periods in the range of 130-170 s. The expected intensity variations from adiabatic acoustic waves with the observed velocity amplitudes (0.8-3.5 km/s) are in good agreement with the measured intensity amplitudes.

2.3. The Shape of the Transition Region - Coronal Loops

There can be little doubt that magnetic fields dominate the structure of the whole solar atmosphere above sunspots, and that the structure is therefore intimately linked to the distribution of magnetic flux tubes in that part of the atmosphere. From the discussion in Section 2.1, it also seems that flux tubes may play an important role in the atmosphere - including the transition region - away from sunspots. We are hence led to conclude that the transition region forms part of such loops that often must extend into the corona. It is this rather complicated transition region that responds to solar flares, and the response may sometimes be seen near the foot-points of loop, sometimes higher up, depending on the thermodynamics of the loops and on the initial energy release.

3. THE TRANSITION REGION DURING FLARES

3.1. Behavior of the O IV / Si IV Transition-Region Plasma - The Flares of October 12 and 30, 1980

The emission in the lines OIV (1401 Å) and SiIV (1402 Å) comes from a plasma at temperatures around 80,000-100,000 K in the transition region. We have studied this emission and its relation to hard X-ray bursts in a disk flare on October 12, 1980 (AR 2725, S13, E18, SB max. 0159 UT), when the UV continuum was also observed, and in a limb flare of October 30, 1980 (AR 2776, N17, E90, C2 max. 1800 UT; no reported optical flare). These two sets of observations show that, even though we can draw some general conclusions as to the nature of transition-region flash phases of small flares, the spatial and temporal evolution of UV impulsive bursts may be quite diversified. In the disk flare of October 12, the impulsive flare brightenings occurred in many small point-like kernels distributed within a confined area. The limb flare of October 30 showed impulsive brightenings at two widely separated (∿ 50 arc s) points on the limb. From these observations and comparison with hard X-ray data from the HXRBS SMM experiment, we draw the following conclusions (Cheng and Tandberg-Hanssen, 1983):

a. There was considerable preflare activity with many UV transient bursts occurring in many small, point-like kernels. The flare later occurred in these active kernels. Figure 3 shows the temporal variation of the SiIV and UV continuum intensities from 0115 to 0205 UT. These intensities are the sums of all pixels in the 1' x 1' field-of-view of the raster. As can be seen from the figure, there were three, large transient brightenings in SiIV before the flare occurred at 0155 UT. The integrated intensities in the raster were rather steady except at the three transient SiIV brightenings. Notice particularly that the UV continuum only shows enhancement during the flare. The three SiIV brightenings occurred at 0124, 0131, and 0147 UT, and there were no detectable hard X-ray emission or repeated optical flares associated with them. Observations obtained with the UVSP show that transient brightenings in transition-zone lines without any accompanying flare activities are, in fact, common occurrences in active regions (Cheng, et al., 1982).

b. There is good correlation between the time evolution of the UV and the hard X-ray bursts when they occur. Individual peaks in the hard X-ray bursts can be identified with individual peaks in the UV light curves of individual flaring kernels.

c. There was impulsive brightening in the UV continuum during the time of the hard X-ray bursts.

d. The UV bursts started to rise a little earlier than the corresponding hard X-ray bursts. Figure 4 shows the SiIV intensities at both points 1 and 2 on the limb during the October 30 flare, separated by 50 arc s. Both emissions started to rise around 0157 UT; at 0159:30 UT the intensities had increased by a factor of 3. The impulsive brightenings at points 1 and 2 seem to have reached maximum at the same time, $\sim$ 1802 UT. The very impulsive hard X-ray burst started at $\sim$ 1800 UT.

3.2. Behavior of the Hot Transition-Region Plasma - The Flares of November 1 and 8, 1980

The hot part of the transition region plasma can be monitored during flares with the UVSP experiment using the OV (1371 Å) and FeXXI (1354 Å) lines. While the OV line emission reveals conditions in the transition-region plasma at $\sim$ 250 000 K, the FeXXI line would be more properly called a coronal line, and is seen in the transition region only during flares.

On November 8, 1980, several fast impulsive-phase bursts were observed with the SMM instruments; see Woodgate et al. (1982). The bursts were seen to be coincident in the OV line and in 25-300 keV X-rays to within the time resolution of 1 s (see Figure 5). This lack of time difference is inconsistent with models for these flares in which the hard X-rays are produced at the top of a loop, followed by the formation of a thermal conduction front which travels to the foot-

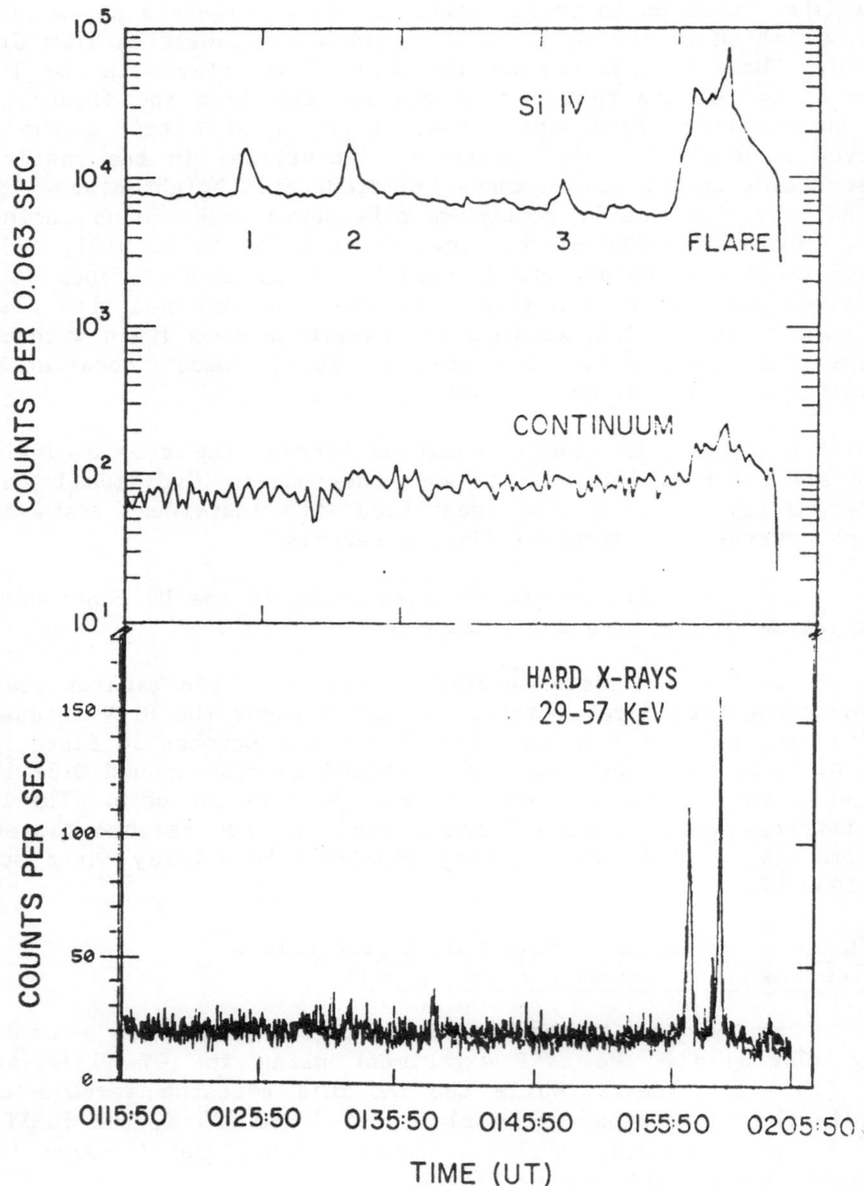

Fig. 3. Upper panel: temporal variations of the integrated SiIV (1402 Å) and the UV continuum emission in the 1'x1' field of view of the UVSP raster during the entire observing period of the 12 October 1980 flare. Lower panel: the corresponding impulsive hard X-ray bursts during the flare. The three SiIV brightenings before the flare are indicated by the numerals 1, 2, and 3.

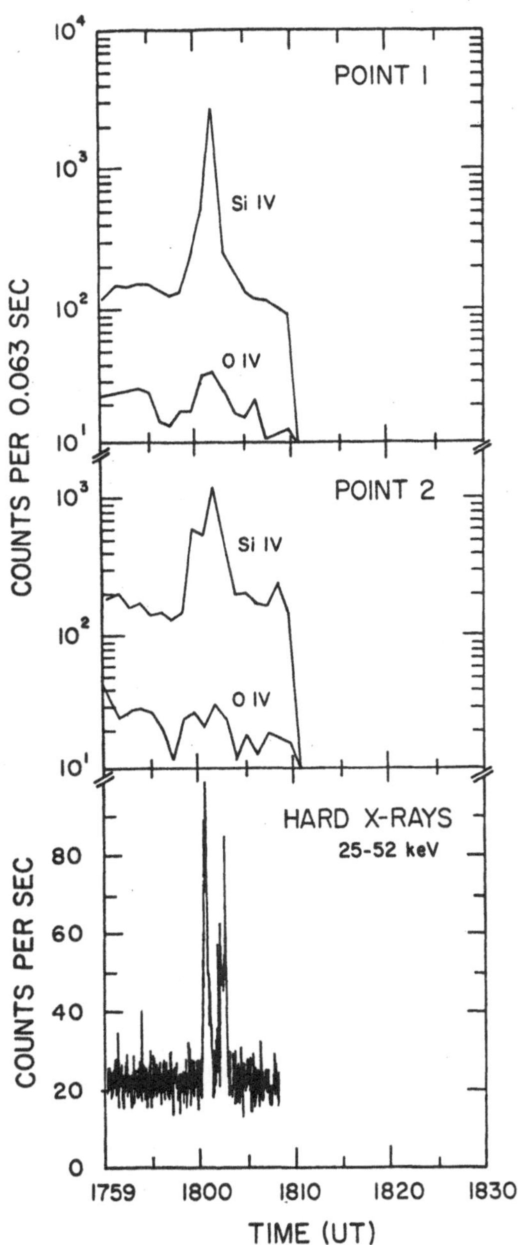

Fig. 4. Temporal variations of the SiIV and OIV intensities at the bright kernels 1 and 2 for the 30 October 1980 flare compared with the corresponding impulsive hard X-ray bursts.

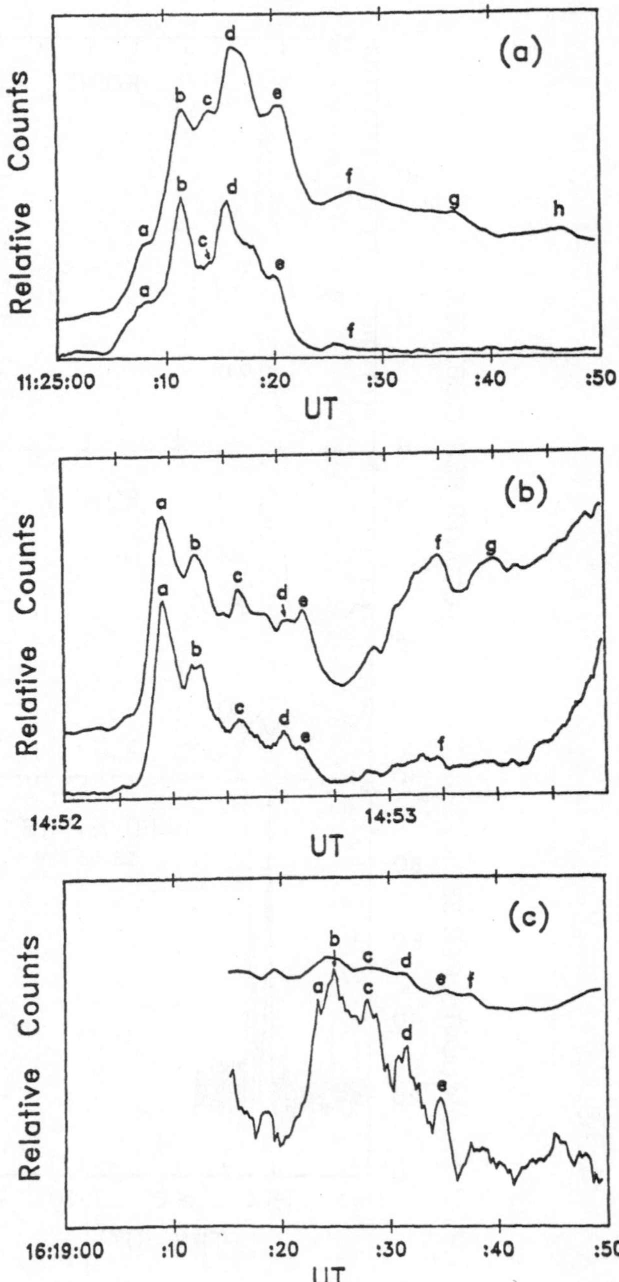

Fig. 5. Relative intensity versus time profiles of hard X-rays (30-85 keV) from HXRBS (lower curve) and OV (1371 Å; T equals 250 000 K) from UVSP (upper curve) for three flares: (a) 11:25 UT; (b) 14:51 UT; and (c) 16:19 UT, on 8 November 1980.

point where a UV burst is produced by heating. On the other hand, models are allowed in which both X-rays and UV are produced at the foot-point, or in which an electron beam transmits energy between loop top and foot-point in less than 1 s.

An exceptionally well observed impulsive-phase flare occurred on November 1, 1980. Due to good Hα, D_3, and magnetic field coverage, we ascertained that the overall magnetic configuration in which the flare occurred was a closed, simple bipolar arch or loop (see Tandberg-Hanssen et al., 1983).

In both soft and hard X-ray flux, the flare became detectable at about 19:16 (see Figure 6). These observations show that there was a gradual onset phase in which the X-ray emission slowly increased until the start of the impulsive phase at about 19:17:30. The impulsive phase produced two hard X-ray spikes; the first peaked at about 19:18:00, the second at about 19:19:15. As the hard X-ray emission subsided after its second peak, the soft X-ray flux continued to rise, reached its maximum at about 19:23, then gradually decayed to background level over the next half hour.

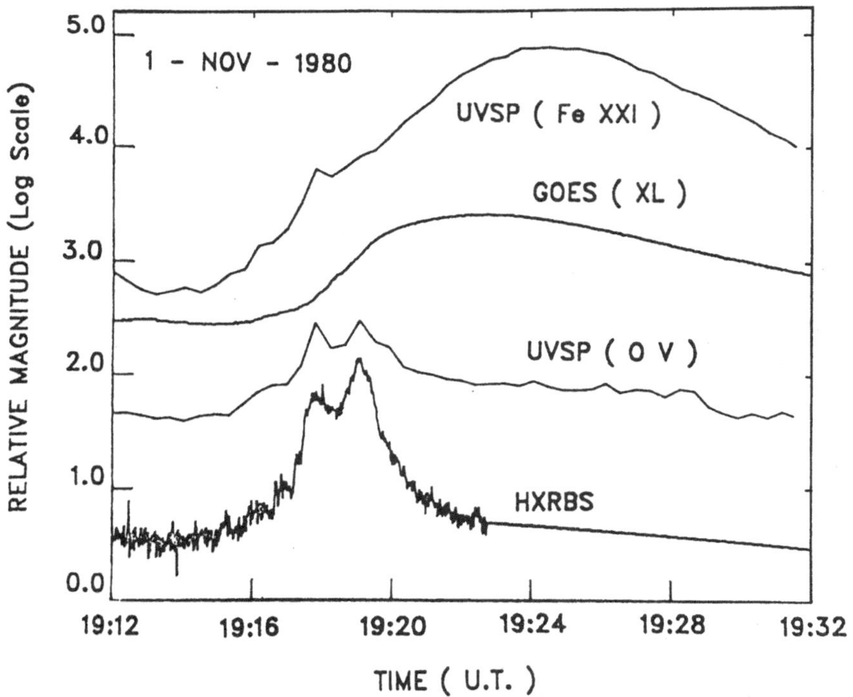

Fig. 6.

The temporal development of the OV 1371 Å emission mimicked the hard X-ray emission, clearly showing the two spikes.

The Hα movie frame in which the flare brightening was first visible was at 19:14:42. In the OV and FeXXI spectroheliogram sequences, the flare brightening was first detectable in the filtergram pair which began at 19:14:54. To within the spatial resolution, the emissions in Hα, OV, and FeXXI began at the same point in each foot of the flare; to within the temporal resolution of the experiment, the two foot-points turned on together, and the onsets of all three emissions were simultaneous.

The D_3 observations show that when the two spikes in X-rays and OV/FeXXI occurred, different areas of the chromospheric flare ribbons (the foot-points) brightened up. This indicates that the impulsive energy release was centered on different field lines in the two spikes. The conclusions to be drawn concerning impulsive-phase, small flares of the type exemplified by the November 1 flare are:

a. The overall magnetic configuration in which the flares occur is a closed, simply bipolar arch or loop.

b. The flares apparently occur spontaneously within the flare arch because (1) both feet of the flare turn on simultaneously and well away from the inversion line, and (2) in case of the November 1 flare, there was no evidence of an emerging bipole or other polarity reversal near either foot.

c. The FeXXI emission came from the feet of the flare well into the impulsive phase, but from over the inversion line between the feet in the decay phase; the OV emission was concentrated over the feet throughout the flare. Hence, through at least the first minute of the impulsive phase, practically all of the flare-heated plasma at $T < 10^7$ K was confined to the feet of the flare loop.

d. For the November 1 flare, the two spikes of the impulsive energy release occurred on different subsets of field lines within the overall flare loop.

3.3. O V and Fe XXI during Flare Flash Phases

The discussion in Section 3.2 indicates how the OV and FeXXI emissions may behave in a "typical" simple, compact flare. However, different flares exhibit a nearly bewildering variety of impulsive-phase behavior, and Figure 7 shows some examples from the UVSP data bank.

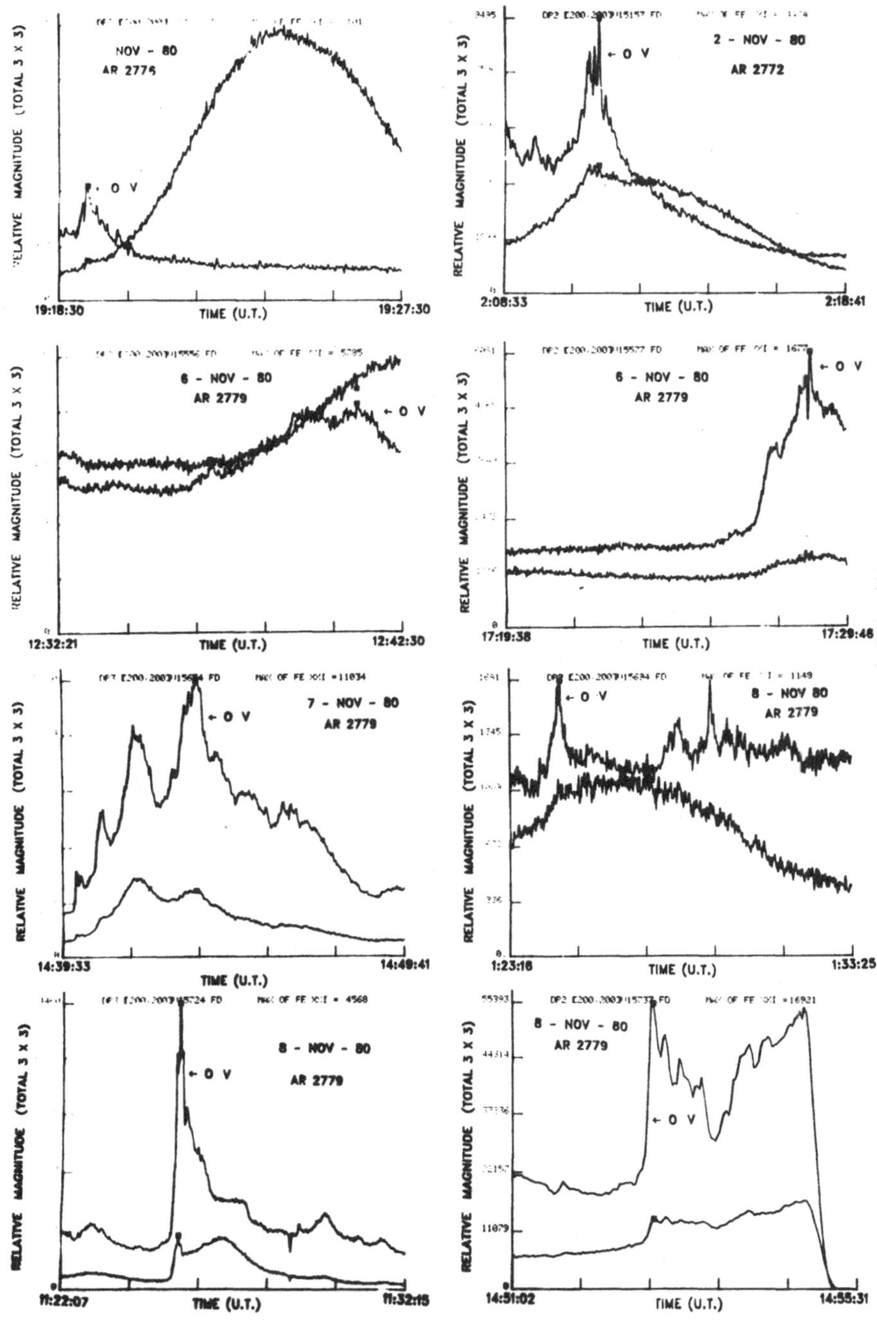
NOV - 80
AR 2775
← O V
RELATIVE MAGNITUDE (TOTAL 3 X 3)
19:18:30
TIME (U.T.)
19:27:30
2 - NOV - 80
AR 2772
← O V
2:08:33
2:18:41
6 - NOV - 80
AR 2779
← O V
12:32:21
12:42:30
6 - NOV - 80
AR 2779
← O V
17:19:38
17:29:46
7 - NOV - 80
AR 2779
← O V
14:39:33
14:49:41
8 - NOV 80
AR 2779
← O V
1:23:16
1:33:25
8 - NOV - 80
AR 2779
← O V
11:22:07
11:32:15
8 - NOV - 80
AR 2779
← O V
55393
44314
33236
22157
11079
14:51:02
14:55:31

Fig. 7.

REFERENCES

R. G. Athay, J. B. Gurman, W. Henze, and R. A. Shine: 1982a,b, Astrophys. J., in press.

R. G. Athay, J. B. Gurman, and W. Henze: 1982c, submitted to Astrophys. J.

E. C. Bruner, Jr.: 1977, Space Science Instrumentation 3, 369.

C.-C. Cheng, E. C. Bruner, E. Tandberg-Hanssen, B. E. Woodgate, R. A. Shine, P. J. Kenny, W. Henze, and G. Poletto: Observations of Solar Flare Transition Zone Plasmas from the Solar Maximum Mission: 1982, Astrophys. J. 253, 353.

C.-C. Cheng and E. Tandberg-Hanssen: 1983, in preparation.

R. F. Donnelly and L. A. Hall: 1973, Solar Physics 31, 411.

K. B. Gebbie, F. Hill, J. Toomre, L. J. November, G. W. Simon, R. G. Athay, E. C. Bruner, R. A. Rehse, J. B. Gurman, R. A. Shine, B. E. Woodgate, and E. A. Tandberg-Hanssen: Steady Flows in the Solar Transition Region Observed with SMM: 1981, Astrophys. J. (Letters) 251, L115.

J. B. Gurman, J. W. Leibacher, R. A. Shine, B. E. Woodgate, and W. Henze: Transition Region Oscillations in Sunspots: 1982, Astrophys. J. 253, 939.

C. Jordan: in Illustrated Glossary for Solar and Solar-Terrestrial Physics, 1977 (eds., A. Bruzek and C. J. Durrant).

E. Tandberg-Hanssen, P. Kaufmann, E. J. Reichmann, D. L. Teuber, R. L. Moore, L. E. Orwig, and H. Zirin: 1983, in preparation.

E. Tandberg-Hanssen, R. G. Athay, J. M. Beckers, J. C. Brandt, E. C. Bruner, R. D. Chapman, C.-C. Cheng, J. B. Gurman, W. Henze, C. L. Hyder, A. G. Michalitsianos, R. A. Shine, S. A. Schoolman, and B. E. Woodgate: Preliminary Observations and Results Obtained with the Ultraviolet Spectrometer and Polarimeter: 1981, Astrophys. J. (Letters) 244, L127.

B. E. Woodgate, E. A. Tandberg-Hanssen, E. C. Bruner, J. M. Beckers, J. C. Brandt, W. Henze, C. L. Hyder, M. W. Kalet, P. J. Kenny, E. D. Knox, A. G. Michalitsianos, R. A. Rehse, R. A. Shine, and H. D. Tinsley: The Ultraviolet Spectrometer and Polarimeter on the Solar Maximum Mission: 1980, Solar Physics 65, 73.

B. E. Woodgate, R. A. Shine, A. Poland, and L. E. Orwig: Rapid Impulsive Phase Bursts in Solar Flares Observed by SMM on November 8, 1980, and Limitations on Models: 1982, in preparation.

DISCUSSION

PRIEST: What velocity fields do you find near prominences and how do they compare with the OSO-8 observations of upflow both sides of a filament and of the Meudon observations of upflow in a filament?

TANDBERG-HANSSEN: We have just started the systematic study of the velocity field around prominences as seen on the disk; hence, it is premature to give a final answer. However, in at least one case we confirmed the OSO-8 observation of upflow on both sides of a filament with typical amplitudes of about 20 km/s.

SVESTKA: Isn't it possible that the maximums of the FeXXI and OV emissions are in different positions so that you miss one of them with your small field-of-view?

TANDBERG-HANSSEN: This is certainly true in some cases. However, before we go to the small field-of-view for rapid scanning sequences, we normally have a larger field-of-view, 3 x 3 arc min, where we can follow the buildup of both FeXXI and OV emissions. In addition, the small field-of-view is 30 x 30 arc sec, which is not really all that small compared to the size of loop footpoints.

BRUECKNER: The OV line at 1370 Å as a diagnostic of the transition zone must be used with caution. This line is allowed; however, its ground state is excited by a spin-forbidden line. The latter case of lines shows a very peculiar behavior during flares. They are often suppressed during the explosive phase. This may reflect very high density conditions during the initial phase. However, Skylab has also observed flares where the spin-forbidden lines did brighten together with the allowed lines.

THE OPTICAL FLARE

H. Zirin
Big Bear Solar Observatory
California Institute of Technology

ABSTRACT. Optical observationd now present considerable information on the flare process. It is always associated with filaments and with simplification of existing magnetic connections, and it arises from the emergence and expansion of new flux.

The optical flare divides into impulsive phase, with multiple flashes along the neutral line, and thermal phase, with two-ribbon expansion. The former bears some resemblance to tearing mode phenomena.

The appearance of loops in emission requires very high densities in those phenomena.

The ratios of the hydrogen lines, the excitation of HeII 4686, and the relation of vertical to horizontal structure all remain to be explained.

1. INTRODUCTION

The great Russian humorists Ilf and Petrov pointed out that "pedestrians must be respected." Pedestrians discovered the law of gravitation, the differential calculus, even the automobile. The optical observers are thus the pedestrians of flare research. Most of what we know about flares today is the result of optical observations. Although it would be most difficult to interpret these without the remarkable radio and X-ray observations reported at this conference, the optical data are still our most powerful tool in understanding what governs the flare. Why? What question do we answer by optical observations? After all, most of us agree (but it is not proven) that the initial flare energy release occurs in very high temperature material invisible in the optical range.

Yet optical observations have the following advantages:

Solar Physics **86** (1983) 173–184. 0038–0938/83/0861–0173 $ 01.80.

1) An immense number of photons are produced, permitting high space and time resolution. A flare occupies hundreds of pixels in optical data - a few in X-ray or radio pictures.

2) High resolution telescopes for optical work are easy and cheap to build.

3) Magnetic fields are easily measured.

4) The flare results from preflare conditions that are invisible to X-rays or radio waves, even with low resolution. The origin of the flare must thus be studied in the optical.

5) Almost all high energy flare effects are reflected in optical effects. With improving understanding, we can pick out different elements to deduce time behavior of HXR, SXR and other radiations.

6) It is quite conceivable that the filament that is part of almost every flare is in fact the seat of the energy release, hence it is possible we already see the flare energy release process in the visible but don't understand it.

2. INFORMATION ON LOCATION AND CIRCUMSTANCES OF FLARES

It must be recalled that almost no data on flare location and morphology existed until the introduction of the magnetograph at Mt. Wilson and high quality Hα cinematography by Ramsey and Moreton at Lockheed. Severny (1958) was the first to worry about the location of flares occurring on neutral lines in the longitudinal fields. The Lockheed observers early recognized the activation of filaments before the Hα flare. So, since 1958-1960 we have known that the flare takes place along the longitudinal neutral line. This is one of the key facts in the flare problem.

The fact that the flare took place at a point where the field was not zero bothered many people and made them dubious of this result. But modern day high resolution magnetograms and Hα grams permit us easily to resolve the problem.

In general, two kinds of longitudinal neutral lines are seen in active regions. Both are marked in Hα by dark features. In one case these dark features run directly from one polarity to the other, and for this reason I called them field transition arches (FTA). They are always perpendicular to the neutral line. In the other kind, the field lines run parallel to the neutral line; these we know as filaments, and this is where flares occur. The reason of course is obvious: perpendicular fibrils involve shorter field lines and much less energy than filaments.

How do optical observations demonstrate this fact?

1) The pre-flare fibrils are seen to run parallel to the neutral line.

2) Measurement of fields in filaments show them to be parallel to the axis.

3) Flares along parallel neutral lines show bright elements in either polarity but displaced along the neutral line.

4) After flares we see loops perpendicular to the neutral line and still later, the parallel fibrils are seen to have been replaced by perpendicular fibrils.

Optical observations have established that, except for occasional small events, flares do not occur away from the general neighborhood of filaments. The parallel boundary is established by sunspot motions, usually connected with newly erupting flux. Because reconnection cannot take place quickly, the expanding flux pushes the old flux aside and a sheared boundary parallel fibril results. It is well established that flares occur as the result of sunspot motions. The prime generator of this is flux emergence in the middle of existing fields. In other cases spot motion occurs for mysterious reasons. For example, the spots of August 1972 were inverted and apparently the p spot was pulled forwards by the unknown Hale-Nicholson force.

A similar event occurred in June 1982. A pair of spots appeared close together in NS alignment on the east limb. Between June 4 and June 6 the following spot (or, more likely, preceding if the group belonged to the other hemisphere) moved rapidly westward, covering 30,000 km in 2 days and culminating in a huge flare on June 6 (Figure 1). The rapid east-west motion produced the sharp east-west neutral line that was the site of the big flare, but many flares also took place on the advancing edge of the moving spot. The flare is the way the magnetic field finally adjusts to the changes produced by spot motion; changes which cannot be accommodated by diffusion but only by instabilities. The process seems quite similar to the process of earthquakes which are also of common interest in California and Japan.

3. THE OPTICAL FLARE: APPEARANCE IN Hα

It did not take very good data for early observers to recognize the two-ribbon flare. Virtually all large flares show this property, the strands starting close together at the neutral line, and moving apart at 5-15 km/sec as the thermal phase of the flare continues. As can readily be seen from pictures, such as those of the August 7, 1982 flare (Zirin and Tanaka, 1973), the two strands are connected by bright Hα loops. X-ray pictures of post-flare loops (Svestka, 1976, Figure 20b) clearly show the association between the post-flare coronal condensation and the bright Hα loops. The appearance of bright Hα loops against the disk is a diagnostic of a definite minimum density and temperature; to my knowledge the values have not been

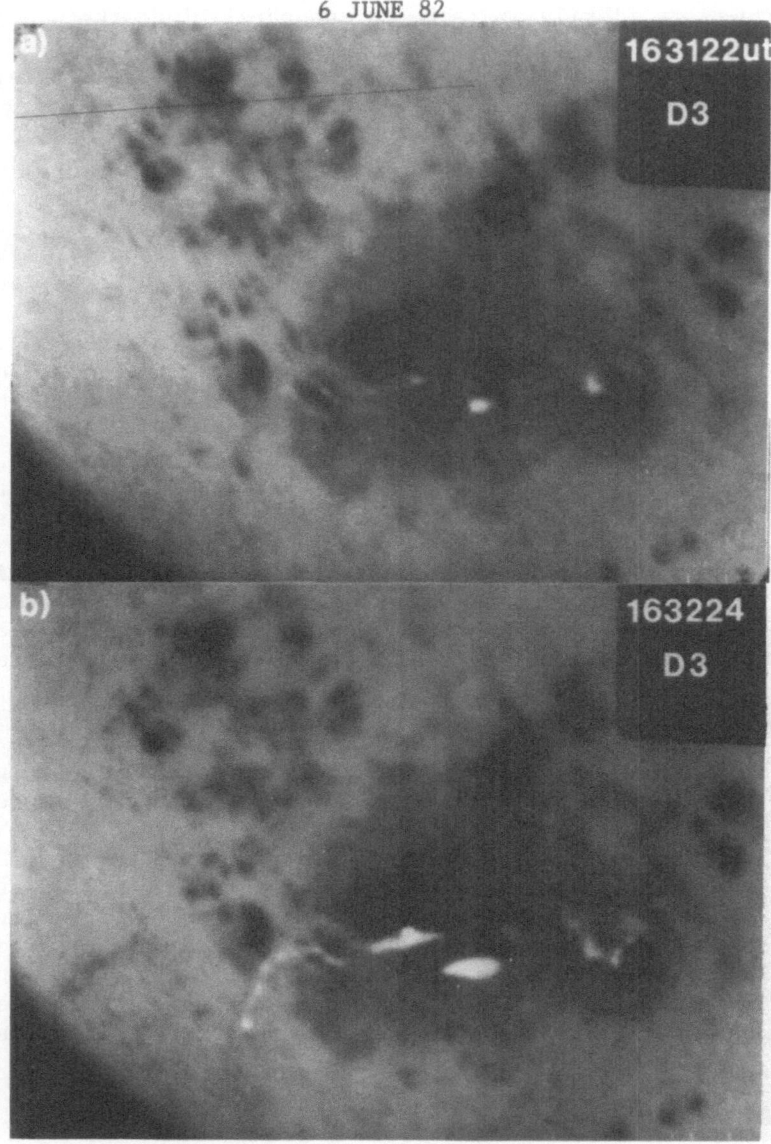

Fig. 1: D3 and Hα images of the June 6, 1982 flare. The large round spot at lower right had been directly S of the large main spot on 6/4; it travelled 30,000 km in two days.
(a) D3 16:31:22 UT -- Impulsive beginning: transient bright kernels appear in the moving spot (p polarity), the large delta spot (p below, f above) and a small f spot to the left.
(b) D3 16:32:24 UT -- Development of the main phase of the flare: although numerous transient flashes still appear, two intense kernels appear in the delta spot where the hot thermal phase is beginning. Field lines that once connected the p spot at right now connect the two intense kernels.

6 JUNE 82

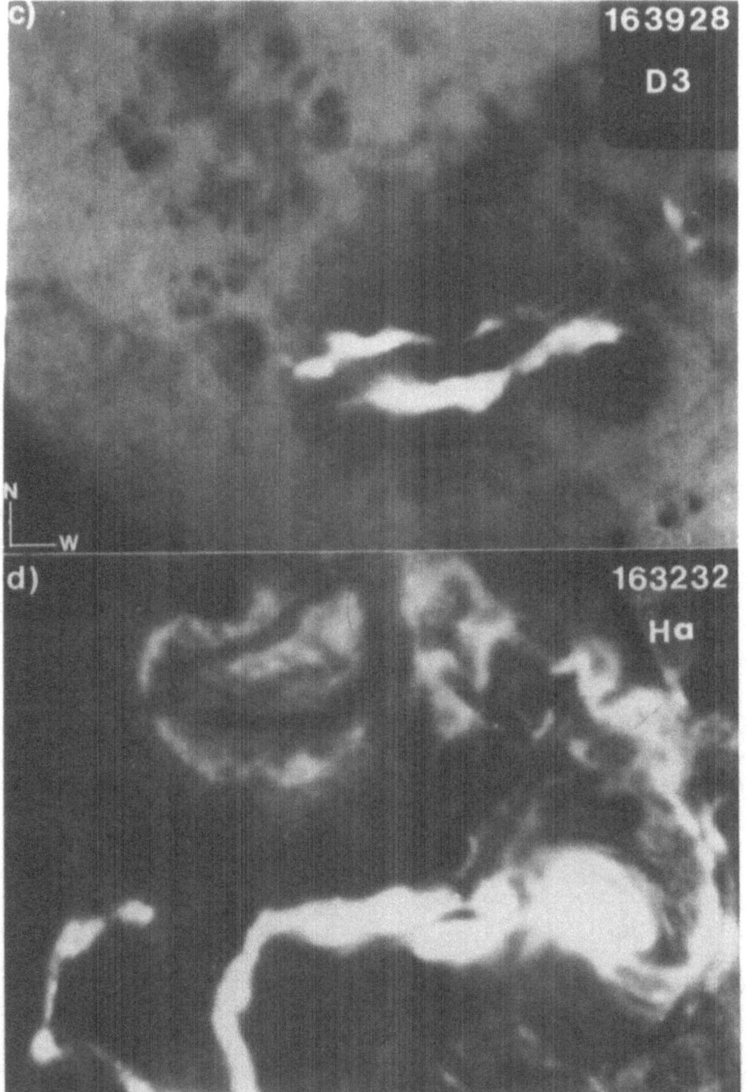

Fig. 1 - continued
(c) D3 16:39:28 UT -- Mainphase: bright loops now connect the two main kernels, which are spreading and elongated; the moving spot at right is also very intense and new emission appears at far right where an associated flare developed.
(d) Hα + 0.5 A 16:32:22 UT -- Note the intense emission over the p spot, corresponding to several small kernels in (b).

calculated. The appearance of Hα and coronal line loops high above the surface, visible on Lyot's early films, is an important fact in understanding the thermal flare process. Our flare models must produce a strong departure from hydrostatic equilibrium in the high density region at the loop tops, where the reconnection in the thermal phase appears to take place.

The two strand phenomenon is a result of the fact that all closed field lines must intersect the solar surface in two points. The rapid spread of the footpoints probably signals the rapid spread of flare energy release from its inception at a low height along the neutral line. The loops trace the field line connection between points of opposte polarity. Whether these connections existed before the flare or are new cannot presently be proven, but, at least in the 6/6/82 flare, it seems clear that the field lines previously ran between the two large spots that appeared together. After the flare the loops show them connecting the largest spots directly across the neutral line.

So much emphasis has been placed on the "two-ribbon flare" that the question is often asked "Where are the other two ribbons?" The Petschek theory, as well as other models of magnetic merging require the involvement of four magnetic poles, and energy from the reconnection area must flow down to each pole and produce surface brightening. In fact, in a complex flare in Hα there are plenty of ribbons (often more than four) because the line is easily excited by any energetic particles that get into a particular line of force; we have many examples of brightening 100,000 km away. Thus the main ribbons must be picked by using less sensitive lines, like D3 or Hα, and a reasonable magnetic connection. For example, two bright D3 ribbons connected by loops must be the two poles whose connection results from the flare; an area connected to the kernels by Hα fibrils can contain one of four ribbons. One important means of determining transverse field links is to study small flares in a region; these usually have 2 bright points which must be at the ends of a single line of force.

But the main phase of most big flares shows two long ribbons connected by an arcade of loops! If we examine the D3 pictures of the 6/6/82 flare (Figure 1), we see in the impulsive phase 11 separate kernels in four general areas, two of each polarity. The neutral line has been so stretched by spot motion that the poles are all aligned roughly along these strands. When the main phase of the flare begins, the peak intensity rapidly spreads from the kernels to the entire string of poles. The spreading apart of the two strands corresponds to the spread of the energy release from the critical neutral line region and is of course matched by the increasing height of loops above the region.

Finally, the three-ribbon flare is an event which is much more common than previously thought. An example is the flare of July 1,

1980 (Zirin and Neidig, 1981). A few minutes after the impulsive peak a third region brightens and remains bright long after the two ribbons near the neutral line fade. Feldman et al. (1982) found a number of cases where the third strand was prominent. The effect is not prominent in Hα but quite marked in D3, and even more prominent in the continuum, judging by the July 1, 1980 event.

4. OTHER LINES

Interesting perspectives on flares may be obtained in other lines, which may be observed with the universal filter or specialized filters. Hβ is particularly interesting because it is not as deep and saturated as Hα but the atmosphere phenomena are easily seen. However, exposures with the universal filter are long because the polaroids have poor transmission in this region.

It is surprising that the most intense Balmer line is Hγ, both in solar and stellar flares, i.e. there is a Balmer increment. Probably this is due to the fact that the lines are very deep (Neidig estimates τ (Hα) = 10^4) and reach the Planck limit. If the temperature where the flare emission occurs is 10 000° the Planck peak will of course be in the blue.

As noted above, D3 is a beautiful line to study flares; no chromospheric background is seen and the emission may be quite intense. With the exception of post flare loops, which are in emission only in the most intense events, all atmospheric phenomena (surges, sprays, active filaments) are in absorption. This is because D3 is seen against the 6 000 degree photosphere, and densities greater than 10^{13} are required to produce sufficient collisional excitation for the line to be seen. Feldman et al. showed that the emission in D3 appears in two ways: scattered transient brightenings along the neutral line during the impulsive phase, and intense brightening of two ribbons on either side of the neutral line during the thermal phase. This behavior allows us generally to separate impulsive and thermal phases and to identify footpoints of the core of the flare. The most intense D3 occurs during the peak of SXR emission from a 2 keV thermal plasma (Figure 1). It is intriguing to compare the transient brightenings along the neutral line with the "islands" that appear in the first phase of tearing mode instability (van Hoven, 1981); each island could generate the particles for each pair of footpoint kernels.

Other lines, such as MgI b and NaD appear virtually identical to D3 but less intense (partly because they are in absorption lines) and without absorption from surges, sprays, etc. Even weak lines such as FeI 5324 and CaI 6103 appear in emission in high resolution images, with intensities about twice the line center (i.e., about 0.2 x the photosphere) and distribution somewhat similar to D3. All these emissions appear to be footpoints, either of streaming impulsive particles or conduction from thermal loops (Figure 2).

9 JULY 78

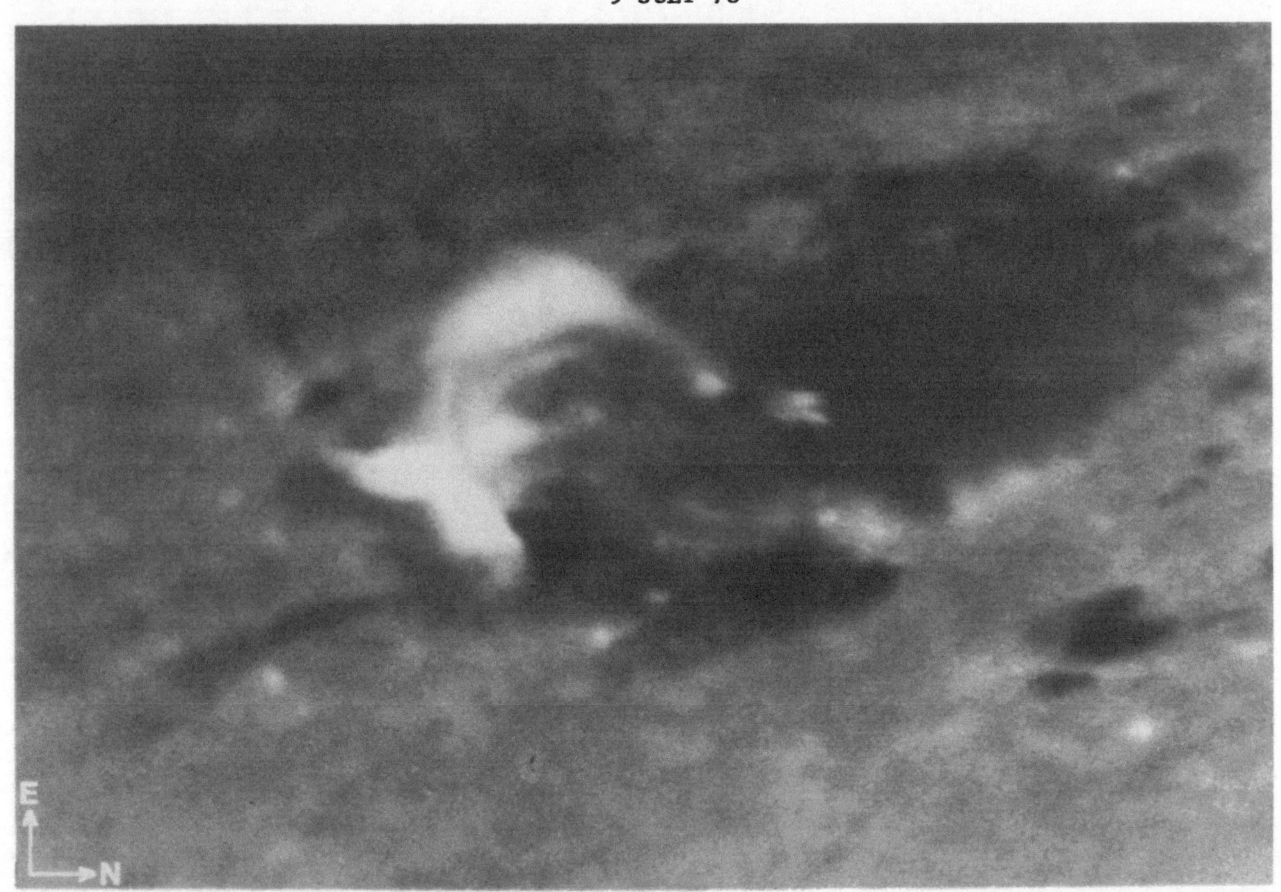

Fig. 2: Hα-1.5Å filtergram of a flare in an active region not far from the east limb. Note the loops connecting the footpoints of the flare.

5. CONTINUUM OBSERVATIONS

The continuum emission from flares is of great interest and represents one of the major energy outputs of the flare. The morphology of the continuum is similar to that described for D3, except that the impulsive kernels are relatively more intense and the emission decays more rapidly. Until recently we thought that white light flares could only occur on the disk and that they were essentially due to heating of the photosphere by streaming particles or conduction. But Harvey's observation (unpublished) of a white light flare above the limb shows that flares can emit continuum comparable in intensity to the photosphere from regions some distance above it. A simple calculation shows a hot very dense loop ($N_e = 10^{14}$, T 100,000°) might produce the observed effect. However as I noted above, most white light flares are observed at footpoints, and we must consider the general case as heating of footpoints by either streaming particles or conduction.

The problem of the blue continuum remains unresolved. A strong increase in flare continuum intensity is observed below 4000 Å in both solar and stellar flares. For example, Zirin and Neidig (1981) found no detectable increase at 5000 Å in flares where the continuum intensity doubled at 3862 Å. Neidig (1982) found I/Iphot of 3.6 at 3610 Å, 0.65 at 4275 Å and 0.4 at 6203 Å. Although Neidig (1982) has shown that some of this effect is due to the drop in the background intensity, he points out this cannot explain kernels where the intensity increases to 3 or 4 times the photosphere in H^-. Opacity at these wavelengths also seems unlikely in view of the known shape of the H^- absorption. Similarly, it is hard to explain the spectral dependence of the white light flare by hot black body emission, as was done in stellar flares by Mochnacki and Zirin (1980). Any hot black body will produce considerable emission in the green which is not observed. The effect of lines appears to me inadequate. If we filled in all the lines we would only increase the intensity at 3800 Å by 1.9 and 4300 Å by 1.5. The metallic lines are in fact narrow (about 0.1 Å wide) and it is hard to understand how they could provide enough intensity to double the continuum at 3610 while producing an only 5% increase at 4275 as occurred in the 1 July 1980 flare. Perhaps there are autoionization continua in this region.

The spectral evidence now favors strong Balmer continuum emission in flares. This conclusion is based on Neidig's data at 3610 Å (1982) and Zirin and Neidig (1981) and the work of Mochnacki and Zirin (1980) who found strong Balmer continuum in stellar flares. My experience with stellar flare spectra is that the Balmer continuum disappears if the detector has low sensitivity below 3800 Å, and I attribute earlier failure (Svestka, 1965) to detect Balmer continuum to that cause.

In summary, the different behavior of the continuum and D3 emission during impulsive and hot thermal events shows that heating at various low levels in the atmosphere is produced by both streaming and conducting electrons at footpoints. The amount transmitted is greatest in the hot thermal phase, and large increases (2 to 4 times) above the photospheric background can occur in D3 and the blue continuum.

6. THE LIMB FLARE

Most flares are seen on the disk and give useful information on the flare location relative to the magnetic fields. But since there is good evidence that the flare energy release takes place above the photosphere, it is important to study the limb flare, particularly in Hα where we have enough opacity to give hope of seeing the flare source.

What we see on the limb bears little relation to the flare on the disk, except for loops and sprays. Two common types of limb flare are (1) a bright condensation (usually the neutral line filament) appears above the limb and blows off in spray, followed by loop condensation, (2) ragged bits of emission, much brighter than the disk in Hα are seen. Except for the post-flare loops, the Hα light curve is short and relatively similar to the hard X-ray time profiles (Zirin, 1979). This is further evidence that (1) the HXR come from the Hα-active region just above the limb and (2) the main flare brightening after the impulsive burst is a low chromospheric phenomenon heated by conduction from the thermal flare.

We do not know exactly what the bits of Hα emission are, although in some cases they clearly match the erupting neutral line filament, in others that is not the case - they brighten and fade without motion and may or may not be loops. In other cases, the flare appears to be a set of low lying loops which grow upward.

7. FOOTPOINTS: A PEDESTRIAN SUBJECT

Any doubt about the connection of loops and footpoints should be removed by Figure 2, a frame in Hα-1.5 Å obtained about 20 minutes after a class 2 flare. The footpoints seen here were emitting in white light 15 minutes earlier; the impulsive flashes were at other points nearby. We clearly see how the bright loops connect the heated footpoints.

Although the role of footpoints is clear, the formation is hard to understand. All indications from optical data are that the heating at footpoints goes much deeper than current models. The He D3 emission requires $T > 15000°$ at $N \sim 10^{14}$. The white light data requires $T > 8000°$ at the photosphere. The HeII data of Linsky et al. (1976) require 85000° at a density ($\sim 10^{12}$) where 304 is optically deep. The fact that much of this heating occurs in the thermal phase is just as surprising as the instantaneous production of the little kernels in the flash phase.

8. THE RIDDLES OF FLARE SPECTRA

I would like to summarize some riddles of the optical flare:

1) The blue continuum.

2) Hγ is the strongest line of the Balmer series. This is possibly explained by black body emission in the Balmer lines at 7000°.

3) The integrated intensity of Hα equals that of Lyα. This could be explained by black body emission in the Balmer lines at 15 000°, which obviously contradicts (2). I do not accept published models which find a set of parameters that fit this observation; we require a more physical explanation of why this happens so often in nature.

4) To what feature of the disk flare does the limb flare correspond?

5) Jefferies, Smith and Smith (1959) found HeII 4686 in the spectra of disk flares. Although the brightness is not great, high loop densities (at least $10^{12.5}$) are required. The 4686 intensity measured by them exceeds the intensity of λ304 measured by Linsky et al. (1976) in a similar flare.

6) The D3 line is observed in emission at footpoints and in loops. The footpoint emission can reach 4 times the photosphere. The loops are at least 10% above the photosphere. Densities of $N_e > 10^{13}$ are required.

7) Post flare loops also appear in emission against the disk in Hα.

Where does the future lie in optical flare observations? Although we have considerable good data, we still need:

1) D3 flares near the limb.

2) Flare images in HeII 4686.

3) Polarization measurements (discussed by Dr. Henoux).

4) Flare spectra of the real kernels.

5) Definitive measurements of the Balmer continuum.

6) Better measurements of the transverse field.

7) Magnetograms in Hα.

This survey has been prepared during my stay at Meudon. I thank Dr. Henoux for his hospitality, and the Ministère d'Etat à la Recherche et à la Technologie for a grant supporting my visit. The research at Big Bear was supported by the National Science Foundation under ATM-8112698 and by NASA under NGL 05 002 034.

REFERENCES

Feldman, U., Liggett, M., and Zirin, H.: 1982, BBSO #0214, submitted to Ap.J.

Jefferies, J. J., Smith, E. V., and Smith, H. J.: 1959, Ap.J. 129, 146.

Linsky, J. L., et al.: 1976, Ap.J. 203, 509.

Mochnacki, S. W., and Zirin, H.: 1980, Ap.J. 239, L27.

Neidig, D.: 1982, submitted to Ap.J.

Severny, A. B.: 1958a, Izv. Krym. Astrofiz. Obs. 19, 71.

Svestka, Z.: 1965, Adv. Astron. Astrophys. 3, 119.

Svestka, Z.: 1976, Solar Flares, Reidel.

van Hoven, G.: 1981, in E. R. Priest (ed.) Solar Magnetohydrodynamics, Gordon and Breach, London.

Zirin, H., and Neidig, D.: 1981, Ap.J. 248, L45.

Zirin, H.: 1978, Solar Phys. 58, 95.

Zirin, H.: 1980, Ap.J. 235, 618.

DISCUSSION

HIEI: Can we identify the loop prominence systems which connect all the Hα bright regions of two-ribbon flares?

ZIRIN: In principle, yes, but complete identification is limited by signal-to-noise ratio. On Sept. 10, 1974, we see loops connecting almost everything.

OBSERVATION OF THE FLARE OF 12 JUNE 1982 BY NORIKURA CORONAGRAPH AND HINOTORI

E. Hiei, T. Okamoto, and K. Tanaka
Tokyo Astronomical Observatory, University of Tokyo,
Mitaka, Tokyo, 181, Japan

ABSTRACT

Flare activity was observed near the limb with two coronagraphs at the Norikura Solar Observatory and the Soft X-ray Crystal Spectrometer (SOX) aboard HINOTORI. A prominence activation occurred and then Hα brightenings were seen on the disk near the prominence. The prominence became very bright and its electron density increased to $10^{12.8}$ cm^{-3} in 1/2 hour. Loop prominence systems appeared above the Hα brightenings about half an hour after the onset of the flare, and were observed in the coronal lines CaXV 5694Å, FeXIV 5303Å, and FeX 6374Å. Shifted and asymmetric profiles of the emission line of 5303Å were sometimes observed, and turbulent phenomena occurred even in the thermal phase. The energy release site of the flare at the onset would be lower than 20 000 km above the solar limb.

1. INTRODUCTION

A limb flare provides information on the height structure of a flare and activity in the corona, which will be observed with a coronagraph. Moreover, we can see Hα brightenings due to different causes such as a) the chromospheric brightening due to energetic particles/heat conduction /soft X-ray irradiation, b) loop prominence systems due to cooling of high temperature plasma, and c) brightening "in situ" of a prominence.

Original energy release site of a flare would be in the coronal region and the limb observation with a Hard X-ray Imaging Telescope and an optical telescope is important for determining the energy release site. However, there were no observations with the instruments aboard HINOTORI except the thermal phase in this limb event.

2. OBSERVATIONS

Active phenomena were observed in the east limb from the morning on 12 June with two coronagraphs at the Norikura Solar Observatory. A prominence became suddenly bright and soon chromospheric flare brightenings

Solar Physics **86** (1983) 185–191. 0038–0938/83/0861–0185 $ 01.05.

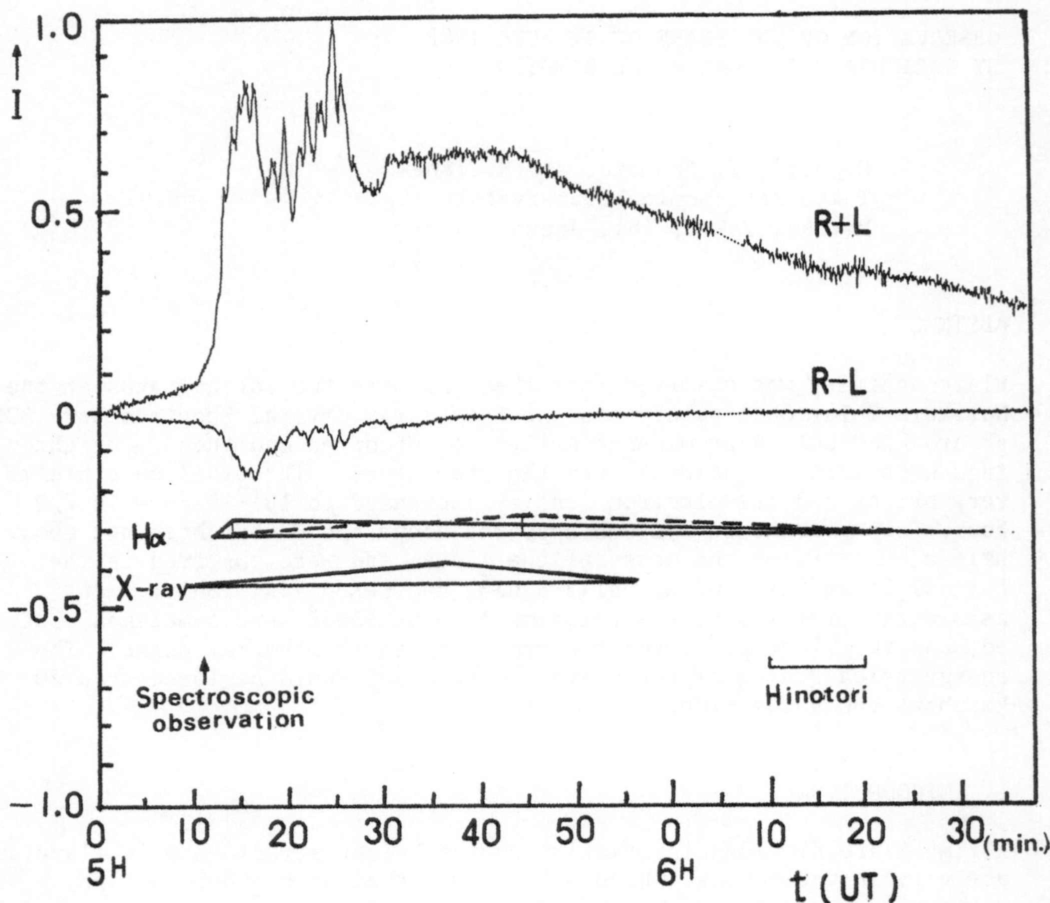

Figure 1. Time profile of radio burst at 9400 MHz observed at Toyokawa. The ordinate is relative intensity. Hα flare, X-ray flare, the start of the spectroscopic observation, and observed time made with HINOTORI are shown.

were seen on the disk near the limb (Kurokawa, 1982). According to "Solar Geophysical Data", two Hα flares occurred at N10E83 from 0512 UT to about 0633 UT; one (sn) reached maximum at 0514 UT and the other (1b) at 0544 UT. The X-ray flare (GOES observation) began at 0509 UT, reaching maximum at 0536 UT and ending at 0556 UT, and was classified as X 3.6. About half an hour after the onset of the flare, loop prominence systems appeared above the chromospheric flare region.

The limb event was spectroscopically observed with a 25cm aperture coronagraph in the spectral region between 3600Å and 6600Å. An achromatic image of the Sun, 81mm in diameter, is formed on a slit of a spectrograph, the linear dispersion of which is 0.5Åmm^{-1}at the 2nd order (Hiei, 1982). The spectroscopic observation was made from 0511 UT to 0740 UT, and the continuum, Balmer lines, metallic lines, and coronal lines CaXV 5694Å, FeXIV 5303Å, and FeX 6374 were observed.

Direct images of the coronal event associated with the limb flare were also taken with the 10 cm aperture coronagraph by using interference filters for 5303Å, 6347Å, D_3 and Hα. The half widths of the filters are 3.28Å, 3.86Å, 3.63Å, 3.0Å, respectively, and solar images (16mm in diameter) were taken with a Nikon motor-drive camera on Kodak 2415 film.

The soft X-ray crystal spectrometer (SOX) aboard HINOTORI (Tanaka et al. 1982) also observed the flare between 0611 UT to 0621 UT.

3. Hα LIMB FLARE

Figure 2 shows part of a spectrum of the limb flare, which was a prominence before the flare onset. The prominence slowly expanded on a time scale of 3 hours, and suddenly became bright about 0510 UT. The tilted emission lines in Figure 2, which are often seen in a limb flare, show a rotating motion as a whole.

The electron density was determined from widths of the Balmer lines.(Suemoto and Hiei, 1959). The derived electron density is $10^{12.2}$ cm^{-3} at 0525 UT and $10^{12.8} cm^{-3}$ at 0534 UT. Time variation of the continua at 3640Å and 3800Å is shown in Figure 3. Balmer jump and emissivity per unit volume of the Balmer continuum depend on the electron density and temperature (Jefferies and Orrall, 1961). The emissivity of the Balmer continuum in the flare loop can be derived from the observed intensity if the length of the loop along the line of sight is known. The flare loop, however, can be composed of filamentary structure and the filling factor is not known. We therefore use the electron density derived from the widths of higher Balmer lines. The derived electron density of $10^{12.8}$ cm^{-3} and intensity ratio of Balmer continuum to the continuum in the Paschen region at 0534 UT give two temperatures; one is 7600 K and the other 11 000 K. The higher temperature, however, is not accepted because it is too high for explaining the continuum in the Paschen region,which is partly due to H^- emission. The emissivity is inferred from the given temperature and electron density, and thus the effective length of the loop along the line of sight is turned out to be about 500 km. The temperature of the loop is almost constant even though the Balmer jump changes very much as shown in Figure 3. The increased electron density could be due to a contraction of the flare plasma.

The coronal lines were no longer seen at the flare loop, and its temperature thus is inferred not to increase up to several million degrees. The flare loop became fainter after 0534 UT, which would be due to a loss of material onto the chromosphere.

4. LOOP PROMINENCE SYSTEMS

The loop prominence systems appeared in the coronal region above the Hα bright points, which were observed at the Hida Observatory. The up-flow material would be expected from some of the bright points.

The direct image of the green line observed with the 10 cm coronagraph showed the loop structures seen from 0545 UT. The loop prominence

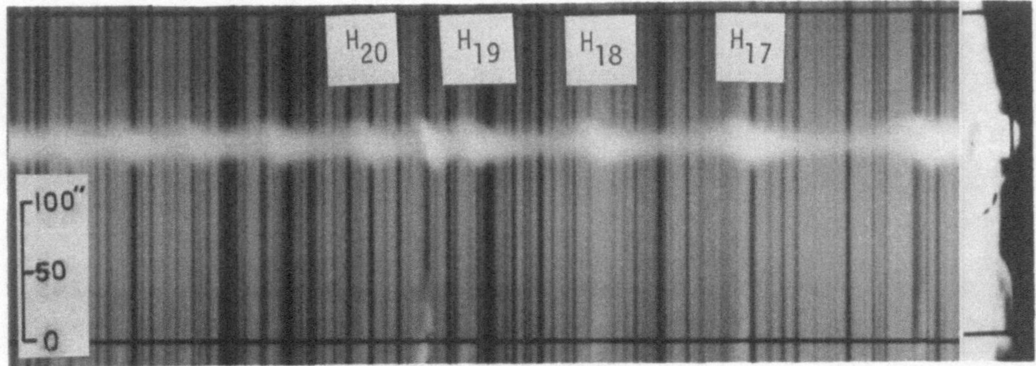

Figure 2. A spectrum of the limb flare taken at 05h18m40s.

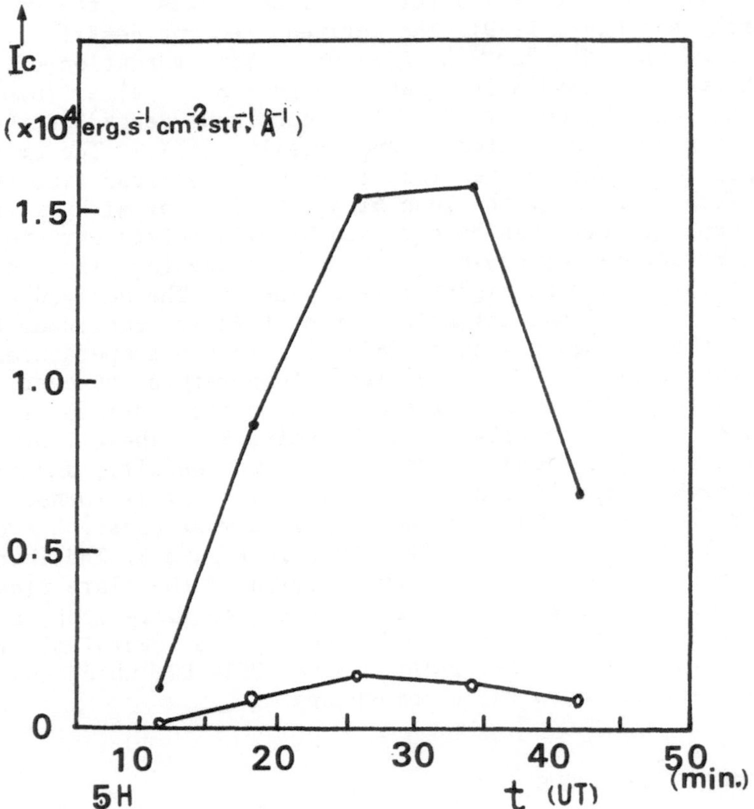

Figure 3. Time variation of the continua at 3640Å (dots) and at 3800Å (circles). The ordinate is intensity of the continuum. Intensity ratio of the dot to the circle gives a value of Balmer jump.

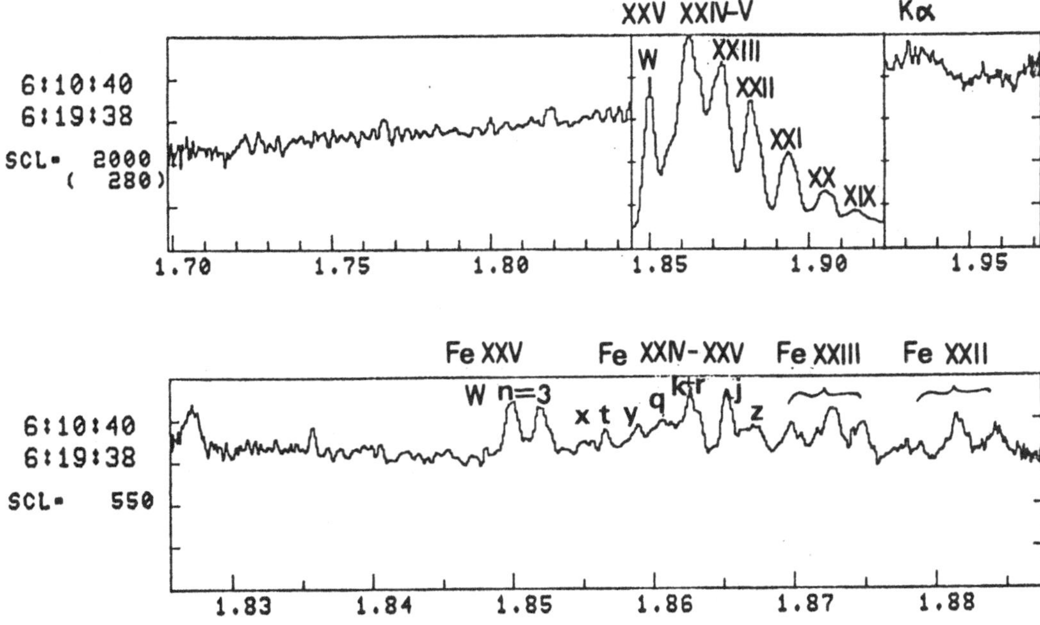

Figure 4. Spectra of X-ray region observed with SOX instrument aboard HINOTORI.

systems in the lower level than 20 000 km from the limb were masked by the occulting disk of the coronagraph and not seen, but the slit jaw pictures of the spectrograph of the 25 cm coronagraph showed the whole features of Hα loops . The Hα loops appeared at 0545 UT, and rising up to the loop was not observed. The coronal lines appeared from 0538 UT.

Three possible cases of flare loops at the onset can be suggested: (i) all flare loops are low, say, below 7000 km; (ii) all flare loops are high and their temperature is too high to be seen in coronal lines originating at several$\times10^6$ K, (iii) the loops could be composed of a mixture of type (i) and (ii). The loop prominence systems appearing half an hour after the onset of the flare can be explained as an expansion of the low level loops in case (i), or by the cooling of the hot plasma in the high level loops in cases (ii) and (iii).

The direct images of green line and Hα line in the earlier phase showed that no appreciable change in the structures in the coronal region above 20 000 km from the limb could be seen. The observation suggests that the energy release site at the onset would be lower than 20 000 km from the solar limb, if the flare loop is vertical to the Hα bright points, and therefore the choice of case (i) would be preferred.

Spectra of the coronal line 5303Å taken in the later phase show a wider profile, which is explained by a turbulent velocity of 30-40 km s^{-1},

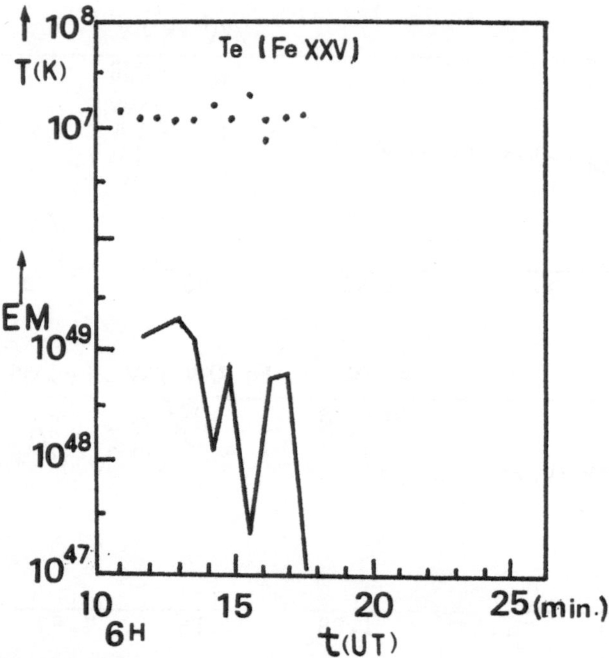

Figure 5. Reduced electron temperature and emission measure from SOX data.

and sometimes a shifted profile, which corresponds to a velocity of about 80 km s^{-1}. The $2x10^6$ K plasma of the flare loop is thus sometimes in a turbulent state even in the thermal phase.

Figure 4 shows X-ray spectra observed with the SOX instrument aboard HINOTORI and Figure 5 time change of the derived electron temperature and emission measure. The emission measure of the $2x10^6$ K plasma, derived from intensity of 5303Å emission line (Jordan, 1966), turns out to be $10^{47.2}cm^{-3}$, smaller by one order of magnitude than the emission measure of the $12x10^6$ K plasma derived from the SOX observation. The result is in consistent with the emission measures in the thermal phase derived by Widing and Spicer (1980).

If, we assume that the flare plasma is of loop structure and a pressure balance is hold between $12x10^6$ K plasma and $2x10^6$ K plasma, then the diameter of hot plasma loop should be larger by 30 times than that of $2x10^6$ K plasma, which means that hot plasma has a capability of making 10 loops of $2x10^6$ K plasma. If part of hot plasma happens to contract to 1/3 of its original diameter of the loop, the density becomes larger by 10 times and cooling times shorter by about 1/100, and 2x106 K plasma and H_α line will be seen.

Faint prominences above the flaring region were seen in the earlier phase, but gradually disappeared when the loop prominence systems began to appear. Presumably the temperature of the coronal region surrounding the prominences became too high for the prominences to maintain H_α line.

ACKNOWLEDGEMENTS

We thank Prof. Moriyama, Dr. T. Watanabe and Mr. K. Akita for providing the SOX data and reducing them, and Mr. M. Irie for supplying the observed data taken with the 10 cm coronagraph.

References

Jefferies, J.T. and Orrall, F.Q.: 1961, Astrophys.J. 134, 747.
Jordan, C. : 1966, Monthly Notices Roy. Astron. Soc. 132, 515.
Hiei, E. : 1982, Solar Phys. 80, 113.
Kurokawa, H. : 1982, in this proceeding.
Suemoto, Z. and Hiei, E. : 1959, Publ. Astron. Soc. Japan 11, 185.
Tanaka, K., Watanabe, T., Nishi, K., and Akita, K.: Astrophys. J. 254, L59.
Widing, K.G. and Spicer, D.S. : 1980, Astrophys. J. 242, 1243.

DISCUSSION

DE JAGER: Very remarkable are the inclined Balmer lines in the spectrum obtained in the early part of the flare. Have these observations been interpreted in terms of a dynamical flare model?

HIEI: The flare spectrum at the initial phase often shows the inclined Balmer and metallic lines. We collect these data and try to interpret the dynamical phenomena of the flare at the onset.

TANDBERG-HANSSEN: In the decay phase you found turbulent velocity. How was this measured?

HIEI: It was measured from the line width of Fe XIV 5303 Å, under an assumption of $T = 2 \times 10^6$ K.

PRIEST: The picture that arose from Skylab observations is of an arcade of hot loops above cool Hα loops during the main phase. The model of Cargill relates the temperature of hot loops and rise speed. Have you calculated these quantities for your event?

HIEI: If the loop prominence system, which appeared about 30 min. later than the onset of the flare, rose up from the surface, then the upward velocity of the loop were to be 10 km/s at the initial phase. We have not compared the heights of the coronal loops at 5303 Å and Hα in this event yet, but I also think that high temperature plasma exists at higher level than low temperature plasma does, as was shown in the 21 Jan. 1974 event of the Skylab observations.

OBSERVATIONS OF A COMPACT SOLAR FLARE ON 1981 SEPTEMBER 7 IN Hα, X-RAY, AND MICROWAVE RADIATIONS

(Extended Abstract*)

MITSUO KANNO, HIROKI KUROKAWA, and THE HINOTORI GROUP
Kwasan and Hida Observatories, University of Kyoto,
Kamitakara, Gifu Prefecture 506-13, Japan

We observed the 1981 September 7 flare (importance 1N and M2.0) with the Hα Lyot filter installed at the 60 cm domeless solar telescope at the Hida Observatory. Hard and soft X-ray observations of the flare were made by the HINOTORI satellite. We also have the microwave data of the flare from Nobeyama (17 GHz), Toyokawa (9.4, 3.75, and 2.0 GHz), and Nagoya (35 GHz).

We have studied the Hα filtergrams of this flare referring to the X-ray and microwave data to find the morphological and evolutionary characteristics of the Hα flare region. We estimated the temporal variations of the temperature and emission measure of the thermal X-ray flare plasma from the data of the soft X-ray monitor between 2 and 12 keV (FLM) aboard HINOTORI. This paper presents the results of these investigations which are used for inferring the physical conditions of the flare plasma in the initial phase. A preliminary report of the results is given in Kurokawa and Kanno (1982, Proc. Hinotori Symp. on Solar Flares, p. 199).

The important features of this flare are: (1) An abrupt Hα brightening occurs initially in a pair of Hα kernels with opposite magnetic polarity. The Hα kernels consist of several bright points of about 1" size especially at the early stage of the flare. (2) The Hα brightening at these bright points is closely associated with slight enhancements of hard X-rays and microwaves prior to the main bursts. (3) The Hα intensity at each bright point decreases as rapidly as a typical spike of hard X-ray and microwave bursts. It seems that the Hα brightening at these bright points is the chromospheric manifestation of an "elementary flare burst" in the overlying coronal flare plasma. We conclude that the Hα brightening at the initial phase is caused primarily by precipitation of nonthermal electrons into the chromosphere which also produce impulsive bursts of hard X-ray and microwave.

---

*Paper to be submitted to Publ. Astron. Soc. Japan.

Solar Physics **86** (1983) 193–194. 0038–0938/83/0861–0193 $ 00.30.

DISCUSSION

DE JAGER: Your observations seem to indicate that the Hα kernel brighttenings start before the X-ray flux increases. This observation would not support the canonical scenario that the Hα brightenings would be caused by a thick-target process, i.e., bombardment of the low chromosphere by energetic particles from higher regions.

KANNO: The position of Hα peaks in the time variation curves may be affected in two ways: First, the time resolution of our Hα observations is about 10 s, so that the time of Hα peaks contains an uncertainty of ± 5 s. Second, the observed Hα intensities may be affected by seeing conditions at this observatory, causing another uncertainty in the position of Hα peaks of ± 10 s. Therefore, I believe that the times of the Hα brightening and an 'elementary impulsive burst' during slight enhancement of the hard X-ray and microwave emissions are in good agreement.

BRUECKNER: Hα line brightenings as flare precursors must not necessarily be interpreted as caused by electron beams, which cannot be seen as X-ray emission. More likely, the Hα brightenings represent a heating of the plasma by magnetic reconnection preceding the catastrophic energy release during the main flare phase. Similar observations have been made in transition zone lines from Skylab.

KANNO: The Hα line brightenings mentioned in this paper are not those of flare precursors. They occurred abruptly at bright points (∿ 1") in the Hα kernels (∿ 10"). On the other hand, Hα line intensities at these bright points were already gradually rising during a period prior to the abrupt Hα line brightening. This gradual Hα rise should be considered as a flare precursor in Hα which can be explained by heating of the chromosphere due to heat conduction from coronal flare plasmas. The coronal flare plasmas are heated by an energy release (not necessarily by magnetic reconnections) preceding the catastrophic energy release during the main flare phase.

It is clearly seen from a comparison of the time profiles between hard X-ray and microwave at 17 GHz, that the abrupt Hα line brightenings mentioned in this paper correspond to an 'elementary impulsive burst' in the coronal flare plasmas, though there appears to be a discrepancy in timing ($\lesssim$ 10 s) between Hα and hard X-rays (see my reply to de Jager's comment). Further evidence for our conclusion comes from the fact that the abrupt variation of the Hα line brightenings has a time scale as short as that of a typical spike of hard X-ray and microwave bursts.

In transition zone lines we also frequently observe abrupt brightenings with a short time scale, in addition to gradual brightenings that are flare precursors.

THE HEIGHT OF Hα FLARE EMITTING REGION

(Extended Abstract*)

Hiroki Kurokawa
Kwasan and Hida Observatories, University of Kyoto,
Kamitakara, Gifu Prefecture 506-13, Japan

Direct and clear observational evidence which can determine the height of an Hα flare emitting region is demonstrated. There have been many attempts to study the vertical structure of the flare with observations of the so-called limb flare. In many cases, however, they can be identified as either flare-associated active prominences, surges or post flare loops. In this paper we show that the top of an Hα flare emitting region is much lower than that of the normal Hα chromosphere, and that Hα flares on the solar limb can not be correctly observed in Hα line center.

The 12 June 1982 flare, classified as X 3.7 in the X ray intensity, occurred very close to the solar limb. This rare chance enabled us to determine the real height of the Hα flare emitting region. Observations were made with the 60 cm Domeless Solar Telescope at Hida Observatory of the University of Kyoto.

The study of the Hα monochromatic images and Hα line profiles of the flare shows the following results. (1) Many bright points in the flare can be seen successively brightening in Hα ± 2.4 Å images, and their abrupt brightening and dimming are closely associated with the impulsive microwave spikes at 17 GHz observed at Nobeyama Solar Radio Observatory. (2) While these flare points very close to the solar limb are brightest in the Hα ± 1.2 Å images, they are significantly obscured by opaque chromospheric structures along the line of sight in Hα line center (Figure 1). Especially the regions A and B in Figure 1(b) are just on the solar limb and can be hardly seen in Hα line center. (3) The top of Hα flare emitting region is about 1800 km high above the photosphere in the light of Hα ± 1.2 Å. (4) The brightest limb feature C in Hα line center (Figure 1(a)), or so-called limb flare, can be identified with a flaring-up phenomenon in an active prominence.

With these results we conclude that the main heated part of an Hα flare is confined to the low chromosphere and much lower than the surrounding chromospheric structures. This conclusion is compatible

* Paper to be submitted to Solar Physics.

Solar Physics **86** (1983) 195–196. 0038–0938/83/0861–0195 $ 00.30.

with evaporation flare models which evacuate the upper chromosphere (Hirayama, 1974; Acton et al., 1982).

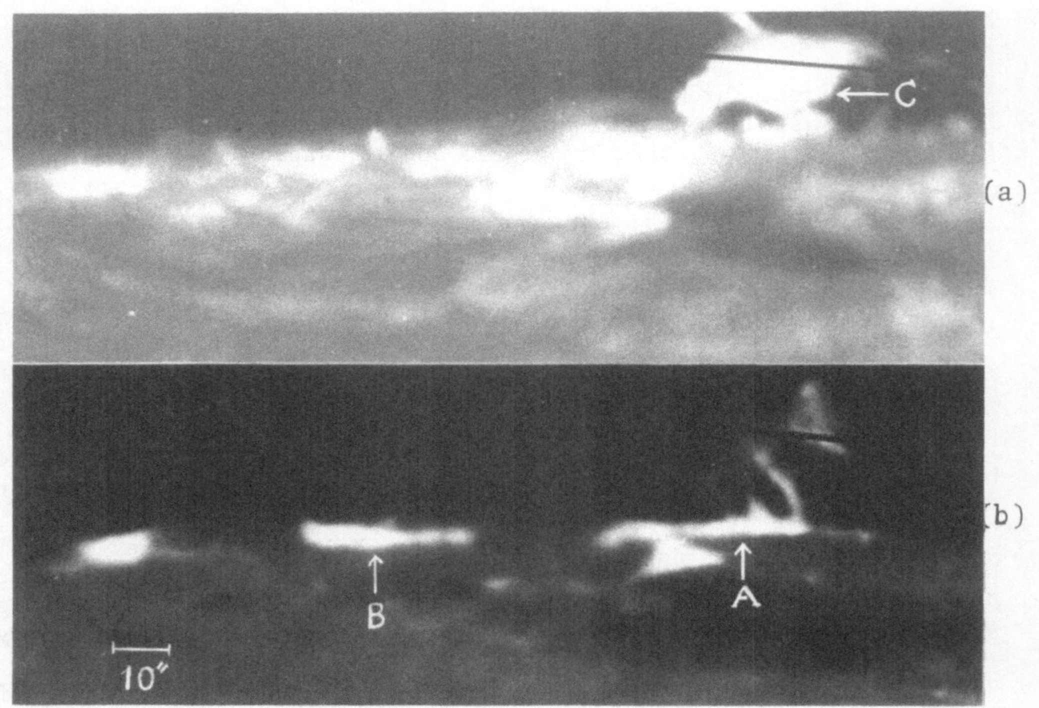

Fig. 1. The 12 June 1982 Hα flare just on the solar limb or very close to it. (a) upper: at 052943 UT in Hα ± 0.0 Å. (b) lower: at 053116 UT in Hα + 1.2 Å. The flare regions very close to the solar limb are very bright in Hα + 1.2 Å, but significantly obscured by the opaque chromospheric structures along the line of sight in Hα ± 0.0 Å. Especially, the regions A and B, just on the solar limb, can be hardly seen in Hα line center.

References

Acton, L. W., Canfield, R. C., Gunkler, T. D., Hudson, H. S., Kiplinger, A. L., and Leinbacher, J. W.: 1982, Astrophys. J. 263, 409.
Hirayama, T.: 1974, Solar Phys. 34, 323.

POSITIONAL MEASUREMENTS ON THE ERUPTIVE LOOP PROMINENCE OF 27 APRIL 1981 AND COMPARISON WITH THE X-RAY SOURCES

Li Su-Cuan
Yunnan Astronomical Observatory, Academia Sinica, Kunming, China

and

Cao Tian-Jun
Purple Mountain Observatory, Academia Sinica, Nanjing, China

ABSTRACT

Using the Hα observational data from Yunnan Observatory, we have made position measurements on the eruptive loop prominence of 27 April 1981, and have compared the results with the positions of X-ray sources obtained by the hard X-ray telescope (SXT) on board the HINOTORI satellite. From the results of measurement and comparison, it is suggested that 1) The two mounds A and C at 0830 UT are extensions of two ribbons in the flare near the limb, which started before 0758 UT. 2) The central positions of two X-ray sources at 0756 UT are just situated at the top of the mound A and the mound C, respectively. The Hα foot-point corresponding to the main source of X-rays was behind the solar limb. The second source of X-rays corresponds to C_1 and C_2. 3) The X-ray sources were probably located near the footpoints of loops.

1. INTRODUCTION

EUV and X-ray observations from space have indicated the appearance of high temperature (up to $3x10^7$ K) kernels during the impulsive phase or the start of the flash phase of a solar flare. They are often situated near the tops of loop structures. The EUV and X-ray emission may spread along the loop from the kernel and an X-ray arcade appears. The emission can also jump to other nearby loops from the initial bright one (Kahler et al., 1975; Brueckner, 1976; Cheng et al., 1975; Vorpahl et al., 1976; Thomas, 1975).

The different X-ray morphology of flares may provide certain observational evidence for different flare models. Likewise, the comparison between the positions of Hα initial bright knots of a flare and of the X-ray sources could give important information for the study of theoretical flare models. But this comparison in positions must be based on accurate positional measurements in Hα.

For this purpose, comparison between the Hα and X-ray positions for the flares occurring near the solar limb is more convenient than that of disc flares, because this comparison may not only generate

Solar Physics **86** (1983) 197–202. 0038–0938/83/0861–0197 $ 00.90.

information on the azimuth of the structure in question, but also the height which is very important for comparison with flare models. These cannot be directly determined for flares on the disk.

For the eruptive loop prominence that occurred on the west limb on 27 April 1981, the most complete Hα observational data were obtained at three observatories in China. At Beijing Observatory and Yunnan Observatory, whole disk images were photographed. These data provide the observational material for this report.

2. DATA AND MEASUREMENTS

The accurate positional information reported in this paper was made by using the Hα observational data obtained at Yunnan Observatory. From the results of the measurement, Hα morphology and precise position (including the height and position of all structures in question) and the time development of the key Hα structures could be determined.

This active region, SESC 3049, was located at about 150L, 15N. Central meridian distance was about 85W at 0044 UT on 27 April, so it was near the west limb during the occurrence of the flare. The first Hα photograph was taken at 0758 UT when some bright structures were appearing on the western limb. After this time, photographs could not be taken until 0830 UT because of clouds. Observations were then carried out continuously from 0830 UT to about 1020 UT.

From the accurately measured positions for the corresponding structures on five frames taken at 0758, 0830, 0831, 0835 and 0845 UT, we may infer the development of these interesting structures and provide the exact position of the Hα flare material to compare with the X-ray source.

The measurements were made using a projected image. The diameter of the solar image was 15.55 mm and was amplifed twenty times on the projection screen. The reading precision on the X- and Y- axes was 0.001 mm and 0.01 mm respectively. The north-south direction was set on the X-axis, with the east-west direction parallel to the Y-axis. With this arrangement we can draw the profiles of interesting structures and measure them directly in terms of the values of X and Y at the center of the disc. At the same time, r and ϕ values of corresponding points are obtained from the X, Y values.

Since the reading error on the X-axis is very small, it should be much smaller than the empirical error, which can be calculated from the measurements. For r, the measured error is about 0.024 mm, corresponding to 2.15×10^3 km, and for ϕ, it is about 0°.12.

3. RESULTS AND DISCUSSION

(1) From the results of the positional measurements in Figure 1, we find that the pre-existing mound A on the west limb was not bright yet at 0758 UT. Since the Hα emission was very weak, it could possibly have been an extension of a plage in the active region. Because

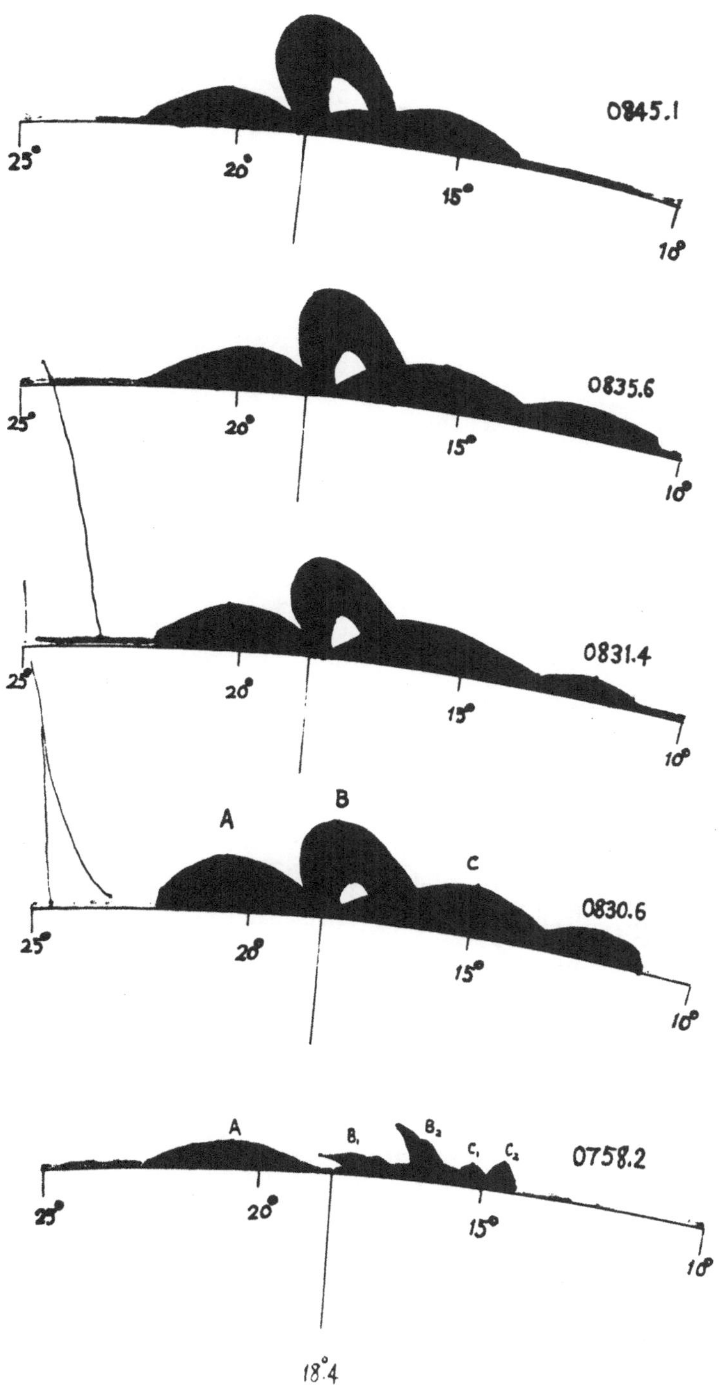
0845.1
25°
20°
15°
10°
0835.6
25°
20°
15°
10°
0831.4
25°
20°
15°
10°
A
B
C
0830.6
25°
20°
15°
10°
A
B₁
B₂
C₁
C₂
0758.2
25°
20°
15°
10°
18.°4

Figure 1

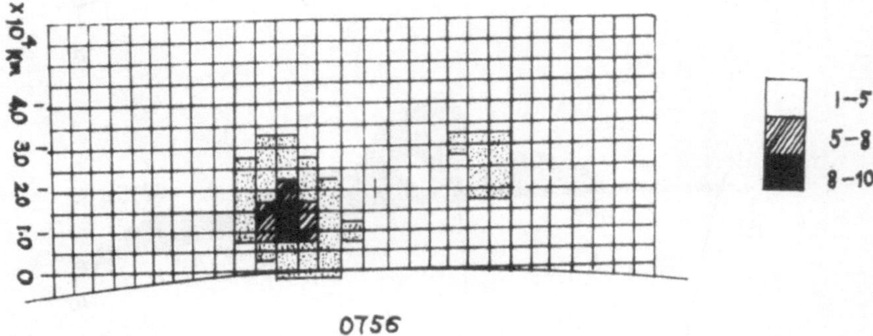

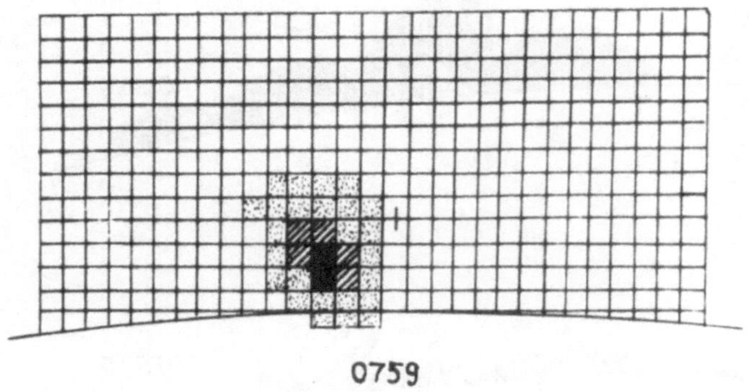

Figure 2

the structures B_1, B_2, and C appeared bright in emission, the flare had just started at this time.

We suggest that the two mounds A and C are extensions of two ribbons in the flare limb. This is consistent with the change in brightness of the two mounds at 0830 UT, half an hour after the beginning, and the morphological variations after that time. It is possible that the mound C developed from the structures C_1 and C_2 at 0758 UT and the mound A was formed by a brightening of the pre-existing plage.

(2) In Figure 1, we indicate the positional angle relative to the solar equator. It is calculated from the azimuth of the rotational axis at that time, and the position angle of other structures can thus be determined. The positions of the X-ray sources obtained at 0756 and 0759 UT on HINOTORI (Tsuneta et al., 1982; Takakura et al., 1983) are given in Figure 2 at the same scale. For convenience of comparison, the position of the calibration point $18\overset{\circ}{.}4$ is marked in these two figures. If we superpose one on the other, the position of the X-ray sources relative to the Hα features can be determined.

The central positions of two X-ray sources at 0756 are just situated at the top of the mound A and above the mound C respectively.

(3) As mentioned above, from the Hα photograph taken at 0758 UT it is found that the first bright knots of the flare had occurred but that the mound A had not brightened yet. However at this time, the X-ray source was situated just above it. We suppose that the Hα foot point corresponding to the main source of X-rays was behind the solar edge. The brightening of mound A at 0830 UT was possibly the upward extension of the foot point. The second source of X-rays at 0756 UT corresponds to C_1, C_2.

(4) From the positional comparison with X-ray sources at 0756 UT, we infer that the X-ray sources were probably located near the footpoints of loops. It is possible that the X-ray emission was caused by high energy electrons descending from a higher atmospheric layer rather than by thermal radiation from an extremely high-temperature plasma.

REFERENCES

Brueckner, G. E.: 1976, Phil. Trans. Roy. Soc. London A 281,443.

Cheng, C. C and Widing, K. G.: 1975, Ap. J. 201, 735.

Kahler, S. W., Krieger, A. S. and Vaiana, G. S.: 1975, Ap. J. (Lett.) 199, L57.

Takakura, T., Tsuneta, S., Ohki, K., Nitta, N., Makishima, K., Murakami, T., Ogawara, Y., Oda, M , and Miyamoto, S.: Ap. J. (Letters), submitted.

Thomas, R. J.: 1975, In S. R. Kane (ed.), I.A.U. Symp. No. 68, Solar Gamma X- and EUV radiation (Dordrecht, Reidel), p. 25.

Tsuneta, S., Ohki, K., Takakura, T., Nitta, N., Makishima, K., Murakami, T., Ogawara, Y., Oda, M and Kondo, I.: 1982, "Hinotori Symposium on Solar Flares" (ISAS, Tokyo), p. 130.

Vorpahl, J. A., Gibson, E. G., Landecker, P. B., Mckenzie, D. L. and Underwood, J. H.: 1976, Solar Physics 45, 199.

DISCUSSION

KIPLINGER: What is the absolute uncertainty in the angle around the limb for the hard X-ray image?

CAO: $\pm$ 0°.1 (Hα map).

TAKAKURA: $\pm$ 1° (X-ray map). It reduces to $\pm$ 17" at the limb in the tangential position.

V. EMISSION PROCESSES AND SOURCE STRUCTURE

Chairmen: T. Takakura and M. R. Kundu

SPATIAL CHARACTERISTICS OF MICROWAVE BURSTS

M. R. Kundu
Astronomy Program, University of Maryland,
College Park, MD 20742

ABSTRACT.

The spatial characteristics of microwave bursts are discussed in relation to impulsive and post-burst phases. The existence of two components - a gradual and another spiky - in the impulsive phase is discussed from the WSRT high time and spatial resolution observations. Using the WSRT data, evidence is presented for homologous flares at 6 cm, from the similarity of their spatial structure and their temporal evolution. Preflare changes in microwave active regions are presented and their interpretations in terms of newly emerging flux at the flare site are discussed. High spatial resolution observations of the structure of microwave flaring loops and their interpretation in terms of arcades of loops as the sites of primary energy release are presented. Theoretical interpretation of the confinement of microwave producing energetic electrons in the coronal part of loops is discussed. The relative timing of the peaks of impulsive hard X-ray and microwave burst is discussed. Possible diagnostics of impulsive phase onsets from cm-λ polarization data are presented, and the role of the emergence of new flux and of the current sheet formed between closed loops in producing impulsive energy release at centimeter wavelengrths are analyzed.

1. INTRODUCTION

In this paper, I shall discuss some spatial properties of microwave bursts, as obtained by the Very Large Array (VLA), the Westerbork Synthesis Radio Telescope (WSRT) and to some extent the Nobeyama 17 GHz interferometer. I shall start with a brief description of the distinctive spatial properties of the impulsive and post-burst phases of a microwave burst. Then I shall successively deal with high time and spatial resolution observations of the impulsive phase, homologous flares, preflare changes in active regions, two dimensional mapping of cm burst sources, and tests of flare models from two dimensional maps in total intensity and polarization, obtained with the VLA.

Solar Physics **86** (1983) 205–217. 0038–0938/83/0862–0205 $ 01.95.

2. IMPULSIVE AND POST-BURST PHASES

The microwave burst has been classified into three basic types: simple or impulsive burst, post-burst, and gradual rise and fall (GRF) bursts (Kundu, 1965; Kundu, 1980). The GRF's will not be discussed here, because our knowledge of them has not changed over the last several years. The spatial characteristics of the impulsive and post burst phases can be summarized as follows (Kundu, 1959; Alissandrakis and Kundu, 1978). At the time of impulsive phase maximum, the source is most compact (size $\lesssim$ 10"); it is strongly polarized and its brightness temperature generally exceeds 10^8 K. After the maximum (that is, in the post-burst phase), the burst core expands, often with a velocity < 30 km/s, up to a size of > 1′ arc; it is unpolarized and the brightness temperature is generally ~ 10^6 K. The bursts occur quite often near the neutral line of the magnetic field, as determined by the polarization maps of 6 cm active regions (Kundu et al., 1977). A typical burst (of intensity $\lesssim$ 10 sfu) is found to be polarized only in the impulsive phase and sometimes only one sense of polarization is observed over the entire extent of the burst source (Fig. 1). This suggests that, if the burst is associated with loop structures, the emission must be associated with one leg of the loop. Kundu and Vlahos (1979) qualitatively explained this result by considering an asymmetrical bipolar magnetic field where the two legs of the loop have dissimilar field strengths; in the post burst phase, the emission comes from the entire loop.

Kai et al. (1982) discussed the change of microwave source characteristics during the transition from the impulsive to the post-burst phase on a statistical basis, using a fairly large number of bursts (peak flux > 50 sfu) observed with the Nobeyama 17 GHz interferometer (spatial resolution ~ 35" arc and time resolution 0.8 sec). They found that the two different phases in the burst, that is, the impulsive phase and the post-burst phase, are distinguished not only by the morphological difference of flux time profiles, but also by the difference of brightness temperatures (10^7 - > 10^9 K vs 10^5 - 10^7 K), degree of circular polarization (0 - 50% vs 0 - 10%), and size (< 5 - 25" arc vs 10 - 70" arc). Almost invariably there is a sudden decrease of circular polarization at the transition between the two phases, typcially from 20 - 30% down to a few precent. The height of the impulsive source is lower than that of the post-burst source. In the post-burst phase, the source shows remarkable expansion and ascension with a velocity of the order of 5 km s^{-1}.

One of the intriguing problems about the impulsive phase is: where does the burst source occur in a flaring magnetic loop, i.e., whether it occurs near the legs of the loop or at the top. According to Kai et al. (1981) the lower height of the impulsive source at 17 GHz than that of the associated post-burst source suggests the source to be located near one or two legs of the loop, in agreement with the interpretation of Kundu and Vlahos (1979). On the other hand, there is strong observational evidence that the impulsive burst source is located in the upper part of the flaring loop (Marsh and Hurford, 1980; Kundu et al., 1982a).

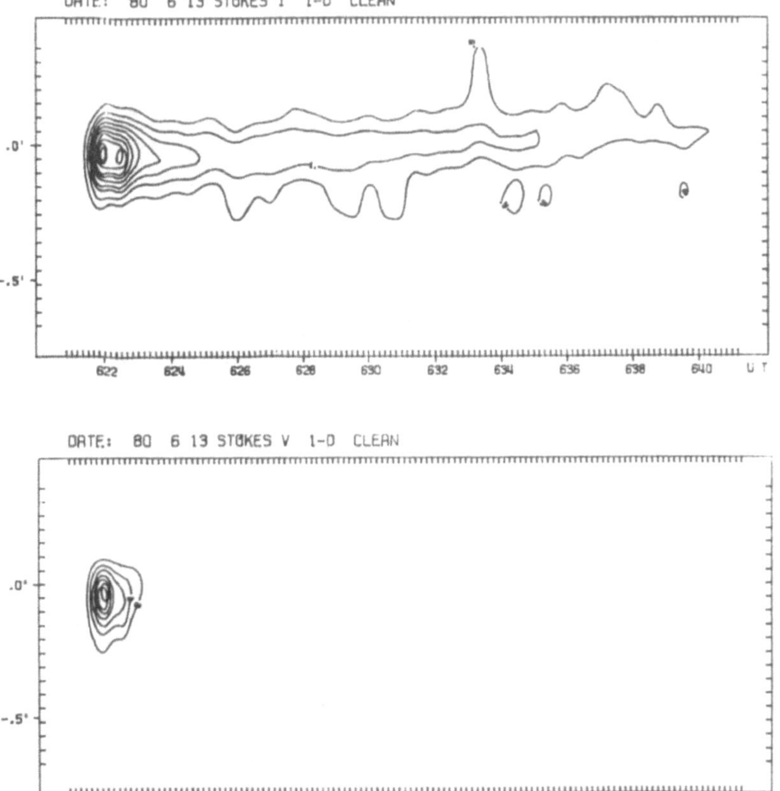

Fig. 1. One-dimensional brightness temperature as a function of position and time for a 6 cm burst observed with the Westerbork Synthesis Radio Telescope (WSRT) with 3" arc resolution. Top: total intensity (I); bottom: circular polarization (V). The contour labels are in arbitrary units of K arcsec. Note that the burst is polarized in the impulsive phase and only in one sense.

3. HIGH SPATIAL OBSERVATIONS WITH 0.1 s TIME RESOLUTION

Using the WSRT, Kattenberg and Allart (1982) studied several 6 cm bursts with up to 3" arc spatial resolution and 0.1 s time resoltuion. They find that often the impulsive phase of the burst consists of a gradual component and a spiky component which is highly variable in time. The spiky component lasts from 10 to 100 s, it is narrow (< 10") and bright ($T_b > 10^8$ K); the gradual component lasts longer (1 - 2 minutes), it is broader (> 15") and fainter ($T_b < 5 \times 10^7$ K). Often the spiky component is less strongly polarized than the gradual component. The gradual component is separated from the spiky component in general by 10" to 100". No real interpretation has been given for the spiky and gradual components of the impulsive phase, although Kattenberg and Allart suggested that an asymmetrical flaring loop or more than one loop may be involved in the production of microwave emission.

4. HOMOLOGOUS FLARES AT 6 cm

In the course of our observing with the WSRT during the SMY period, Alissandrakis and Kundu (1983) detected two sets of three homologous flares out of more than 60 bursts recorded at 6 cm with 3" resolution. One set of three homologous flares is shown in Fig. 2. As can be seen from this figure, the one-dimensional contour maps (I) for the three flares are rather similar, although the corresponding polarization maps (not shown here) are different. In Fig. 3, are drawn the position lines of the three burst sources, including the component sources on a Kitt Peak National Observatory (KPNO) magnetogram. It is clear that both component sources for each of the three flares on May 26 occurs near some neutral line and the three corresponding components occur close together. These observations provide, in our view, the first example of similar spatial structure and position of homologous flares at centimeter wavelengths.

5. PREFLARE CHANGES IN ACTIVE REGIONS

It has been known for many years that the flare build-up manifests itself at centimeter wavelengths (2-6 cm), in the form of increased intensity and increased polarization of the active region. With the availability of arc-second resolution using the VLA it has been possible to study the nature of this build-up from two-dimensional synthesized maps over short periods before the start of a flare. For a hard X-ray associated impulsive centimeter burst observed on June 25, 1980 (Kundu, Schmahl, and Velusamy, 1982a), we produced several 15-minute synthesized maps in total intensity (I) and polarization (V). Figure 4, shows the central 1.'1 x 1.'6 regions of 15 minute synthesis maps over the period 14:45-15:45 UT. As can be seen from these Figures, the region is very complex, consisting of numerous components, many of which are bipolar. These oppositely polarized components could be the footpoints of magnetic loops. These components have brightness temperatures of 6-9 x 10^6 K during the hour before the flare. The burst of 25 June, 1980, starting at 1550 UT, was located close to the neutral line of these oppositely polarized regions near B. The burst maximum is identified with a "+" and the burst extent averaged over the period 1551-1600 UT, is shown by the dotted contour in the last map.

There was a definite trend for the active region undergoing brightness and polarization changes. The central component B (see Fig. 4) intensified at 1515-1530 UT and increased in polarization slightly. In the last map (1530-1545) several new components appeared with polarization of 40-80%. However, the most remarkable feature is the change of the sense of polarization of component B; also, the component on the northern side of B greatly increased in polarized intensity, with polarization of 80-90%. This might imply the emergence of a flux of reverse polarity at coronal levels. (The

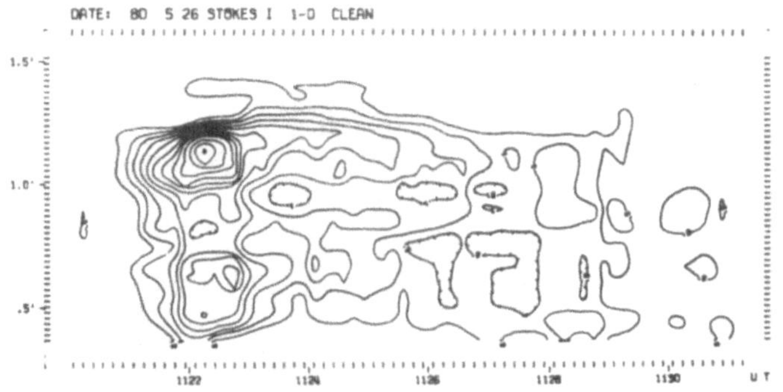

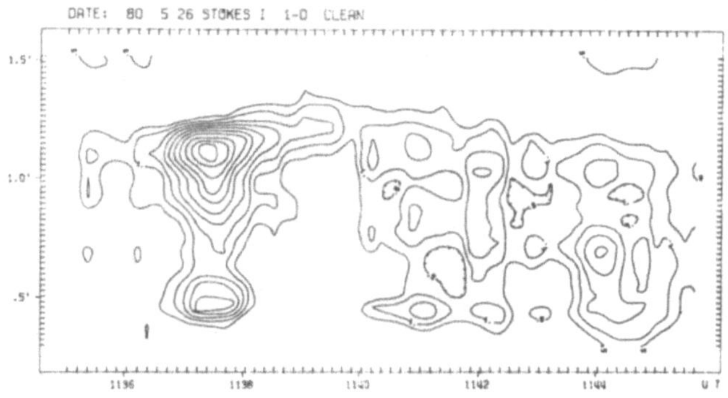

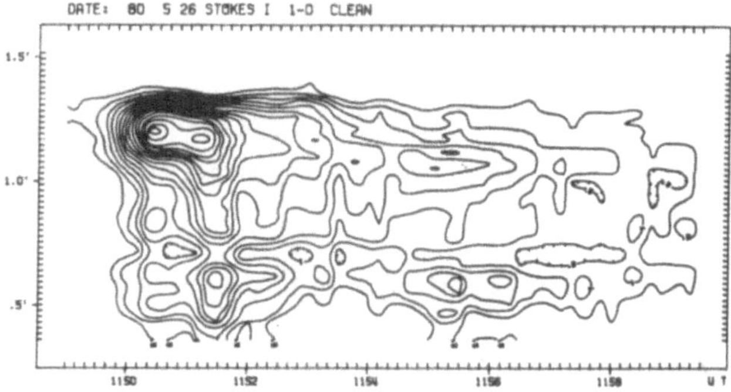

Fig. 2. One-dimensional brightness temperture as a function of position and time for a set of three homologous 6 cm bursts observed on May 26, 1980, using the WSRT with 3" arc resolution. Only the total intensity (I) contours (in arbitrary units of K arc sec) are shown here.

photospheric magnetograms show little or no change.) We believe that this may be caused by the expansion of a pre-existing flux tube in which twisting increases its coronal magnetic field; at the same time there must be some heating of the loop. Alternately one can conceive of a previously existing loop of opposite polarity which was not observable at 6 cm due to its weak magnetic field becoming observable due to a sudden onset of currents in the loop.

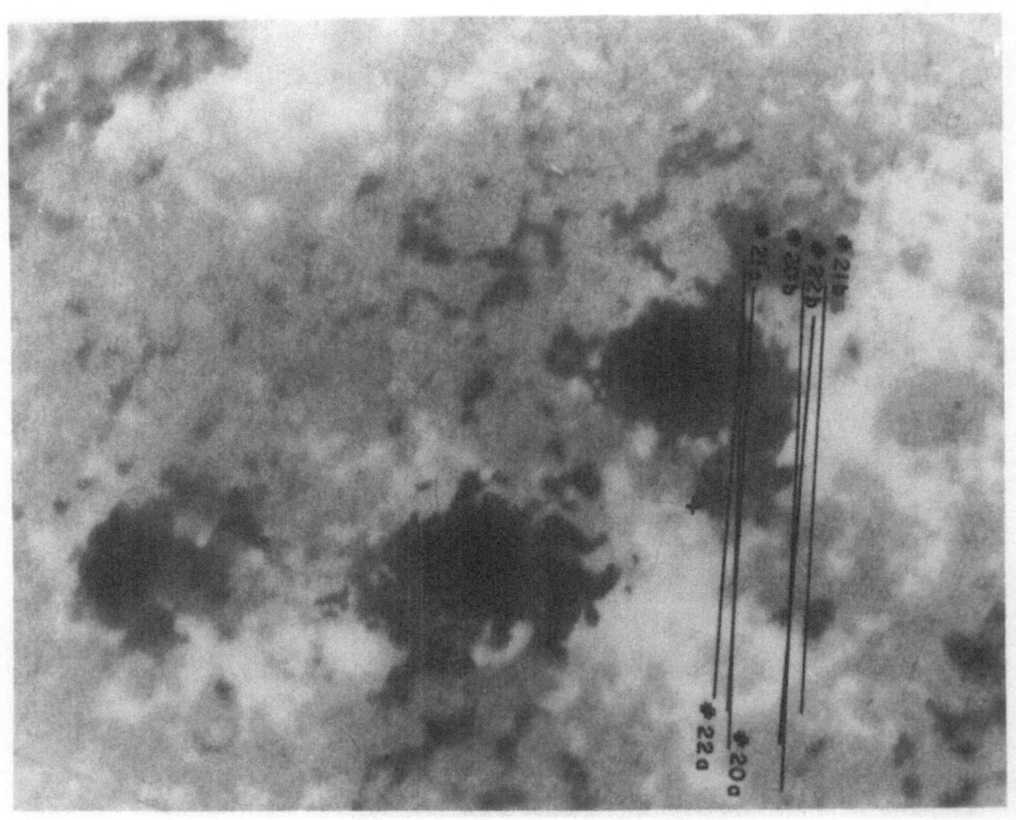

Fig. 3. One-dimensional positions of the 6 cm bursts of Fig. 2, drawn on the Kitt Peak magnetogram.

6. TWO-DIMENSIONAL MAPPING OF CENTIMETER BURST SOURCES

Marsh et al. (1980), using the VLA at 6 cm with up to 1" spatial resolution and 10 s time resolution, showed that the burst source is located between two Hα kernels, close to the magnetic neutral line, in agreement with the results of Alissandrakis and Kundu (1978). Marsh and Hurford (1980) also used the VLA at 2 cm and 1.3 cm to obtain images with 1" x 0".75 resolution showing a compact source (size ~ 2", $T_b \sim 5 \times 10^8$ K) correlated spatially between Hα kernels. The post-impulsive microwave emission came from a larger elongated source delineated by the magnetic field lines joining these kernels. They placed the energy release region near the top of the magnetic arch, with an upper limit of 2" in size.

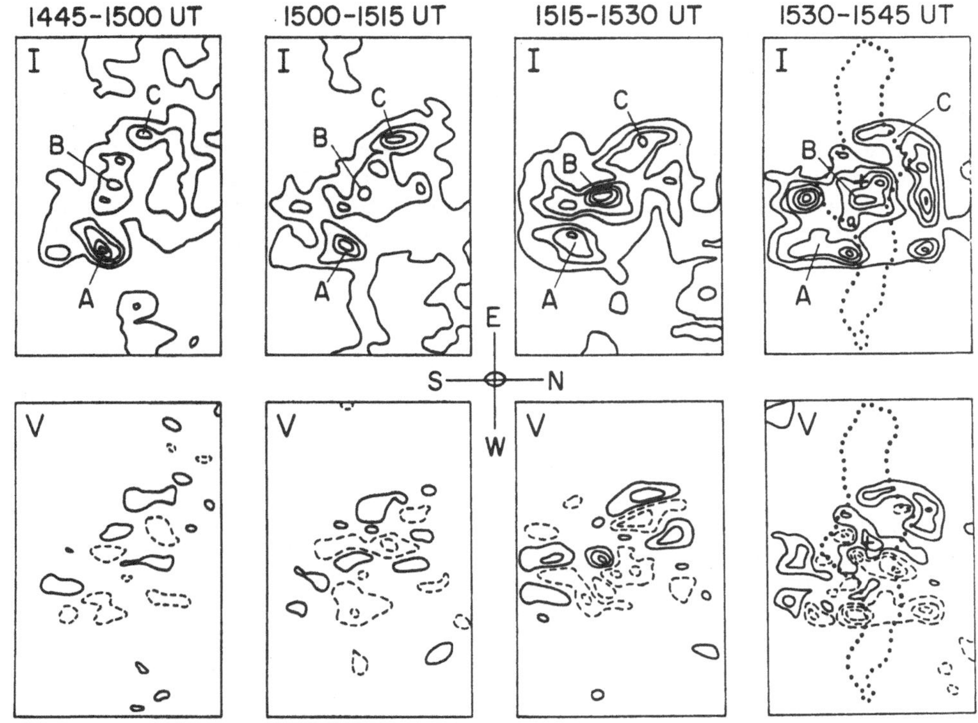

Fig. 4. Central 1:1 x 1:6 regions of pre-flare (15^m) maps of I and V, 15:00-15:45 UT, 25 June 1980. The beam size is 4" x 6". The first contour is 1×10^6 K and the contour interval is 2×10^6 K. Note the reversal of polarity in component "B" in the last map, close to the subsequent burst peak (marked by a +). The dotted outline shows the extent of the burst source in the impulsive phase, 1551-1600 UT.

7. TESTS OF FLARE MODELS

Using the VLA, Kundu and his collaborators studied several impulsive bursts at 6 cm and tried to interpret their results in terms of existing flare models. This is possible because the VLA data provide information on magnetic field structure near the sites of energy release.

(i) Arcades of loops as sites of primary energy release

As one can see from the samples of 10 second synthesized maps (Fig. 5), obtained with a resolution of 1" x 2", the 6 cm burst source associated with the impulsive phase of a flare observed on 25 June, 1980, consists of several bipolar component sources, the most intense of which is located near the center. Several components, including the strong central source, have almost the same finite extent in total intensity and polarization (~ 5" x 10"); the neutral line (the line of zero polarization at 6 cm) passes through the region of total intensity and divides it into two regions of opposite polarity across the shorter extent of the region. Most of the Hα emission occurs near the footpoints of the loops. The component sources appear in the form of arcades of loops, and the 6 cm emitting region occupies a substantial portion (5"-10") on both sides of the flaring loop. There are, however, cases where the two opposite polarities correspond to two distinct, well resolved total intensity peaks, with minimal emission near the center.

It is clear that the loop or arcade model of flares involves the release of energy in the coronal part of a magnetic loop. This energy release (possibly through magnetic reconection brought on by a tearing mode instability - Spicer, 1977) impulsively heats the plasma in the upper part of the loop. A nonthermal tail of high energy electrons (energies up to a few hundred keV) may also be produced at this time. The hot plasma is confined between a pair of conduction fronts which propagate down the legs of the loop with a velocity near the ion sound speed (Brown, Melrose, and Spicer, 1979; Vlahos and Papadopoulos, 1979). Electrons with velocities greater than approximately three times the electron thermal speed in this region, however, are not confined by the conduction fronts and escape to the lower part of the loop. Holman, Kundu and Papadopoulos (1982) showed that, when the electron gyrofrequency exceeds the plasma frequency, the escaping electrons are unstable to the generation of electrostatic plasma waves which scatter the particles in pitch angle to a nearly isotropic distribution. They showed that this scattering can enhance the microwave emission from the upper part of the loop because more electrons are trapped in the upper part of the loop, and the scattered electrons have a higher mean pitch angle. In a complimentary study, Petrosian (1982), on the other hand, computed the variation of gyrosynchrotron intensity and spectrum along a closed (semi-circular) magnetic loop. Using an isotropic particle distribution and a uniform (little variation from top to footpoints) magnetic field, he showed that microwave emission originated predominantly from the upper part of the loop and that a loop would appear larger in the optically thick regime than in the optically thin regime, consistent with the result that microwaves at high frequencies come from smaller volumes than those at lower frequencies.

The impulsive microwave and hard X-ray emissions from the June 25, 1980 flare have been studied in more detail by Holman, Kundu, and Dennis (1982). They have obtained time profiles for the four major bipolar regions contributing to the 6 cm emission from the flare and find that the light curves are consistent with the regions flaring in unison (to within the 10 sec time resolution of the VLA) rather than sequentially or in an uncorrelated manner. There is no indication of any change in the magnetic field structure in the emitting regions to within the 2" spatial resolution of the observations. The maximum 6

(a)

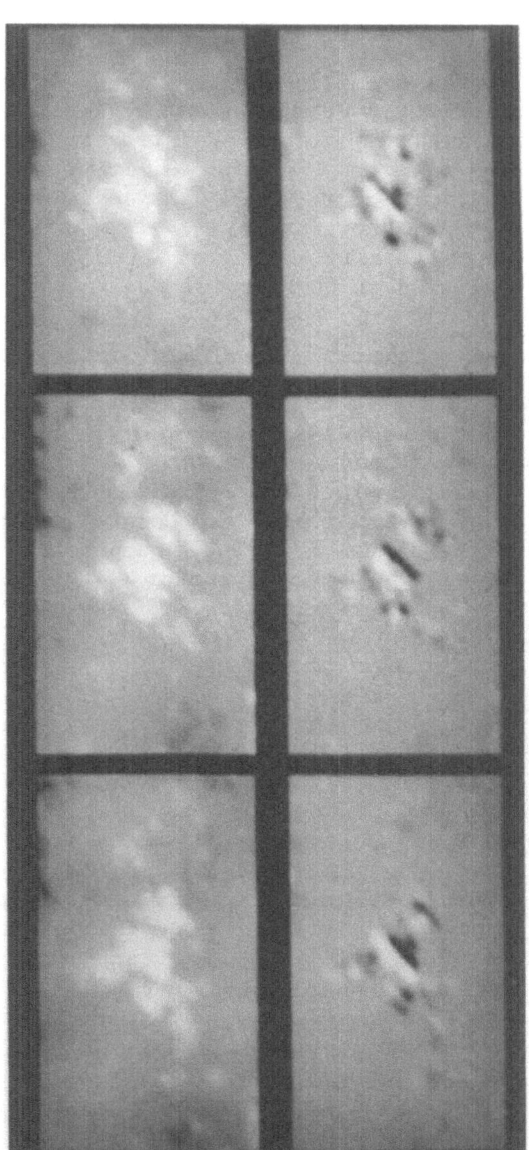

Fig. 5. (a) A sequence of three 10 s snapshot maps in I (left) and V (right) produced during the impulsive phase of a burst observed on June 25, 1980, with the Very Large Array. The synthesized beam is 1"x2". Note the strong bipolar component in the midst of other bipolar components, indicating arcades of loops. (b) The arcade of loops for the main source is illustrated in the set of diagrams, from left to right, total intensity (I), circular polarization (V) and schematic diagram of the arcade.

(b)

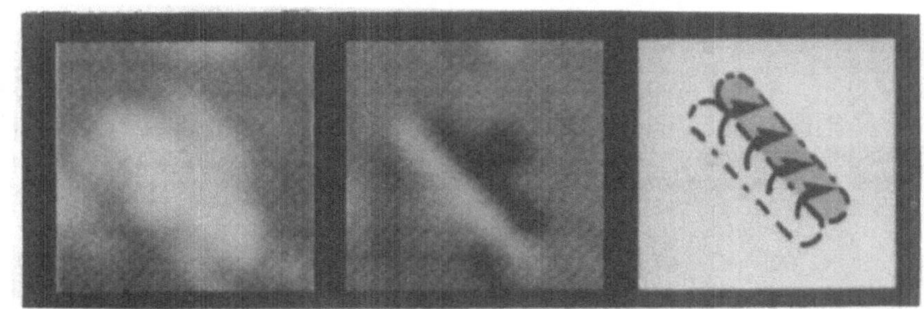

cm flux occurs 1.4 min after the maximum in the integrated 28-498 keV X-ray emission, and coincides with a secondary peak in the X-ray light curve. This occurs because the X-ray spectrum is harder at the time of the 6 cm flux maximum. A computation of the power flux of > 25 keV electrons, assuming thick target bremsstrahlung for the X-ray emission, shows no significant increase of the electron energy flux at the time of the 6 cm emission maximum. This indicates that in this secondary X-ray peak already energetic electrons were accelerated to higher energies, with the total energy in > 25 keV electrons approximately conserved. These observations also indicate that the emission at the time of the 6 cm maximum is nonthermal.

(ii) <u>Current sheet between closed loops and centimeter impulsive bursts</u>

Kundu (1981) and Kundu et al. (1982b) discussed a set of 6 cm VLA observations (resolution ~ 2") that pertains to changes in the coronal magnetic field configurations that took place in the various stages of the evolution of a flare (observed on 14 May 1980). The burst appeared as a gradual component on which was superimposed a strong impulsive phase (duration ~ 2 minutes) in coincidence with a hard X-ray burst. The pre-flare region (Fig. 6) showed intense emission with peak $T_b \sim 10^7$ K extended along a neutral line situated approximately in the east-west direction. A burst source of intense emission with $T_b \sim 4 \times 10^7$ K appeared initially (Fig. 6). The most remarkable feature of the burst source evolution was that an intense emission extending along the north-south neutral line (line of zero polarization) appeared, just before the impulsive burst occurred. This north-south neutral line must be indicative of the appearance of a new system of loops, possibly due to reconnections. In the 20 seconds preceding the impulsive peak ($T_b \sim 1.1 \times 10^9$ K) the arcade of loops (burst source) changed and ultimately developed into two strong bipolar regions or a quadrupole structure (at 19:19:55 UT, Fig. 6) whose orientations were such that near the loop tops the field lines were opposed to each other. This quadrupole field configuration is reminiscent of the flare models in which a current sheet develops at the interface between two closed loops. The impulsive energy release must have occurred near one of the centrally located neutral lines. The bright compact bipolar source is obviously related to the region of energy release by some kind of magnetic reconnection of the field lines originating from the two bipolar regions between which this compact region is located. Soon after the maximum, a loop-like structure with a lower brightness temperature ($< 3 \times 10^8$ K) developed around the compact source. The footpoints of the loop have predominantly opposite circular polarization. The two ribbon Hα flare associated with the burst occurred near the footpoints of the loop and the intense compact source was located near the top of this loop.

Kundu et al.'s (1982b) study has provided several examples of burst maps produced with the VLA, which appear to provide support for one or the other of the different classes of flare models. Thus, in

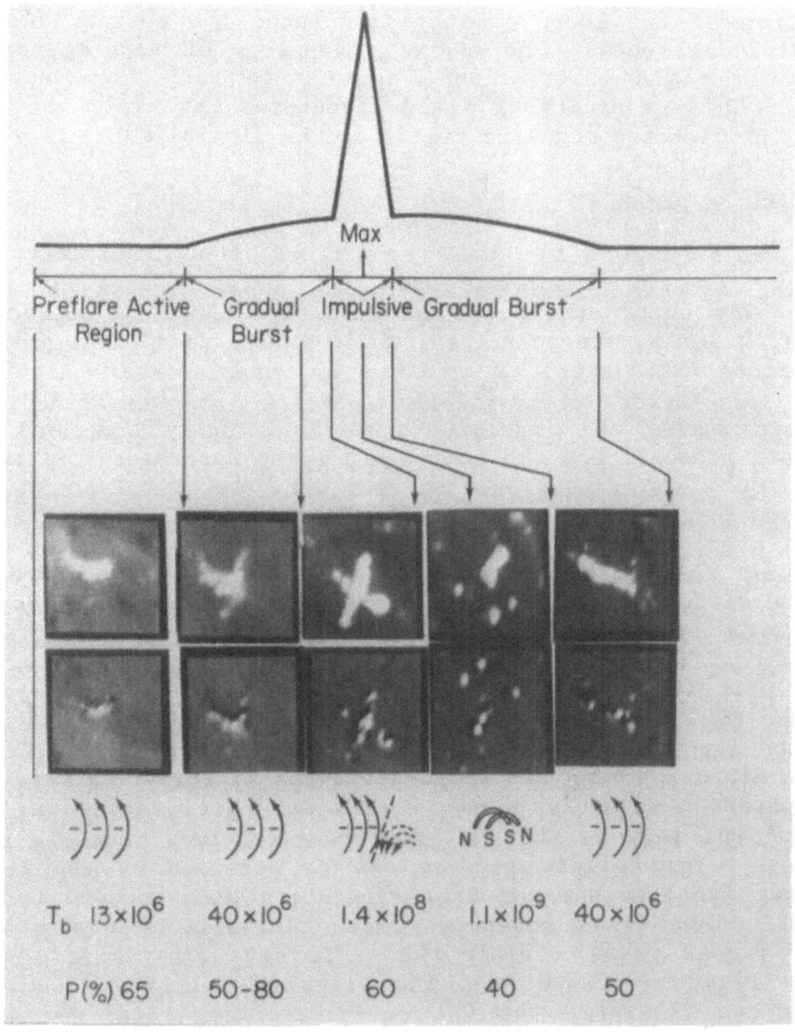

Fig. 6. Preflare active region and burst source maps for the 14 May 1980 burst at 6 cm. Each map was synthesized from data taken during a time interval appropriate to the observed 6 cm flux. Preflare 6 cm map 18:06-18:45 UT, $T_b(max) \sim 13 \times 10^6$ K; gradual phase of burst, 18:59-19:14, $T_b(max) \sim 40 \times 10^6$ K; last 5^m before impulsive phase, 19:14-19:19, $T_b(max) \sim 1.4 \times 10^8$ K; peak of the impulsive phase 19:19:55-19:20:05, $T_b(max) \sim 1100 \times 10^6$ K; (Note the remarkable quadruple structure); gradual phase of burst, 19:21-19:30, $T_b(max) \sim 40 \times 10^6$ K.

some cases, an isolated loop model is favored, in that the emission from bipolar loops originates from a large part of the loop surrounding its top. In a unique case, a quadrupole structure reminiscent of two bipolar loops in close proximity was believed to be responsible for the triggering of a flare. It is possible that a neutral sheet developed at the interface between two loops or that the interaction of two loops greatly influenced the current flow within the individual loops. The magnetic topology of flaring regions, as inferred from VLA observations, appears to vary from one case to another. This diversity of field structures may imply the diversity of ways in which the magnetic fields can be dissipated in flares.

8. CONCLUDING REMARKS

In this review we have summarized the distinctive spatial properties of the impulsive and post burst phases of microwave bursts. The burst source in the impulsive phase is compact ($< 10''$ size), $T_b > 10^8$ K, and polarized, while in the post burst phase, it is more diffuse ($\gtrsim 1'$ size), $T_b \sim 10^6$ K and unpolarized. High time and spatial resolution observations indicate the existence of two components during the impulsive phase - a spiky component and the other more gradual, the two components being separated spatially by as much as 100". Two-dimensional mapping with the VLA of preflare active regions at 6 cm with resolution of a few arc seconds has indicated in one case a polarization reversal of a component source where the burst originated; this is interpreted as due to changes in a pre-existing flux tube either due to twisting or to onset of currents. We conclude this review by noting that most of the microwave positional data indicate that the primary release of flare energy takes place in the coronal part of a magnetic loop. A nonthermal tail of high energy electrons ($E > 100$ keV) is produced at this time and some of these electrons are trapped near the top of the loop by scattering and magnetic mirroring, and are responsible for microwave emission via the gyrosynchrotron mechanism. The hard X-rays are produced at the foot points of the loop by thick target bremsstrahlung from the streaming electrons. This simple picture can be modified by the loop field structure, its symmetry and its orientation with respect to the line of sight. Thus it is possible to see microwave sources at the foot points in some cases. Since the radio data provide information on magnetic field structure near the sites of energy release, we have attempted to interpret the VLA two-dimensional total intensity and polarization maps in terms of existing flare models. The VLA data suggest that in some cases an isolated loop model is favored in that the emission from bipolar loops originate from a large part of the loop surrounding its top. They also suggest that a neutral sheet between closed loops may trigger the impulsive onset of a flare as indicated by the development of a quadrupole structue or by the change in the orientation of a neutral plane associated with the emergence of new flux. The magnetic topology of flaring regions, as inferred from VLA observations, appears to vary from one case to another. This diversity of field structures may imply the diversity of ways in which the magnetic fields can be dissipated in flares.

REFERENCES

Alissandrakis, C. E. and Kundu, M. R.: 1978, Ap. J. 222, 342.
Alissandrakis, C. E. and Kundu, M. R.: 1983, in preparation.
Brown, J. C., Melrose, D. B., and Spicer, D.: 1979, Ap. J. 228, 592.
Holman, G. D., Kundu, M. R., and Papadopoulos, K.: 1982, Ap. J. 257, 354.
Holman, G. D., Kundu, M. R., and Dennis, B. R.: 1983, in preparation.
Kai, K., Kosugi, T., and Nakajima, H.: 1982, 75, 331.
Kattenberg, A. and Allart, M.: 1982, Ap. J., in press.
Kundu, M. R.: 1959, Ann. d'Ap. 22, 1.
Kundu, M. R.: 1965, "Solar Radio Astronomy" (John Wiley-Interscience, New York).
Kundu, M. R., Alissandrakis, C. E., Bregman, J. and Hin, A.: 1977, Ap. J. 213, 278.
Kundu, M. R. and Vlahos, L.: 1979, Ap. J. 232, 595.
Kundu, M. R.: 1981, Proc. SMY Symposium Workshop, Crimea, p. 24.
Kundu, M. R., Schmahl, E. J., and Velusamy, T.: 1982a, Ap. J. 253, 963.
Kundu, M. R., Schmahl, E. J., Velusamy, T. and Vlahos, L.: 1982b, Astron. Astrophys. 108, 188.
Marsh, K. and Hurford, G. J.: 1980, Ap. J. 240, L111.
Marsh, K, Hurford, G. J. and Zirin, H.: 1980, Proc. I.A.U. Symp. No. 86 "Radio Physics of the Sun" (ed. M. R. Kundu and T. E. Gergely), p. 191.
Spicer, D.: 1977, Solar Phys. 53, 305.
Vlahos, L., and Papadopoulos, K.: 1979, Ap. J. 233, 717.

DISCUSSION

KAI: I have got an impression from your recent review article that you are not completely inclined toward the idea that the microwave source lies at the top of magnetic loops. Is there any evidence against it?

KUNDU: Indeed, there are some examples in a paper by myself, Schmahl and Velusamy (Ap.J. 253, 963, 1982). There are cases of two emission regions at 6 cm coinciding with Hα flaring kernels with minimum emission in between. One can probably take this as an evidence that the microwave emission can occur at the footponts of loops.

DUAL FREQUENCY OBSERVATIONS OF FLARES WITH THE VLA

George A. Dulk and Timothy S. Bastian
Department of Astro-Geophysics, University of Colorado,
Boulder, Colo., U.S.A.

and

Gordon J. Hurford
Solar Astronomy, California Institute of Technology,
Pasadena, Calif., U.S.A.

ABSTRACT. We describe observations of three flares made at 5 and 15 GHz with the VLA, two subflares near the limb on 1981 November 21 and 22, and an M7.7 flare on 1981 May 8. Even though the time histories of the November flares indicated simple impulsive bursts, the VLA observed no 5 GHz radiation at all from one flare, and from the other, the 15 GHz radiation emanated from a source which was smaller, lower and displaced from the 5 GHz source. Without the spatial information, we would have derived incorrect results from the assumption that photons of different energy (both at X-ray and radio wavelengths) arose from one homogeneous volume.

The 1981 May 8 flare was intense and complex, having two or more sources at both 5 and 15 GHz. Prior to the peak of the flare, the sources grew in size to > 20" to 40", after which they were not visible to the VLA; only (weak) subsources could be seen. These were located between or at the edge of the Hα ribbons and the two hard X-ray sources imaged by the Hinotori. Highly polarized, bursty radiation observed at Toyokawa at 1 and 2 GHz, indicated that an electron-cyclotron maser operated during the flare. We derive 360 to 660 gauss as the maximum field strength in flaring loops.

1. INTRODUCTION

We have observed several solar flares simultaneously at 5 and 15 GHz using the VLA. The quality of the observations and ease of interpretation depends greatly upon the configuration of the VLA when the observations were made. With a widespread array such as "A" or "B", there are frequently too few antenna pairs with short spacings, i.e., low resolution. As a result, there are large numbers of replicas on the maps, making it difficult to determine where the source lay and, in some cases, the large scale structure of the flare is resolved out, i.e., is not visible on the maps. The "C" array is perhaps optimum for dual frequency observations as there are an adequate number of short spacings and the resolution, typically 1", is sufficient for most flares.

Solar Physics **86** (1983) 219–226. 0038–0938/83/0862–0219 $ 01.20.

Here we describe observations made on two occasions: 1981 November 21/22 and 1981 May 8. The former are the better because the "C" array was used; however the latter (made with the "B" array) are of special interest because the flare was intense and was imaged in hard X-rays by the Hinotori spacecraft.

2. THE FLARES OF 1981 NOVEMBER 21 AND 22

These were small subflares located near the limb, at about 73° and 84° longitude respectively. A detailed description of the observations and interpretation is given by Dulk, Bastian and Kane (1983). Here we summarize the results.

On 21 November radiation was observed at 15 GHz and not at 5 GHz, which is peculiar because non-coherent emission mechanisms produce more low frequency radiation than high and, even if optically thick at 5 GHz, the source should have been observable with the VLA. On 22 November two sources were observed, having nearly identical impulsive flux histories, but displaced from one another by about 12". One, "Source A", emitted only 15 GHz radiation while the other, "Source B", was some 2 to 5 times brighter at 5 GHz than at 15 GHz. Both sources were located at a low altitude, $\lesssim$3000 km; in projection they appeared at the same height as, or below the Hα features. Comparison of Hα pictures from Big Bear with the 15 GHz VLA maps showed that, for the source on 21 November and source "A" on 22 November, the radiation had to pass over a large sunspot enroute to Earth. We believe that 15 GHz radiation was little attenuated while the 5 GHz radiation was gyroresonantly absorbed by the background plasma at the 3rd harmonic of the gyrofrequency; this implies that the field strength at $\lesssim$3000 km altitude was ~600 G. The 15 GHz radiation probably arose from gyrosynchrotron emission from electrons of E~0.5 MeV in a weak, nonthermal tail embedded in a magnetic field of about 200 G (Marsh et al. (1981) reached similar conclusions about a different flare). The degree of circular polarization was ~0.5 RH, indicating that the field was inclined about 45° to the line of sight.

The second source, "B", with a brightness temperature at 5 GHz of $\approx 10^7$ K, probably originated from the thermal, soft X-ray plasma. Its polarization varied with time, changing from RH to LH and then partly back to RH; we attribute this to the emission arising in a field of ~200 G, initially with a viewing angle of $\approx$70°, then $\approx$110°, and then back to $\approx$80°. In this interpretation the changing polarization was due to the source brightening at different locations along the line of sight, and was not due to any temporal variations in the field strength or topology.

Concluding this section, the main result of these dual frequency observations is that what appeared to be simple, impulsive flares on light curves were in fact complicated in their spatial and polarization structures. Also, the heights of the radio sources were unexpectedly low.

3. THE LARGE FLARE OF 1981 MAY 8

Hinotori observations of this flare have been described by Ohki et al. (1982). It was a 2B, M7.7 flare located at N10 E36, starting at about 2205 UT, with impulsive peaks some 20 min later, and lasting more than three hours altogether. Intense type II (shock-related) bursts and type IV continuum (from a stationary source) were recorded at Culgoora starting at 2233 UT. Figure 1 shows the time histories of the 25-400 keV X-rays from SMM (Ohki et al. show a similar plot from Hinotori) together with 15 GHz flux profiles from Owens Valley and the VLA. Taking account the linear vs. log scales, the X-ray and Owens Valley profiles are similar, as usual, but the VLA profile differs.

We note that the VLA and Owens Valley profiles are essentially identical in the early stages of the flare, until about 2220 UT, and that the VLA profile represents the flux as seen by an interferometer with fringe spacing 15".4 (the smallest antenna spacing) whereas the Owens Valley profile comes from a single antenna with a beamwidth of 3'.1. Thus we attribute the decreasing flux after 2220 UT as seen by the VLA to an increase in source size to 15" and larger. Comparing the two flux profiles in Figure 1, we conclude that, by 2230 UT, at least 90% of the 15 GHz flux arose from a source larger than 15". Because the source was invisible to the VLA, we have no information on its actual size or its position. Only the positions of small substructures can be found after about 2220 UT.

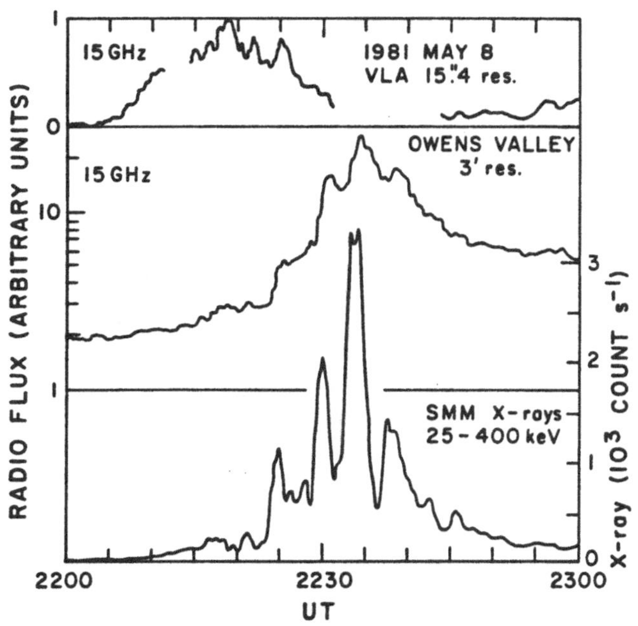

Figure 1. Flux density profiles of the 1981 May 8 flare observed at 15 GHz by a close-spaced VLA interferometer (top) and a 27 m antenna at Owens Valley (center). At the bottom is the hard X-ray profile from SMM.

Figure 2 shows the radio spectrum and polarization at four times as recorded by the frequency agile receiver at Owens Valley. The spectra are relatively flat, with unusually strong emission both at low and high frequencies. The spectral slope at low (optically thick) frequencies is $\lesssim 1$, indicating that the source size was an increasing function of wavelength. At high frequencies, the shallow slope indicates the presence of a significant nonthermal tail and electrons of E~1 MeV. At all times the polarization was relatively weak, $\lesssim 0.2$; the change of sense from LH to RH at ~3 GHz may reflect a change from dominantly o-mode to x-mode.

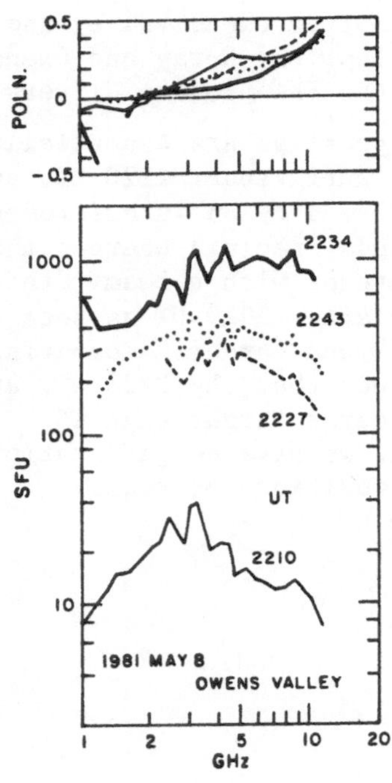

Figure 2. Spectra of the radio flux and polarization at four times as observed with 1-18 GHz frequency agile receiver at Owens Valley.

Figure 3 shows drawings of the sunspots, Hα flare, line of photospheric field reversal, and 15 GHz radio source positions at two different times. The Hα and magnetic features were drawn from excellent pictures from Big Bear Observatory, courtesy of H. Zirin. During the rise phase at 2211 UT, the 15 GHz source size was ≈3"× 6" and it was located very close to one of the three bright Hα patches. Possibly it was close to one footpoint of a very large loop connecting two of the Hα kernels. Because of equipment problems and saturation of the VLA, the last 15 GHz image obtained before the peak was at 2228 UT, at which time most of the flux came from a source larger than 15". Three small bright patches were seen by the VLA; these are shown relative to Hα and magnetic features (at 2233 UT) on the lower half of Figure 3. One of the 15 GHz patches appears to be superimposed on a bright Hα feature while the others are located between the bright features. It is at this time that Hinatori obtained the hard X-ray images published by Ohki et al. (1982). The 17-40 keV X-rays arose from two sources, a weak one near the Hα feature "A" and an intense one near "B". The latter was large, ≈40" in extent, and the eastmost

15 GHz feature was located on its edge. It is unfortunate that we are unable to determine whether the large 15 GHz source coincided with the hard X-ray source "B" or whether it was concentrated between the two hard X-ray sources.

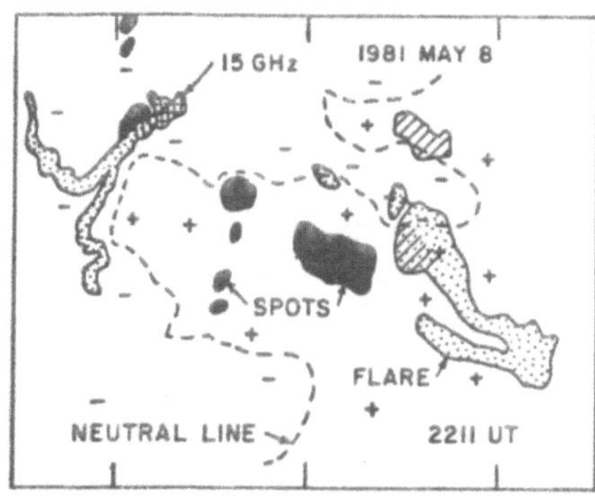

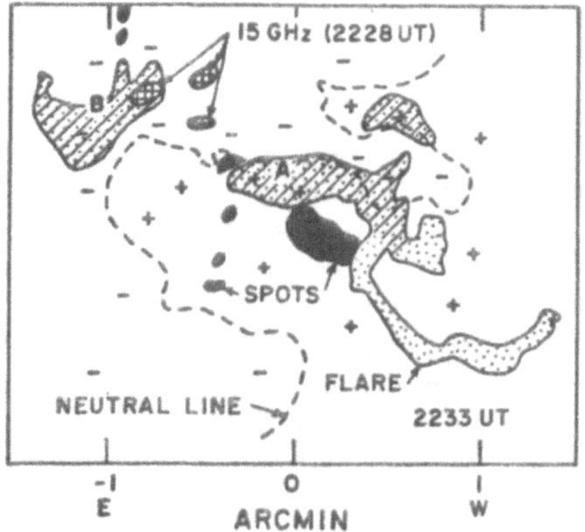

Figure 3. Tracings of the sunspot positions (black), Hα flare position (stippled, with hatched areas showing of brightest Hα), magnetic fields (dashed neutral line and +, - showing polarity), and 15 GHz source positions (cross-hatched). Top: during the rise phase at 2211 UT. Bottom: near the peak at 2233 UT (except for radio positions at 2228 UT) and with "A" and "B" indicating the two hard X-ray source positions from Hinatori (Ohki et al., 1982).

VLA observations were made at 5 GHz simultaneously with those at 15 GHz. Similarly, the 5 GHz flux density derived from the closest antenna pair (≈40" fringe spacing) decreased after ≈2220 UT, showing that this source also became very large. Examination of the dirty maps showed that there were at least two and probably three subsources; however, because of replication on the maps we have so far been unable to locate the sources unambiguously.

Figure 4 shows the flux profile at 1 GHz from Toyokawa, courtesy of S. Enome. This remarkable profile shows very fast time structure, $\lesssim$ 0.5 s, and that the bursty radiation was 100% circularly polarized. (There is a weak, nearly unpolarized component also present, the counterpart of the higher frequency microwave burst.) The bursty component is closely associated with the hard X-ray and higher frequency peaks, possibly more with the leading edges than the peaks themselves. The 2 GHz profile from Toyokawa shows a similar bursty component, although not so strong or prevalent. This bursty component must have arisen from a coherent emission mechanism. Melrose and Dulk (1982a) worked out the details of how an electron-cyclotron maser operates, and argued that a maser at the 2nd harmonic of the gyrofrequency could account for such bursts. However, more important for flare theory are the consequences of masers at the first harmonic. These can contain a significant fraction of the flare energy, as much as the electrons contributing to hard X-ray bursts (Melrose and Dulk, 1982b). As this radiation cannot escape, it can rapidly heat a large volume of surrounding plasma. Also the maser can cause electrons to precipitate rapidly from magnetic traps and cause bright hard X-rays to come from footpoints. In the case of the 8 May flare, we can use the theory and observations to estimate the maximum magnetic field strength in flaring loops: taking the radiation as being emitted at $\omega = 2\Omega_e$ and the fact that Toyokawa observed maser-like bursts at 2 GHz but not at 3.7 GHz, we find that $360 \lesssim B_{max} < 660$ G.

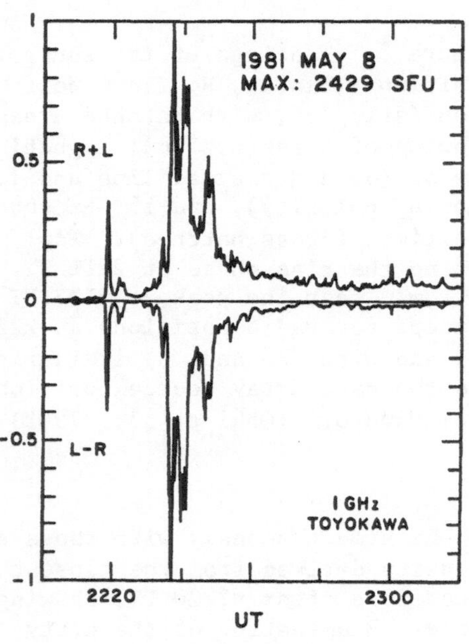

Figure 4. Flux density (Stokes parameter I) and polarization (Stokes parameter V) profiles at 1 GHz as recorded at Toyokawa (courtesy S. Enome). The nearly exact symmetry of I and V indicates that the dominant, bursty component was approximately 100% circularly polarized.

We thank the organizers of this symposium, NSF, and JSPS, for the opportunity for one of us (G.A.D.) to attend. We are grateful to the VLA personnel for their assistance with the VLA data, to H. Zirin for the Big Bear pictures, to M. Liggett for accurately locating the Hα features for comparison with the VLA, to S. Enome for the Toyokawa data, and to R. Stewart for the Culgoora data. This work has been supported in part by NASA's Solar Terrestrial Theory and Solar Heliospheric Physics programs (grants NAGW-91 and NSG-7287 to the University of Colorado).

REFERENCES

Dulk, G.A., Bastian, T. and Kane, S.R.: 1983, "Two-Frequency Imaging of Microwave Impulsive Flares near the Limb," to be submitted to Astrophys. J.

Marsh, K.A., Hurford, G.J., Zirin, H., Dulk, G.A., Dennis, B.R., Frost, K.J. and Orwig, L.E.: 1981, Astrophys. J. 251, 797-806.

Melrose, D.B. and Dulk, G.A.: 1982a, Astrophys. J. 259, 844-858.

Melrose, D.B. and Dulk, G.A.: 1982b, Astrophys. J. (Letters) 259, L41-45. L41-45.

Ohki, K., Tsuneta, S., Takakura, T., Nitta, N., Makishima, A., Murakami, T., Ogawara, Y., Oda, M. and Miyamoto, S.: 1982, Hinatori Symp. on Solar Flares, (Tokyo: Inst. of Space and Astronautical Sci.), p. 102.

DISCUSSION

RAMATY: In a gyrosynchrotron source which is optically thick at low frequencies and optically thin at high frequencies, the polarization should change with frequency from the thick to the thin regimes. Was this effect observed?

DULK: I have not seen such a reversal. Usually you do not see the maser part of a source at, say, 5 GHz as at 15 GHz , sizes of sources at the two frequencies are 10" - 15" and 2" - 3", respectively.

KUNDU: We have one case of simultaneous observation of a burst at 2 and 6 cm with the VLA (c-configuration). The positions and structures at both wavelengths were the same, implying that we are dealing with the same loop producing 2 and 6 cm emission. The polarization was the same at both wavelengths around the peak of the impulsive phase.

ENOME: As regards Ramaty's question on polarization reversal, such a reversal with respect to frequency in a single source has been observed only in the recent high-resolution measurements.

KUNDU: In one event we saw a source of the same size and at the same location at 5 - 15 GHz. But polarization reversal was uncertain. The optical depth at 5 GHz was about unity.

DULK: I will show an example of an Owens Valley spectrum which shows a change from optically thick to thin at 5 GHz and, at approximately the same frequency, from weak LH to moderate RH.

KUNDU: In the case of the Nov 21 - 22 event which was rather weak (∿ 8

SFU), do you think that the rapid change of polarization at 5 GHz could be related to rapid change of the neutral plane? If so, could there be uncertainty in polarization changes due to inherent weakness of the burst?

DULK: A change of polarization could be caused by a time-changing field, as you suggest. An alternative possibility, which I prefer, is to have the source centroid shift slightly, along the line of sight, from a location where $\theta \simeq 70^\circ$ to one where $\theta \simeq 110^\circ$.

ENOME: The burst on Nov. 21-22, 1981 should be classified as a GRF, since $T \lesssim 3 \times 10^6$ and it is associated with soft X-ray emission of tens of keV.

DULK: Temporal behavior suggested an impulsive burst, with a duration only one minute or a few minutes. The 15 GHz radiation could not have come from soft X-ray plasma; it required electrons of 0.3 to 1 MeV or so, e.g., in a weak but flat non-thermal tail.

DISSIPATIVE THERMAL MODELS FOR IMPULSIVE MICROWAVE BURST DELAYS

John C. Brown
Department of Astronomy, University of Glasgow, G12 8QW, U.K.

ABSTRACT. Microwave emission is analysed for a source heated by magnetic dissipation. As the temperature rises and the field falls, the resulting variation in microwave optical depth results in time delays between emission features at different frequencies. A single such source can, however, only explain events exhibiting small delays ($\approx$ 1sec) and with lower frequencies peaking later. Continuous production of many such heated regions (multiple kernel model) can, however, explain long delays and produce either advancement or retardation of lower frequencies.

1. RESULTS

Solar microwave bursts sometimes exhibit delays in emission peaks between different frequencies in the tens of GHz range, these delays ranging from seconds or less (Kaufmann et al., 1982) up to tens of seconds (Wiehl, Schöcklin, and Magun, 1980) and can involve either advancement or retardation of low frequencies. In a more detailed paper (Brown et al., 1982) we have reviewed these observations and considered their possible interpretation in terms of thermal source models involving magnetic dissipation. Here we summarise the main results of this study.

We find that the bulk of microwave emission from a single small heated kernel occurs in a time short enough for energy losses by conduction connection and radiation to be negligible as a first approximation. Then for an initially cool plasma of density n and initial field B_o, the electron temperature $T(t) = 10^8\ T_8$ rises with time t according to

$$T\ (t) \simeq 1{\cdot}92 \times 10^8 \frac{B_2^2}{n_{10}} (1-b^2)\ (K), \tag{1}$$

where $B_2 = B\ /(10^2$ gauss), $n_{10} = n/(10^{10}\mathrm{cm}^{-3})$ and $b = B(t)/B_o$ defines

Solar Physics **86** (1983) 227–230. 0038–0938/83/0862–0227 $ 00.60.

the relative time variation of the declining field. Then, using the Dulk et al. (1979) expressions for gyrosynchrtron emissivity and absorption coefficient for a thermal plasma the flux density observed at time t, if the source has typical dimension 10^9 L_9 cm, should vary with b(t) according to

$$S_\nu(\mathrm{sfu}) \simeq 135 \frac{L_9^2 \, B_2^2}{n_{10}} \nu_{10}^2 (1-b^2) \times$$

$$\times \left[1-\exp\left\{ -2.22\times10^{-4} \frac{L_9^2 \, B_2^{23}}{n_{10}^6} \nu_{10}^{-10} b^9 (1-b^2)^7 \right\} \right], \qquad (2)$$

where ν_{10} is the frequency in 10 GHz units.

Equation (2) has the property that S_ν (b) peaks at b = 0.6 at high (optically thin) frequencies but for smaller and smaller b at lower and lower (optically thick) frequencies. Since, according to the model, b decreases monotonically with time, frequency dependent delays can be explained but only in the sense of retardation of low frequencies. The typical delay time can be roughly estimated by assuming b to decline according to

$$\frac{db}{dt} \simeq -\frac{b}{t_o} \quad \text{where } t_o \simeq \frac{L}{v_A} = 4.62 \, L_9 \frac{n_{10}^{1/2}}{B_2} \ (\mathrm{s}) \qquad (3)$$

with v_A the Alfvén speed.

Taking a typical impulsive burst spike to last a few seconds, to have a peak flux of some hundreds of s.f.u. at a spectral peak in the 10-20 GHz range, and delays of about one second over this range, we find from Equations (2) and (3), when inverted, that the model parameter values necessary are:

$$L \simeq 5.8\times10^8 \ \mathrm{cm}; \quad n \simeq 5.6\times10^9 \ \mathrm{cm}^{-3}; \quad B \simeq 220 \ \mathrm{gauss}. \qquad (4)$$

For these parameters, the microwave flux density variations at various frequencies predicted by Equations (2) and (3) are shown in Figure 1 together with the variations in the optical depth τ_{15} at 15 GHz. (Optical depths at other frequencies scale according to $\tau_\nu \sim \nu^{-10}$).

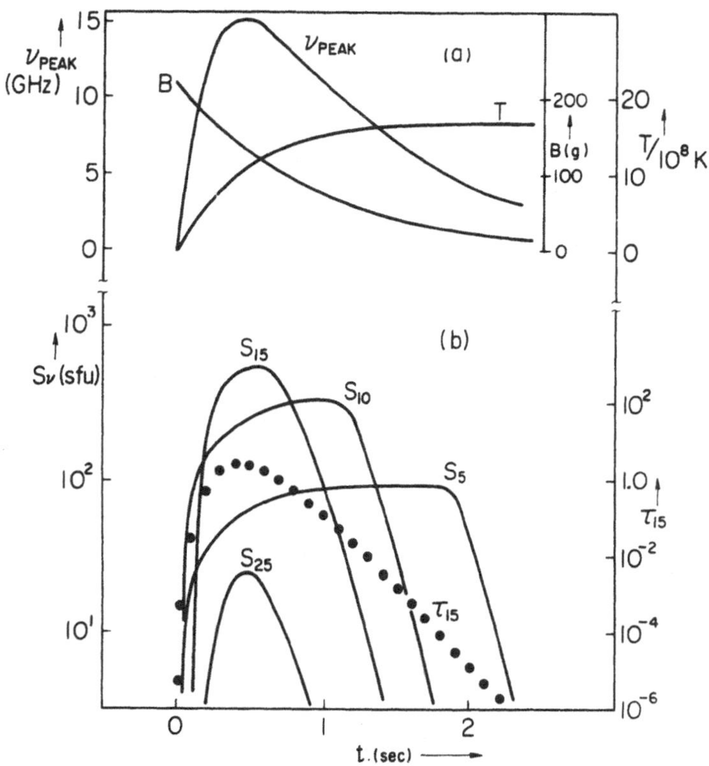

Figure 1

(a) Time variation of magnetic field B and temperature T in typical dissipative model and the resulting variation in microwave spectrum peak frequency ν_{peak}.

(b) Time variations in microwave flux densities S_ν at various frequencies ν predicted by the model. Also shown is the variation in optical depth τ_{15} at 15 GHz.

It is not possible in this way, however, to explain very long low ν delays (tens of seconds) in large events with high ν_{peak} values such as that of 1978 July 11 observed by Wiehl (1980) which, according to Equations (2) and (3) would require meaninglessly high fields ($B_o \gtrsim 2400$ gauss) and densities ($n \gtrsim 2 \times 10^{13} cm^{-3}$). Nor can such a single heated region explain retardation of low frequencies. We have therefore extended the above concepts to include the possibility of continuous production of many short lived source kernels (conceivable as tearing mode islands) throughout the microwave burst duration, similarly to the proposal by Brown, Craig and Karpen (1980) for hard X-ray bursts.

We find that, depending on the time variation of the distribution of kernel production rate over kernel size and optical depth, it is feasible in this multikernel model to obtain either advancement or retardation of low frequencies, and also to explain long delays for physically reasonable parameters. Furthermore, such distributions of kernel production parameters can give rise to microwave burst spectra which are less steep both at low and high frequencies than isothermal source spectra ($\sim \nu^2$ and $\sim \nu^{-8}$ respectively, cf. Dulk et al., 1979) and so closer to typical spectra observed (MacKinnon and Brown, 1982).

REFERENCES

Brown J.C., Craig I.J.D. and Karpen J.T.: 1980, Solar Phys. 67, 143.

Brown J.C., MacKinnon A.L., Zodi A. and Kaufmann P.: 1982, submitted to Astron. Astrophys.

Dulk G.A., Melrose D.B. and White S.: 1979, Astrophys. J. 234, 1137.

Kaufmann P., Costa J.E.R. and Strauss F.M.: 1982, submitted to Solar Phys.

MacKinnon A.L. and Brown J.C.: 1982, preprint.

Wiehl H.J., Schöchlin W.A. and Magun A.: 1980, Astron. Astrophys. 92, 260.

Wiehl H.J.: 1980, preprint.

DISCUSSION

ACTON: How do you obtain a high frequency (ν) time delay from an ensemble of sources which individually produce only low ν delays?

BROWN: In a steady state production of kernels, each of very short duration, there is a fixed ν_{peak} depending on the distribution of B, n, and L over the kernels. If one then allows this distribution to very via the production function, ν_{peak} can be made to shift either up or down in time, i.e. the mean kernel properties have to vary suitably in time.

LATE PHASE GRADUAL ENHANCEMENTS IN MICROWAVES AND HARD X-RAYS OF THE 6 NOVEMBER 1980 FLARE

K. Kai, H. Nakajima and T. Kosugi
Tokyo Astronomical Observatory, University of Tokyo
and
S.R. Kane
Space Sciences Laboratory, University of California, Berkeley

ABSTRACT. Gradual enhancements which repeatedly appeared 20-50 min after the onset of an impulsive burst were observed in microwaves and hard X-rays. The observed characteristics of the gradual enhancements are (i) the similarity of time profiles in both the frequency range of 17 GHz to 160 MHz and the energy range of 40 to $\sim$400 keV, (ii) the low peak frequency of the microwave spectrum ($\sim$4 GHz), and (iii) the flat X-ray spectrum with the power-law exponent of $\sim$1.7. It is suggested that the radio and X-ray gradual enhancements are due to a single population of energetic electrons which extend high in the corona (at least 1.4×10^5 km). Nevertheless the 17 GHz source is likely to be at a low altitude, not much higher than the preceding impulsive burst source.

1. INTRODUCTION

Gradual enhancements which occur tens of minutes after the onset of microwave impulsive bursts are called gradual components or extended components. Bursts of this type are of particular interest from a view point of particle acceleration, since they are often accompanied by energetic particles which are measured in the interplanetary space. As a possible mechanism of the energetic particle production the concept of two-phase acceleration has been proposed (de Jager, 1969); electrons are accelerated to 10 - 100 keV energies at the impulsive phase and then further accelerated up to relativistic energies at a later phase by coronal shock waves which manifest themselves as type II bursts at metre wavelengths. Wild, Smerd and Weiss (1963) and de Jager (1969) suggest that these energetic electrons give rise to gradual components at a late phase. It has been revealed by recent hard X-ray and gamma ray observations (see, e.g., Chupp, 1982) that energetic electrons and protons are produced without the classic two-phase acceleration process. Nevertheless the particle acceleration which takes place at a late phase in the corona seems to be still of particular importance as an acceleration process which is distinctive from the acceleration in the impulsive phase.

Gradual hard X-ray bursts were studied first by Frost and Dennis

Solar Physics **86** (1983) 231–236. 0038–0938/83/0862–0231 $ 00.90.

(1971), subsequently by Hudson (1978) and Hudson et al. (1982). These three gradual bursts had hard spectra with power-law exponent of ∿2. Hudson and Hudson et al showed that by limb occultation gradual bursts were purely of coronal origin. On the other hand Kane et al (1982) have concluded from a stereoscopic observation that the main source of a gradual hard X-ray burst lay in the lower atmosphere. Although microwave bursts which show both impulsive and late phase gradual enhancements are much rarer than simple impulsive bursts, they seem to form a distinctive category of microwave bursts. The burst on 6 November 1980 is likely to be a typical example of the composite bursts. From the analysis of the 6 November 1980 burst we intend to show (1) that gradual enhancements at a late phase appear widely at cm to m wavelengths and in hard X-rays possibly due to a common, non-thermal electron population, and (2) that inspite of the wide spread in frequency the microwave source lies low in the corona.

2. OBSERVATIONS

The hard X-ray and 17 GHz radio observations reported here were made with the ISEE-3 X-Ray Spectrometer (Anderson et al.,1978; Kane et al., 1982) and the 17 GHz polarimeter and interferometer at Nobeyama (Nakajima et al., 1980), respectively. Additional radio measurements were made with the polarimeters at Toyokawa (S. Enome, private communication) and with the radiospectrometers at Nobeyama. The Hα-flare observations were made at Mitaka (F. Moriyama, private communication).

The observed characteristics of the gradual enhancements which appear in the late phase of the 6 November 1980 burst can be summarized as follows;

(1) Gradual enhancements occur at 17 GHz repeatedly even ∿50 min after the onset of the impulsive burst (Figure 1). The source is rather small in size (∿20") and moderately circularly polarized (∿30 % RH). Those characteristics definitely distinguish the enhancements from post-burst increases.

(2) The microwave spectrum has a peak at ∿4 GHz which is significantly lower than the peak frequency of the impulsive burst (∿17 GH).

(3) The gradual enhancements show similar time profiles in a wide frequency range of 17 GHz down to at least 160 MHz (Figure 2).

(4) The gradual enhancements appear also in hard X-rays with a time profile similar to that at 17 GHz (Figure 1). The X-ray spectrum (Figure 3) is significantly harder ($\gamma \simeq 1.7$ above 80 keV) than that of the impulsive burst ($\gamma \simeq 3.2$). The harder component appears in coincidence with the gradual enhancements.

(5) The sources of the gradual enhancements are located ∿24" - 32" west (toward the disk centre) of the impulsive burst source at 17 GHz. The associated Hα-flare shows a complex structure extending over ∿ 1ʹ.5 area. However, we can identify the bright region as marked by A in Figure 4 with the impulsive burst.

(6) The source height can be estimated from a comparison of the radio position with the position of the associated Hα-flare region. The height of the impulsive burst source thus obtained is (12 $\pm$ 10)"

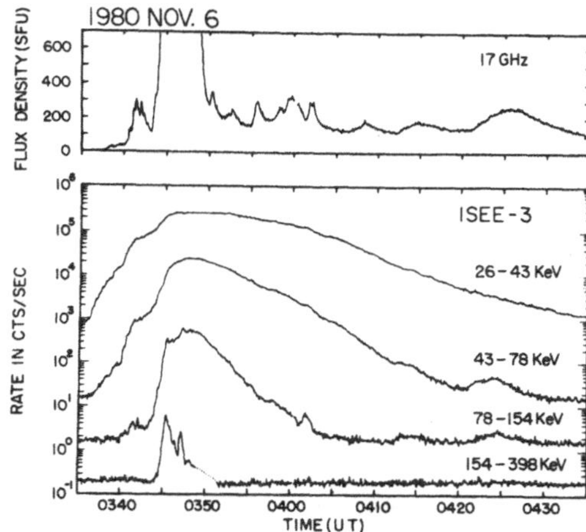

Fig. 1. Time profiles of the microwave-X-ray burst on 6 November 1980. Here, 17 GHz observations were made with the polarimeter at Nobeyama. The strongest part of the impulsive burst is truncated in order to emphasize gradual enhancements in the late phase. The peak flux density is 6000 sfu. X-ray measurements were made with the X-ray spectrometer aboard the ISEE-3 spacecraft (Anderson et al., 1978; Kane et al., 1982). Note successive gradual enhancements at the late phase between 0405 and 0435 UT both at 17 GHz and in the 43-78 and 78-154 keV channels.

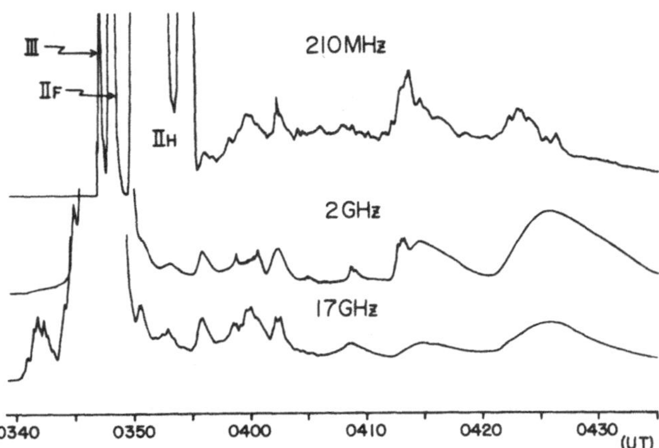

Fig. 2. Comparison of time profiles recorded at three different frequencies; 17 GHz, 2 GHz, and 210 MHz. Flux density scales are arbitrary. The 2 GHz data was obtained at Toyokawa (courtesy by Dr Enome). Below 150 MHz strong spiky storms which may be classified as stationary type IV burst obscure the gradual enhancements.

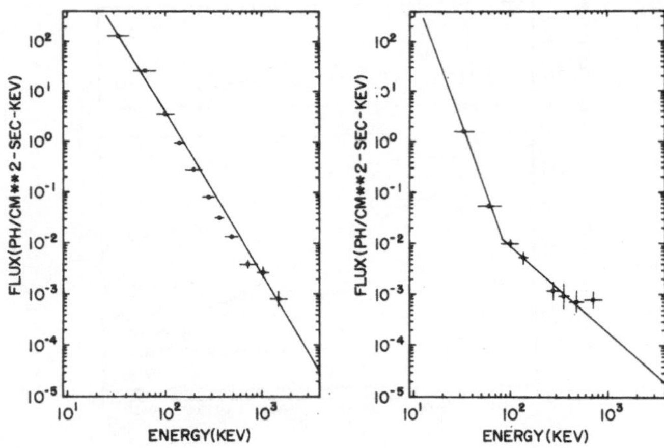

Fig. 3. X-ray spectra taken at 0345 UT (a) and 0425 UT (b). Fitting with a single power-law gives an exponent of 3.2 for (a). Fitting with two power-laws gives an exponent of 5.3 below 80 keV and that of 1.7 above 80 keV.

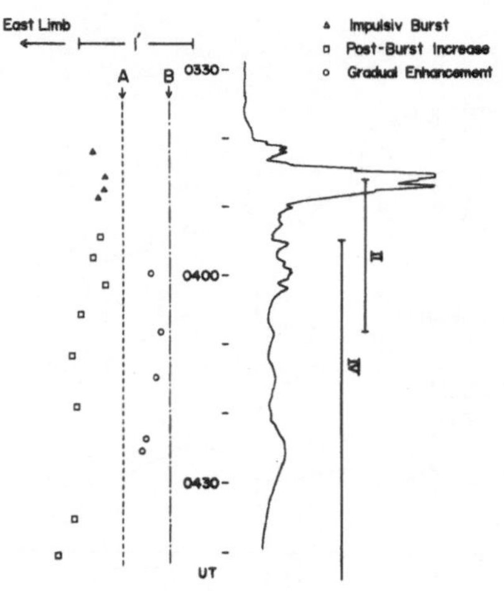

Fig. 4. Time plots of centre positions of 17 GHz sources. Position measurements were made with the one-dimensional (east-west) interferometer at Nobeyama (Nakajima et al., 1980). The east limb is on the left. There is uncertainty of $\sim\pm 10''$ left in the absolute position determination. The mark A indicates the centre position ($\sim 12' 35''$ east of the disk centre) of Hα-flare bright region (10" - 15") which brightens strongly during the impulsive burst. The mark B indicates the position of the west end of Hα-flare bright regions which covers part of the preceding sunspot. The Hα-flare was observed at Mitaka and the pictures were provided by Dr Moriyama.

above the photosphere. On the other hand the gradual enhancement source is 15" - 7" east of the position of the preceding sunspot marked by B in Figure 4. Therefore we have no evidence that the gradual enhancement source is much higher than the impulsive burst source at 17 GHz.

3. INTERPRETATION

Gradual enhancements observed at a wide frequency range appear to be due to a common single population of high energy electrons. This was suggested earlier by Hudson et al. (1982) from the simultaneity of a hard X-ray gradual burst and a metre wave flare-continuum, and subsequently by Klein et al. (1982) from the similarity of time profiles between hard X-ray, microwave and metre wave flare-continuum and moving type IV bursts. It is clear that the high energy electrons must extend to an altitude of at least 1.4×10^5 km ($\sim$200") above the photosphere, which corresponds to the 200 MHz local plasma level on a standard model. The 160 MHz continuum source at 0410 UT is located at an altitude of $\sim 4.5\times10^5$ km (Svestka et al., 1982). On the other hand the 17 GHz source is located at a much lower altitude. Therfore, the source of the gradual enhancements must have a <u>columnar</u> structure which extends from low to high corona. The structure is probably a large scale magnetic loop. The 17 GHz source may be near a foot point of the loop where magnetic fields are strong with the line of force directed towards the photosphere.

The mechanism of the particle acceleration which takes place in the late phase of a solar flare is still not well understood. It is not clear whether the first and second phase accelerations are physically interconnected or they happen independently. In the present event the impulsive and gradual sources are separated by a projected distance of $\sim$30". We need further detailed studies of the 6 November 1980 flare and other events of the same categoty in order to clarify the relationship between the first and second phase acceleration.

We thank Drs S. Enome and F. Moriyama for providing us respectively polarimeter records obtained at Toyokawa and Hα-flare data obtained at Mitaka, and Messrs. M. Miyazawa and Y. Shiomi for assistance in reduction of Hα-flare data.

REFERENCES

Anderson, K.A., Kane, S.R., Primbsch, J.H., Weitzman, R.H., Evans, W.D., Klebesadel, R.W., and Aiello, W.P.: 1978, IEEE Trans. GE-16, 157.

Chupp, E.L.: 1982, in Gamma Ray Transients and Related Astrophysical Phenomena, ed. by R.E. Lingenfelter, H.S. Hudson and D.M. Worrall, p.363.

De Jager, C.: 1969, in Cospar Symposium in Solar Flares and Space Research, ed. by C. de Jager and Z. Svestka (North Holland Publishing Co.), p.1.

Frost, K.J. and Dennis, B.R.: 1971, Astrophys. J. 165, 655.
Hudson, H.S.: 1978, Astrophys. J. 224, 235.
Hudson, H.S., Lin, R.P. and Stewart, R.T.: 1982, Solar Phys. 75, 245.
Kane, S.R., Fenimore, E.E., Klebesadel, R.W., Laros, J.G.: 1982, Astrophys. J. Lett. 254, L53.
Klein, L., Anderson, K., Pick, M., Trottet, G., Vilmer, N., and Kane, S.R.: 1982, Solar Phys. (in press).
Nakajima, H., Sekiguchi, H., Aiba, S., Shiomi, Y., Kuwabara, T., Sawa, M., Hirabayashi, H., Kosugi, T., and Kai, K.: 1980, Publ. Astr. Soc. Japan 32, 639.
Svestka, Z., Dennis, B.R., Pick, M., Raoult, A., Rapley, C.G., Stewart, R.T., and Woodgate, B.E.: 1982, Solar Phys. 80, 143.
Wild, J.P., Smerd, S.F., and Weiss, A.: 1963, Ann. Rev. Astron. Astrophys. 1, 291

DISCUSSION

HUDSON: The gradual source has characteristics strikingly similar to those of the limb-occulted events observed by OSO-5 (1969 March 30) and OSO-7 (1971 Dec.14 and 1972 Aug. 2); late acceleration and hard spectrum. What was the hard X-ray spectrum of this event, and how was it classified in meter-wave observations?

KANE: There are some similarities between this event and the 1969 March 30 event observed by OSO-5. However, there are several differences, too: (1) The Nov.6, 1980 flare was not occulted. (2) The additional acceleration described by Dr. Kai occurred later in the gradual phase.

KUNDU: With regard to the columnar source for 17 GHz, and 160 MHz, was its meter-wave burst stationary? (I'd expect it to be stationary). If you consider the entire source to be one loop, its thickness has to be rather big, extanding from a few thousand km to ∿ 200 000 km. Do you think more than one loop is involved? I think the event is like a flare continuum.

KAI: I have no position measurements at meter wavelengths. We consider that the entire source is a single loop, because the time profile at 160 MHz is similar to that at 17 GHz. The late enhancements might be a flare continuum as you suggest, but we need further confirmation.

SHORT-PERIOD PULSATIONS OBSERVED SIMULTANEOUSLY BY X-RAY AND RADIO WAVES

S.S. Degaonkar*†, T. Takakura*, P. Kaufmann**, J.E.R. Costa**, K. Ohki***, and N. Nitta*

*Department of Astronomy, University of Tokyo, Tokyo 113, Japan

**Instituto de Pesquisas Espaciais, C.P.515, 12200, Saõ José dos campos, SP, Brazil

***Tokyo Astronomical Observatory, University of Tokyo, Mitaka, Tokyo 181, Japan

†on leave of absence from Physical Res. Lab., Ahmedabad, India

ABSTRACT

From simultaneous high-time-resolution observations of solar X-rays from Hinotori and the millimeter waves at Itapetinga Radio Observatory in Brazil during a solar flare on November 4, 1981 at 1827 UT, short period ($\sim$300 ms) pulsations have been detected in five time intervals of 2 s each. Both a cross-correlation analysis between X-rays and microwaves and a Fourier analysis were made to verify the significance of the quasi-periodic pulsations. The cross-correlation is significant but the pulsations could not be periodic oscillation.

(A more detailed version of essentially the same paper was published by T. Takakura, P. Kaufmann, J. E. R. Costa, S. S. Degaonkar, K. Ohki, and N. Nitta in Nature 302, 317 (1983).)

DISCUSSION

KUNDU: If the 200 - 300 ms periodicity is due to bouncing of electron clouds in a flux tube $\sim 10^4$ km, then with an instrument like the VLA having 10 s integration time, you'd expect to see the whole tube emitting at cm wavelengths. Am I not correct?

TAKAKURA: The modulation degree in radio waves is very small, 1% or so. Furthermore, the duration of the quasi-pulsation was 2 - 3 s in each time section. Accordingly, we may not expect to see the whole tube.

Solar Physics **86** (1983) 237–238. 0038–0938/83/0862–0237 $ 00.30.

PRIEST: Although the lengths of loops are typically 10 000 km, their widths are on the order of 1000 km, and so short-period pulsations could be due to an MHD effect, namely, fast magnetoacoustic waves propagating across such a coronal flux tube. Alternatively, there could be tearing mode islands 1000 km long, and you should not rule out an MHD interpretation.

TAKAKURA: One problem in the former case is that the modulation degree of X-ray pulsations is as high as 10%. In order to have such a value, the whole X-ray source must be composed of only 10 flux tubes or less, even if the modulation degree is about 100 % in one tube. However, we agree that an MHD interpretation can not be ruled out.

FAST TRANSIENTS IN HARD X-RAY SOLAR FLARES

Alan L. Kiplinger[1,2], B.R. Dennis[2], A. Gordon Emslie[3], K.J. Frost[2], and L.E. Orwig[2]

[1] Applied Research Corporation, Landover, Maryland
[2] Goddard Space Flight Center, Greenbelt, Maryland
[3] Dept. of Physics, University of Alabama in Huntsville, Huntsville, Alabama

ABSTRACT

We present the results of a search for fast spikes in 5483 hard X-ray solar flares as observed with the Hard X-Ray Burst Spectrometer on the Solar Maximum Mission (SMM). Hundreds of fast spikes with durations of less than 1 second have been detected at time resolutions of 128 ms and 10 ms. Fast spikes have been detected with rise and decay times as short as 20 ms and with widths as short at 45 ms that represent the fastest hard X-ray variations yet seen from the Sun. The observations of such fast variations place new constraints on the physical nature of the source.

(A slightly modified version of this contribution can be found in Ap. J. Letters 265, L99.)

DISCUSSION

DE JAGER: Have you subjected your material to a probability analysis, such as the following: Assuming an average flare duration of 100 s, your 5000 flares would contain 5 x 10^6 time intervals of 0.1 s. What fraction of these would, by mere chance, deviate from the background level by more than 3σ? (In addition, the chance of these fictitious bursts would be larger if the time intervals considered would be taken smaller than 0.1 s).

KIPLINGER: Although a statistical test such as you describe is worthy of consideration, it has not been conducted on the present set of 159 flares believed to contain fast spikes. The 3σ criterion cited serves only as a very conservative lower limit to the significance of a single 128 ms sample. Flares with a single 128 ms sample 3σ above the slow

Solar Physics **86** (1983) 239–240. 0038–0938/83/0862–0239 $ 00.30.

component were generally ignored since most fast spikes have longer durations and most of the 159 events have several significant peaks. Some short duration spikes have statistical significances as high as 20σ.

HURLEY: (1) Data from our experiments aboard the Prognoz and Venera spacecraft seem to indicate that as one looks at higher energies, fluctuations become faster. (2) I believe that the e-folding times are physically more significant than the times to maxima, and in addition, are more reliably derived from the data, since it is often difficult to define the absolute minima. Could you give an idea of the e-folding times involved here?

KIPLINGER: (1) It is also very obvious from our data that the impulsive behavior of flares is enhanced at higher X-ray energies. (2) The shortest rise and decay e-folding times observed are on the order of 12 ms.

HARD X-RAY DYNAMIC SPECTRUM OF FLARES OBSERVED BY HINOTORI

N.NITTA and T.TAKAKURA
Department of Astronomy, University of Tokyo
Bunkyo-ku, Tokyo, 113, Japan

K.OHKI
Tokyo Astronomical Observatory, University of Tokyo
Mitaka, Tokyo, 181, Japan

M.YOSHIMORI
Department of Physics, Rikkyo University
Toshima-ku, Tokyo, 171, Japan

ABSTRACT. Time variations of the hard X-ray spectrum in solar flares are observed by the hard X-ray spectrometer (HXM) aboard the Hinotori satellite. With a new presentation of the dynamic spectrum we have studied the differences between impulsive and gradual hard X-ray bursts. In the impulsive events a "bent" spectrum up to some hundred keV persists at least until the main peak. In the gradual events, on the other hand, a power-law spectrum augmented by a low-energy excess is dominant.

1. INTRODUCTION

This paper surveys the hard X-ray spectral variations of solar flares observed by Hinotori. We find in the literature at least four distinct components in the solar hard X-ray emission; a new thermal component (Lin et al., 1981); emission from second-step-accelerated electrons (Bai and Ramaty, 1979; Bai et al., 1983); coronal gradual sources (Frost and Dennis, 1971; Hudson, 1978; Hudson et al., 1982); and the usual impulsive component (Kane and Anderson, 1970; Crannell et al., 1978; Kane et al., 1980). Due to this complexity in the nature of hard X-ray emission and because from a theoretical point of view thermal and nonthermal models do not necessarily correspond to an exponential and a power-law spectrum respectively (cf. Smith and Auer, 1980; Brown, 1974), the hard X-ray spectrum itself may no longer play an important role in the determination of the best flare model. However we hope to clarify the situation by considering its time variation throughout an event.

In this report we present a new description of the hard X-ray spectrum, the "dynamic spectrum", for different types of bursts observed by Hinotori hard X-ray spectrometer (HXM), the description of which can be found elsewhere (Ohki et al., 1982a). "Dynamic spectrum" is a sequential plot of evolving spectrum with time on a third axis. This clearly displays the spectral variations throughout an event with

Solar Physics **86** (1983) 241–246. 0038–0938/83/0862–0241 $ 00.90.

good counting statistics.

In the next section we describe systematic differences between impulsive and gradual events, using these displays. Section 3 extends the HXM spectrum into the gamma-ray range ($h\nu \lesssim 1$ MeV). The differences between the two types of burst are also seen in these regions. In section 4 comparisons of our results with previous observations are made and some implications are considered.

2. IMPULSIVE AND GRADUAL BURSTS

For discussion let us call a burst "impulsive" if it has spike(s) of duration (FWHM) $\lesssim$ 1 minute and total duration $\lesssim$ 10 minutes. This type of burst is the most common in the Hinotori observations. We show an example in Figure 1. To supplement the qualitative information in the dynamic spectrum we also present time profiles accompanied by comments on spectral fits (exponential or power-law) and best-fit parameters.

We have studied more than 20 impulsive events. These include events with a wide range of intensity, duration, hardness, and number and width of spikes. Still they tend to have remarkable common features.

(a) Crowded regions that occasionally appear in the dynamic spectrum (Figure 1 (a)) correspond to the spikes in the time profile. Just after such a region, the spectrum often changes from an exponential to a power-law form (Figure 1 (b)).

(b) After the spectrum changes to a power-law, the slope becomes steeper and steeper. The scatter found in the high energies (Figure 1 (a)) in the dynamic spectrum toward the end of a burst is due mostly to the statistical fluctuations and indicates the depletion of photons of $h\nu \gtrsim 150$ keV.

(c) The spectrum is hardest at peaks of spikes.

"Gradual" bursts are those without sharp spikes and of duration $\gtrsim$ 20 minutes. These include some of the largest bursts that Hinotori observed, and the total number is small (less than 10). An example is found in Figure 2. We notice first that the time profiles in lower energies are different from those in higher energies. Our tentative double spectral fitting (exponential plus power-law) shows that in lower-energy channels most of photons come from a thermal plasma of $T = 1.2 \times 10^8$ K and EM = 2×10^{48} cm^{-3}. Though other gradual bursts have different T and EM, the lower-energy excess is the common feature. At high energies the power-law spectrum with small and decreasing γ is persistent. This is most vivid toward the end of the burst.

We must mention here that there are of course bursts which belong to neither category. One of the example is the event of 17 July 1981 (Ohki et al., 1982b), a gradual one but with extremely soft spectrum.

3. RELATION TO GAMMA-RAY ENERGIES

The consistency between the flux in the higher-energy channels of HXM and the lower-energy channels of SGR (Solar Gamma-Ray spectrometer, for details see Okudaira et al., 1981) has been checked in seven different bursts. However for the fine analysis of spectral variations the accuracy is still insufficient (sometimes a factor 2 discrepancy) due largely to the different deconvolution techniques between HXM and

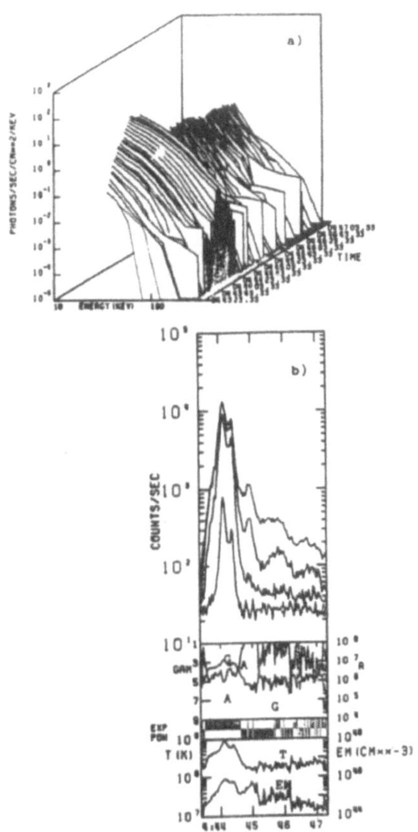

Fig. 1. The impulsive event of 15 October 1981.
(a) Dynamic spectrum: time goes up and to the right.
(b) (up to down) Time profiles in four energy bands (PC 1, PC 2+3, PC 4+5, PC 6+7, i.e. 33-44 keV, 44-75 keV, 75-169 keV, 169-399 keV). Variation of A and γ in $I(\varepsilon)=A\varepsilon^{-\gamma}$. Indication of better spectral fit (exponential plus power-law). Temperature and emission measure in thermal fit.
Integration time is 2 seconds. Spectral fitting in (b) is made using PC 2 through PC 7

SGR. In Figure 3 we use the same events as in Figures 1 and 2 to compare the spectral behaviour in the range of $h\nu \gtrsim 100$ keV. We find that while in the event of 15 October 1981 the flux above 300 keV changes in phase with that at lower energies, the gradual burst of 7 October 1981 shows a different pattern. That is, the spectrum above 500 keV hardens as the flux at 100 keV decreases.

Another point to be noticed is that the spectrum above 100 keV is much steeper ($\gamma \gtrsim 4$) in impulsive bursts than in gradual bursts ($\gamma \sim 2$). Actually only in very few impulsive bursts can we detect photons beyond 1 MeV.

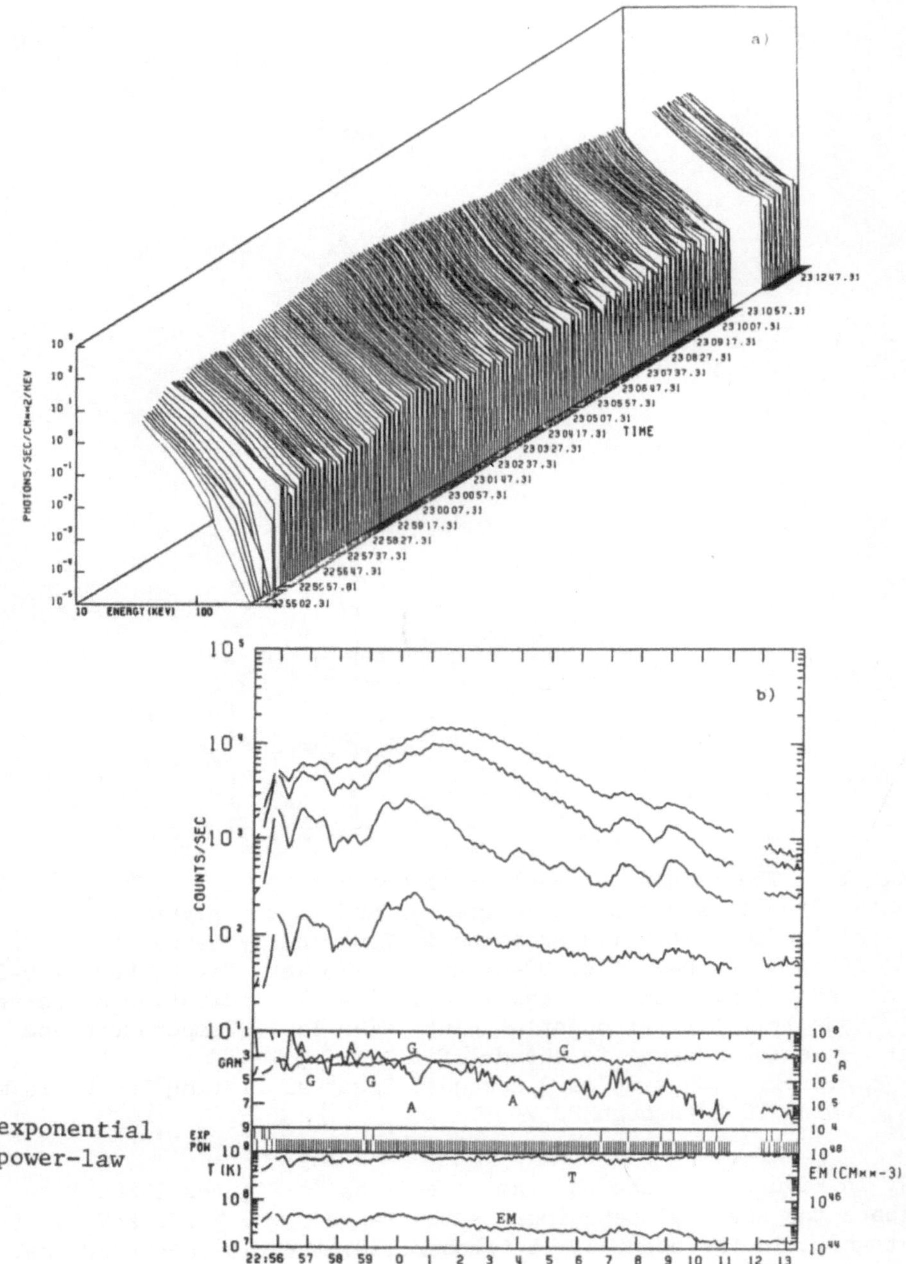

Fig. 2. The gradual burst of 7 October 1981. The same as Fig. 1. But the integration time is 5 seconds, and the spectral fit is made using only PC 4 through 7.

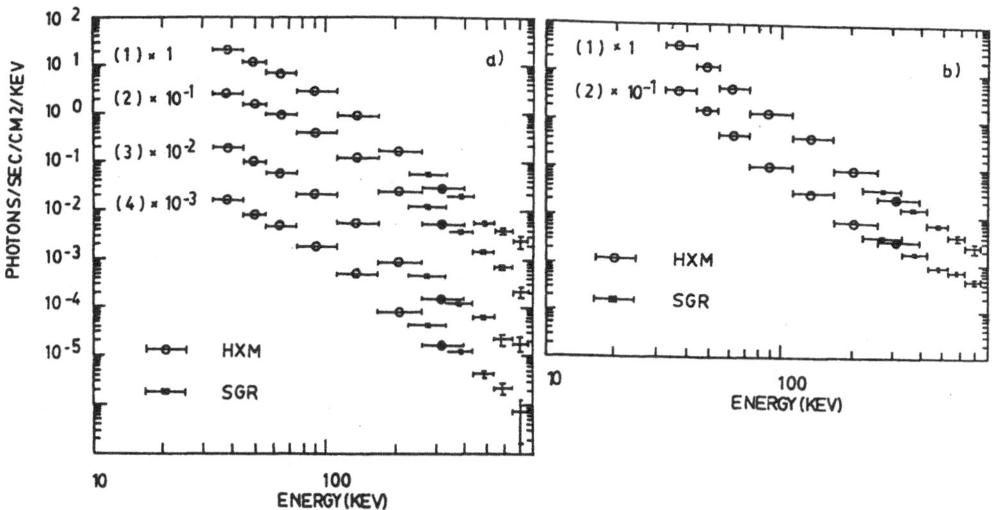

Fig. 3. X-ray and gamma-ray spectra for the events of 15 October (a) and 7 October 1981. Time intervals are 0443:57-0444:05 (1), 0444:05-0444:13 (2), 0444:13-0444:21 (3), 0444:21-0444:29 (4) in (a), and 2300:05-2300:13 (1), 2300:41-2300:49 (2) in (b).

4. DISCUSSION

Our results are generally consistent with previous observations. The bent spectrum seen at least until the peaks of impulsive bursts has been reported to be very common (Kane and Anderson, 1970; Crannell _et al_., 1978; Elcan, 1978). However when gamma-ray data are taken into account an isothermal spectrum does not explain the higher-energy part ($h\nu \gtrsim 200$ keV) if there is only one electron population as suggested by the event of 15 October 1981. The spectral softening previously observed in the decay phase of small bursts with a single spike (Kane and Anderson, 1970) is also seen in our larger impulsive events. The extent to which this spectral softening is due to the contribution from the thermal ($T \sim 3 \times 10^7$ K) plasma is now being studied. Also in the decay phase of impulsive bursts the spectrum fits a power-law better than an exponential. But contrary to the case of gradual bursts the spectrum is very steep and shows complete loss of photons above 150 keV.

Turning to gradual bursts one of the most remarkable feature is that time profiles in lower energies are not similar to those in higher energies. If we attribute these low-energy photons to some thermal component the temperature should be between 4×10^7 K and 1.4×10^8 K, varying with events, a little higher than that of thermal emission found by Lin _et al_. (1981). This excess in lower energies appears not only after the peak, but also in some cases from the beginning.

Another common feature in gradual bursts is that the spectrum in higher energies hardens throughout an event. The energy-dependent time delay (Ohki _et al_., 1983), which has not been found in impulsive bursts to an accuracy of about 1 second, will result from this spectral hardening.

Gradual events seem to belong to the so-called extended bursts (Hoyng et al., 1976). They are often situated in the corona as seen at around 20 keV (Takakura et al., 1983; Ohki et al., 1983), but more detailed comparisons with other wavelengths will unveil what makes differences essentially between impulsive and gradual bursts.

ACKNOWLEDGEMENTS

The deepest thanks of the authors are due to Prof. M.Oda, Prof. Y. Ogawara, Dr. T.Murakami and Dr. K.Makishima for their collaboration and encouragement throughout every stage of the HXM experiment. We are also grateful to Dr. H.S.Hudson for valuable advice. We wish to thank Prof. Y.Tanaka of ISAS, Prof. I.Kondo of ICR, University of Tokyo and Prof. K.Tanaka of TAO, University of Tokyo for their general management of the Hinotori project.

REFERENCES

Bai, T. and Ramaty, R.: 1979, Ap. J. 227, 1072.

Bai, T., Hudson, H.S., Pelling, R.M., Lin, R.P., Schwartz, R.A. and von Rosenvinge, T.T.: 1983, Ap. J. (preprint).

Brown, J.C.: 1974, in G.Newkirk, Jr. (ed.), 'Coronal Disturbances', IAU symp. 57, 395.

Crannell, C.J., Frost, K.J., Matzler, C., Ohki, K. and Saba, J.L.: 1978, Ap. J. 223, 620.

Elcan, M.J.: 1978, Ap. J.(Letters) 226, L 99.

Frost, K.J. and Dennis, B.R.: 1971, Ap. J. 165, 655.

Hoyng, P., Brown, J.C. and van Beek, H.F.: 1976, Solar Phys. 48, 197.

Hudson, H.S.: 1978, Ap. J. 224, 235.

Hudson, H.S., Lin, R.P. and Stewart, R.T.: 1982, Solar Phys. 75, 245.

Kane, S.R., Crannell, C.J., Datlowe, D., Feldman, U., Gabriel, A., Hudson, H.S., Kundu, M.R., Matzler, C., Neidig, D., Petrosian, V. and Sheeley, Jr.,N.R.: 1980, in P.A.Sturrock (ed.), 'Solar Flares', Corolado Associated Univ. Press, 187.

Kane, S.R. and Anderson, K.A.: 1970, Ap. J. 162, 1003.

Lin, R.P., Schwartz, R.A., Pelling, R.M. and Hurley, K.C.: 1981, Ap. J.(Letters) 251, L 109.

Ohki, K., Tsuneta, S., Takakura, T., Nitta, N., Makishima, K., Murakami, T., Ogawara, Y., Oda, M. and Miyamoto, S.: 1982a, 'Hinotori Symposium on Solar Flares', ISAS, 102.

Ohki, K., Nitta, N., Tsuneta, S., Takakura, T., Makishima, K., Murakami, T., Ogawara, Y., Oda, M. and Miyamoto, S.: 1982b,ibid.,69.

Ohki, K., Takakura, T., Tsuneta, S. and Nitta, N.: 1983, in these proceedings.

Okudaira, K., Hirashima, Y., Yoshimori, M. and Kondo, I.: 1981, in the 17th ICRC (Paris) SH-1, 11.

Smith, D.F. and Auer, L.H.: 1980, Ap. J. 238, 1126.

Takakura, T., Tsuneta, S., Ohki, K. and Nitta, N.: 1983, Ap. J. (Submitted).

A FLARE MODEL DEDUCED FROM HINOTORI AND MILLIMETERWAVE INTERFEROMETER OBSERVATIONS

Kin-aki Kawabata, Hideo Ogawa, and Ikuro Suzuki
Department of Physics and Astrophysics, Nagoya University
Chikusa-ku, Nagoya 464, Japan

ABSTRACT. The γ-ray and white-light flare of 13 May, 1981 is used for a study of spatial distributions of energetic electrons and high-temperature plasma.

1. INTRODUCTION

Systematic studies of flares over a wide range of the spectrum, from γ-rays through the radio band, are of great importance for understanding of physical processes involved in solar flares. However, we have only a small chance for having complete data from simultaneous observations in such a wide range of the spectrum. On 13 May 1981, a big flare with importance 2B or 3B and X 1.5 (Solar Geophysical Data) occurred near the east limb (E58N09). The flare was accompanied by γ-ray line emissions and by a white light flare (Hiei et al., 1982). The flare started at 0345 UT, reached maximum development at 0412 UT, and ended at 0612 UT, The flare was observed by the Hinotori X-ray imaging telescope (SXT) in soft (5-10 keV) and hard (17-40 keV) X-rays, by the 35 GHz fanbeam interferometer at Nagoya, by an Hα-filtergraph at Mitaka, and by a spectrograph at Mt. Norikura. These observations give us various kinds of information on spatial distributions of energetic electrons and high temperature plasma. Observations of the flare were also made by the other instruments on Hinotori, and in a wide range of radio frequencies by a radiospectrograph, polarimeters and radiometers (Solar Geophysical Data). The event is the one of the best observed big events and gives us a chance to carry out a systematic study.

X-ray images obtained by SXT show that both of soft (5-10 keV) and hard (17-40 keV) X-ray sources are located at the top of an arcade of magnetic loops which was observed by an Hα-filtergraph in the decaying phase of the flare (Tsuneta et al., 1983). Both X-ray sources were stationary and had an elongated shape with a size of 60 000 km in length and 28 000 km in width. The electron temperature and the emission measure of the soft X-ray source had been already enhanced at the start time of observations and these remained nearly the same values throughout the observations (Hiei et al., 1982). The observations indicate that the heating and the material supply to the soft X-ray source were finished at the beginning of the observations. The electron temperature, emission measure and the electron density were 2×10^7 K, $3 \times 10^{49}\text{cm}^{-3}$, and 3×10^{10} cm^{-3}, respectively (Tsuneta *et al*., 1983). Nitta *et al*. (1983)

Solar Physics **86** (1983) 247–252. 0038–0938/83/0862–0247 $ 00.90.

have shown that the low energy part of hard X-rays also has been already enhanced at the beginning of observations and they have mentioned a spectral hardening starting at 0412 UT. Kawabata et al. (1982) have shown that the time history of 35 GHz radio emission resembles that of the γ-rays higher than 720 keV. In the present article, we describe some results deduced from the 35 GHz interferometer observations combined with these investigations from Hinotori observations by various authors.

2. 35 GHz OBSERVATIONS

Figure 1 illustrates a comparison of the hard X-ray image (Tsuneta et al., 1983) and the 35 GHz fanbeam brightness. As is seen in the Figure 1 the 35 GHz radio source is composed of two components. One of them can be identified with the X-ray source at the top of arcade. Comparing with an Hα photograph (Mitaka) and a sketch of sunspots (Mitaka), the other source can be identified with a sunspot (Takakura et al., 1982). The radio brightening at the top of the arcade can be first recognized at about 0400 UT and remained weak until 0413 UT. The radio component started to increase in brightness at about 0413 UT, and at the same, the other component was also enhanced. Comparing with the dynamic spectrum of X-rays, the radio enhancement can be identified with a hardening of the X-ray spectrum.

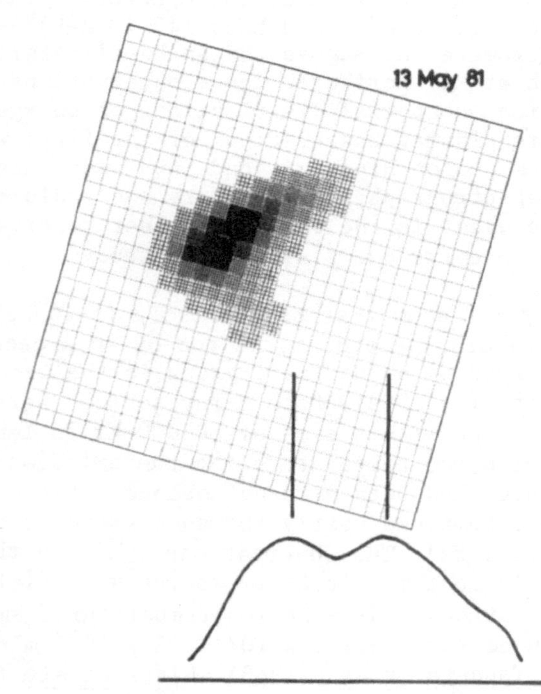

Figure 1. X-ray map (17-40 keV) and fanbeam brightness distribution at 35 GHz. X-ray map was supplied by Mr. S. Tsuneta.

3. INTERPRETATIONS

Because a rough estimation from gyro-synchrotron theory shows that only 1 % of the energetic electrons in number produce sufficiently strong 35 GHz emissions near the sunspot region, we concentrate our discussions only on the radio emission from the top of the arcade.

The free-free emission at 35 GHz computed from the emission measure and the electron temperature from the soft X-ray emission-line spectrum is about 15 s.f.u.. As is shown later, the magnetic field strength at the top of the arcade is about 50-60 gauss and then the gyro-resonance emission from the thermal plasma can be ignored. Similarly, gyro-resonance emission from the electrons producing hard X-rays can be also ignored until 0412 UT, because the hard X-ray spectrum is soft. The computation is in good agreement with the radio flux densities observed at 35 GHz before 0413 UT. Thus we conclude that the 35 GHz emission at the top of the arcade before 0413 UT is due to free-free transitions in the plasma with a temperature of 2×10^7 K.

Computations of the 35 GHz radio flux densities due to the gyro-synchrotron theory have been done for various assumed field strengths from the electron spectrum extrapolated from the hard X-ray region at an angle of 60° from the magnetic field direction. Because the ion density

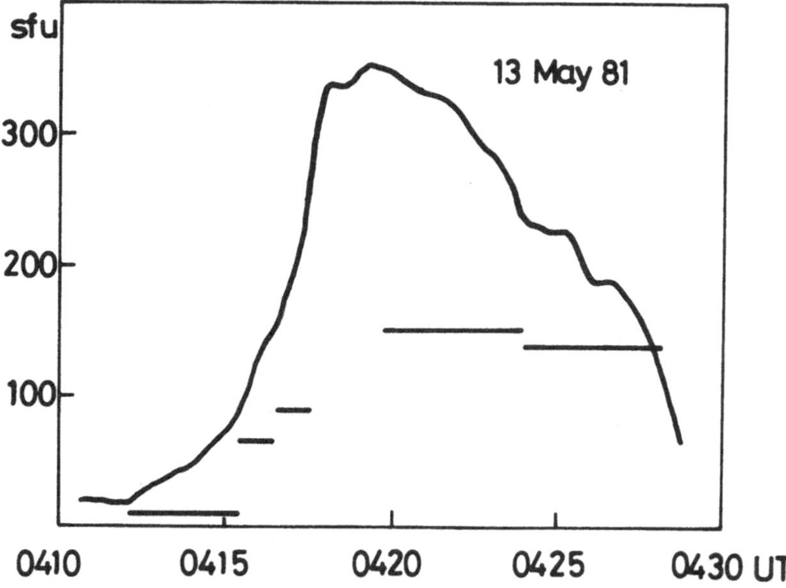

Figure 2. Time history of the flux density at 35 GHz. About a half of the flux density comes from the top of the arcade. Horizontal bars indicates computed flux for the source at the top of arcade by the gyro-synchrotron theory.

of the background plasma is known from the spectroscopic and imaging observations, the only unknown parameters are the field strengths and the angle between the field direction and the line of sight. The time history of the computed flux densities for a field strength of 63 Gauss is shown in Figure 2. If we take into account that about half of the radio emission comes from the source at the top of the arcade, the computed time history of flux density reproduces the observed one well. In this computation, most of the 35 GHz radiation is emitted from electrons having an energy of several MeV. The computation is compatible with the one-to-one correspondence between γ-ray bursts and 35 GHz bursts (Kawabata et al., 1982).

In this computation, we have put a low energy cut-off of 100 keV and high energy cut-off of 6.5 MeV. The low energy cut-off is in practice not important, because 35 GHz radiation from electrons with energy less than 100 keV are not significant. The high energy cut-off has been put equal to the highest observed γ-ray energy somewhat arbitrarily. When we put a high energy cut-off as high as 10 MeV or more, we obtain the best fit at 50 gauss.

When the field direction is at random, the median value of the angle between the field direction and the line of sight is 60°. Then the computation at the angle of 60° gives a reasonable typical value of the field strength. When the angle is close to 90°, the required field strength becomes somewhat weaker, say 50 gauss.

Sakurai(1983) has computed the potential field of the active region that produced the 13 May 1981 flare from magnetograph data, obtaining a field strength of 50 Gauss at the X-ray source. The field strength for the interpretation of 35 GHz flux densities is thus very close to the potential-field approximation. The agreement may suggest that the magnetic field lines in the radio emitting region do not carry large currents and are fairly simple.

4. CONCLUSIONS

During the observations on Hinotori, the electron density of the ambient plasma was $3 \times 10^{10}\ cm^{-3}$ (Tsuneta et al., 1983) as was already described. At his electron density, the ionization loss time of 20 keV, 100 keV, 300 keV, and 4 MeV electrons become 0.5, 5, 30, and 350 seconds, respectively. Thus it is evident that the electrons must have been continuously accelerated. The ionization loss time is less than the travel time over the scale lengths of the X-ray source for electrons below 30 keV. The electrons below 30 keV are not able to escape from the source region. Thus the confinement of the electrons observed by hard X-ray imaging telescope can naturally be interpreted. The 35 GHz radio observations indicate that electrons with an energy of several MeV are trapped in the source. The spectral hardening in the decaying phase mentioned by Nitta et al. (1983) can be interpreted by ionization loss.

The radio brightness at the sunspot may be due to the precipitation of electrons in the loss-cone; if so the time history of radio brightness may be a manifestation of the rate of particle acceleration. The time delay of 3 minutes observed at hard X-ray and γ-ray energies and in the 35 GHz flux density implies that it takes 3 minutes to accelerate electrons up to 300 keV.

ACKNOWLEDGEMENT

Authors would like to thank to Prof. T. Takakura, Dr. N. Ohki, Dr. Y. Ogawara, Dr. M. Yoshimori, Mr. S. Tsuneta, Mr. Nitta and to all the members of the Hinotori Group at the Tokyo Astronomical Observatory and the Institute of Space and Astronautical Science for their supply of X-ray maps, X-ray and γ-ray data before publication. They would also like to thank to the Institute of Plasma Physics for the use of their computer.

The solar interferometer at 35 GHz used in the present work was constructed under the support by the Grant-In-Aid for Scientific Research 942026 and 343003. Observations and the reduction of radio data have been carried out by support from Grant-In-Aid for Special Project 56110004 and by Grant-In-Aid for Scientific Research 565401.

REFERENCES

Hiei,E.,Takanka,K.,Watanabe,T.,and Akita,K.:1982,Proc.Hinotori Symposium on Solar Flares,Institute of Space and Astronautical Science, Tokyo, p.208.

Kawabata,K.,Ogawa,H,,Takakura,T.,Tsuneta,S.,Ohki,K.,Yoshimori,M.,Okudaira,K.,Hirasima,Y.,and Kondoh,I.:1982, Proc. Hinotori Symposium on Solar Flares,Institute of Space and Astronautical Science, Tokyo,p.168.

Nitta,N.,Takakura,T.,Ohki,K.,and Yoshimori,M.:1983,These Proceedings.

Sakurai,T.:1983,These Proceedings.

Takakura,T.,Ohki,K.,Tsuneta,S.,Nitta,N.,Makishima,K.,Murakami,T., Ogawara,Y., and Oda,M.: 1982, Proc. Hinotori Symposium on Solar Flares, Institute of Space and Astronautical Science, Tokyo, p.142.

Tsuneta,S.,Takakura,T.,Nitta,N.,Ohki,K.,Ogawara,Y.,Murakami,T., Makishima,K., and Oda, M.:1983,These Proceedings.

DISCUSSION

KUNDU: I'd like to note that, just as you observe 35 GHz emission when the hard X-ray spectrum hardens, we find a similar result. We've studied the relationship between 6 cm emission and hard X-rays (28 - 498 keV) as observed by Frost experiment. We find that 6 cm flux corresponds not so well with the integration flux (28 - 498 keV), but very well with the highest energy channel. This corresponds also with the time when the X-ray spectrum hardens.

KAWABATA: In our cooperative work with Hinotori SGR group, we have shown a one-to-one correspondence between Hinotori γ-ray events and 35 GHz radio events with peak flux densities higher than 300 sfu. The time history of 35 GHz bursts resembled that of γ-ray bursts. The results have been published in Hinotori Symposium, ISAS, Jan. 1982.

HUDSON: What is the energy of electrons producing 35 GHz under these conditions? (A few MeV - an extrapolation is necessary). Isn't there enough information from hard X-rays, γ-rays, and mm-wave to determine the overall electron specrum quite well? Does the same spectral distribution extend over the whole 10 keV - 10 MeV range? Is there any possibility of observing synchrotron energy losses?

KAWABATA: From the radio spectrum in the optically thin part, we can obtain the electron energy spectrum. In order to specify the electron energy and number of electrons, we require magnetic field strengths.

SIMNETT: (A) At what frequency did the microwave spectrum peak on May 13? (B) Regarding your general statement regarding maximum frequency of the microwave emission, there is no physical reason why it should be less than, or on the order of, 35 GHz. In fact, during the last solar maximum, e.g., July 8, 1968, the maximum frequency was observed to exceed 71 GHz.

KAWABATA: (A) For the 13 May 1981 event, no observations have been made at 17 GHz during its main phase. Hence we cannot give a meaningful answer. However, 17 GHz observations give generally the highest peak flux density in most of the big or moderate events. (B) This is the results of observations. We have never observed an event with a spectral peak above 35 GHz whenever the data at 17 GHz are available. I agree with you on the point that we should investigate why the spectral peak were always below 35 GHz. It is a still unresolved problem.

ACTON: There appears to be an offset between the most intense X-ray location from the SXT map and the most intense 35 GHz emission from your radio scan. What is the explanation of this?

KAWABATA: The accuracy in absolute position is estimated to be ± 7" from our own data. Intercalibrations between the data from Nobeyama Observatory and our data have given a better agreement with a discrepancy of a few arc sec. Thus we feel the accuracy will be better than ± 7".

Unfortunately, we had trouble with our computer on 13 May 1981 and so no digital data are available for the main part of this event. We determined the location from the pen record taken for quick-look data. Thus the accuracy was not so good and is estimated to be ± 15" in peak-to-peak value. The SXT map also has a certain amount of error. Thus we can not claim a discrepancy between the most intense part of the SXT map and the radio position.

RELATION BETWEEN HARD X-RAY SPECTRA AND ELECTRON ENERGY SPECTRA

Ikuro Suzuki and Kin-Aki Kawabata
Department of Physics and Astrophysics, Nagoya University,
Chikusa-ku, Nagoya 464, Japan

ABSTRACT. The relationship between hard X-ray spectra and energetic electron spectra in solar X-ray bursts is investigated, and a simplified cross-section for bremsstrahlung which is applicable to the region of mildly relativistic energies is proposed. Using the proposed cross-section, we solve an integral equation to obtain the electron energy spectrum. The validity of the proposed cross-section is checked by comparing the spectrum calculated by the exact Bethe-Heitler formula. A good agreement between two calculated spectra is obtained up to 10 MeV energy with an accuracy of 20 %.

1. INTRODUCTION

The relation between bremsstrahlung X-ray spectra and the source electron energy spectra has been solved for nonrelativistic energies by Brown (1971,1975). For thin-target X-ray production in the nonrelativistic energy range, Brown (1971) has given the relation $N(E) \propto E^{0.5} I(E)$ for a power-law X-ray spectrum, where N(E) is the electron erergy spectrum emitting the X-rays and I(E) is the X-ray energy spectrum. On the other hand Ramaty et al. (1975) calculated X-ray spectra by using the exact Bethe-Heitler cross-section. Assuming a power-law electron spectrum, they obtained $N(E) \sim E^{0.5} I(E)$ in the nonrelativistic region, going over to $N(E) \sim E^{-0.1} I(E)$ in the ultrarelativistic region.

In this paper we investigate the relation between two spectra in order to get a simple and useful formula in the entire energy range <10 MeV.

2. ELECTRON ENERGY SPECTRUM AND X-RAY ENERGY SPECTRUM

At the Earth, the X-ray flux is given by the equation

$$I(\varepsilon,t) = \frac{n}{4\pi R^2} \int_{\varepsilon}^{\infty} c\beta(E)\sigma(\varepsilon,E)N(E,t)dE \quad , \tag{1}$$

where n is the average value of the number density of thermal protons, R=1AU, $\beta(E)=v(E)/c$, $\sigma(\varepsilon,E)$ the bremsstrahlung cross-section, ε and E photon and electron kinetic energy respectively.

Jackson (1962) gives a bremsstrahlung cross-section applicable even

Solar Physics **86** (1983) 253–257. 0038–0938/83/0862–0253 $ 00.75.

in the relativistic case,

$$\sigma(\varepsilon,E) = \frac{16}{3}\frac{\alpha r_0^2}{\varepsilon\beta(E)^2}\ln\left(\frac{2m_0 v(E)^2}{(1-\beta(E)^2)\varepsilon}\right) , \tag{2}$$

where

$$\beta(E) = \frac{\{2E/m_0c^2(1+E/2m_0c^2)\}^{1/2}}{1+E/m_0c^2} , \tag{3}$$

$\alpha=e^2/\hbar c$ is the fine structure constant, $r_0=e^2/m_0c^2$ the electron classical radius. As the logarithmic factor varies slowly, we may deal with the factor nonrelativistically. If we replace $\beta(E)^2$ in the logarithmic factor by 0 and $v(E)$ by an average value of the initial state and the final state in an emission process, we obtain the modified cross-section

$$\sigma(\varepsilon,E) = \frac{16}{3}\frac{\alpha r_0^2}{\varepsilon\beta(E)^2}\ln\frac{1+\sqrt{1-\varepsilon/E}}{1-\sqrt{1-\varepsilon/E}} . \tag{4}$$

This is the same as the conventional expression of the Bethe-Heitler cross-section in the nonrelativistic case, if we replace the factor $1/\beta^2$ by $m_0c^2/2E$. Therefore in the following calculation we use the cross-section (4) with $\beta(E)$ given by Equation (3).

Using Brown's (1971) method, we obtain the solution of Equation (1) as

$$N(E,t) = \frac{4R^2}{nK}\left(\frac{m_0}{2}\right)^{1/2}\frac{(1+E/2m_0c^2)^{1/2}}{1+E/m_0c^2}\int_E^\infty d\varepsilon\,\frac{\frac{\partial}{\partial\varepsilon}\varepsilon\frac{\partial}{\partial\varepsilon}\varepsilon I(\varepsilon,t)}{\sqrt{\varepsilon-E}} , \tag{5}$$

where $K=(8/3)\alpha r_0^2 m_0c^2$. Equation (5) agrees with Brown's (1975) result in the nonrelativistic limit of $E/m_0c^2\ll 1$. For a conventional power-law X-ray spectrum,

$$I(\varepsilon,t) = A(t)\varepsilon^{-\gamma(t)} , \tag{6}$$

we obtain

$$N(E,t) = \frac{4R^2}{nK}\left(\frac{m_0}{2}\right)^{1/2}(\gamma(t)-1)^2 B(\gamma(t)-1/2,1/2)\times$$
$$\times\frac{(1+E/2m_0c^2)^{1/2}}{1+E/m_0c^2}A(t)E^{-\gamma(t)+0.5} , \tag{7}$$

where $B(x,y)$ is the Beta function. If we express $A(t)E^{-\gamma(t)}$ in terms of $I(E,t)$, the ratio $N(E,t)/I(E,t)$ becomes

$$\frac{N(E,t)}{I(E,t)} = \frac{4R^2}{nK}\left(\frac{m_0}{2}\right)^{1/2}(\gamma(t)-1)^2 B(\gamma(t)-1/2,1/2)\frac{\{E(1+E/2m_0c^2)\}^{1/2}}{1+E/m_0c^2} . \tag{8}$$

In the nonrelativistic region ($E/m_0c^2\ll 1$) the energy dependence of the ratio N/I is proportional to $E^{0.5}$, on the other hand in the ultrarelativistic region ($E/m_0c^2 \gg 1$) the ratio approaches $E^{0.0}$. The reason for this change in the energy dependence comes from the fact that $\beta(E)$ becomes $(2E/m_0c^2)^{0.5}$ in the nonrelativis ic approximation and 1 in the ultrarelativistic limit. Therefore the electrons with a simple power-law spectrum $E^{-\delta}$ emit the X-rays with spectra $E^{-\delta-0.5}$ in the nonrelativistic region and $E^{-\delta}$ in the ultrarelativistic region.

3. COMPARISON OF OUR TREATMENT WITH EXACT TREATMENT

In order to examine our treatment we compare the X-ray spectrum calculated by the proposed cross-section (4) with that calculated by the exact Bethe-Heitler cross-section. For the sake of definiteness we assume a power-law spectrum for energetic electrons,

$$nN(E) = CE^{-\delta} \qquad \text{for } E>E_c \quad , \tag{9}$$

where E_c is a low energy cutoff. The exact cross-section for electron-proton bremsstrahlung is given by (Koch and Motz, 1959)

$$\begin{aligned}\sigma(\varepsilon,E) = \alpha r_0^2\frac{P'}{\varepsilon P}\{\frac{4}{3} - 2TT'(\frac{P^2+P'^2}{P^2P'^2}) + \frac{kT'}{P^3} + \frac{k'T}{P'^3} - \frac{kk'}{PP'} + \\ +L\left(\frac{8}{3}\frac{TT'}{PP'} + \frac{\varepsilon^2(T^2T'^2+P^2P'^2)}{(m_0c^2)^2P^3P'^3} + \frac{\varepsilon/m_0c^2}{2PP'}(k\frac{TT'+P^2}{P^3} - \right. \\ \left. -k'\frac{TT'+P'^2}{P'^3} + \frac{2\varepsilon TT'}{m_0c^2P^2P'^2}\quad)\right)\} \quad ,\end{aligned} \tag{10}$$

where $T = E/m_0c^2+1$, $T' = T - \varepsilon/m_0c^2$, $P = (T^2-1)^{1/2}$, $P' = (T'^2-1)^{1/2}$

and $L = 2\ln\dfrac{TT'+PP'-1}{\varepsilon/m_0c^2}$, $k = \ln\dfrac{T+P}{T-P}$, $k' = \ln\dfrac{T'+P'}{T'-P'}$.

Inserting Equations (9) and (10) into Equation (1), the solid line in Figure 1 is calculated. On the other hand using equations (4) and (9) the dotted line in Figure 1 is calculated. The two lines show close agreement in the energy range from 10 keV to 10 MeV, with a difference of about 20 % in the region of energy over 100 keV. The disagreement in the range over 1 MeV may be due to the relativistic effect in the logarithm.

However as the discrepancy is minor, our treatment is useful for applications in astrophysical problems. Since a one-to-one correspondence in solar flares exists between γ-ray bursts and millimeter-wave bursts (Kawabata et al., 1982), the correspondence indicates that the same electrons emit γ-rays by bremsstrahlung mechanism and millimeter waves by gyrosynchrotron mechanism simultaneouly. Applying our method to solar flares, the relation between γ-ray bursts and millimeter-wave bursts may be made clear.

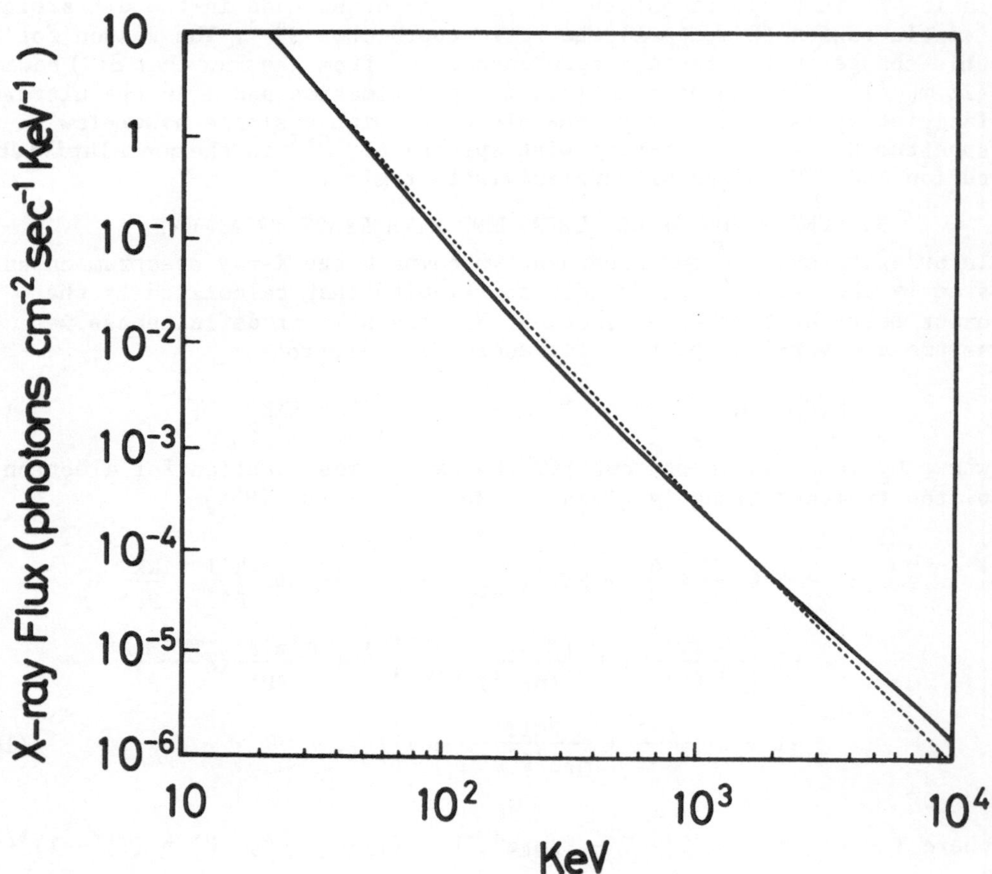

Fig. 1. X-ray emission from electrons with $nN(E) = 2.83 \times 10^{47} E(\text{keV})^{-2.5}$ $\text{cm}^{-3}\text{keV}^{-1}$ at the Earth. A solid line is obtained from the exact Bethe-Heitler cross-section (10) and using the proposed cross-section (4) a dotted line is calculated.

4. DISCUSSION

In this paper we consider only electron-proton bremsstrahlung. Electron -electron bremsstrahlung is negligible in the nonrelativistic region, and it contributes only 35 % in the total bremsstrahlung above 0.5 MeV (Bai and Ramaty, 1976). Therefore our treatment is approximately good up to 10 MeV energy. Since most of solar X- and γ-ray bursts occur in the energy range from 1 keV to 10 MeV, our method is useful for data analyses of solar bursts.

REFERENCES

Bai, T. and Ramaty, R.: 1976, Solar Phys. 76, 343.
Brown, J. C.: 1971, Solar Phys. 18, 489.
Brown, J. C.: 1975, in S. R. Kane (ed.), "Solar Gamma-, X-, and EUV Radiation", IAU Symp. 68, 245.

Jackson, J. D.: 1962, "Classical Electrodynamics", Chp. 15, Jhon Wiley & Sons, New York.

Koch, H. W. and Motz, J. W.: 1959, Rev. Mod. Phys. 31, 920.

Kawabata, K., Ogawa, H., Takakura, T., Tsuneta, S., Ohki, K., Yoshimori, M., Okudaira, K., Hirashima, Y., and Kondo, I.: 1982, Proc. Hinotori Symposium on Solar Flares, p. 168. Institute of Space and Astronautical Science, Tokyo.

Ramaty, R., Kozlovsky, B., and Lingenfelter, R. E.: 1975, Space Sci. Rev., 18, 431.

DISCUSSION

EMSLIE: At relativistic energies, the bremsstrahlung cross-section is highly anisotropic. Thus the modification to Brown's (1971) formulae due to anisotropy is probably larger than the 15% obtained for the direction-integrated cross-section.

SUZUKI: As I consider the region of energy $\lesssim 5$ MeV the effects due to anisotropy are not important. My formula of the direction-integrated cross-section gives a good approximation to several MeV.

A COMPARISON OF HIGH-ENERGY EVENTS IN THE QUIET SUN WITH SOLAR FLARES

G.E. Brueckner
E.O. Hulbert Center for Space Research,
Naval Research Laboratory, Washington, DC 20375, U.S.A.

ABSTRACT. Observations of high-speed coronal clouds (OSO-7), flare ejecta (Skylab) and high-energy jets (HRTS) are compared. It is possible that the same physical mechanism - an expanding loop - which is responsible for the high speed jets (400 km sec^{-1}, 2.5 x 10^{26} ergs) can also account for the high-speed coronal clouds (1300 km sec^{-1}, 4 x 10^{30} ergs), which were correlated with a flare-connected spray. Field strength of 15 gauss and 2500 gauss are required for the jets and the sprays, respectively.

1. INTRODUCTION

This paper is trying to develop scaling laws which are to provide boundary condition for mechanism to explain the high-velocity mass ejections from the solar surface. Sprays are the most energetic (~10^{32} ergs) phenomena of this class. They are well-observed in Hα. However, there are no observations of the magnetic configuration which may lead to their acceleration. On the other end of the scale there are the jets, which have been found recently in the quiet Sun (Brueckner, 1980). Because they are also high-velocity events (~400 km sec^{-1}) they may be caused by a similar mechanism which provides much less energy. Although jets can only be observed in the ultraviolet part of the spectrum, the few high-resolution observations carried out with NRL's High Resolution Telescope and Spectrograph (HRTS) experiment during sounding rocket flights provide many detailed information about the physical process which lead to their acceleration. The observations of three typical events (spray, flare-ejecta and jets) are summarized in the next paragraph. From these parameters the physical condition for the jet mechanism are derived and finally an attempt is made to scale this mechanism up to explain high-energetic sprays.

Solar Physics **86** (1983) 259–265. 0038–0938/83/0862–0259 $ 01.05.

2. OBSERVATIONS

(a) High-energetic clouds in the corona caused by a spray.

The NRL white-light coronagraph on OSO-7 observed a string of four high velocity clouds (Fig. 1) on December 14, 1971 at 0407, 0418 and 0430 U.T. moving outward at 6-8 R_o with projected velocities between 950-1100 km sec^{-1} (Brueckner, 1974). When traced back to the solar surface taking into account only gravitational deceleration the clouds coincide within an error of one minute with four discrete type II bursts observed between 0241 and 0257 by Culgoora. If one uses Newkirk's (1961) coronal streamer model, the type II bursts show a drift velocity of 1600 km sec^{-1}, which can be interpreted as the shock waves caused by the individual ejecta. It must be concluded that the acceleration of the clouds took place below the

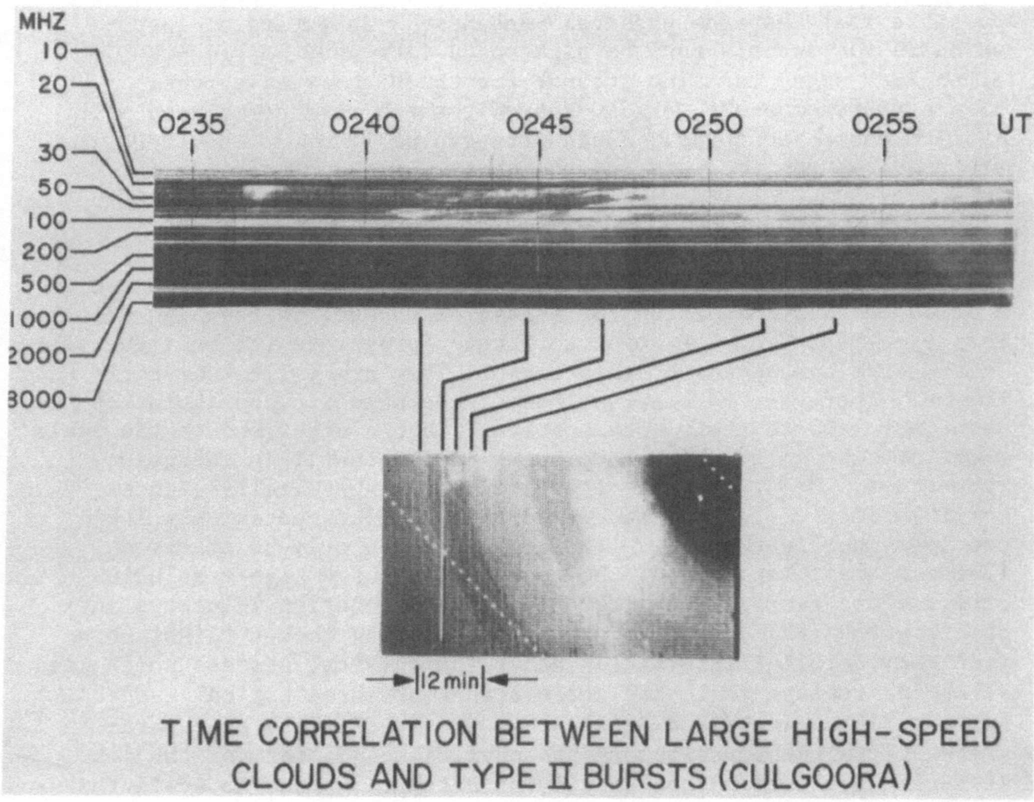

Fig. 1: OSO-7 white-light coronagraph observations of high-speed coronal clouds compared with type III and type II bursts caused by the same event (Dec. 14, 1971). The high-speed clouds are located between 6 and 9 $R_\odot$.

100 MHz plasma frequency level (1.3 R_o in Newkirk's model) and no additional energy was supplied above this level. From the dimension and density a total mass of 8 x 10^{14} g can be determined for each cloud which carries at 1300 km/sec a kinetic energy of 4 x 10^{30} ergs. The event was also accompanied by a huge coronal blast in the south-east quadrant which is resolved into numerous smaller clouds (Figure 1). The bulk of the energy (10^{32} ergs) and the mass (10^{16} g) is contained in this blast. Assuming that the strong type III burst at 0236.5 (Figure 1) signals the explosive phase of the flare, the clouds traveled 4.5 min to reach the 100 MHz (1.3 $R_\odot$) level. This leads to a minimum value of 5 km sec^{-2} for the acceleration. The physical properties of the clouds are listed in Table I.

(b) High-energy jets. A string of four high-energy jets were observed in a time sequence of ultraviolet spectra during the second rocket flight of the HRTS instrument (Brueckner, 1980; Bartoe and Brueckner 1982). Their appearance in the C IV line at 1548 Å lines is shown in Figure 2. If one measures the time pro-

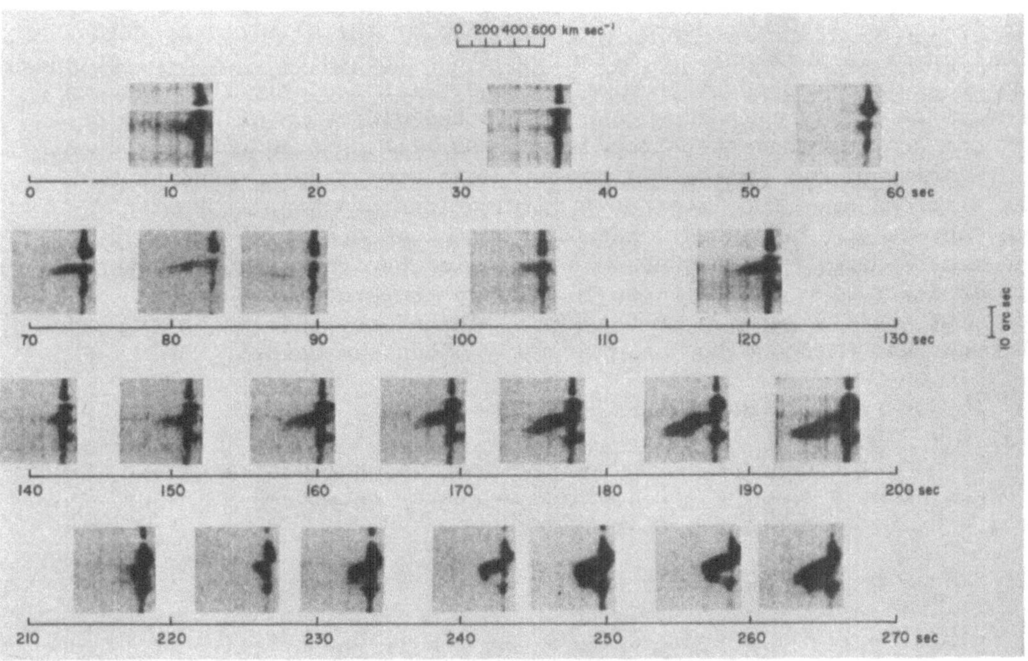

Fig. 2: Time sequence of high-resolution C IV 1548 Å spectra (HRTS). (HRTS). Blue is to the left, the spectrograph slit is vertical. Doppler velocities up to 400 km sec^{-1} are observed.

gression of the maximum velocity one obtains a constant acceleration of 5 km/sec for all four events. The Doppler velocity field of the two last events shows two discretely separated blueshifted components which merge at the maximum velocity. It is therefore very likely that the acceleration takes place in an exploding loop. The material at the top of the loop perpendicular to the field lines experiences the strongest acceleration. The acceleration continues when the event disappears out of the slit. Therefore, 400 km/sec is their minimum velocity. The disappearance of the jet can be explained either as a result of strong proper motions which move the plasma cloud out of the field of view or its heating to coronal temperatures during the acceleration. The jets can be seen simultaneously in lines from Ly-α (20 000K) to N V (200 000K). During their acceleration the number density shifts from cooler to hotter ions. There are no coronal lines in the spectral regime of the HRTS experiment which are strong enough to show emission from the jets. The physical properties of the jets are summarized in Table I.

Figure 3 shows a flare-related high-speed ejecta photographed with the NRL spectroheliograph on-board Skylab. The sequence in four different spectral lines covers the explosive phase of a large flare. The direction of dispersion of the slitless spectroheliograms is vertical. At A, B and C one recognizes vertical streaks, which must be caused by Doppler motions. Particularly the discrete streak at "a" is aligned with a gap in the direction of dispersion. If one assumes that the streak is the Doppler shifted gap, one obtains a velocity of 400 km sec^{-1}. However, one must keep in mind that in slitless spectroheliograms proper motions and Doppler shifts can not be distinguished. The size of the ejecta is 4" or 3×10^8 cm, comparable with the jets seen in the quiet Sun, however their density is at least 10^{11} and they can be seen in temperatures between 80 000K (He II) and 2.5 MK (Fe XVI). Therefore, this event falls in between the jets and the massive spray discussed before.

3. ACCELERATION MECHANISM AND SCALING LAWS

If the acceleration of the jets is caused by magnetic forces, then $v = v_A$, which implies an energy density $E = 1/2\ \rho v^2$ and a field strength $B = (E \times 8\pi)^{1/2}$.

With $N_e = 7 \times 10^9$ cm^{-3}, $v = 4 \times 10^7$ cm sec^{-1} one obtains

$$E = 9.3 \text{ ergs cm}^{-3} \text{ and } B = 15 \text{ gauss.}$$

The size of the exploding loop is $\ell \sim 3 \times 10^8$ cm. Magnetic fields in the order of 15 gauss or more over a length of 3×10^8 cm have been observed in the quiet Sun in emerging flux regions (Harvey and Martin, 1973; Wiehr, 1970). A magnetic flux of

$$F_{mag} = B \times A = 1.4 \times 10^{18}\ Mx$$

is needed to drive a single jet, assuming an area $A = 9 \times 10^{16}\ cm^2$. The assumption is made that the efficiency of the process is close to unity. (The simultaneous heating of the gas to transition zone or coronal temperatures requires only a small fraction of the energy needed for the acceleration.) If the spray is accelerated in a similar size loop ($\ell \sim 3 \times 10^8$ cm), a density $N_e = 1.8 \times 10^{13}$ and a magnetic field strength of $B = 2.5 \times 10^3$ gauss are required to explain the observed velocity $v = 1.3 \times 10^8\ cm\ sec^{-1}$ and the total mass of 8 x 10 g. Fields of 2.5×10^3 gauss are common in active regions near big sunspots where large sprays originate.

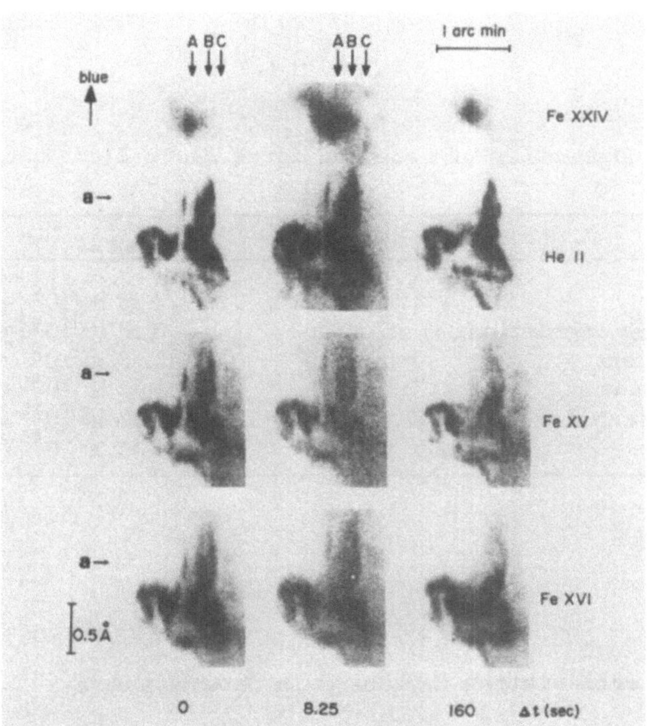

Fig. 3: Explosive phase of the June 15, 1973 flare observed with NRL's slitless spectroheliograph on-board Skylab. Streams of highly Doppler-shifted material are observed at A, B and C. The streak at "a" seems to be originating in the gap below it. The wavelength shift corresponds to a Doppler shift of 400 km sec^{-1}. The Fe XXIV plasma is located above the emission gap in He II and does not show strong Doppler motions.

The detailed observations of the accelerating velocity field in the expanding loop show that the acceleration of the material is accompanied by a stretching of the loop. Mullan and Ahmad (1982) have proposed that the ballooning instability accounts for the acceleration of material in magnetic unstable loops. However, the differential pressure between the interior of the loop and its surrounding corona which is responsible for the ballooning instability is too small to account for the observed acceleration of the jets.

It is much more likely that Lorentz forces beneath the solar photosphere are responsible for the rapid expansion of the loop which eventually results in a kink instability at its top (see Figure 2) and a subsequent reconnection at the feet of the loop.

TABLE I

Observed Properties of Jets and Large Flare Ejecta (Sprays)

	Jets	Sprays	
Max. velocity	4×10^{7}	1.3×10^{8}	cm/sec
Density (during acceleration)	7×10^{9}	1.8×10^{13}	cm^{-3}
Size, large event	3×10^{8}	3×10^{8}	cm
Energy, large event	2.5×10^{26}	4×10^{30}	ergs
Mass, large event	3.1×10^{11}	8×10^{14}	g
Acceleration	5×10^{5}	$>5 \times 10^{5}$	cm sec^{-2}

TABLE II

Accelerating Mechanism in Jets and Sprays

	Jets	Sprays	
Columnar density	2×10^{18}	2.5×10^{21}	cm^{-2}
Force	1.7	2×10^{3}	dyne
Kinetic energy density	9	2.5×10^{5}	ergs cm^{-3}
Field strength	15	2.5×10^{3}	gauss
Alfvẽn speed	4×10^{7}	1.3×10^{8}	cm sec^{-1}
Area	9×10^{16}	9×10^{16}	cm^{2}
Magnetic flux	1.4×10^{18}	2.3×10^{20}	Mx

REFERENCES

Brueckner, G.E.: 1980, Proc. XVII General Assembly IAU, D. Reidel Pub. Co., Dordrecht, Holland.

Brueckner, G.E.: 1974, Proc. IAU. Symp. #57, D. Reidel Pub. Co., Dordrecht, Holland.

Brueckner, G.E.: 1976, Phil. Trans. R. Soc. Lond. A 281, 443.

Harvey, K.L., and Martin, S.F.: 1975, Solar Physics 32, 389.

Mullan, D.J. and Ahmad, I.A.: 1982, Solar Physics 75, 347.

Newkirk, G.N., Jr.: 1961, Ap. J. 133, 983.

Wiehr, E.: 1970, Solar Physics 15, 148.

PRE- AND POST-FLARE X-RAY VARIATIONS IN ACTIVE REGIONS

Zdenek Svestka[1,2] and Aert Schadee[1]
[1]Space Research Laboratory of the Astronomical Institute
Beneluxlaan 21
3527 HS Utrecht, The Netherlands

[2]Center for Astrophysics and Space Science
University of California San Diego
La Jolla, CA 92093, U.S.A.

ABSTRACT

Extremely low background noise of the HXIS experiment aboard the SMM made it possible to detect > 3.5 keV X-ray emissions from non-flaring active regions which are $10^3 - 10^4$ times weaker than the X-ray flux from flares. Short-lived X-ray bursts and long-lived X-ray enhancements of various intensities seem to characterize active regions in different phases of their development. After major two-ribbon flares, giant X-ray arches are seen in the corona, slowly decaying for many hours after the flare end. Associated with these arches appear to be quasi-periodic flare-like variations of purely coronal nature.

1. SHORT- AND LONG-LIVED VARIATIONS

The principal aim of the Hard X-ray Imaging Spectrometer (HXIS) aboard the SMM was the imaging of the impulsive phase of flares in hard X-rays (> 16 keV) and the subsequent development of the thermal flare in the 3.5 - 30 keV X-rays. These tasks were fulfilled successfully, as the reader can see in the contributions presented by Duijveman, De Jager, and Simnett elsewhere in these proceedings. However, somewhat to our surprise, HXIS made it also possible to study X-ray emission from non-flaring active regions. These observations demonstrate that even in the absence of flares there must be areas in the active region corona with temperature in excess of 6×10^6 K!

These emissions, not associated with flares, are usually very weak. The cause that HXIS has been able to detect them is its extremely low background. In the lowest energy channel, covering the 3.5 - 5.5 keV range, the average background noise is only 4.56×10^{-3} counts per second, i.e. about one count every four minutes, in each individual pixel of the HXIS coarse or fine field of view (FOV). For extensive sources, this is of particular advantage in the coarse pixels which cover a relatively large area of 32" x 32". If, per example, there is a source, emitting twice the average background in one pixel for half an hour, we get 24 counts instead of the average background of 8 counts after 30 minutes

Solar Physics **86** (1983) 267–277. 0038–0938/83/0862–0267 $ 01.65.

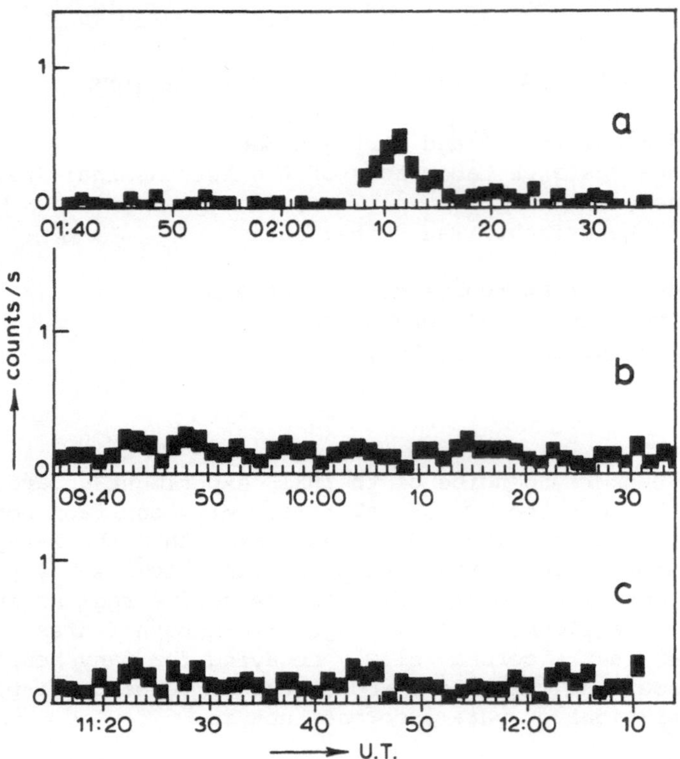

Figure 1 Time-development records of two typical weak X-ray brightenings on May 20, 1980, in AR 2456. (a) A short-lived source (counts per second from 6 pixels of the coarse FOV); (b) A long-lived source (counts per second from 17 pixels of the coarse FOV); (c) the same long-lived source one orbit later.
The maximum flux in (a) is ∿ 0.1 counts/second per pixel; the average flux in (b) is ∿ 0.01 counts/second per pixel. Energy range 3.5 - 5.5 keV, instrumental background 0.0046 counts/second per pixel.

of integration. Thus we detect an X-ray source with 16 ± 6 counts, with statistical significance of 2.7 σ, in spite of the fact that the source produced only one X-ray count every two minutes! However, we will not be able to see the source in the fine FOV, unless the source is so small that it fills only one or two 8" x 8" pixels. (Already for 2 pixels the statitical significance of the source discovery decreases to 1.6 σ.)

In higher energy channels of HXIS the background is higher (8.94×10^{-3} and 9.44×10^{-3} counts per seconds, respectively, in the 5.5 - 8.0 keV and 8.0 - 11.0 keV bands) and the flux weaker so that we must restrict the imaging to the lowest 3.5 - 5.5 keV energy band. One should note, however, that even this range of energies (corresponding to 2.25 - 3.54 Å) significantly exceeds the energies imaged in X-rays aboard Skylab; there the images correspond to wavelengths $\gtrsim$ 10 Å even in the hardest filters used. Thus temperatures in excess of $2.5 - 3.0 \times 10^6$ K

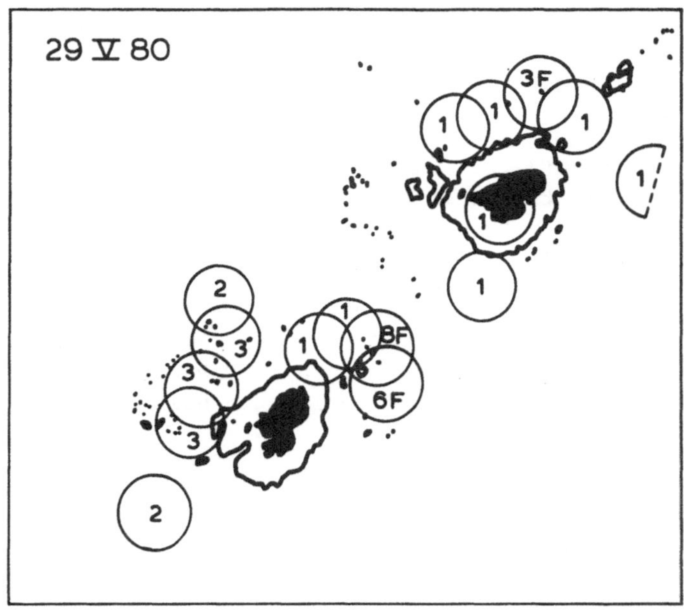

Figure 2. A summary of the occurrence of short-lived X-ray brightenings during one day (29 May 1980) in the active region complex 2466/69/70. Each circle marks the position of an X-ray source (with spatial resolution of 32") and the number inside each circle gives the number of brightenings recorded in the source during May 29. Letter F means that one or more of the brightenings in that source reached the intensity of a subflare.

were high enough to produce recognizable emission in the Skylab pictures. In the case of detectable HXIS sources, the coronal temperature must be higher than 6×10^6 K, unless chromospheric densities are involved in the emission. Because of the weakness of the sources we were usually unable to determine the actual temperature and emission measure in them. But a few stronger events ($\gtrsim 0.2$ counts/s per pixel within 3.5 - 5.5 keV) made it possible to compare the counts in the 3.5 - 5.5 keV and 5.5 - 8.0 keV energy bands. For them we have found temperatures between 6.6×10^6 and 6.8×10^6 K, and emission measures in the range from 6×10^{47} to 1.5×10^{48} cm^{-3}.

There are (at least) two different kinds of the weak X-ray emissions in non-flaring active regions (cf. Figure 1):

(1) Short-lived bursts, with durations similar to the duration of subflares (2 - 30 minutes), but generally with only $\sim$0.1 - 1.0 counts/s per pixel at maximum flux which is a flux about 10^3 - 10^2 times weaker than in a flare.

(2) Long-lived enhancements that last for one or more orbits of the SMM spacecraft, up to 10 hours. They are usually still weaker than the short-lived bursts, with about 0.01 counts/s per pixel of the coarse FOV, with very little and slow fluctuations. Their flux is thus $\sim 10^4$

times weaker than in a flare.

There are active regions which produce tens of the short-lived bursts per day with greatly variable intensity. An example is shown in Figure 2: in the activity complex AR 2466/69/70, on May 29, 1980, HXIS recorded 38 X-ray enhancements, of which four were classified in Hα as components of two flares. Six other brightenings were classified in Hα as subflares. All the others emitted 0.1 $\simeq$ 1.0 counts/s. Altogether, from May 24 through May 29, HXIS recorded 111 short-lived X-ray brightenings in this region, i.e. 20 events per day, on average. In contrast to that, only 9 such events per day were seen, on average, in the AR 2456 during May 19 - 22, 1980, and they were all very weak ($\sim$ 0.1 c/s). This region produced also 25 long-lived X-ray sources during the same period, of which one example is shown in Figure 3. Essentially all the time at least one long-lived source could be seen in half-an-hour integrations of > 3.5 keV X-rays.

Starting from $\sim$13 UT on May 20, the most common position of the long-lived X-ray enhancements in AR 2456 was close to the position of the major flare which occurred in the region at 20:52 on May 21. (Figure 3 shows one of these sources). Thus one may be inclined to suggest that these sources were some kind of precursors to the flare, starting $\sim$32 hours before the flare. But, as a matter of fact, a long-lived source in a very similar position was seen again by HXIS at 09:30 UT on May 22, i.e. 12 hours after the flare onset. Thus, though probably related, these very weak long-lived sources can hardly be called 'flare precursors' in the common sense of this word.

Long-lived X-ray sources were also present in the active complex AR 2466/69/70 and they were even more intense there than in AR 2456. It is, however, more difficult to analyze them because of the superposition of the multitude of short-lived sources in > 3.5 keV X-rays.

Some relation to the flare activity is indicated: AR 2456 was an old mature region, which during 3 days produced one major flare and only three subflares; it was the site of only rare and weak short-lived X-ray bursts, but essentially all the time some parts of the region, mostly along the filament channel, were sources of long-lived X-ray emission above 3.5 keV. On the other hand, the complex AR 2466/69/70 was a set of developing sunspot groups, with many elementary flux emerging regions; it produced many subflares and frequent short-lived bursts of > 3.5 keV X-rays with wide spectrum of peak intensity. Thus, the X-ray 'signature' of an active region might be indicative of the kind of flares one may expect in it.

We do not know the physical nature of these phenomena. According to Gaizauskas (private communication) many of the short-lived X-ray bursts in AR 2466/69/70 were related to newly emerging flux regions (but note that the spatial resolution of 32" in X-rays does not permit 100%-sure associations). At least some of the long-lived enhancements may be related to filament activations. But more comparisons with chromospheric and transition-region data are needed to confirm these tentative associations.

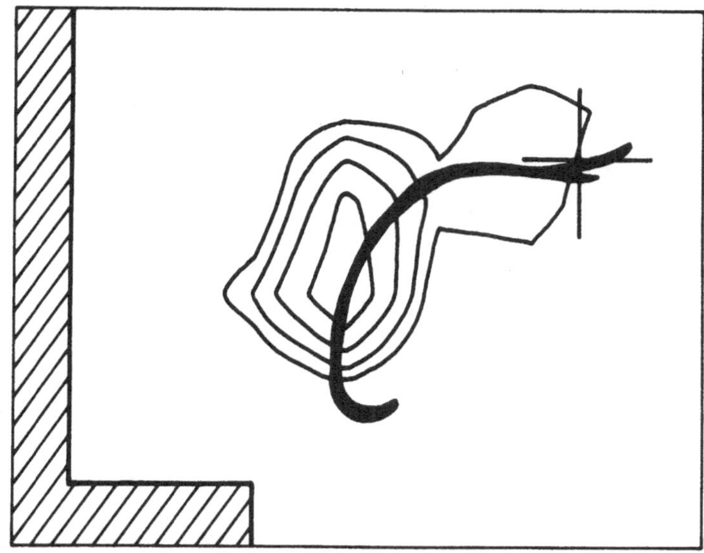

Figure 3. 3.5 - 5.5 keV image of a long-lived X-ray source in AR 2456 on May 21, 1980. The dark AR filament is shown here for a reference. The contours correspond to counts exceeding the background by factors of $2^{n/2}$ for n = 1,2,3, and 4, after 1575 seconds of integration, from 03:15:45 through 03:42:00 UT. The source was seen first in the integrations made from 00:04 through 00:30 UT and last from 07:58 through 08:28 UT. During its eight-hour existence it was also the site of a short-lived burst from 04:04 through >04:09 UT with maximum flux of 0.12 counts/s per pixel. The average intensity of the image presented here was 0.025 counts/s per pixel of the coarse FOV. The center of HXIS FOV is marked by a cross and the boundary of the coarse FOV is hatched.

2. POST-FLARE ARCHES

After the major two-ribbon flare of 20:52 UT on 21 May 1980 in AR 2456, HXIS recorded a long-lived X-ray source of unusually large extent (Figure 4) and with characteristics greatly different from all the other sources. It obviously was the image of a giant arch extending high into the corona above the active region (Svestka et al., 1982a). HXIS saw it first at 03:07 UT (SMM did not look at AR 2456 for two orbits before) and could follow it until 08:50, when it became too weak. As the insert in Figure 4 shows, this arch was the bottom part of a type I noise storm recorded at Culgoora.

A very similar event, but by an order of magnitude more intense, was seen on 6 November 1980 (Svestka et al., 1982b). In Figure 5 we can see the arch extending above the eastern solar limb over an active region that produced one of the most important flares of the present cycle six and a half hours earlier. On the right-hand side of Figure

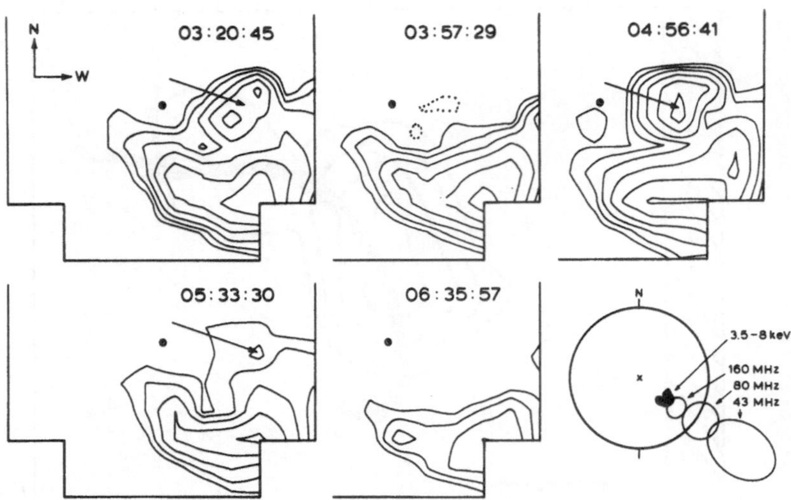

Figure 4. X-ray contours ($2^{n/2}$ times background for $n = 1,2,...6$) of the long-lived source observed on 22 May 1980 after the major flare of 21 May, 20:52 UT. The given times are the mean times of 1525 s integrations. The position of the flare is marked by dotted contours in the image at 03:57:29. The source is a giant coronal arch above AR 2456. The insert shows the relative positions of this X-ray arch and three images of the stationary type I noise storm as seen at Culgoora at 03:20 UT.

5 we again compare the X-ray image with Culgoora radio images. It was night at Culgoora when the X-ray image was made. But both the stationary type IV burst at the time of the flare (above) and the following type I noise storm next morning (below) were clearly above the X-ray arch.

We suggest that the Kopp and Pneuman (1976) model of the post-flare loops is capable of explaining the formation and existence of such arches: The field lines are distended upwards during the filament eruption, and afterwards they subsequently reconnect starting from below and proceeding upwards (Figure 6b). However, since the field is sheared, a field line never reconnects with the same field line with which it was connected before (Anzer and Pneuman, 1982). There is always another field line closer to it, with which the reconnection is preferably accomplished. The circle and arrow in Figure 6b demonstrate this situation. As the result we get a system of field lines above, which become interconnected along the $H_{||} = 0$ line (Figure 6c). Each reconnection process energizes particles, some of these particles are trapped in the upper spiral fields, thermalize there, and so build up the plasma arch. In this way we may obtain exactly the structure of the arch we observe in X-rays.

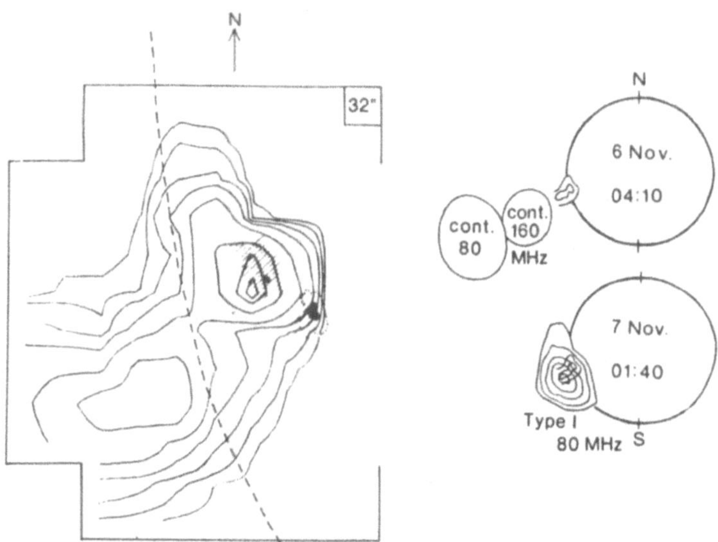

Figure 5. LEFT: The coronal arch imaged by HXIS in 3.5 - 5.5 keV X-rays with spatial resolution of 32" at 09:53 UT on November 6, 1980. The X-ray flux was integrated for 7 minutes. The X-ray image is projected on a sketch of the Hα picture of the active region - hatched areas indicate the Hα plage. The largest spots are also indicated. The dashed line is the solar limb. RIGHT: The relative positions of this X-ray arch and images of the stationary type IV radio burst (above) and type I noise storm (below) as seen by Culgoora.

3. CORONAL FLARE-LIKE VARIATIONS

Figure 4 shows that the top of the arch on May 22 was slowly, continuously decaying. However, a region below the arch, close to the flare site but not quite identical with it, showed striking variations in X-ray brightness. This region is indicated by arrows in Figure 4. The region was bright at 03:20 UT, disappeared at 03:57 UT, reappeared and became even brighter than the top of the arch at 04:56 UT, was weakly seen at 05:33 UT, and disappeared again at 06:35 UT.

Similar variations were also recorded below the arch of 6 November, 1980. In this case these variations were so strong that also the GOES-2 satellite of NOAA could image them: they started exactly at 10:00 UT and seem to occur quasi-periodically with an apparent period of about 20 minutes for about 4 hours.

All these variations could be seen in various coronal lines by other SMM experiments and, rather surprisingly, also on metric waves at Culgoora (22 May, 80 MHz), and Nançay (6 November, 169 MHz). In the transition layer, however, the O V line shows only a very small response to the coronal variations, and below it, in the chromospheric Hα line, high resolution Ramey photographs do not show any brightness variation

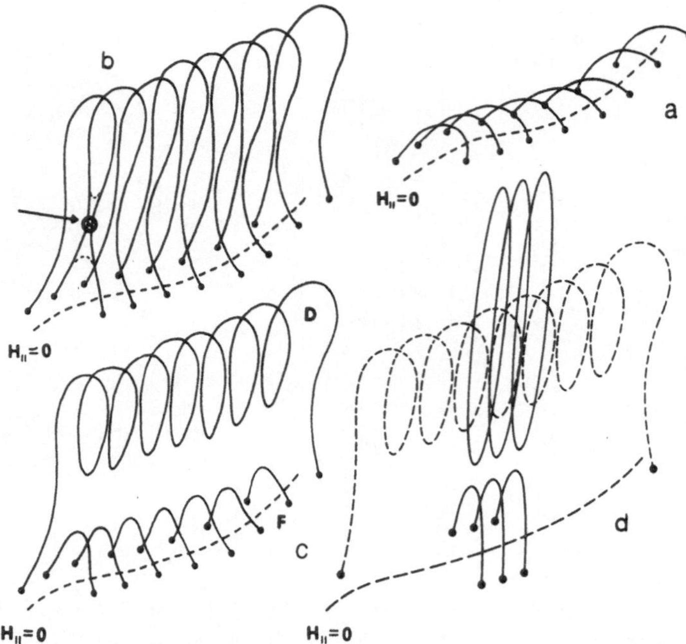

Figure 6. If the magnetic field is sheared (a) and is distended upwards during the filament eruption (b), reconnection occurs between neighbouring field lines (arrow in (b)). As a result we get less sheared flare loops below (F in (c)) while the upper disconnected loops (D in (c)) become interconnected along the $H_{\|} = 0$ line in the form of an arch. Later on, when non-sheared field lines reconnect, isolated field lines are being produced around the arch (d).

at all (Svestka et al., 1982b).

Still, the observed quasi-periodic X-ray enhancements on 6 November were definitely so strong that they should have heated the transition layer and produced at least subflares in Hα (Svestka et al., 1983). Therefore, we suggest that there may be only one reason why no heat at all was transported through the transition layer to the chromosphere: namely, that the field line structure, associated with these variations, was completely disconnected from the lower atmospheric layers. Thus we suggest the situation sketched in Figure 6d:

One can imagine that rare field lines still reconnect many hours after the flare (cf., e.g., Nolte et al., 1979), but these field lines are anchored far from the $H_{\|} = 0$ line and the shear of such field lines is small (e.g., Zirin and Tanaka, 1973). Therefore, the topological interconnection, indicated in Figure 6c, may not be accomplished anymore, and instead of it isolated closed field lines, encircling the arch, are being produced. An accumulation of still reconnecting field lines above some spatially limited part of the active region might create a plasmoid encircling the X-ray arch. One can suppose then that particles are accelerated high in the corona in the radio noise storm region and stream along the closed field lines below the arch, where they excite the corona. No heat, of course, can be conducted downwards, since there are no connecting field lines going through the transition layer to the chromosphere.

All these puzzling pre-flare and post-flare phenomena are being studied further, using the data obtained aboard the SMM by HXIS and other experiments, and comparing them with high-resolution ground-based observations.

ACKNOWLEDGEMENTS

We thank C. de Jager for stimulating discussions which have led to the use of the illustrative sets of $2^{n/2}$x background line contours. The research described in this paper was made possible by the support of the Netherlands Committee for Geophysics and Space Research of the Royal Netherlands Academy of Arts and Sciences. NASA provided the SMM facilities at the GSFC. The US National Science Foundation contributed to the trip expenses of ZS to Japan.

REFERENCES

Anzer, U. and Pneuman, G.W.: 1982, Solar Phys. 79, 129.
Kopp, R.A. and Pneuman, G.W.: 1976, Solar Phys. 50, 85.
Nolte, J.T., Gerassimenko, M., Krieger, A.S., Petrasso, R.D. and Svestka, Z.: 1979, Solar Phys. 62, 123.
Svestka, Z., Stewart, R.T., Hoyng, P., van Tend, W., Acton, L.W., Gabriel, A.H., Rapley, C.G. and 8 co-authors: 1982a, Solar Phys. 75, 305.
Svestka, Z., Dennis, B.R., Pick, M., Raoult, A., Rapley, C.G., Stewart, R.T. and Woodgate, B.E.: 1982b, Solar Phys. 80, 143.
Svestka, Z., Schrijver, J., Somov, B., Dennis, B.R., Woodgate, B.E., Fürst, E., Hirth, W., Klein, L. and Raoult, A.: 1983, Solar Phys., in press.
Zirin, H. and Tanaka, K.: 1973, Solar Phys. 32, 173.

DISCUSSION

KUNDU: I find some similarity between your events and the events observed by AS&E, namely X-ray (2 - 54Å) brightening in filament cavities after filament disruption. As you may know, there also was some correlation between such X-ray brightening events and storm radiation sources. In fact, Dave Webb and myself concluded in a paper that storm-producing electrons (10 - 100 keV) may be responsible for heating (by collisional processes) the X-ray plasma.

SVESTKA: These observations really may be related, as we have stated in

the published paper. However, you and Webb have studied enhancements immediately following filament eruptions, whereas we see long-lived arches existing for 10 hours or more. These could not be seen by the AS&E experiment on Skylab because of high background, I suppose.

HIEI: Would it be possible for your model to explain the 'gradual' hard X-ray enhancement of, say, the May 13, 1981 event?

SVESTKA: No. I think that what you see are the tops of the flare loops at an altitude of about 50 000 km, at the time of the flare. What I described was an arch at 150 000 to 200 000 km altitude, lasting for many hours after the flare.

PRIEST: Are your short bursts related to X-ray bright points?

SVESTKA: I don't think so. I had in mind emerging flux in an active region. We studied bright points outside active regions for two days in cooperation with ground-based observatories and the UVSP experiment. They detected five newly formed X-ray bright points (elementary active regions) in the HXIS field of view, and two of them flared. HXIS could not image any of them, even when they were flaring. Thus the temperature or emission measure was too low.

DOSCHEK: What do you think are the physical characteristics of the pre-flare loop that becomes the site of soft X-ray emission from $\sim 2 \times 10^7$ K plasma? That is, what are the peak temperature in the preflare loop and the loop pressure?

SVESTKA: You apparently have in mind a compact flare. I know nothing about the loop prior to the flaring, because we see the loop (if at all) only in the coarse field of view with poor spatial resolution, and only in the lowest energy channel. Thus we have no means for obtaining the physical characteristics.

ACTON: Your HXIS 'microflares' may be related to what we see with FCS in many active regins and call 'Mg XI flares'. These are not seen in Si XIII or S XV. Thus, the peak temperatures are on the order of 6×10^6 K. The duration of these events is on the order of 10 minutes and they are not very impulsive. Some active regions produce many of these -- others none at all.

SVESTKA: These really seem to be events identical to our 'short-lived X-ray variations'. We see similar characteristics to theirs.

BRUECKNER: The existence of pockets of 10 MK plasma is confirmed by HRIX observations of Fe XXI. Size of the emission centers is 5" - 8". Since no particular brightening can be seen in their vicinity in transition zone lines, these emission centers seem to be isolated from their surroundings. The lifetime is longer than the duration of a rocket flight (see Brueckner, 1981, in Physics of Active Regions, F. J. Orrall, Ed.)

UCHIDA: Concerning your mechanism of the production of the helical field by reconnection, I would like to point out that I have proposed similar one in the Cambridge Symposium in 1979 (Proc. IAU Symp. No. 91, p 303).

IMAGING OF IMPULSIVE SOLAR FLARE PHENOMENA

A. Duijveman and P. Hoyng
Astronomical Institute
Space Research Laboratory
Beneluxlaan 21
3527 HS UTRECHT, The Netherlands

ABSTRACT. We review some recent advances in our understanding of impulsive solar flare phenomena obtained through new hard X-ray and radio imaging instruments (the Solar Maximum Mission and Hinotori satellites, the VLA and VLBI).

1. INTRODUCTION

Solar flares have been studied extensively during the last solar maximum and many new instruments have been developed. There was a worldwide coordination of observations in the framework of the Solar Maximum Year (Svestka and de Jager, 1980). This review is directed towards the results of instruments imaging in hard X-rays and microwaves, with emphasis on the impulsive phase of flares. Sections 2 and 3 concentrate on the interpretation of hard X-ray data. Section 4 also includes microwave data. Finally in Section 5 we discuss the implications of the presence of a hot component ($T \sim 4 \times 10^7$ K) for the flare energy balance.

2. HARD X-RAY IMAGING

Before the launch of SMM on February 14, 1980 and of Hinotori on February 21, 1981 already some information on the height structure of solar flares in hard X-rays was available from stereoscopic observations with two spacecraft (PVO and ISEE-3; Kane *et al.*, 1979). Actual imaging in hard X-rays became possible with the Hard X-ray Imaging Spectrometer (HXIS) aboard the Solar Maximum Mission satellite (SMM) and the Solar hard X-ray Telescope (SXT) aboard Hinotori.

HXIS images simultaneously in six energy bands in the energy range 3.5 - 30 keV with a time resolution between 1.5. and 7 sec, depending on the mode of operation. Its spatial resolution is 8" x 8" over a FOV 2'40" (van Beek *et al.*, 1980).

SXT consists of two bigrid modulation collimators and images in the energy range of typically 17 - 40 keV. Its time resolution is about 4 s, while its angular resolution is claimed to be as high as 7" x 7" in the case of good counting statistics (Makishima, 1982).

On the theoretical front, before the launch of these instruments, two basic models for the sources of the impulsive hard X-rays had been developed and these can now be observationally tested. In one model the hard X-rays originate from non-thermal energetic electrons emitting

Solar Physics **86** (1983) 279–288. 0038–0938/83/0862–0279 $ 01.50.

bremsstrahlung when colliding with the background ions. The fast electrons, if accelerated in a magnetic loop, precipitate at the footpoints of the loop and emit most of the hard X-rays there (thick target model, Brown, 1976). In the second model the source of the hard X-rays is thermal ($T \gtrsim 10^8$ K). The hot source, presumably created at the top of the loop, expands along the loop until it hits the transition zone and chromosphere (Brown, Melrose and Spicer, 1979; Smith and Lilliequist, 1979; Smith and Harmony, 1982). This alternative was studied because it was in principle able to alleviate the extremely high acceleration efficiency required by the thick target model (Smith, 1980), though it must be realized that a thermal model not always has an energetic advantage over a non-thermal interpretation (Kiplinger, private communication).

For at least three flares HXIS appears to confirm the non-thermal, thick target interpretation. The relevant observations are summarized in Figures 1,2 and 3. Figure 1 shows contour plots obtained by integration of HXIS images over the main hard X-ray spike of the impulsive phase. The bottom row of hard X-ray images (16 - 30 keV) is made up of bright points, labelled A,B and C. The top row of soft X-ray images (3.5 - 5.5 keV) is very different in character. It does not show the

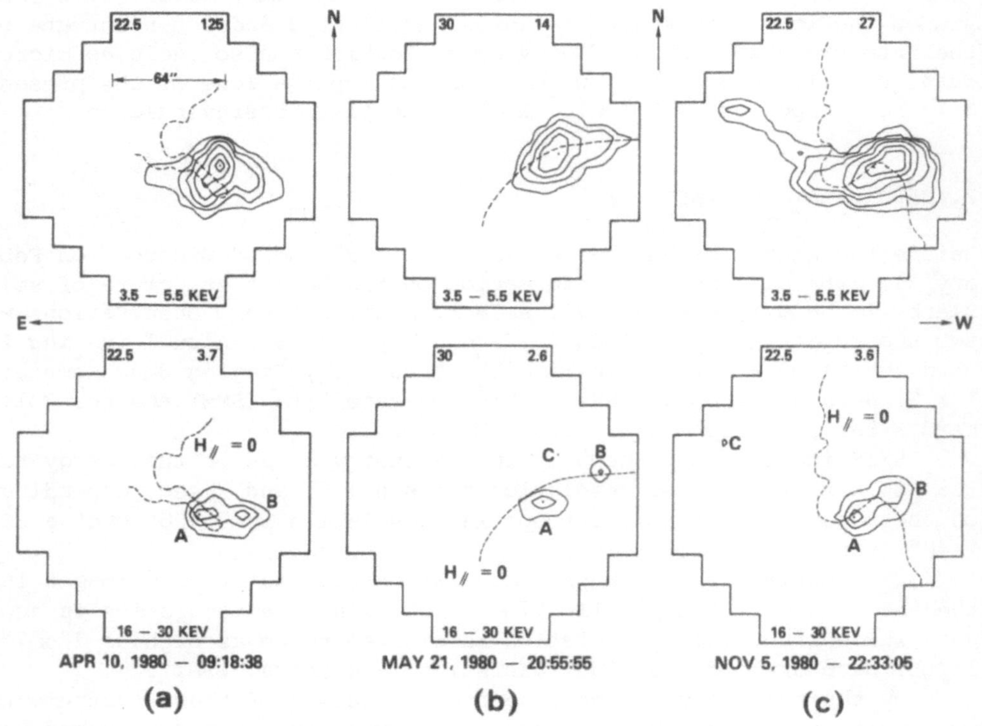

Figure 1. HXIS contour plots in soft and hard X-rays for three flares. The integration is over the impulsive spike. The dashed lines indicate the neutral line. The hard X-ray bright spots are labelled A, B and C (Duijveman et al., 1982).

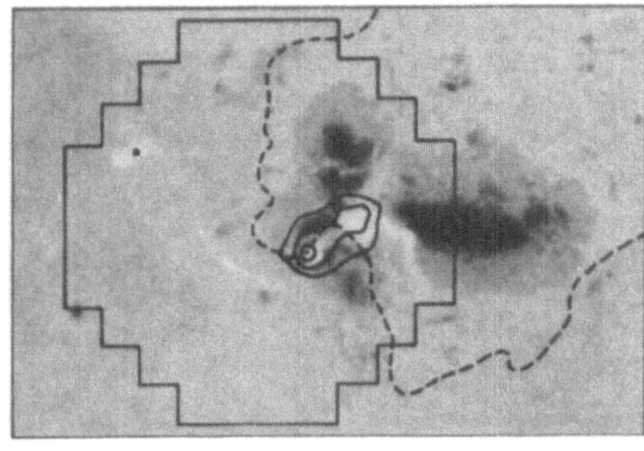

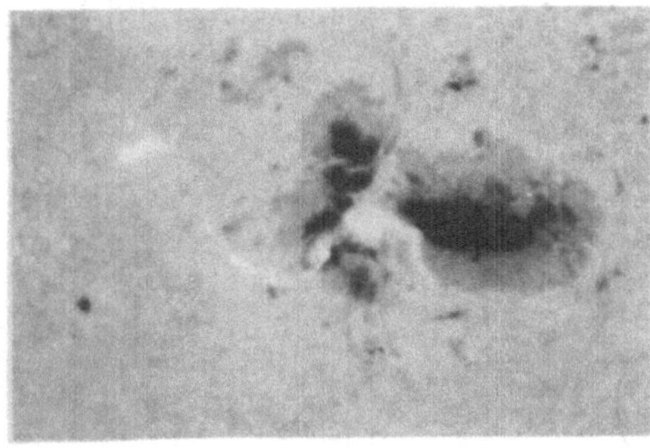

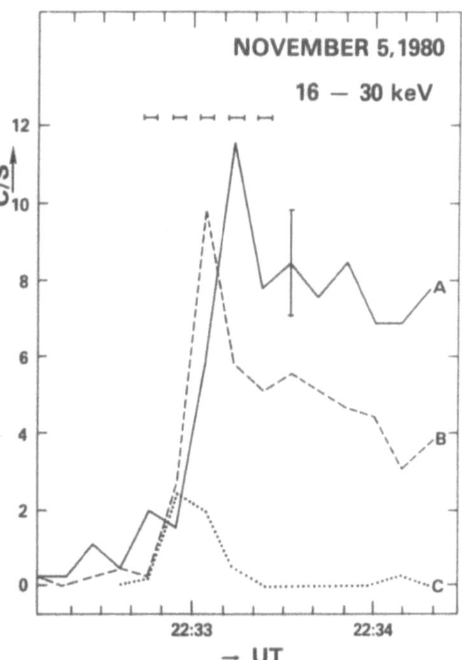

Figure 2. (left): Overlay of the hard X-ray image with offband Hα (+ 1Å) for the November 5, 1980 flare (Duijveman et al., 1982).

Figure 3. (right): Time profiles for the hard X-ray bright spots A,B and C of the November 5, 1980 flare (Duijveman et al., 1982).

multiple source structure displayed by the hard X-ray images. The bulk of the soft X-ray emission comes from in between the hard X-ray bright points.

Figure 2 shows for the November 5, 1980 flare of Figure 1c that the hard X-ray bright points are also bright in Hα, i.e. at the location of the hard X-ray bright points also the deeper layers of the atmosphere are affected. This leads to the conclusion that the hard X-ray brightenings are located at the footpoints of loops. Note in this respect that the bright point C of the November 5, 1980 flare in Figure 1 is actually located at the end of a large loop-like structure in soft X-rays. Ultraviolet observations (UVSP-Woodgate *et al.*, 1980; Woodgate *et al.*, 1983) confirm that the transition region is affected at the location of the hard X-ray brightenings.

Figure 3 shows that the hard X-ray brightenings B and C of the November 5, 1980 flare start simultaneous to within about 5 s (for the April 10 and May 21 flares, see Duijveman *et al.*, 1982).

The observed simultaneity of the hard X-ray bright points is most straightforwardly explained in terms of non-thermal electron beams. The speed of the beam electrons is amply sufficient to account for the observed simultaneity, while the Alfvén or conductive speeds are too low. Three remarks are relevant at this point:

a. The presence of electron beams is inferred only during the initial intense hard X-ray spike(s), hence during a small fraction of the total time the flares are visible in hard X-rays. Duijveman *et al.* (1982) analyzed the acceleration efficiency and found a lower limit of about 20%. This remains a theoretical enigma.
b. No trace of an expanding hot source was observed.
c. HXIS observed only a few flares which were sufficiently strong and spatially extended above 16 keV so that the above analysis could be made (Duijveman *et al.*, 1982; Harrison *et al.*, 1983).

3. QUASI-THERMAL FLARES; CONFLICT BETWEEN HXIS AND SXT?

The HXIS results on beams should not be generalized! If one accepts the above analysis, then for a few flares strong electron beams occur during

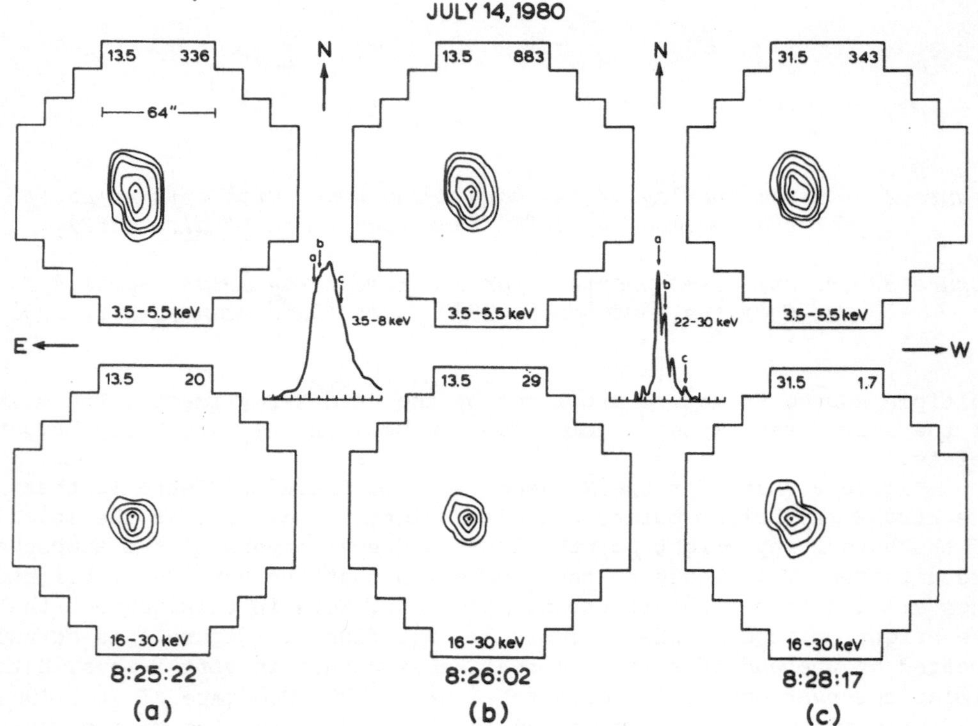

Figure 4. HXIS contour plots in soft and hard X-rays for the July 14, 1980 flare. Soft and hard X-ray images coincide at all times, i.e. there are no double sources. Light curves show the soft and hard X-ray time development of the flare.

the hard X-ray spikes. However, the multiple hard X-ray sources of the flares discussed in Section 2 merge into quasi-thermal single sources ($\geqslant$ 16 keV), spatially coinciding (or nearly so) with the 3.5 - 5.5 keV sources. This happens all within what is commonly referred to as 'the impulsive phase', though 'early thermal phase' might be a better name. An example is given by the May 21, 1980 flare. After the occurrence of the hard X-ray spikes it shows a coronal hot source ($T \approx 4 \times 10^7$ K) with a lifetime much larger than the duration of the spikes. We refer to Section 5 for details.

Secondly, many flares imaged by HXIS above 16 keV never show double sources, even if they are quite strong in hard X-rays. An example is the flare of July 14, 1980. For this flare soft and hard X-ray images coincide at *all* times (Figure 4). At the first (main) hard X-ray spike the plasma has a quasi-thermal, two-temperature structure ($T \approx 2 \times 10^7$ K and 4×10^7 K) similar to that observed in the early thermal phase of the flares in Section 2. Apparently for this type of flare a phase with strong electron beams cannot be confirmed.

The SXT observations reported so far (Ohki *et al.*, 1982; Tsuneta *et al.*, 1982; Takakura *et al.*, 1982 and publications in these proceedings) do not show the occurrence of well-separated hard X-ray sources. Instead, SXT generally finds evidence for slowly varying *coronal* hard X-ray sources. Double sources have been observed in a few cases only, and they are usually displaced with respect to the Hα kernels. An explanation could be that SXT observed a different class of flares as it operated one year later in the solar cycle. A gradual change of the hard X-ray character of flares with time has been seen by HXRBS on SMM (Kiplinger, private communication). The SXT events appear to be rather gradual and lack the large hard X-ray spikes which were characteristic for the flares discussed in Section 2. It could be that the SXT hard X-ray sources are similar to the hot sources observed by HXIS during the early thermal phase.

4. MICROWAVE IMAGING

A major development in microwave imaging was the construction of the Very Large Array (VLA) that allows high resolution (subarcsec) imaging in four wavelength bands between 1.3 and 21 cm. Its time resolution in the snapshot mode is 10 s.

Typical results for the impulsive phase microwave sources at high frequencies ($\gtrsim$ 15 GHz) may be summarized as follows (Marsh and Hurford, 1980; 1982):

- the microwave source is located in between Hα patches (Figure 5);
- the source is compact ($\approx 2''$) and most likely located at the top of the loop (above the neutral line);
- the source is highly polarized (typically about 60%) and the sense of the polarization changes rapidly over the source. This is indicative of synchrotron radiation. Note that after the impulsive phase often the whole loop becomes visible in microwaves and the polarization decreases to values of about 10%. This is interpreted as thermal bremsstrahlung.

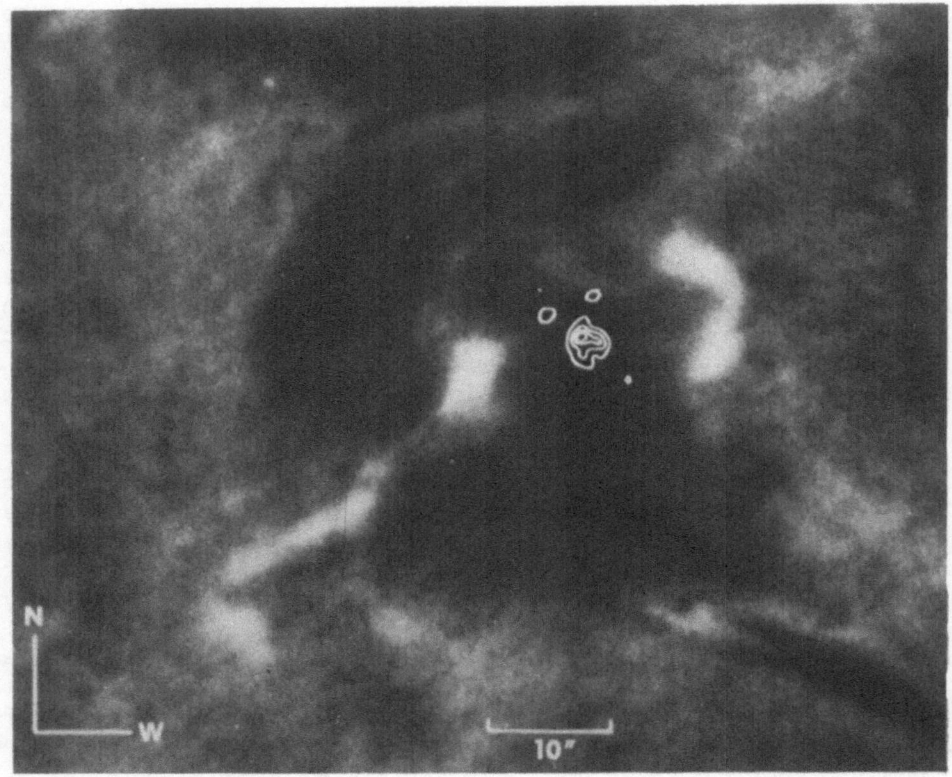

Figure 5. The position of a compact microwave source with respect to Hα brightenings for a flare on September 9, 1979 at about 16:11 UT. The microwave source was observed by VLA at 22.5 GHz (after Marsh and Hurford, 1980).

Several possible explanations for the above impulsive microwave sources have been given:

- Holman *et al.* (1982) proposed that energetic electrons with energies $\gtrsim$ 50 keV are confined at the top of the loop by electrostatic turbulence generated by the fast electrons themselves;
- the compact size of the source and the rapid change in the sense of the polarization could be the combined result of the directivity of the synchrotron radiation (due to a rapid change in direction of the magnetic field vector) and the restricted dynamic range of the VLA;
- Melrose and Dulk (1982) suggested that the scattering of electrons into the loss cone is enhanced near the footpoints due to the action of an electron-cyclotron maser. The top of the loop is brighter because the scattering into the loss cone is not enhanced there.

Kundu *et al.* (1982) made VLA observations at a wavelength of 6 cm. They studied the development of the magnetic field by studying the polarization structure of the microwave sources before, during and after the flare and found evidence for new bipolar loops emerging just before the flare. This supports the emerging flux models of Heyvaerts *et al.* (1977). 1D observations with the Westerbork Synthesis Radio Telescope (WSRT) at 6 cm indicate the presence of double sources, co-spatial with Hα brightenings. The stronger microwave source corresponds with the weaker hard X-ray source. Kattenberg *et al.* (1983) explain this in terms of an asymmetric loop model.

A very interesting new development is a recent observation by Kuijpers *et al.* (1982) using VLBI at 18 cm. Evidence was found for an (unresolved) source smaller than 40 km with a brightness temperature of about 10^{12} K. The observations were made with radiotelescopes at Onsala, Sweden, and Dwingeloo, the Netherlands (baseline 620 km). The source was not flare related.

HXIS and VLA have a joint data set for one flare on November 5, 1980 which has recently been analyzed (Hoyng *et al.*, 1983). The authors conclude that the data exclude a common origin of microwaves and hard X-rays from a purely thermal plasma. However, there is a fair agreement with the thick target picture. In this picture the electrons generating the microwaves at the top of the loop by gyrosynchrotron radiation are free streaming toward the footpoints where they produce the hard X-ray burst.

We should mention here that the diagnostics of the parameters of the flaring plasma has been refined considerably through the availability of new imaging instruments. Temperatures and densities were determined by a variety of methods (see e.g. Gabriel *et al.*, 1981; Phillips *et al.*, 1982; Duijveman *et al.*, 1982). From the joint VLA-HXIS data set for the November 5, 1980 flare, Hoyng *et al.* (1983) determined the plasma $\beta = 8\pi nkT/B^2$ and found $\beta \simeq 0.01$.

5. HIGH TEMPERATURE COMPONENT

There is evidence for the frequent occurrence of a hot thermal plasma component during the early thermal phase of flares. For example Lin *et al.*, 1981 found a 34×10^6 K component from measurements with a balloon-borne array of detectors with very high spectral resolution ($\lesssim$ 1 keV). Other examples are given by the April 10, 1980 (Machado *et al.*, 1982), May 21, 1980 (Hoyng *et al.*, 1981), April 30, 1980 (de Jager *et al.*, 1983) and July 14, 1980 (Section 3) flares. After the impulsive spikes, the May 21, 1980 flare shows a single hard X-ray source which is spatially nearly coincident with the soft X-ray source. A two-temperature analysis leads to a cool component characterized by $n_c = 15 \times 10^{10}$ cm^{-3}, $T_c = 2 \times 10^7$ K and a hot component $n_h = 2 \times 10^{10}$ cm^{-3}, $T_h = 4 \times 10^7$ K.

The interesting point is that the hot plasma is almost collisionless. The mean free path of an electron in the hot source is about 5000 km compared to an observed size of the source of about 10^4 km. Heat losses from this hot source can therefore not be described by classical formulae. It can be shown that the hot source can be maintained against conductive heat losses only by non-classical dissipation processes (Duijveman, 1983). The source is observed to exist for about 10 minutes.

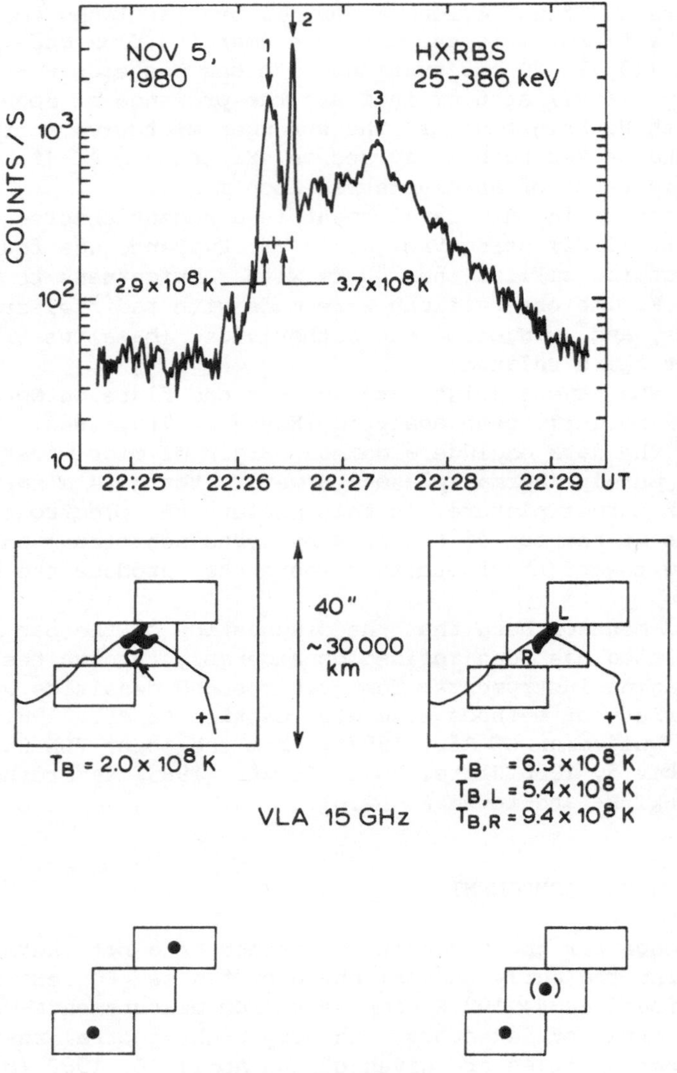

Figure 6. (Top): the hard X-ray time profile of the November 5, 1980 flare as observed by the Hard X-ray Burst Spectrometer on board SMM; this flare occurred in the same active region and 7 min. prior to the one shown in Figure 1-3.
(Middle): the location of the 15 GHz microwave source at the time of peaks 1 and 2, respectively (20% contour Stokes I); the source is strongly polarized at the edges (R = Right, L = Left circularly polarized).
(Bottom): a set of three 8" x 16" pixels from HXIS; a dot indicates a significant 16 - 30 keV flux in excess of a single temperature fit (after Hoyng et al., 1983).

It is estimated that approximately 10^{31} ergs are necessary to maintain the source over its lifetime. This is an order of magnitude larger than the estimate for the energy in the beam electrons during the impulsive phase (Duijveman *et al.*, 1982). The hot source plays therefore an important role in the energy balance of the flare.

ACKNOWLEDGEMENT

The development and construction of HXIS was sponsored by the Netherlands Committee for Geophysics and Space Research (GROC) of the Royal Netherlands Academy of Arts and Sciences (KNAW), and the Science Research Council of the United Kingdom.

REFERENCES

Brown, J.C.: 1976, *Phil. Trans. R. Soc. Lond. A281*, 473.
Brown, J.C., Melrose, D.B. and Spicer, D.S.: 1979, *Ap. J. 228*, 592.
De Jager, C., Machado, M.E., Schadee, A., Strong, K.T., Svestka, Z., Woodgate, B.E., van Tend, W.: 1983, *Solar Phys.* (in press).
Duijveman, A., Hoyng, P. and Machado, M.E.: 1982, *Solar Phys. 81*, 137.
Duijveman, A.: 1983, *Solar Phys.* (in press).
Gabriel, A.H., Acton, L.W., Culhane, J.L., Phillips, K.J.H., Wolfson, C.J., Rapley, C.G. and 9 co-authors: 1981, *Ap. J. 244*, L147.
Harrison, R.A., Simnett, G.M., Hoyng, P., Lafleur, H. and Van Beek, H.F.: 1983, *Solar Phys.* (in press).
Hoyng, P., Duijveman, A., Machado, M.E., Rust, D.M., Svestka, Z., Boelee, A., De Jager, C., Frost, K.J., Lafleur, H., Simnett, G.M., Van Beek, H.F. and Woodgate, B.E.: 1981, *Ap. J. 246*, L155.
Hoyng, P., Marsh, K.A., Zirin, H. and Dennis, B.R.: 1983, *Ap. J.* (in press).
Holman, G.D., Kundu, M.R. and Papadopoulos, K.: 1982, *Ap. J. 257*, 354.
Kane, S.R., Anderson, K.A., Evans, W.O., Klebesadel, R.W. and Laros, J. : 1979, *Ap. J. 233*, L151.
Kattenberg, A., Allaart, M., De Jager, C., Schadee, A., Schrijver, J., Shibasaki, K., Svestka, Z. and Van Tend, W.: 1983, *Solar Phys.* (in press).
Kundu, M.R., Schmahl, E.J., Velusamy, T. and Vlahos, L.: 1982, *Astron. Astrophys. 108*, 188.
Kuijpers, J., Tapping, K.F. and Graham, D.: 1982, IAU Coll. No. 71, Catania, Italy.
Lin, R.P., Schwartz, R.A., Pelling, R.M. and Hurley, K.C.: 1981, *Ap. J. 251*, L109.
Machado, M.E., Duijveman, A. and Dennis, B.R.: 1982, *Solar Phys. 79*, 85.
Makishima, K.: 1982, *Proceedings of Hinotori Symposium on Solar Flares* (Tokyo, January 27 - 29, 1982).
Marsh, K.A. and Hurford, G.J.: 1980, *Ap. J. 240*, L111.
Marsh, K.A. and Hurford, G.J.: 1982, *Ann. Rev. Astron. Astrophys. 20*, 497.

Melrose, D.B. and Dulk, G.A.: 1982, *Ap. J.* *259*, L41.

Ohki, K., Tsuneta, S., Takakura, T., Nitta, N., Makishima, K., Murakami, T., Ogawara, Y., Oda, M. and Miyamoto, S.: 1982, *Proceedings of Hinotori Symposium on Solar Flares* (Tokyo, January 27 - 29, 1982).

Phillips, K.J.H., Leibacher, J.W., Wolfson, C.J., Parkinson, J.H., Fawcett, B.C., Kent, B.J., Mason, H.E., Acton, L.W., Culhane, J.L. and Gabriel, A.H.: 1982, *Ap. J.* *256*, 774.

Smith, D.F. and Lilliequist, C.G.: 1979, *Ap. J.* *232*, 582.

Smith, D.F. 1980, *Solar Phys.* *66*, 135.

Smith, D.F. and Harmony, D.W.: 1982, *Ap. J.* *252*, 800.

Svestka, Z. and De Jager, C.: 1980, *Space Science Reviews* *26*, 317.

Takakura, T., Ohki, K., Tsuneta, S., Nitta, N., Makishima, K., Murakami, T., Ogawara, Y. and Oda, M.: 1982, *Proceedings of Hinotori Symposium on Solar Flares* (Tokyo, January 27 - 29, 1982).

Tsuneta, S., Ohki, K., Takakura, T., Nitta, N., Makishima, K., Murakami, T., Ogawara, Y., Oda, M. and Kondo, I.: 1982, *Proceedings of Hinotori Symposium on Solar Flares* (Tokyo, January 27 - 29, 1982).

Van Beek, H.F., Hoyng, P., Lafleur, B. and Simnett, G.M.: 1980, *Solar Phys.* *65*, 39.

Woodgate, B.E., Tandberg-Hanssen, E.A. and 12 co-authors: 1980, *Solar Phys.*, *65*, 73.

Woodgate, B.E., Shine, R.A., Poland, A.I. and Orwig, L.E.: 1983, *Ap. J.* (in press).

DISCUSSION

BROWN: What fraction of the total hard X-rays (16 - 30 keV) recorded by HXIS, or inferred by HXRBS, in the impulsive phase comes from the foot-point pixels?

DUIJVEMAN: More than about 25%.

KUNDU: In a friendly manner, I'd like to correct some dates in your presentation. Radio astronomers have used spatial resolution of 6" with WSRT since 1974, and if you consider resolution of 20", the Japanese radio astronomers have been using it since the 1960's. With regard to gyrosynchrotron masering, the idea has been around since 1979 (Holman, Eichler, and Kundu, 1980); it has been revived recently in a big way by Dulk and Melrose, and by Vlahos, Papadopoulos and Sharma.

DUIJVEMAN: Thank you for this comment. Actually I meant that high resolution, 2D imaging in microwaves was not possible before 1979.

DOSCHEK: You mentioned the presence of hot emission in the late impulsive phase. It is interesting that Widing has already found with the Skylab data that Fe XXIV emission at 1.5×10^7 K was, in 3 out of 4 cases, from a small region at the top of a loop.

DUIJVEMAN: Please note that the hot source at the top of a loop observed by HXIS (for example, in the May 21, 1980 flare) has a much higher temperature ($T \sim 4 \times 10^7$ K).

TEMPERATURE STRUCTURE OF SPATIALLY RESOLVED HARD X-RAY FLARES

G.M. Simnett
Department of Space Research,
University of Birmingham, England

ABSTRACT. High resolution X-ray images are used to study the temperature structure and evolution of two spatially resolved, but compact, solar flares. Both flares developed within a magnetic loop whose footpoints were separated by typically 15000 km, and involved primary energy release at one footpoint. This was followed by transfer of chromospheric material into and around the loop. The flares involved total energies differing by over an order of magnitude, and they follow different evolutionary paths because of this.

1. INTRODUCTION

The motivation of this work is to help establish the site of the primary energy release in flares. The closest we can approach in time is that of the impulsive hard X-ray bursts so it is reasonable to suppose that if we determine where in the flare region these occur, and also if we can infer the three dimensional magnetic field structure, then we should be able to put limits on the position that the energy release occurred. Current hard X-ray observations do not have sufficient sensitivity combined with spatial resolution to yield directly what is required - a situation that will not change before the next solar maximum - and therefore it is important to understand how to make best use of the available data. This paper discusses two flares observed by the Hard X-ray Imaging Spectrometer (HXIS) (van Beek et al, 1980) on the Solar Maximum Mission (SMM) concentrating on the evolution of the flare plasma from the flash phase through the decay.

Interpretation is less ambiguous if compact events are studied, and we have chosen the flares of 14:56 UT, May 7, 1980 and 22:38 UT, July 5, 1980. These flares, including Hα observations, have been discussed by Acton et al. (1982) and Harrison et al. (1983) respectively. They concluded that both in the impulsive phase and in the several minutes following this phase, the bulk of the hot plasma was confined

Solar Physics **86** (1983) 289–299. 0038–0938/83/0862–0289 $ 01.65

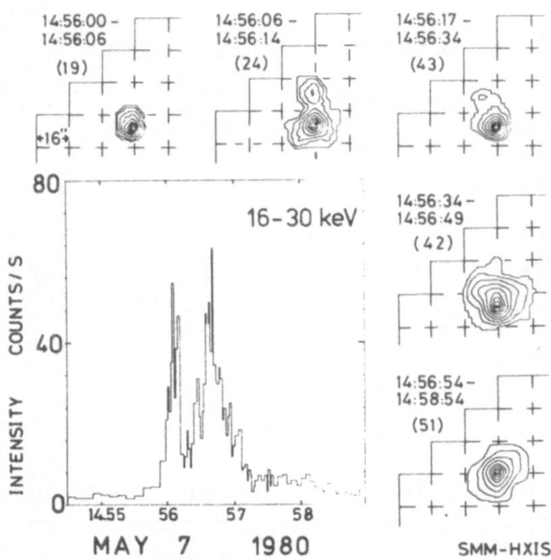

FIGURE 1

The hard X-ray intensity-time profile for the May 7 event. The time resolution during the set of impulsive spikes is 1.5 s. The contours are from the HXIS fine field of view, and have been corrected for the triangular response of the collimator. Each 16" x 16" square encompasses four individual pixels.

within a compact magnetic loop linking bright kernels seen in Hα. This will therefore be assumed in the discussion here. There are substantial increases in material within the loop following the impulsive phase, which is attributed to chromospheric evaporation (Acton et al., 1982). In both events there is a substantial hard X-ray halo around the main flare loop, consistent with photospheric backscatter from a high altitude source as discussed by Brown et al.(1975) and Bai and Ramaty (1978).

THE 14:56 MAY 7 1980 EVENT AT S23W12

Figure 1 shows the intensity-time profile of the 16-30 keV hard X-ray emission, together with spatial contour maps at five times during the impulsive phase. There are two main sets of impulsive bursts, the first set, with a harder spectrum, ending at 14:56:14 UT and the second set with a peak intensity around 14:56:40 UT. We shall ignore the fine structure and refer to them as the first and second bursts. During the onset phase, the emission is concentrated in one pixel, reference no.196. As the first burst progresses, there is evidence for a second patch of emission about 16" to the north or NNE. During the peak of the second burst, 14:56:34-14:56:48 UT the centroid is still in the same position but the emission is relatively diffuse. This is unlikely to be an instrumental effect, such as leakage in the HXIS collimator, as this would be worse during the first burst which had both a harder spectrum and higher intensity above 28 keV (Acton et al., 1982). In the decay, 14:56:54-14:58:54 UT, the centroid of emission moved north by 7" to a point approximately midway between the two hard X-ray patches at 14:56:00-14:56:14 UT. This point is imaged primarily by pixel 197 and it is interpreted as the top of the magnetic loop containing the flare plasma. The separation of the loop foot-

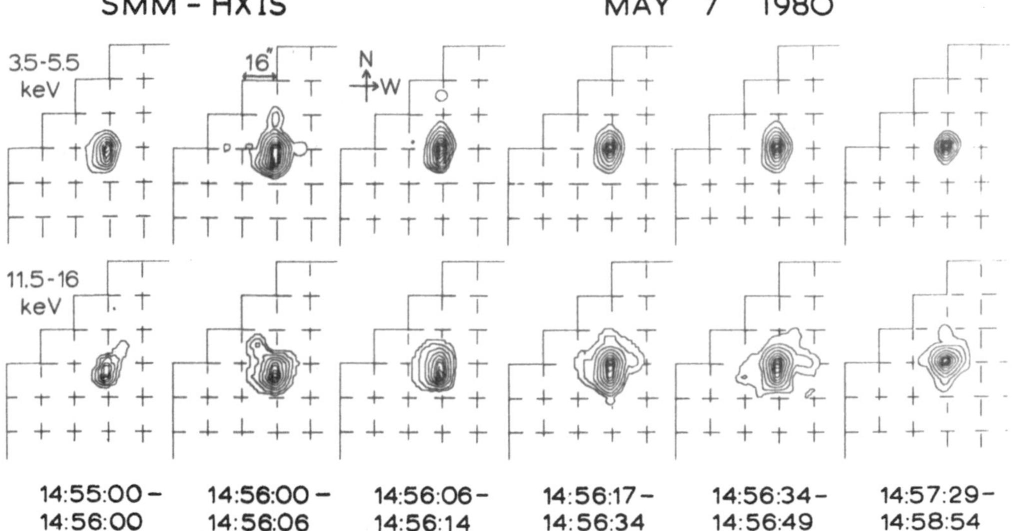

FIGURE 2. The 3.5-5.5 keV and 11.5-16 keV intensity contours throughout the flare. Note that after 14:56:06, the higher energy contour is always more diffuse than that at lower energy, even in the decay phase after 14:57:29 UT.

points is $\approx$16", which compares with a distance of 14" between the edges of the bright H kernels (Acton et al., 1982).

Figure 2 shows the evolution of the 3.5-5.5 keV and 11.5-16 keV images, starting with the pre-impulsive phase emission between 14:55:00 and 14:56:00 UT. Even prior to 14:55:00 UT there was emission from an area a few arc seconds to the east of the loop containing the main flare. We could speculate that this is activity in a separate loop which later interacts with the main loop, triggering the flare. Even during the first impulsive burst the 3.5-5.5 keV emission is primarily at the looptop, although at higher energies it is biased to the south footpoint. Above 22 keV it is almost exclusively within pixel 196; in the first burst 18 photons were detected from this pixel compared with <4 in any of the 8 surrounding pixels, a contrast >4.5. The triangular response of the HXIS collimator predicts that for a single uniformly illuminated pixel, the contrast between it and the adjacent and diagonally adjacent pixels is 16:3:1. In the 3.5-8.0 keV band, a sum of 9 images over the first burst produced a contrast between pixel 196 and the three pixels to the south of [281/(2, 5, 0)]. These results demonstrate that even in a hard event, leakage in the collimator is negligible compared with the imaged intensity. During the first bursts there is again weak evidence for emission from a point N or NNE. Although individual contours are of low statistical significance the appearance of a feature in this position in several independent data sets supports the hypothesis that this is the site of the northern footpoint. The 3.5-5.5 keV emission is clearly centered at the looptop early in the impulsive phase, but not until the end of this phase, after 14:57 UT, does the centroid of the 11.5-16 keV emission shift decisively to the same point.

Table I gives the temperature and emission measure as a function of time for pixel 197, which images the looptop. We have performed a 2-temperature analysis wherever possible. It is significant that there is no 2-temperature fit during the pre-flash period. This clearly indicates some non thermal processes occurring before the main trigger; the region is not simply "heating up" prior to the main event. The apparent sudden cooling at the end of the impulsive phase probably indicates that we are wrong to try to fit the data with a thermal model, despite the good fit; during the impulsive phase there will be additional contributions from non thermal bremsstrahlung which will diminish or cease at the end of the impulsive phase. At no time during the flare will a two temperature model fit the emission from pixel 196, the southern footpoint. During the second impulsive burst, the 16-30 keV data alone require a temperature of 8.3×10^7 K, and an emission measure of $2.7 \times 10^{45} cm^{-3}$, if we adopt a thermal interpretation. We should not be too surprised at the lack of a fit, as a substantial part of the emission from this pixel will be from the looptop because of the collimator response. If we add to this some genuine footpoint emission, perhaps from downflowing material, this will then distort the spectrum.

Apart from the hard X-rays above 16 keV the data are consistent with the majority (>90%) of the emission coming from pixel 197, which images the looptop. We noted above the collimator response, which dictates that for a uniformly illuminated pixel, 50% of the emission will be from that pixel, and 50% from the eight surrounding pixels. Any time the emission from a single pixel exceeds 50% of the sum of the group of 9, then it is consistent with a source less than 8" x 8". Throughout the maximum soft X-ray phase, and through the decay, the 3.5-5.5 keV emission from pixel 197 is always >50% of the 9 pixel sum, with a peak value of 59% at flare maximum. This indicates a source size at the looptop <<8", probably no more than 5", or 4000 km. If we assume the loop is a uniform semicircular tube, the length imaged by pixel 197 is ∿7000 km, leading to a volume, V of $9 \times 10^{25} cm^3$.

From this estimate of V we may determine the density of the loop from the emission measure $\int_V n_e^2 dV$, of the low temperature component. Table I shows that the emission measure at the higher temperature is never more that 3% of the soft component; therefore we are justified in neglecting it. In Table I we show the density n_e and thermal energy content E as a function of time. The nature explanation for the density increase is chromospheric evaporation, and this is discussed in depth by Acton et al.(1982). We have suggested that the majority of the flare is imaged by this pixel; if we accept that, then the emission measures would be increased by a factor 1.6, with corresponding increases in density and energy content of 1.25.

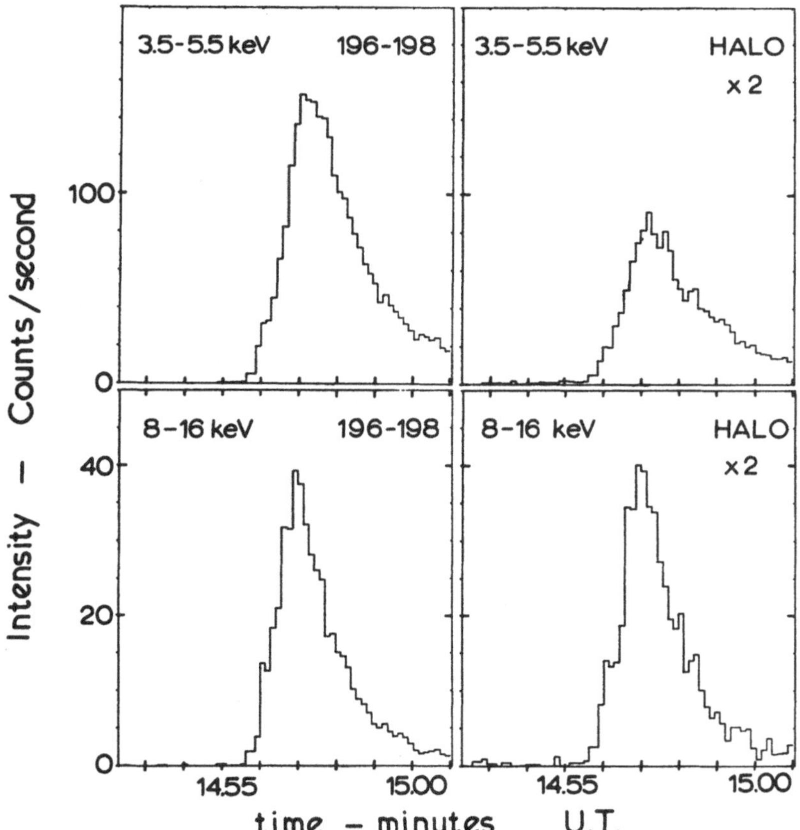

FIGURE 3. A comparison between the intensity-time profiles at 3.5-5.5 keV and 8-16 keV for the main flare region, pixels 196-198, and the halo surrounding it, on May 7, 1980. Note that the intensity of the halo profiles has been multiplied by 2.

The loop extends approximately N-S, imaged by pixels 196(S), 197 (top) and 198(N). These are defined as the primary flare pixels, and the 12 pixels which surround them are the "halo". Figure 3 compares the intensities in the flare and halo pixels in two energy bands, 3.5-5.5 keV and 8-16 keV. It is clear that throughout the event, the spectrum of the halo is significantly harder than that of the flare. This is consistent with photospheric albedo, as predicted by Brown et al. (1975) and Bai and Ramaty (1978). This will be discussed quantitatively in a later paper.

THE 22:38 UT, JULY 5 1980 EVENT, AT N28W29.

This event has remarkable similarities to the May 7 flare. The main flare occurred in a compact magnetic loop of almost identical dimensions to that on May 7. There was evidence for chromospheric evaporation, and during the soft X-ray flare maximum the bulk of the X-ray emission was from the half of the loop linking the flaring footpoint to the looptop. There were two sets of impulsive hard X-ray bursts, with the first being considerably harder in spectrum than the second. The main difference was the scale of the energy release, classified in soft X-rays as M8.9 compared with C7 for the May 7 flare - a factor 15 more intense. The evolution of this flare is somewhat different to that of May 7, but this is not surprising if over an order of magnitude more energy is released in a structure of similar size.

The intensity-time histories of the four HXIS 8" x 8" pixels within which the main flare loop was imaged are shown in Figure 4a, b for the 3.5-5.5 keV and 11.5-16 keV energy bands respectively. These four pixels are arranged geometrically with N at the top and W to the right. Detailed analysis of the 16-30 keV images (Harrison et al., 1983) showed that the magnetic loop stretched from the SW corner of pixel 272 to the southern part of pixel 273. After allowing for foreshortening, they put the separation of the footpoints equivalent to 19" at disc center, or 14 000 km, a distance consistent with the separation of the Hα flare kernels. The loop thickness was estimated by Harrison et al at 4500km (diameter).The site of the impulsive hard X-ray bursts was at the southern end of the loop, in pixel 272. The first impulsive burst, at 22:39 UT, caused a rapid increase in soft X-ray emission, peaking at 22:40:15 UT in pixel 272, after which it started to decline. The features are seen best in the 11.5-16 keV data. Throughout this temporary decline, the intensity in pixel 273 increases, reaching maximum almost in coincidence with the minimum in pixel 272. The first maximum in pixel 273 is reached ∿100 s after the onset of the first impulsive burst, corresponding to a velocity around the loop of 200 km/s if indeed the energy all originates in the southern footpoint. A transfer of hot material around the loop would account for these observations.

The second impulsive burst started at 21:41:25 UT. The 3.5-5.5 keV intensity following the second energy release is confined this time almost totally to the southern part of the loop, extending to the looptop. Throughout the long monotonic rise to maximum the intensity at the other end of the loop is falling. Eventually, well into the decay, the emission from pixel 273 becomes dominant. To summarise, in the aftermath of the first impulsive event, pixel 273, at the northern footpoint, rapidly became the dominant center of emission; after the second burst, the energy transfer is considerably slower.

This behaviour can be interpreted as follows. When the energy associated with the first impulsive burst is released at the southern footpoint, a large amount of chromospheric material is "evaporated" into the loop with high bulk kinetic energy. It rapidly fills the loop, and in the process transfers the majority of the X-ray emitting plasma into the northern half of the loop. After the second impulsive burst, again with the energy release at the southern footpoint, more chromospheric material is evaporated only now the loop is still filled

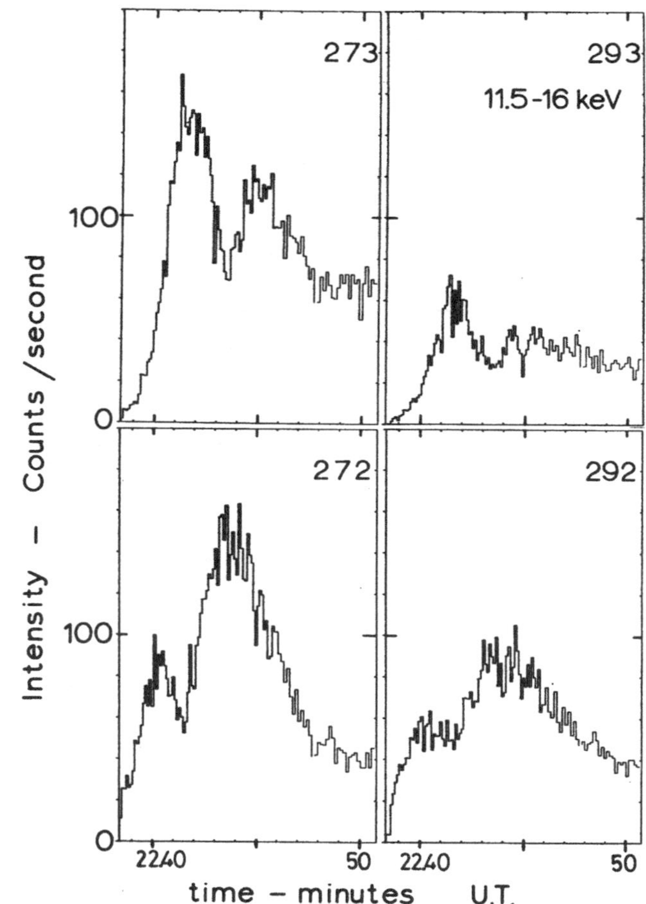

FIGURE 4a The intensity-time profiles for the four pixels imaging the July 5, 1980 flare, in the 3.5-5.5 keV energy band.

FIGURE 4b As 4(a), except in the 11.5-16 keV energy band.

TABLE I. Time resolution of the emission measure Y, temperature T, electron density n_e and energy content E of the principal pixel 197 in the May 7, 1980 event. The data have been analysed assuming a 2-temperature model. It is probable that the majority (>90%) of flare emission is from the volume imaged by this pixel, in which case the emission measures should be increased by 1.6 and the density and energy content by 1.25.

Time : UT	$T_1 \times 10^6$ K	$Y_1 \times 10^{47}$ cm^{-3}	$T_2 \times 10^6$ K	$Y_2 \times 10^{48}$ cm^{-3}	$n_e \times 10^{11}$ cm^{-3}	E Energy in soft X-ray plasma 10^{28} erg.
14:55:00 - 56:00	no 2-temperature fit					
14:56:00 - 56:06	36	0.2	7.1	4.4	2.2	5.9
14:56:06 - 56:14	33	0.4	7.0	5.9	2.6	6.6
14:56:17 - 56:34	28	1.4	7.3	7.2	2.8	7.8
14:56:34 - 56:49	29	2.1	7.2	15	4.1	11
14:56:49 - 57:29	21	8.7	6.6	30	5.8	14
14:57:29 - 58:54	22	4.0	6.6	31	5.9	14
14:58:54 - 00:31	23	0.9	6.4	19	4.6	11
15:00:31 - 01:22	23	0.3	6.4	10	3.3	8.0

TABLE II. Temperatures and emission measures at three phases of the July 5, 1980, flare using a 2-temperature model. Where the low temperature component is omitted, the residual counts, after subtraction of the high temperature component, did not fit a single temperature. All times are UT. The density n_e is calculated under the assumption of a simple compact loop as discussed in the text (after Harrison et al, 1983).

	22:39:50 - 22:40:27			22:42:54 - 22:43:31			22:44:44 - 22:45:21		
Pixel	$T \times 10^6$ K	$Y \times 10^{48}$ cm^{-3}	$n_e \times 10^{12}$ cm^{-3}	$T \times 10^6$ K	$Y \times 10^{48}$ cm^{-3}	$n_e \times 10^{12}$ cm^{-3}	$T \times 10^6$ K	$Y \times 10^{48}$ cm^{-3}	$n_e \times 10^{12}$ cm^{-3}
	30	2.2		24	12.3		25	7.1	
272	7.0	195	1.4	5.3	2300	4.7	6.4	790	2.7
	30	1.4		27	3.7		26	4.2	
292	6.7	130	1.4	5.3	560	2.8			
	30	1.8		26	3.5		26	5.1	
273	7.5	79	0.8	7.3	8.4	0.8	7.2	160	1.2
	33	0.5		26	1.4		26	1.8	
293	6.7	54	1.0						

to a high density with the material evaporated in the first burst. This inhibits further rapid transfer of energy and material throughout the loop, so the subsequent evolution is relatively slow. However, in this flare the energy density is sufficiently high that before the plasma can cool significantly, there is clear evidence of more transfer of material to the northern part of the loop.

Temperatures and emission measures for three different periods in this event are summarized in Table II (after Harrison et al., 1983). The derived densities are calculated for the low temperature component. It is evident for the period 22:39:50 - 22:40:27 UT, covering the first burst, that material rapidly fills the entire loop. Remember that the pre-flare densities are likely to be at least an order of magnitude lower. In the second period, 22:42:54 - 22:47:31 UT, the density in the southern part of the loop has risen to $4.7 \times 10^{12} cm^{-3}$, while in the northern part there has been little change. Following the decay from soft X-ray maximum, the temperature has increased, the density in the southern part of the loop has decreased, whereas in the northern part it has increased. We have seen a transfer of material around the loop, and also a loss of material, presumably back to the chromosphere.

As in the May 7 event, this flare was surrounded by an X-ray halo which is consistent with photospheric albedo.

DISCUSSION

We have presented high resolution X-ray data for two compact flares. Both flares follow almost identical courses, summarised as follows:

1) An energy release occurs near one end of a compact magnetic loop, which accelerates the electrons responsible for the >50 keV X-ray emission, and by inference lower energy electrons, and probably low energy (∿MeV) protons.

2) The bulk of the energy flow is downwards towards the dense chromosphere, where it both heats the cool, dense material and causes it to move rapidly upwards into the magnetic loop.

3) The expansion continues until the bulk kinetic energy of the material is dissipated. In the case of the weak, May 7 flare, this is before the material has got significantly beyond the looptop. Note, however, that, as a fraction of the total emission in the 3.5-8.0 keV band, the intensity from the northern footpoint, imaged by pixel 198, is greatest at the end of the second impulsive burst, having risen monotonically from 5.2% at flare onset to 11.9%.

4) The upward motion of the evaporated material increases the pressure on the magnetic loop, triggering a fresh energy release.

The time delay between the onsets of the two main impulsive bursts coupled with the velocity of the upward moving material, may give a clue to the distance along the loop that the energy release/particle acceleration occurs.

5) The scenario outlined above repeats intself, only now the density in the loop is enhanced, and the bulk of the hot, X-ray emitting plasma which has been produced through dissipation of the impulsive phase energy tends to remain in the southern part of the loop.

6) The flares decay through normal conductive and radiative losses. As the physical dimensions are comparable, but the energy content differs by a factor of 15, we see significantly further along the evolutionary path in the July 5 flare before the flare decays. Harrison et al (1983) point out that in some cases, before the simple evolutionary process can be completed, some other disruptive force may occur, causing quite a different conclusion. In the case of the July 5 flare, before the decay had progressed far, the plasma broke out of the containing flare tube. In other energetic events, there may be multiple loops involved, etc., which distort the overall picture. However, we believe this basic picture of flare development, with two major energy releases within the flare loop,the second triggered by the effect of the first, is applicable to a significant fraction of flares.

REFERENCES

Bai, T. and Ramaty, R. 1978, Ap.J. 219, 705

Brown, J.C., van Beek, H.F. and McClymont, A.N. 1975, Astr.Ap. 41,395

Acton, L.W., Canfield, R.C., Gunkler, T.A., Hudson, H.S., Kiplinger, A.L. and Leibacher, J.W. 1982, Ap.J. (in press)

Harrison, R A., Simnett, G.M., Hoyng, P., La Fleur, H. and van Beek, H.F. 1983, Solar Phys. (in press)

Van Beek, H.F., Hoyng, P., La Fleur, H. and Simnett, G.M. 1980, Solar Phys. 65, 39.

DISCUSSION

SVESTKA: How do you get the high electron densities and what are the uncertainties involved?

SIMNETT: The densities are derived from the emission measure = $\int n_e{}^2 dV$. The emission measures are derived from the HXIS "count-rate prediction program". The size of the X-ray images places quite close limits on the dimensions of the flaring loop; even the altitude of the July 5 flare can be estimated as the loop is projected onto the photosphere due to the N28W29 location of the flare. Therefore the volume is known to within a factor of 2 - 4. The July 5 flare was so compact and so intense that densities well in excess of $10^{12} cm^{-3}$ are required. There are systematic uncertainties in the calculation of emission measure. Comparisons with XRP on SMM have suggested that HXIS emission measures could be high by close to an order of magnitude. Efforts are being made to resolve this.

GENERAL ASPECTS OF HARD X-RAY FLARES OBSERVED BY HINOTORI: GRADUAL BURST AND IMPULSIVE BURST

K. OHKI*, T. TAKAKURA**, S. TSUNETA**, and N. NITTA**
* Tokyo Astronomical Observatory, Mitaka, Tokyo 181, Japan
** Department of Astronomy, University of Tokyo, Tokyo 113, Japan

ABSTRACT. We survey here the observational results on five gradual and four impulsive type events from the hard X-ray imaging (SXT) and spectrometer (HXM) instruments on the Hinotori satellite. A set of differences are clearly recognized between the gradual and impulsive type bursts. These are: (1) Hard X-ray images show the existence of a large coronal source for each gradual burst and a wide variety of source structures for impulsive bursts. (2) The source heights of the impulsive bursts appear to be low. (3) All gradual bursts show power-law spectra while impulsive bursts show exponential thermal spectra at least before the maximum phase. (4) Energy-dependent peak delays are observed only in gradual bursts. From these differences we suggest that two different acceleration and emission mechanisms are involved with these two kinds of hard X-ray bursts.

1. INTRODUCTION

It has long been believed that two distinct phases of flare phenomena exist during the course of large events accompanied by intense type II and IV radio bursts. A typical event occurred on March 30, 1969 (Frost and Dennis, 1971). The two phases were called the impulsive phase and the second phase,respectively. In this paper, we present some 'pure' gradual events in which only a second phase seems to occur from the beginning of the event, without any impulsive phase superposed on it. We propose here to call such an event a gradual burst.

In this paper, we survey all of the gradual bursts found so far in the Hinotori data. The number is only five among more than a hundred intense bursts observed with the imaging experiment since the launch. This fact indicates the very rare occurrence of this type of event. One of the most prominent features is the existence of coronal hard X-ray sources confirmed directly with the imaging observations (Takakura et al., 1982; Tsuneta et al., 1982; Ohki et al., 1982). Although there were several observations of coronal hard X-ray sources isolated by limb occultation (e.g. Frost and Dennis, 1971; Hudson, 1978; Hudson et al., 1982; Kane et al., 1982), direct observations of the main sources in the corona have been obtained for the first time with the SXT. These events all show long enduring, gradual time profiles and are definitely

Solar Physics **86** (1983) 301–312. 0038–0938/83/0862–0301 $ 01.80.

located in the corona at some distance above the photosphere.

On the other hand, many short-duration bursts, the so-called impulsive bursts, were also imaged. These show a fairly wide variety of image types, one of which is a double source structure that was observed also by the X-ray imaging instrument on the SMM satellite (Hoyng et al., 1981). For simplicity, we tentatively use the terms, 'gradual' and 'impulsive' only as an indication of the burst duration. If a burst has a longer duration than, say, ten minutes, we call it a 'gradual' burst; otherwise it is an 'impulsive' burst. In the later sections of this paper we argue that the differences between these gradual and impulsive bursts lie not only in the duration but also in the hard X-ray image and in the spectrum, namely in the physical nature of the sources. For impulsive events, we show here only four strong events selected from many. Energy spectra of these strong events extend to higher energy channels with good counting rates statistics. More impulsive events are described elsewhere from the imaging point of view (Takakura et al., 1983) and from the spectral point of view (Nitta et al., 1983).

2. GRADUAL HARD X-RAY BURSTS

As mentioned in the previous section, five gradual bursts were observed so far. In Figures 1 and 2 we selectively show two typical examples of gradual bursts. The other three bursts are May 8 (Ohki et al., 1982), May 13 (Tsuneta et al., 1983), and April 27 (Takakura et al., 1983) event in 1981. All of them are intense bursts even among the one hundred large events observed with the SXT. All these events lasted more than 17 minutes, the limit of the Hinotori data recorder for one flare.

The images of these five bursts can be subdivided into two kinds. One is a loop-like structure (April 1, May 13, and Oct 7, 1981 events), and the other kind is a double source structure (May 8 and April 27, 1981 events). Two loop-like events (April 1 and May 13) occurred near the limb. The entire loop structure, however, still lay within the disc. If (as is most likely) we are observing loops, a considerable amount of emission must originate at coronal heights. On the other hand, the Oct 7, 1981 event shown in Figure 3 occurred just on the limb. We interpret the main source as a cross-sectional view of a loop top in the corona and the subordinate source as one of the two foot points, with another foot point being occulted by the limb. Since the line-of-sight path length within the loop is longest at the top of the loop, the X-ray surface brightness of the main source may be much brighter than the foot points.

The other two gradual bursts (May 8 and April 27) seem to have somewhat different source structures, appearing to be double. It seems difficult to explain the double source of the May 8 event as two separate loops. As explained in Ohki et al. (1982), at least the main source of the May 8 event seems to be fairly high. The two sources in these two events may be the two mirror-points at a considerable height within a large coronal loop trapping the accelerated electrons which emit hard X-rays through a thick-target trap mechanism (Kane et al., 1980). It is possible, though very speculative, that the double source of April 27,1981 event is a side view of the May 8 double source, namely

GRADUAL BURST

APR 1, 1981 EVENT

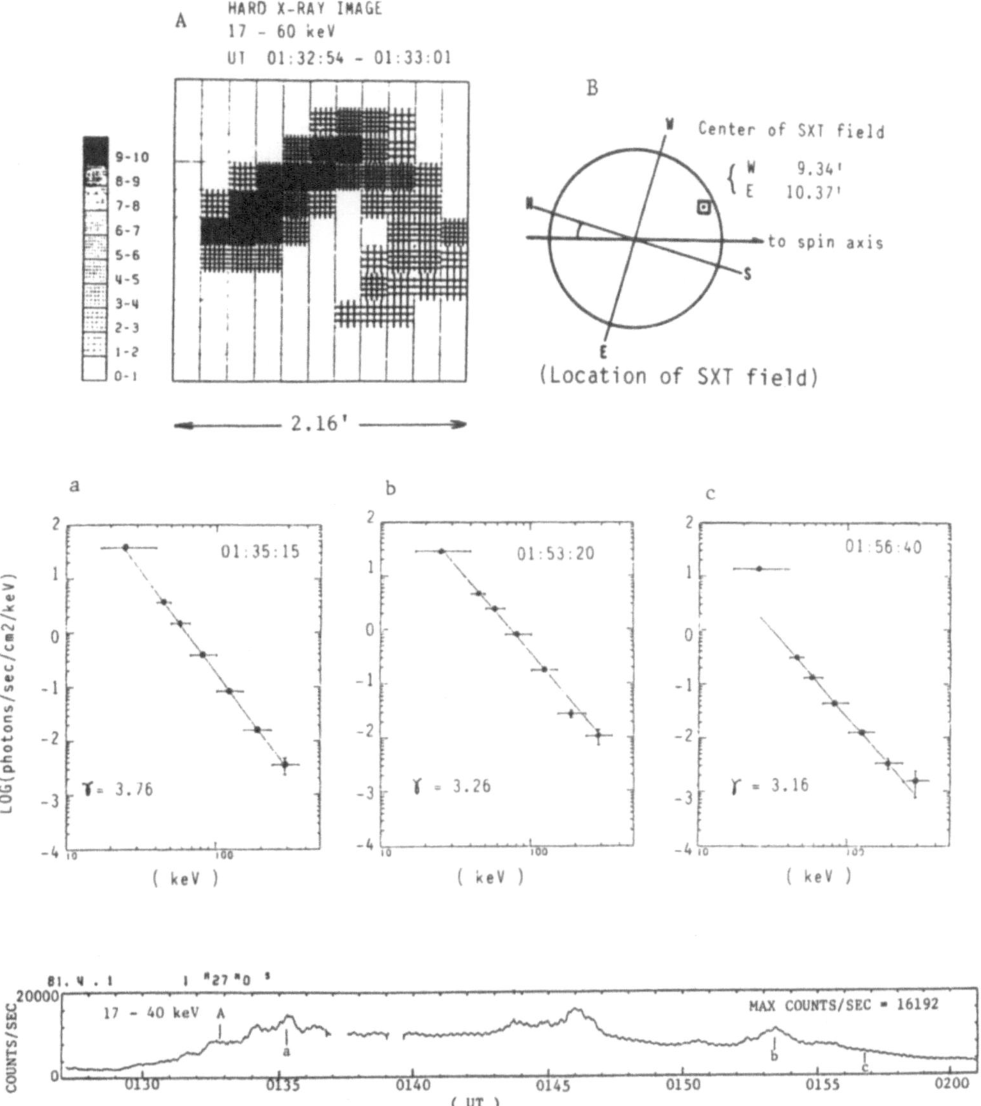

Figure 1 - (A) Hard X-ray image of the 1981 April 1 event at the time indicated in the time profile at the bottom; (B) location of the SXT field of view on the solar disc is shown; (a),(b),and (c) hard X-ray spectra at times indicated in the time profile.

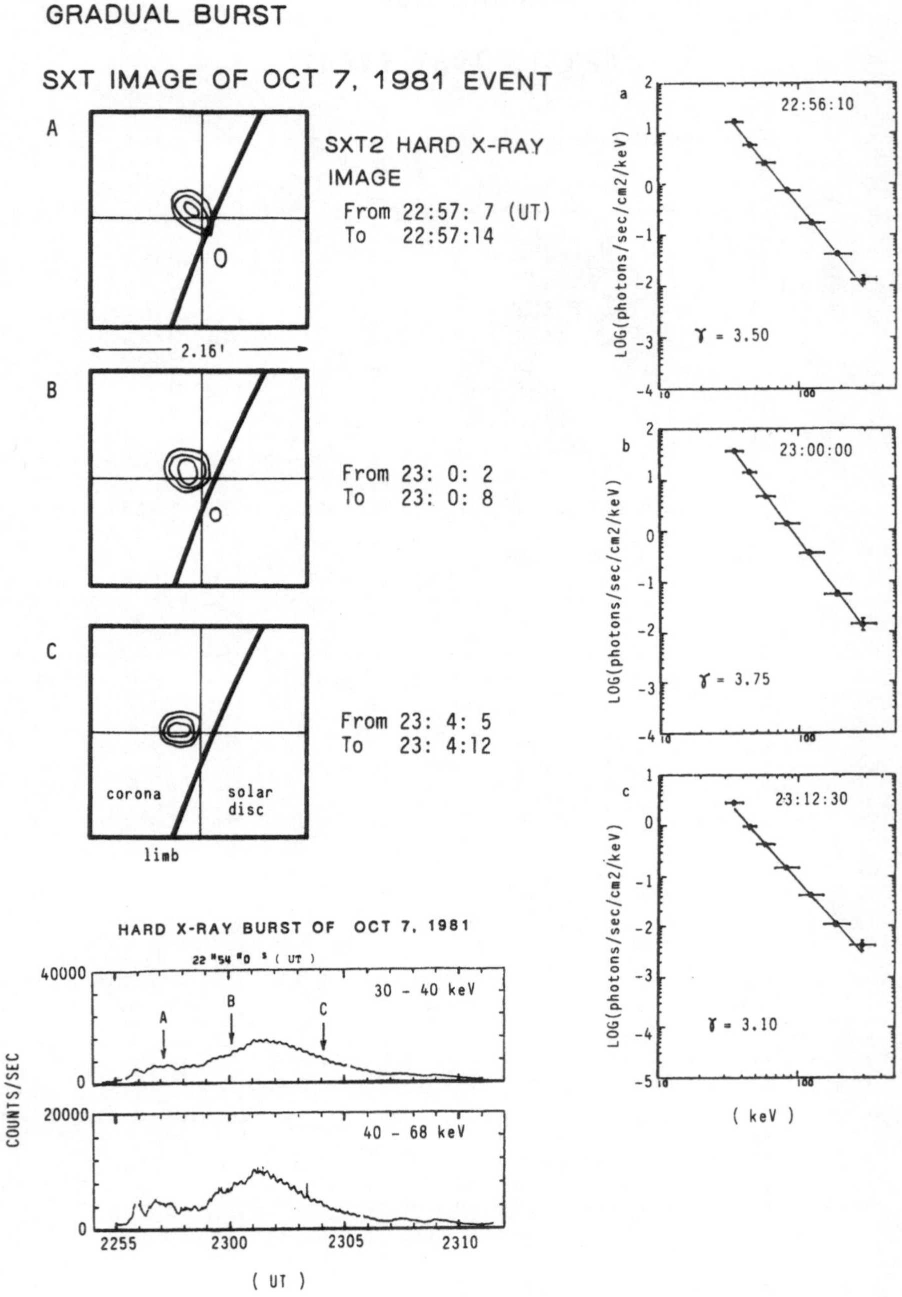

GRADUAL BURST

SXT IMAGE OF OCT 7, 1981 EVENT

A

SXT2 HARD X-RAY IMAGE
From 22:57: 7 (UT)
To 22:57:14

2.16'

B

From 23: 0: 2
To 23: 0: 8

C

From 23: 4: 5
To 23: 4:12

corona
solar disc
limb

HARD X-RAY BURST OF OCT 7, 1981

22 h 54 m 0 s (UT)

30 - 40 keV

40 - 68 keV

COUNTS/SEC

(UT)

a 22:56:10 γ = 3.50

b 23:00:00 γ = 3.75

c 23:12:30 γ = 3.10

LOG(photons/sec/cm2/keV)

(keV)

Figure 2. - Hard X-ray images at three different times of the October 7, 1981 limb event shown with relation to the solar limb position. Hard X-ray spectra at three times are also shown.

the two coronal sources of April 27 may be connected by a very large coronal loop in which accelerated electrons are mirrored at the heights of the two sources.

The spectral characteristics of these five gradual bursts are rather simple. They all have basically power-law spectra from the very beginning through the end of the events. This fact also supports the above interpretation of gradual bursts in terms of a thick-target non-thermal trap model. The only exception is the channel 1 flux in later phases in most events. This channel has some amount of excess flux above the power-law line extension from the higher energies in the later phases of most gradual bursts. This excess is probably due to a different thermal component (Lin et al., 1981) than the power-law component at higher energies. The channel 1 excess, however, appears only in a later phase, so that the loop-like coronal source in the main phase may have a non-thermal origin as well as the emissions at higher energies. In order to have a conclusive result about this problem of the emission mechanism for the loop-like gradual bursts, further detailed studies of the spectra and images are necessary.

3. IMPULSIVE HARD X-RAY BURSTS

Most of the impulsive bursts observed with hard X-ray imaging (SXT) seem to have rather simple, small sources. One example is shown in Figure 3. In this case, the flare occurred at S13,W15, and the hard X-ray source was located on the brightest Hα patches. The hard X-ray spectrum at the impulsive peak fits an isothermal model spectrum clearly better than a power-law, as shown in Figure 3a1 and 3a2.

In Figure 4 and 5, we show two examples of very intense impulsive bursts which occurred just on the limb. Both bursts have gamma-ray line emissions (Yoshimori et al., 1982). The most important result from these limb events is that the hard X-ray source height of the impulsive burst has been directly measured for the first time.At least the main sources of these two limb events are very low in the solar atmosphere as compared with the source heights of the gradual bursts. We conclude that the major part of impulsive hard X-radiation is emitted at altitudes less than 10 arcseconds above the photosphere. A weak source, however, appeared high in the corona around the maximum phase of these two events.

In Figure 6, images of both the double source and the loop-like single source which appeared in 1981 October 15 event are schematically illustrated as well as the corresponding spectra of both phases. As can be seen in Figure 6a1, the spectrum of the double source shows a very good fit to a thermal spectrum with a temperature 6.51×10^8 K, while the spectrum for the single loop-like source is completely different. The latter spectrum is a steep power law with a high energy tail over 100 keV as shown in Figure 6b2. As shown in the figure, the components of the double source in the panel (A) of the Figure are separated by the neutral line, while the loop-like source in the panel (B) cross the neutral line with one foot point anchored in each polarity.

4. ENERGY DEPENDENT PEAK DELAY

In some large flares with sufficient count rates at high energies,

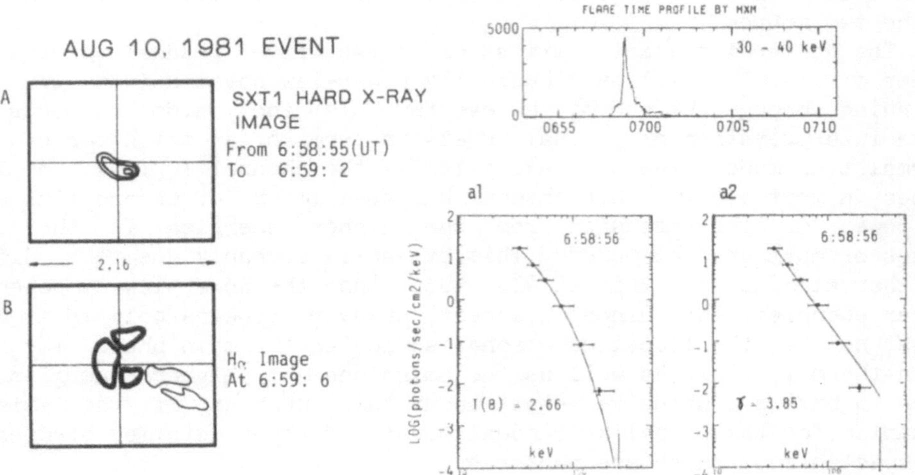

Figure 3. - (A) Hard X-ray image observed during the August 10, 1981 impulsive burst. (B) A sketch from the Hα image at a time near the hard X-ray peak taken at Tokyo Astronomical Observatory. (a1) A hard X-ray spectrum with isothermal model fitting at the X-ray peak. (a2) The same as (a1) except with power-law model fitting. Note that this spectrum fits better to the isothermal spectrum.

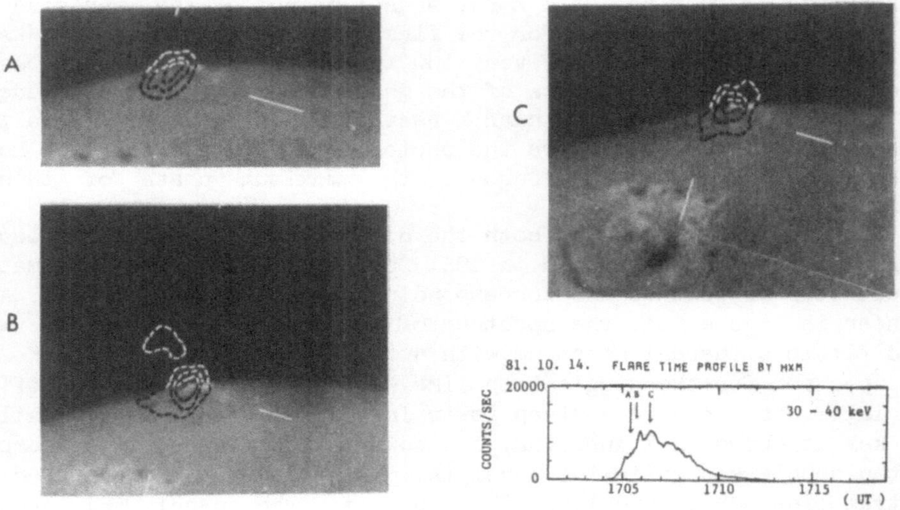

Figure 4. -- Three hard X-ray images superposed on Hα images near the times A, B, and C. Note that the main X-ray source appears to be always very low.

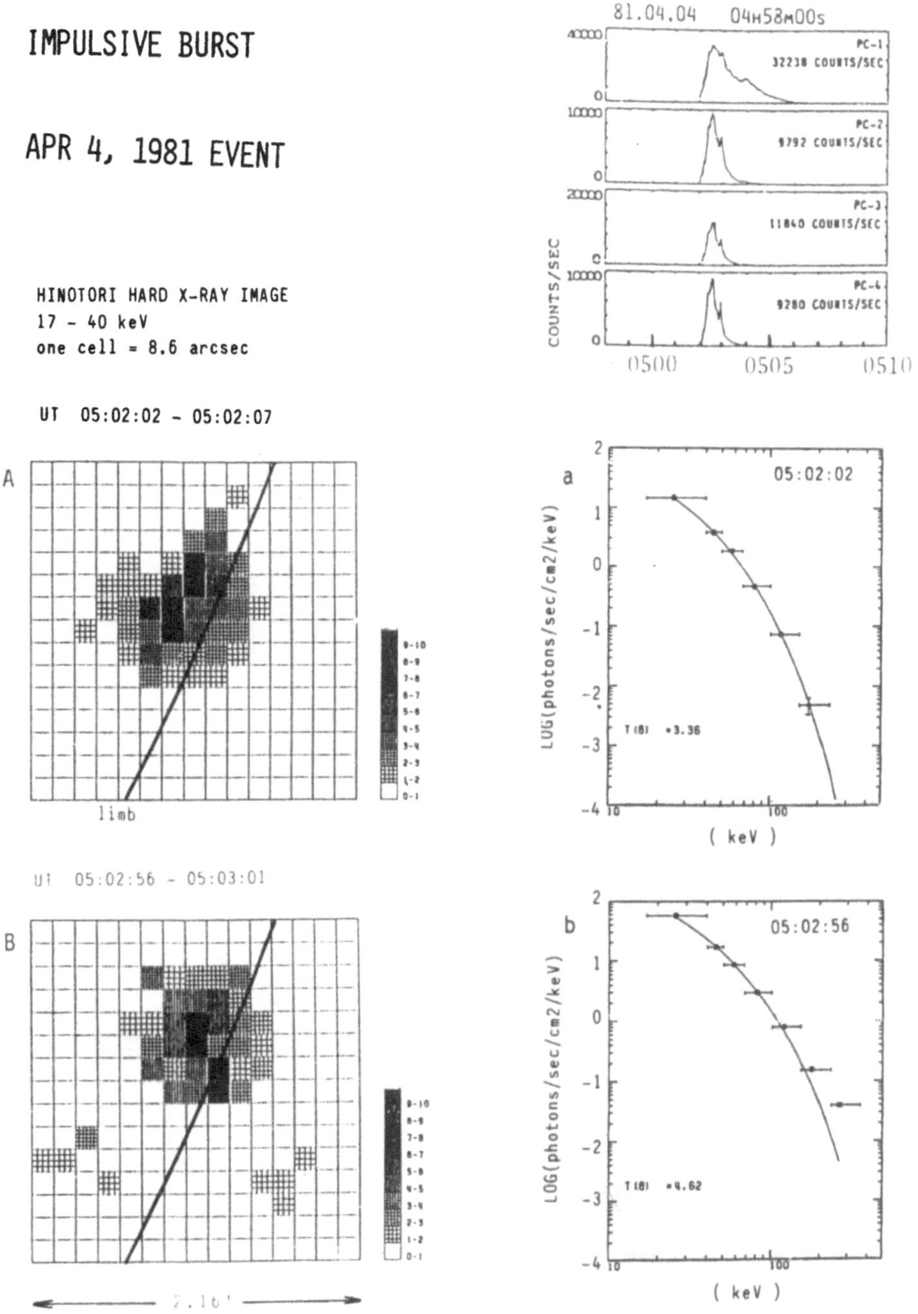

IMPULSIVE BURST

APR 4, 1981 EVENT

HINOTORI HARD X-RAY IMAGE
17 - 40 keV
one cell = 8.6 arcsec

UT 05:02:02 - 05:02:07

A

limb

UT 05:02:56 - 05:03:01

B

2.16'

81.04.04 04H58M00S

PC-1 32238 COUNTS/SEC

PC-2 9792 COUNTS/SEC

PC-3 11840 COUNTS/SEC

PC-4 9280 COUNTS/SEC

COUNTS/SEC

0500 0505 0510

a 05:02:02

T(8) =3.36

LOG(photons/sec/cm2/keV)

(keV)

b 05:02:56

T(8) =4.62

LOG(photons/sec/cm2/keV)

(keV)

Figure 5. - Hard X-ray images of 1981 April 4 impulsive burst at times A and B, with corresponding spectra. The solid line in each image panel indicates the solar limb.

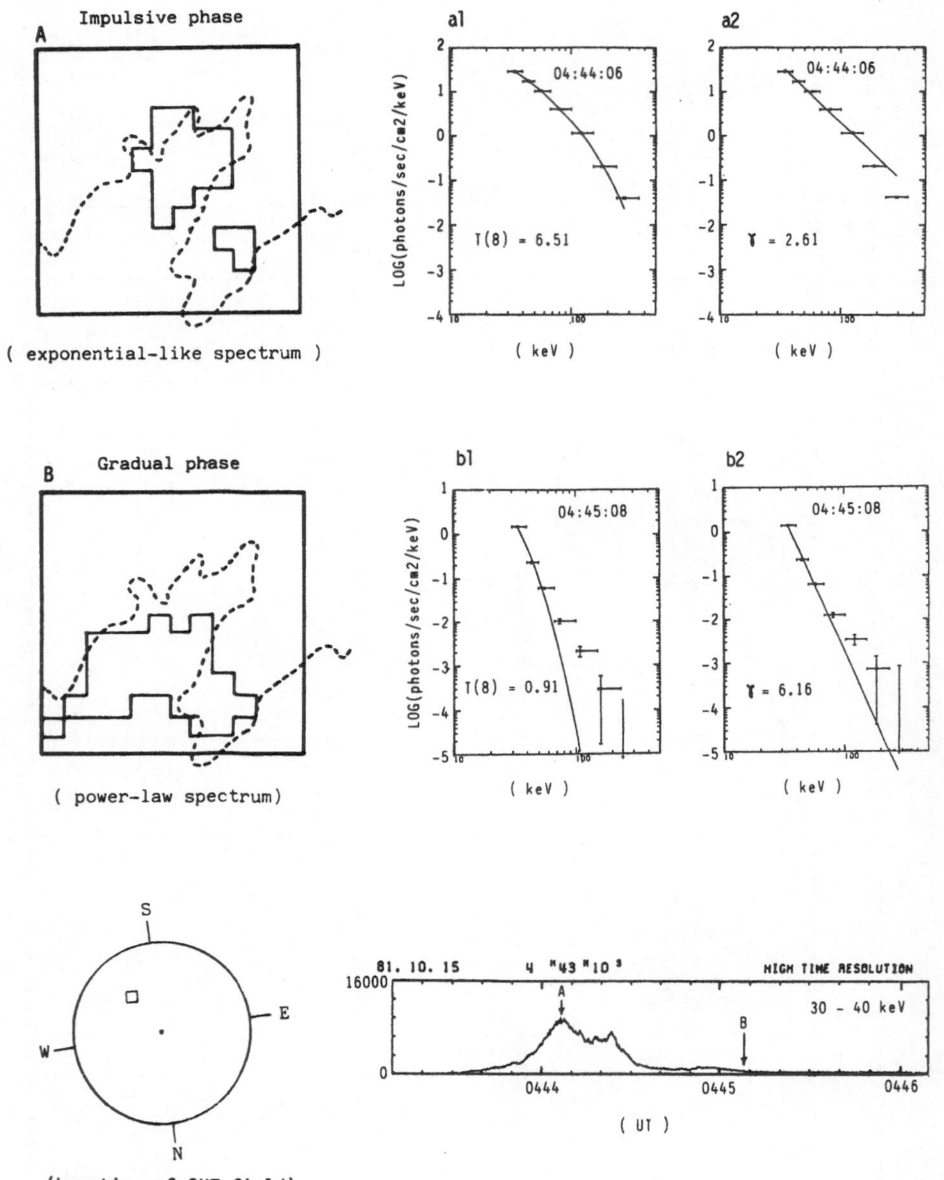

Figure 6. - (A) Schematic hard X-ray image of 1981 Oct 15 event is shown by solid lines along with the magnetic neutral line as a dashed line. (B) The same as above concerning a gradual decay phase. (a1) and (a2) Isothermal and power-law model fitting of the impulsive phase spectrum. (b1) and (b2) The same model fitting of the gradual phase.

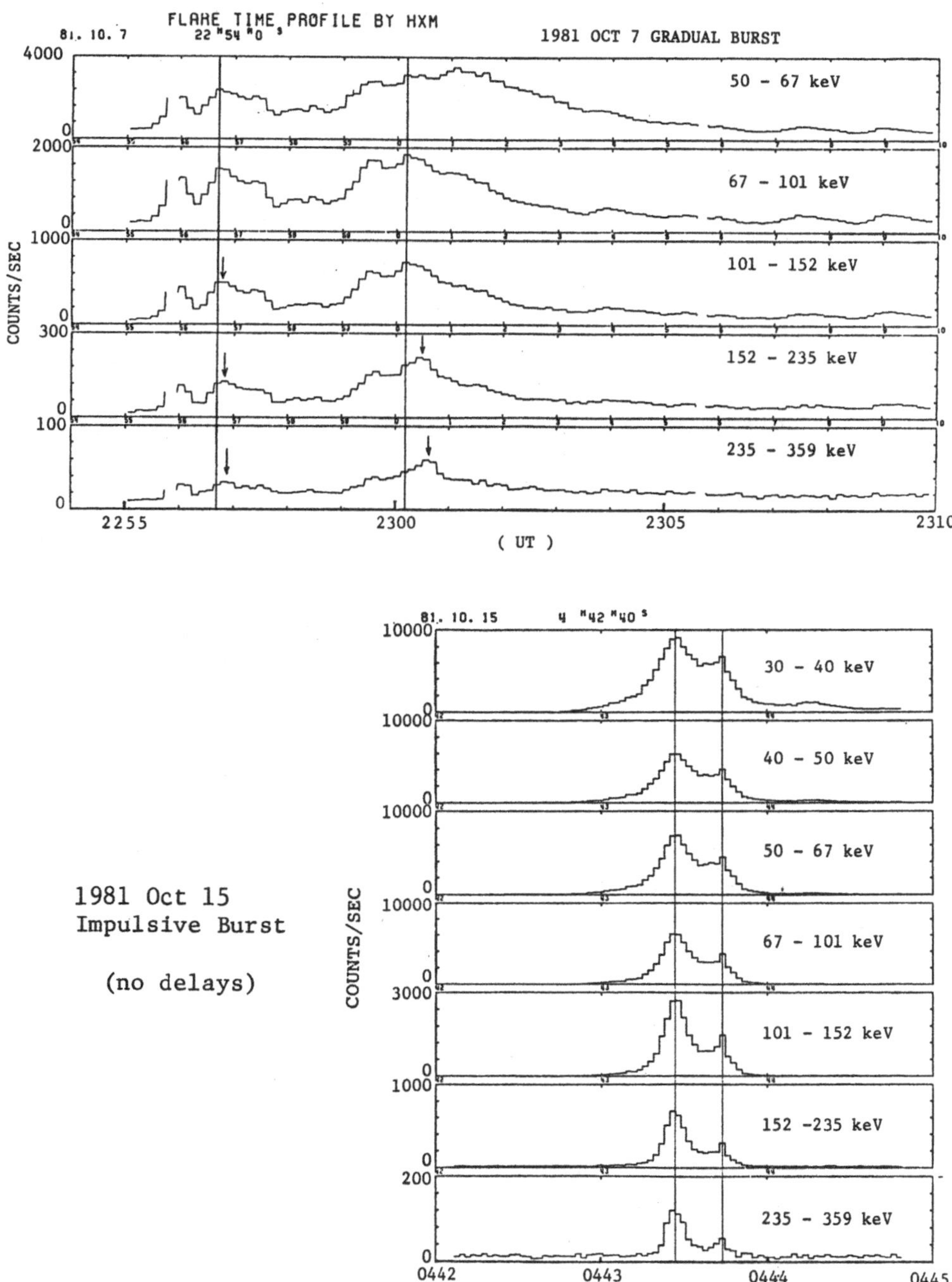

Figure 7. - Search for energy-dependent peak delays for the gradual burst of Oct 7, and the impulsive burst of Oct 15, 1981. Vertical lines near peaks indicate time references.

energy dependent peak delays have been reported by Bai and Ramaty (1979) and Vilmer et al. (1982). We have checked all the gradual bursts observed with Hinotori for this peak delay phenomenon, and find that all gradual bursts show more or less of a delay in peak times. On the other hand, we have also checked many impulsive bursts to search for such delays. At least 20 impulsive bursts with exponential spectra at their peaks were examined, with null results.

In Figure 7, we present one typical example of a gradual burst and that of an impulsive burst. The Oct 15, 1981 impulsive event has no delays from channels 1 through 7. On the other hand, the peaks in the 1981 October 7 event apparently delayed in the higher energy channels.

5. SUMMARY

Although a wide variety of observed characteristics exists among the nine events presented in this paper, there is a distinct set of differences between the gradual bursts and impulsive bursts analyzed so far. Those differences are sammarized in Table I.

TABLE I: Summary of Hard X-ray Bursts Observed with Hinotori

	Gradual Burst	Impulsive Burst
Image	A. Large Loop B. Double source (not low)	A. Single Source (small or extended) B. Double or more sources
Height	high ($\gtrsim 10^4$ km)	low (at least main source)
Spectrum	power-law is prevailing	exponential (early phase) power-law (late phase)
Energy Dependent Peak Delay	Some delays	No delays

From these differences between gradual and impulsive bursts, we suggest that two different kinds of acceleration and emission processes should be responsible for gradual and impulsive bursts. One candidate for the gradual burst emission is a thick target emission by nonthermal electrons trapped in a large coronal loop. In this case, it takes a longer time for electrons to be accelerated to higher energies. On the other hand, all the electrons are accelerated rapidly in the impulsive bursts and precipitated onto the chromospheric region to emit X-rays through either a thick target process or a thermal emission process from an extremely high temperature plasma with several hundreds millions degrees.

ACKNOWLEDGEMENTS

The hard X-ray experiments (SXT and HXM) on Hinotori are a collaboration involving the Institute of Space and Astronautical Science and the Tokyo Astronomical Observatory. The deepest thanks of the authors are due to Prof. M.Oda, Prof. Y.Ogawara, Dr. T.Murakami, and Dr. K. Makishima of ISAS for their collaboration and encouragement throughout every stage of this experiment. We also wish to thank Prof. H.Zirin of the Big Bear Solar Observatory for supplying the Hα photographs.

REFERENCES

Bai, T. and Ramaty, R.:1979, Ap.J. 227, 1072

Frost,K.J. and Dennis,B.R.:1971, Ap.J. 165, 655.

Hoyng, P., Duijveman, A., Machado,M.E., Rust, D.M., Svestka, Z., Boelee, A., de Jager, C., Frost, K.J., Lafleur, H., Simnett,G.M., van Veek, H. F., and Woodgate, B.E.: 1981, Ap.J. Letters 246, L155.

Hudson, H.S.: 1978, Ap.J. 224, 235.

Hudson, H.S., Lin, R.P., and Stewart, R.T.:1982, Solar Phys. 75, 245.

Kane, S.R., Crannell, C.J., Datlowe, D., Feldman, U., Gabriel, A., Hudson, H.S., Kundu, M.R., Matzler, C., Neidig, D., Petrosian, V., and Sheeley, Jr., N.R.: 1980, chapter 5, Solar Flares, ed. P.A. Sturrock, p. 187.

Kane, S.R., Fenimore, E.E., Klebesadel, R.W., and Laros, J.G.: 1982, Ap. J. 254, L53.

Lin, R.P., Schwartz, R.A., Pelling, R.M., and Hurley, K.C.:1981, Ap.J. Letters 251, L109.

Nitta, N., Ohki, K., and Takakura, T.: 1983, in this proceedings.

Ohki, K., Nitta, N., Tsuneta, S., Takakura, T., Makishima, K., Murakami. T., Ogawara, Y., and Oda, M.: 1982b, proc. of Hinotori Sympo. on Solar Flares, p. 69.

Ohki, K., Tsuneta, S., Takakura, T., Nitta, N., Makishima, K., Murakami, T., Ogawara, Y., and Oda, M.: 1982a, proc. of Hinotori Sympo. on Solar Flares, p. 102.

Takakura, T., Ohki, K., Tsuneta, S., Nitta, N, Makishima, K., Murakami, T., Ogawara, Y., and Oda, M.: 1982a, proc. of Hinotori Sympo. on Solar Flares, p. 142.

Takakura, T., Ohki, K., Tsuneta, S., and Nitta, N.: 1983, in this proceedings.

Takakura, T., Tsuneta, S., Ohki, K., Nitta, N., Makishima, K., Murakami, T., Ogawara, Y., Oda, M., and Miyamoto, S.: 1983, submitted to Ap. J. Letters.

Tsuneta, S., Ohki, K., Takakura, T., Nitta, N., Makishima, K., Murakami, T., Ogawara, Y., Oda, M.,and Kondo, I.: 1982, proc. of Hinotori Sympo. on Solar Flares, p. 130.

Vilmer, N., Kane, S.R., and Trottet, G.:1982, Astr. Ap. 108, 306

Yoshimori, M., Okudaira, K., Hirashima, Y., and Kondo, I.:1982, proc. of Hinotori Sympo. on Solar Flares, p. 230.

DISCUSSION

ACTON: How do you explain the delay of high energy peaks in the gradual type burst?

OHKI: The most probable explanation seems to be the delay in the acceleration of higher energy electrons. This is because the hard X-ray source density (for example, of the Oct. 7, 1981 event) is expected to be so large that spectral hardening by collisions in a trapped region seems difficult.

HENOUX: When you see three X-ray sources, what is the speed of propagation of the thermal plasma from the double source to the third source?

OHKI: I presume you are talking about the Oct. 15, 1981 event. In that event, the distance between the initial double source and the later single source is about 30 arc sec (20 000 km). On the other hand, it is about 30 seconds from the peak time of the double source to the onset of the third single source. Therefore, the propagation speed of the exciting agency, if it exists, is about 700 km/s.

VERTICAL STRUCTURE OF HARD X-RAY FLARE

S. Tsuneta, T. Takakura, and N. Nitta
Department of Astronomy, University of Tokyo
Bunkyo-ku, Tokyo 113, Japan

K. Ohki
Tokyo Astronomical Observatory, University of Tokyo
Mitaka, Tokyo 181, Japan

K. Makishima, T. Murakami, M. Oda, and Y. Ogawara
Institute of Space and Astronautical Science
Komaba Meguro-ku, Tokyo 153, Japan

ABSTRACT. This paper presents studies of the vertical structure of hard X-ray flares for two contrasting examples. The 1981 May 13 flare contained a coronal hard X-ray source which was located above 50000 km above the photosphere. On the other hand, the 1981 July 20 flare had a chromospheric double source structure in the initial phase. Electrons in this case were able to stream freely from the corona to the chromosphere.

1. INTRODUCTION

Spatially resolved hard X-ray image can provide vital clues to discriminate among various models of solar flares. Imaging hard X-ray telescope (SXT) aboard the Japanese HINOTORI spacecraft, which was launched on 1981 February 21 by the Institute of Space and Astronautical Science, can image the solar flare in hard X-ray (17-40 keV) and soft X-ray (5-10 keV) ranges with high spatial and time resolutions. The SXT instrument has high signal to noise ratio in the hard X-ray range even for relatively small flares and has whole sun field of view, using the technique of the rotating modulation collimator (Oda, 1965; Oda _et al._, 1976). It has also a capability for determining the absolute location of hard X-ray flares on the solar disk with an accuracy of 5 - 15 arcseconds, depending on the attitude stability of the spacecraft, location of the flare on the solar disk, and location of the spacecraft in orbit. The angular resolution of the image is about 15 arcsec x 15 arcsec over a whole sun in the case of good counting statistics. Same resolution can be obtained even for medium size flare at the expense of the time resolution. Maximum time resolution is about 6 - 10 sec for two channel mode (hard and soft X-rays) and about 3 - 5 sec for single channel mode (hard X-rays alone), depending on the spin period of the spacecraft.

Solar Physics **86** (1983) 313–321. 0038–0938/83/0862–0313 $ 01.35.

The brief description of the instrument is given by Makishima (1982) and Takakura _et al_. (1983). For details on the data analysis system and instument performance, we refer to the paper by Tsuneta (1983).

During 20 months operation of the spacecraft from 1981 February to 1982 October, more than 100 flares with intensities enough to synthesize a hard X-ray image have been observed, some of which occurred near the limb. This paper concerns with the vertical structure of hard X-ray flares for two contrasting examples with emphasis on the results obtained with SXT.

2. OBSERVATION OF THE 1981 MAY 13 FLARE

May 13 event is a two-ribbon 2B/X1.5 flare. Figure 1 shows hard X-ray time profile of this event obtained with the hard X-ray spectrometer, HXM, in four energy channels. According to Solar Geophysical Data, this flare started about 30 minutes before the HINOTORI observation. The data during this period is missing due to the limitation of the onboard data storage capacity. The time variation was very smooth without _any_ impulsive component even in higher channels. Spectral fitting over the range from 40 keV to 200 keV shows that the flare photon spectrum is well represented by a single power law with a power index of about 5 at the start of observation and about 3 in the decay phase.

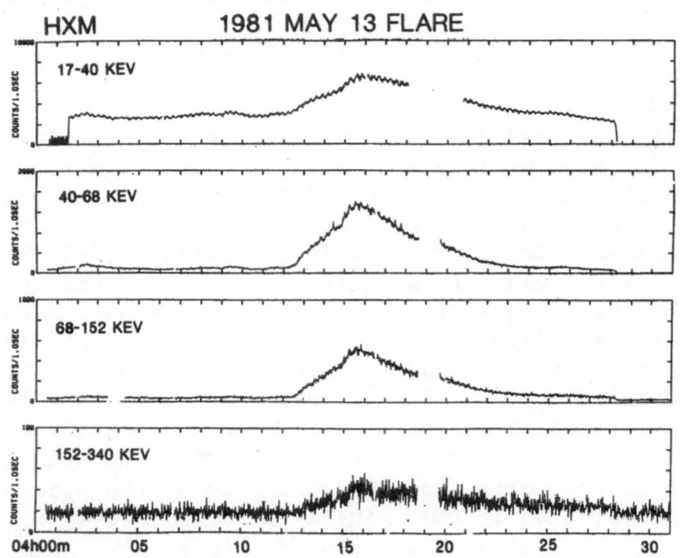

Figure 1. Hard X-ray time profile of the 1981 May 13 flare in four energy channels (channel 1(17 - 40 keV), channel 2 + 3 (40 - 68 keV), channel 4 + 5 (68 - 152 keV), channel 6 + 7 (152 - 340 keV). All data points are plotted with 1.0 second time resolution. Shallow modulation in the lower channel with period of about 15 sec is due to spin rotation of the spacecraft.

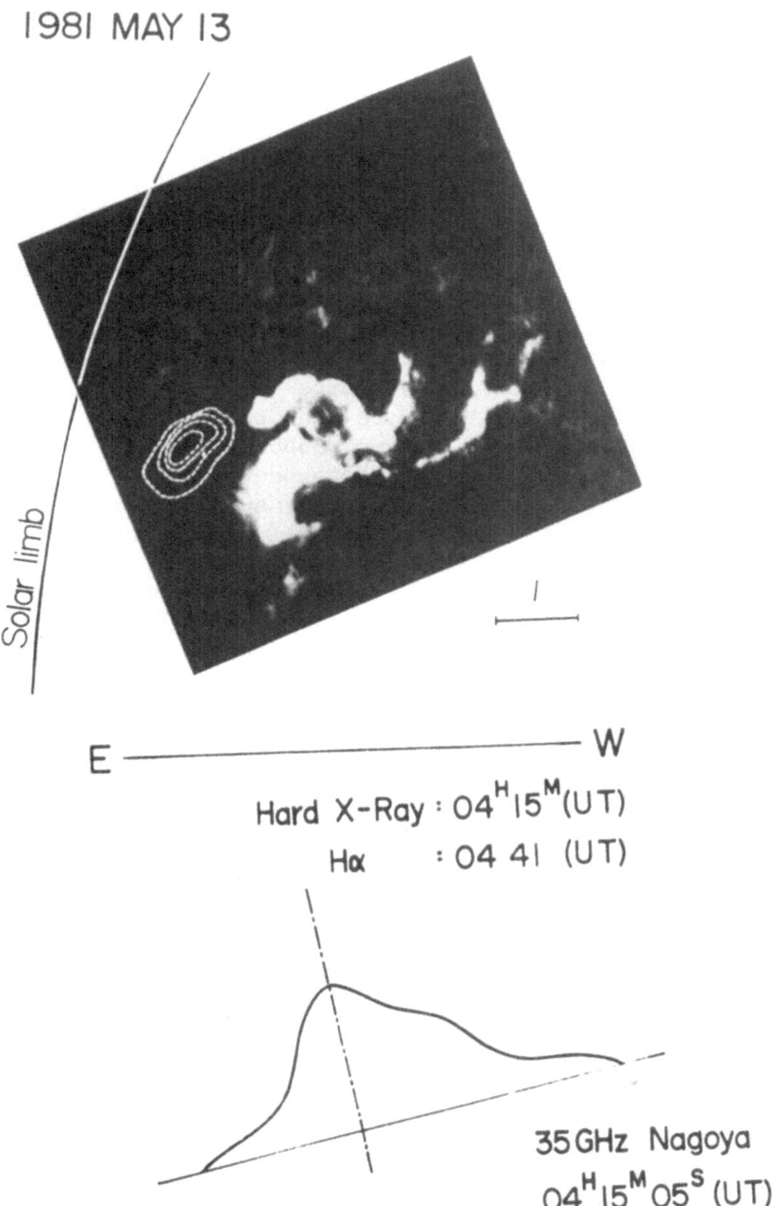

Figure 2. Overlay of the hard X-ray image (17 - 40 keV) near the peak of hard X-rays with Hα photograph taken about 30 minutes after hard X-ray observation for the May 13, 1981 flare (after Tsuneta et al., 1983). The contours indicate the levels at 80, 60, 40 and 20 % of the peak intensity. Microwave brightness at 35 GHz is also shown by the courtesy of K. Kawabata and H. Ogawa at Nagoya University, Japan.

Figure 2 shows the hard X-ray image from 17 to 40 keV at 04h15m UT overlaid on the Hα photograph taken about 30 minutes after the hard X-ray observation. Hard X-ray image is synthesized by Maximum Entropy Method, MEM from one-dimensional scans in about 15 different directions (Tsuneta, 1983). The global structure of the Hα flare did not change during this period. The hard X-ray source is clearly displaced from the well developed two-ribbon flare by about one arcminute toward the limb. This displacement is much larger than the combined positional error in the overlay.

The post-flare loop prominence system connecting the two ribbons appeared about 15 minutes after the hard X-ray peak (Hiei, private communication). The location of the top of loop or archade coincide in position with the hard X-ray source, which is confirmed by Sakurai (1983). The displacement of the hard X-ray source would be due to the projection effect of a coronal hard X-ray source near the limb. If the hard X-ray source was located just above the two ribbons, the height of the X-ray source is about 50000 km above the photosphere.

The soft X-ray (5 - 10 keV) maps were available over the hard X-ray peak for about 4 minutes. The soft X-ray image coincided well in position and size with the hard X-ray image.

The observed displacement of the hard X-ray source persisted throughout 17 minutes of SXT imaging observation from 04h01m to 04h18m over the peak of hard X-rays. We can hardly notice the time variations of the size and source structure such as expansion of the source or the motion toward the limb.

Also shown in Figure 2 is a one-dimensional microwave brightness distribution at 35 GHz (Kawabata _et al._, 1983). The center of the strong microwave source coincides well in position with the hard X-ray source.

Detailed description and analysis of the event are described in the paper by Tsuneta _et al._ (1983), and Tsuneta (1983).

3. OBSERVATION OF THE 1981 JULY 20 FLARE

Two ribbon 1B/M5.5 flare occurred near the limb (S26W75) on 1981 July 20. The characteristics of this flare present a strong contrast to those of the May 13 flare. Hard X-ray time profile in Figure 3 shows some spikes with a duration of about 10 sec in contrast to the smooth time profile of the May 13 flare. Figure 4 shows hard X-ray (17 - 40 keV) images at 13h15m45s and 13h24m48s. Maps are synthesized by MEM. The curve in the map indicates an optical limb. This flare exhibits a double source structure with a distance of about 50000 km lasting for about 6 minutes till around 13h20m (Figure 4(a)).

Figure 5 shows an example of raw one dimensional scans and those derived from the synthesized image of the Figure 4(a). The fit seems to be good. We also check the image by fitting the two Gaussian sources with 12 independent parameters such as source locations, sizes and their peak intensity to the raw data such as shown in Figure 5. The fitting is excellent and the obtained parameters are consistent with the MEM result. Accordingly there is no doubt that the double source structure shown in Figure 4 is real in spite of its relatively low intensity. The intensities of those sources are highly unbalanced. No expanding source in the corona between the double sources was observed. After around

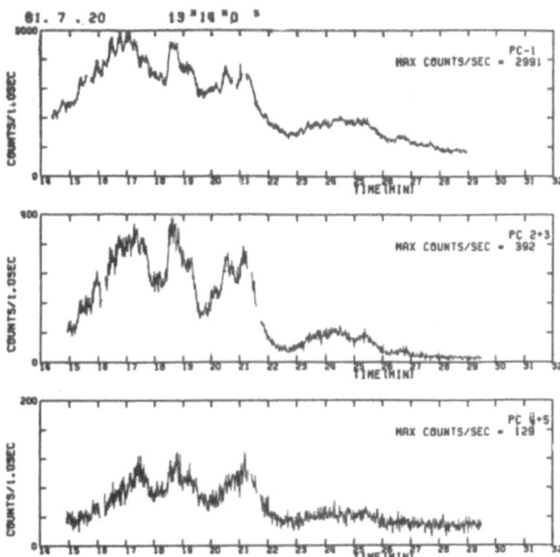

Figure 3. Hard X-ray time profile of the 1981 July 20 flare in three energy channels. From above, Channel 1 (17-40 keV), channel 2+3 (40-68 keV), channel 4+5 (68-152 keV). All data points are plotted with 1.0 sec time resolution.

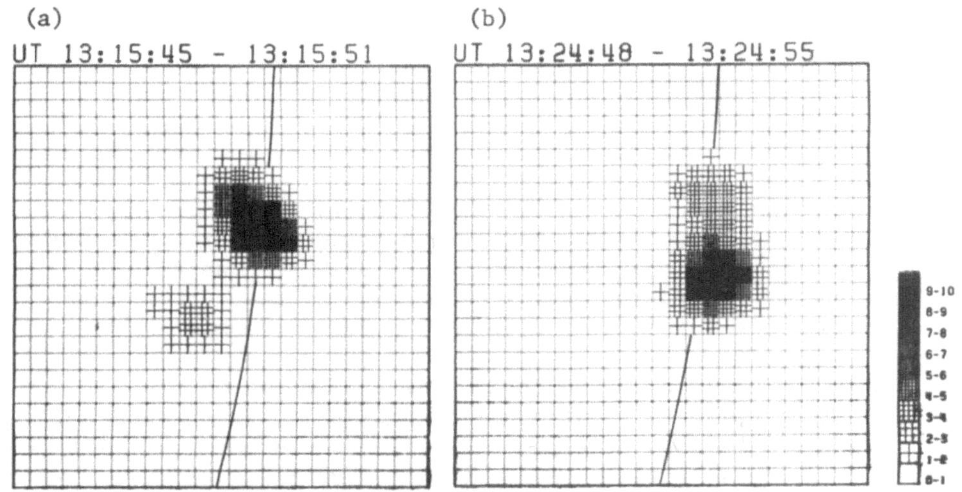

Figure 4. Hard X-ray images of the 1981 July 20 flare at 13h15m45s (a) and at 13h24m48s (b). The curve in each figure is solar optical limb. Pixel size is 7.62 arcsec and FOV is about 3 arcmin. The intensity in each pixel is shown in ten degrees of shading.

13h20m, the loop-like structure, which seems to connect the both footpoints in the corona appeared.

Comparison of the hard X-ray maps with the Hα picture shows that each hard X-ray source in the initial phase coincides in position with each Hα bright region (courtesy of K. Tanaka and H. Zirin). Near the end of the hard X-ray flare, a loop prominence system connenting the two ribbons can be seen. From this comparison, we conclude that the double X-ray source corresponds to the footpoints of a magnetic tube arched over the two-ribbon flare.

Detailed description and analysis of the event are described in the paper by Tsuneta (1983).

4. DISCUSSION

The two examples presented in the preceding section have different characteristics. The July 20 flare had a chromospheric double source which coincided in position with Hα two ribbons. The hard X-ray intensity in the coronal part between the double source is below 10 % of that of chromospheric source. The time profile shows some spiky structure with time scale of several seconds.

The source structure and its time evolution in hard X-rays of the July 20 flare seem similar to those observed by HXIS (Duijveman and Hoyng, 1983 and references therein). The present flare is particularly similar to the 1980 May 21 flare, in that the initial multi source emission coincident in position with Hα brightings was followed by the loop-like emission. The disk flare observed by HXIS would appear very similar to the present flare if occurred on the limb. The duration, during which the double source can be seen was, however, much longer in the present case. In these events electrons are able to stream freely from the corona to both legs of magnetic tube. For the July 20 flare, the lower cutoff energy of the electron beam had to be larger than about 10 keV in order to sustain stable electron beam against the return current instability (Tsuneta, 1983).

On the other hand, The May 13 event had a stable coronal hard X-ray source with an estimated altitude of 50000 km during 17 minutes of imaging observation over the hard X-ray peak. The hard X-ray intensity on the Hα two ribbons is below 10 % of the coronal peak intensity. The time profile of hard X-rays was very smooth even in energies above 100 keV.

Similarly HXIS aboard the Solar Maximum Mission observed a coronal loop-like component with temperature of 4×10^7 K, which appeared near the end of the impulsive phase (Duijveman and Hoyng, 1983 and references therein). Futhermore Lin _et al_. (1981) identified a hot component with a peak temperature of 3.4×10^7 K. Such a high temperature component may constitute a class of flare plasma and be located in the coronal loop.

The parameters of the (cool) plasma located near the loop top are determined and found to be $n = 3 \times 10^{10}$ cm^{-3}, $T = 2 \times 10^7$ K, and plasma $\beta = 8\pi\ nkT/B^2$ is nearly equal to 0.5 throughout the observation for the May 13 event (Tsuneta, 1983).

Figure 5. Examples of raw one-dimensional scans and computed ones (dark dot) obtained from the synthesized image shown in Figure 4 (a).

In order to understand the nature of the hard X-ray source stably located near the top of the coronal loop structure, it is decisively important to estimate the contributions of these cool (2×10^7 K) and hot ($3\text{-}5 \times 10^7$ K) thermal plasma as well as non-thermal power-law component to the hard X-ray image. This kind of work is now in progress with the aids of hard X-ray spectrometer and soft X-ray crystal spectrometer aboard the HINOTORI spacecraft.

ACKNOWLEDGEMENTS

The authors thank Professors E.Hiei and H.Zirin for kindly providing us the Hα picture and to Professor K.Kawabata and Dr. H.Ogawa for providing us the interferometer data of the May 13 flare at Nagoya University.

The highly successful operation of the HINOTORI spacecraft is the results of the several years of intense work by many people. We present our special thanks and appreciations to the members of the HINOTORI project team headed by Professor Y.Tanaka of the Institute of Space and Astronautical Science (ISAS) and by Professor I.Kondo of the Institute of Cosmic Ray Research, University of Tokyo and ISAS.

REFERENCES

Duijveman, A., Hoyng, P.: 1983, in these proceedings.

Kawabata, K., Ogawa, H., Suzuki, I.: 1983, in these proceedings.

Lin, R.P., Schwarts, R.A., Pelling, R.M. and Hurley, K.C.: 1981, Ap.J. 251, L109.

Makishima, K.: 1982, in Proc.Hinotori Symposium on Solar Flares (published by ISAS).

Oda M.: 1965, Applied Optics 4,143.

Oda M., Muranaka, N., Matsuoka, M.,Miyamoto, S.,Ogawara, Y.: 1976, Space Science Instrumentation 2, 141.

Sakurai, T.: 1983, in these proceedings.

Takakura, T., Tsuneta, S.,Ohki, K., Nitta, N., Makishima, K., Murakami, T., Ogawara, Y., Oda, M. and Miyamoto, S.: 1983, to appear in Ap.J. (Letters).

Tsuneta, S.: 1983, Hard X-ray imaging of solar flares with the imaging X-ray telescope aboard the HINOTORI satellite, Ph.D Thesis. University of Tokyo, Japan.

Tsuneta, S., Ogawara, Y., Murakami, T., Makishima, K., Oda, M., Takakura, T., Nitta, N., Ohki, K., Tanaka, K.: to be submitted to Ap.J. (Letters).

DISCUSSION

HUDSON: It is very remarkable to have an Hα flare without a corresponding soft X-ray burst. In fact, this would be unprecedented if the flare were an important one. Could you please clarify the relationship between the X-ray burst and the Hα time history?

TSUNETA: The two ribbon flare started at 03:30 UT. Hard X-ray images are available between 04:02 And 04:18 UT and soft X-ray images are available between 04:14 And 04:18 UT. During this period, soft X-ray source coincided in position and size with hard X-ray source.

EMSLIE: With reference to your interpretation of the May 13 event, it seems difficult to postulate a non-thermal population of electrons confined in a given region for 17 minutes when their thermalization time for the quoted source density of 3×10^{10} cm^{-3} is less than 1 s. For reasonable source volumes and electron injection rates, the source would attain a temperature large enough for thermal hard X-ray emission in a time much less than 10 s.

TSUNETA: As the emission measure and temperature deduced from Fe XXV lines did not vary, and the soft X-ray source size was nearly constant during observation of hard X-rays, the total energy content of the thermal plasma emitting soft X-rays near the loop top also did not vary. The dominant energy loss mechanism is the conduction loss to the transition region, which balances energy input near the loop top.

SIMNETT: Regarding the May 13 event, I have difficulty in understanding the confinement mechanism for the non-thermal electrons in a loop such as you describe. In 17 min such non-thermal electrons would need to be reflected in excess of 10 times without escaping into the lower part of the loop. In view of scattering of electrons, such complete confinement is surprising. Is it possible that you could have somehow misinterpreted your alignment?

TSUNETA: There is no possibility of mis-alignment. The combined error in the overlay is about 10 - 15 arc sec in the radial direction and 15 - 20 are sec in the azimuthal direction along the solar limb. The one arc min displacement of the hard X-ray source is much larger than this radial error. Furthermore, microwave observation by Kawabata and Ogawa showed that microwave source was located near the loop top.

HARD X-RAY IMAGES OF IMPULSIVE BURSTS

T. TAKAKURA*, K. OHKI**, S. TSUNETA*, and N. NITTA*
* Department of Astronomy, University of Tokyo, Tokyo 113, Japan
** Tokyo Astronomical Observatory, University of Tokyo, Mitaka 181, Japan

ABSTRACT. A morphological study is made for the hard X-ray images (25-50 keV) of nine impulsive bursts observed by Hinotori. Most of them revealed single sources, either extended or compact, during the whole duration of the bursts. The sources of all of four spike bursts in the present sample are compact. After the main phase of the impulsive bursts, generally the source size becomes smaller accompanying a shift of position. The X-ray source size is much greater than that of the Hα kernel in two events out of three. Four possible explanations for the X-ray source to be single are suggested. One of these is the strong electric field along the magnetic field as demonstrated to be produced at the decay of force-free current.

1. INTRODUCTION

The hard X-ray images obtained by the Solar Maximum Mission for a few impulsive bursts or the impulsive phase were reported as being double (Hoyng et al., 1981; Machado et al., 1982; Duijveman and Hoyng, 1982). On the other hand, the hard X-ray images of impulsive bursts observed from Hinotori appear as single sources in most cases of the more than ten events so far analyzed, except for the double source of the October 15, 1981 (Takakura et al., 1982; Ohki et al., 1982). The aim of the present report is to provide a morphological study of hard X-ray images (25-50 keV) of the impulsive bursts obtained by Hinotori.

2. SPIKE BURSTS

In order to see the hard X-ray images of the simplest elementary bursts, four spike bursts having the durations of 10 to 15 s and moderate intensities were selected. The hard X-ray time profiles and source images obtained for two of these are shown in Figure 1. The images are single and compact. Similar images have been obtained for the other spike bursts, 1981 August 4 (06:57 UT) and August 13 (07:16 UT).

The contours for the X-ray images in all figures in the present paper

Solar Physics **86** (1983) 323–331. 0038–0938/83/0862–0323 $ 01.35.

are at fractions of 0.88, 0.68, 0.48, 0.28 and 0.08 of the maximum brightness.

The absolute position of the image is unfortunately not reliable enough in these short events to be able to compare it with that of the Hα flare kernel or neutral line.

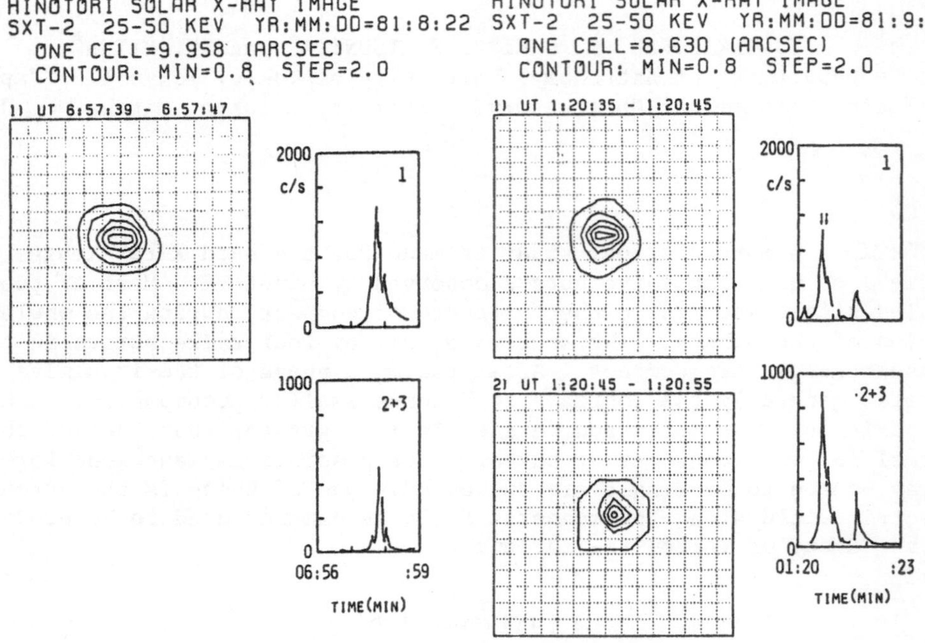

Fig. 1. Time profiles and images of two hard X-ray spike bursts. The nominal energy ranges for the time profiles are 30-40 keV (upper panel) and 40-67 keV (lower panel), respectively.

3. IMPULSIVE BURSTS

Five impulsive hard X-ray bursts with durations of one to two minutes having moderate to high intensities of about 3 to 30 counts cm^{-2} s^{-1} keV^{-1} at 30 keV have been selected:

3.1. Burst of September 7, 1981

The time profiles and X-ray images of this burst are shown in Figure 2. The images are single and compact. The associated Hα flare was observed at the Hida Observatory (Kanno et al., 1982; Kurokawa and Kanno, 1982). The center of the X-ray image nearly coincides with the brightest central kernel in Hα.

3.2. Burst of November 4, 1981

The time profiles of this burst are shown in Figure 3(a) and the X-ray images are shown in Figure 4. The images are single except for the image in panel 2. The source is extended at the main peak and the decrease in size in the later phase is remarkable, accompanying a small

HINOTORI SOLAR X-RAY IMAGE
SXT-2 25-50 KEV YR:MM:DD=81:9:7
ONE CELL=8.630 (ARCSEC)
CONTOUR: MIN=0.8 STEP=2.0

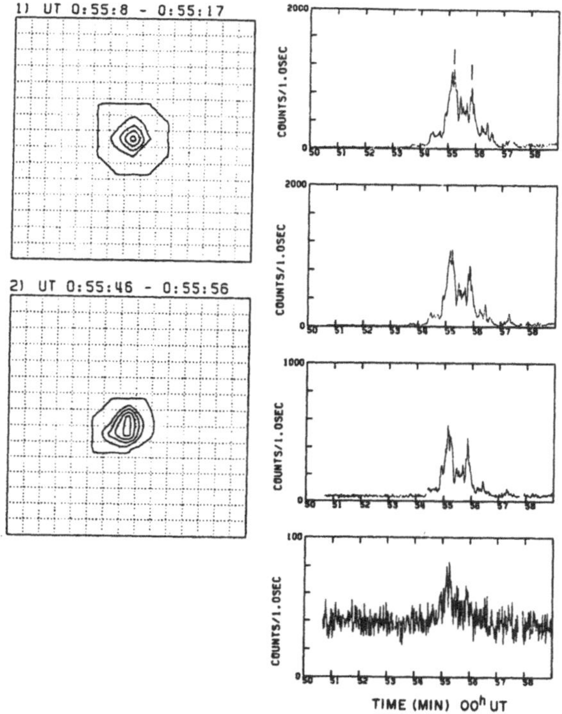

Fig. 2. Time profiles and images of impulsive X-ray burst of Sept. 7, 1981. The nominal energy ranges (in keV) for the time profiles from upper to lower panels are 30-40, 40-67, 67-152 and 152-359, respectively.

shift in position of the centroid.

Note that sub-second pulsations in microwaves and X-rays were observed during the main peak of this event as presented by Degaonkar et al. (1982).

3.3. Burst of September 13, 1981
The time profiles are shown in Figure 3(b) and the X-ray images are shown in Figure 5. The images in panels (1) to (4) show some extensions which could be a subordinate source. In its later phase the source became smaller and the position shifted about 30 arcsec as shown in panel (5).

The associated Hα flare was observed at the Big Bear Observatory (H. Zirin, private communication) and a rough sketch is illustrated in the last panel of Figure 5. The determination of the coordinates of this map was made by Dr. K. Tanaka. The centroid of X-ray source is near the east end of the Hα flare and its source size is much greater than any of the Hα bright points.

3.4. Great burst of August 11, 1981
The time profiles of this burst are shown in Figure 3(c) and the X-ray images are shown in Figure 6. All images are extended without any appreciable change in global structure.

The associated Hα flare was observed at the Tokyo Astronomical Observatory at Mitaka (TAO) by the patrol camera. A rough sketch of the Hα flare patches is shown in the last panel. In a comparison with the X-ray images, we can see that the centroid of the X-ray images nearly coincides with one of the bright portions of the Hα flare patches, but the size of the X-ray image is much greater than that of the kernel. The centroid of one-directional brightness distribution at 17 GHz observed at the Nobeyama Solar Radio Station of TAO is also shown as a dashed line R in this figure (H. Nakajima, private communication).

One notable characteristic of this burst is a remarkable time lag of the microwave peak from the X-ray (30-40 keV) main peak. The delay is about 5 sec at 17 GHz (Nobeyama) and increases with decreasing frequency to be 7.5, 13.5, and 23 sec at 9.4, 3.75 and 2 GHz respectively as observed at the Research Institute of Atmospherics, Toyokawa.

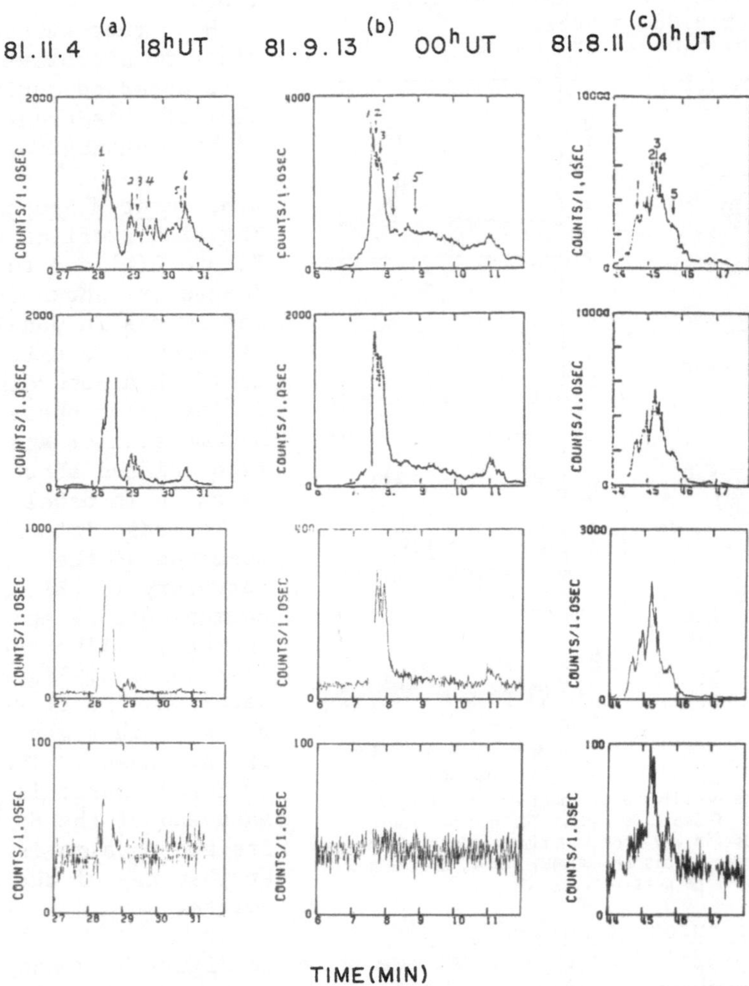

Fig. 3. Time profiles of three impulsive X-ray bursts. The energy ranges are same as given in the caption to Figure 2.

3.5. Great burst of September 15, 1981

The time profiles and X-ray images of this burst are shown in Figure 7. This burst is the largest event in the present analysis. Accordingly images are available from a very early time in the increasing phase. As shown in the figure the source is single throughout the duration of the burst, showing no appreciable change in global structure during the increase to the maximum phases, but a small shift in the centroid can be seen. The decrease in size in the later phase is remarkable. This characteristic is common in the impulsive bursts as described previously for Figures 4 and 5. Another common feature in these bursts is that the better spectral fit changes from exponential during the increasing and main phases to power-law at the later phase.

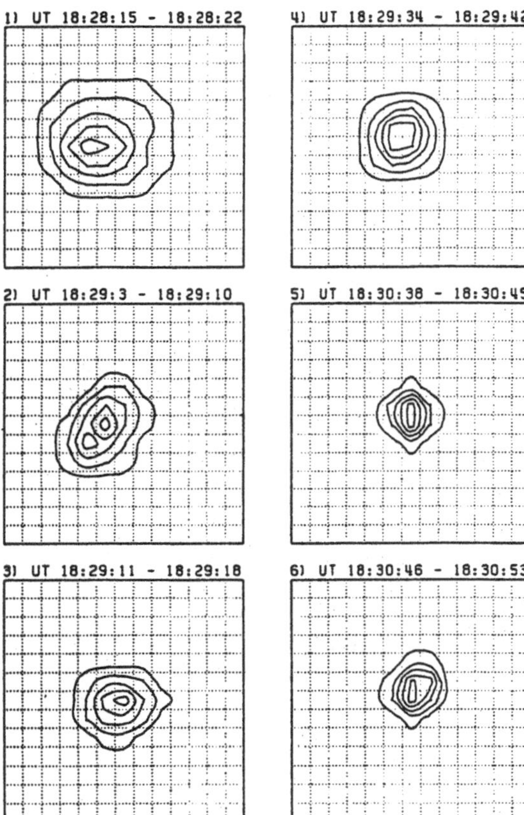

Fig. 4. X-ray images of impulsive X-ray burst of Nov. 4, 1981. Time profiles are shown in Figure 3(a).

4. DISCUSSION

In most of the impulsive hard X-ray bursts, the source images derived so far appear to be single. What would be the reason for this? Four possibilities can be imagined.

(1) The hard X-ray source is near the top or is a large portion of the coronal loops. This situation may occur when the initial density of loop is sufficiently high for the 30 keV electrons to be stopped in the loop.
(2) The separation of two sources, which are the foot points of a single loop, is too small to be resolved in some weak flares.
(3) Magnetic mirrors in the legs of a single loop are asymmetric.
(4) The acceleration of electrons is due to an induced electric field parallel to the force-free magnetic field. Such an electric field can be created during the decay of a force-free current due to anomalous ohmic loss. The condition that the induced electric field $\mathbb{E}$ be parallel to the magnetic field $\mathbb{B}$,

$$\mathbb{E} = \beta\, \mathbb{B}\ , \tag{1}$$

is given by the following as a demonstration.

If β is constant spatially,

$$\frac{\partial\, \mathbb{B}}{\partial t} = -\operatorname{rot}\, \mathbb{E} = -\beta \operatorname{rot}\, \mathbb{B}\ . \tag{2}$$

If we use the condition of force-free current,

$$\mathbf{j} \sim \operatorname{rot}\, \mathbb{B} = \alpha\, \mathbb{B}\ ,$$

Equation (2) reduces to

$$\frac{\partial B}{\partial t} = -\alpha\, \beta\, B\ . \tag{3}$$

Therefore, if the magnetic field starts to decay exponentially from $t = 0$

as

$$\mathbb{B} = \mathbb{B}_0 \exp(-\int \alpha\beta\, dt) \;, \qquad t \geq 0 \;,$$
$$= \mathbb{B}_0 \qquad\qquad\qquad\quad , \qquad t < 0 \;, \tag{4}$$

keeping the force-free condition (for simplicity), the induced electric field is always parallel to the magnetic field so that the electron beam streams into one leg of the magnetic loop. Even for $\partial B/\partial t = 0.1$ gauss s^{-1} (10^{-5} Volt/m^2) and for the radius of magnetic loop of 10^3 km, E is of the order of 10 Volt m^{-1}.

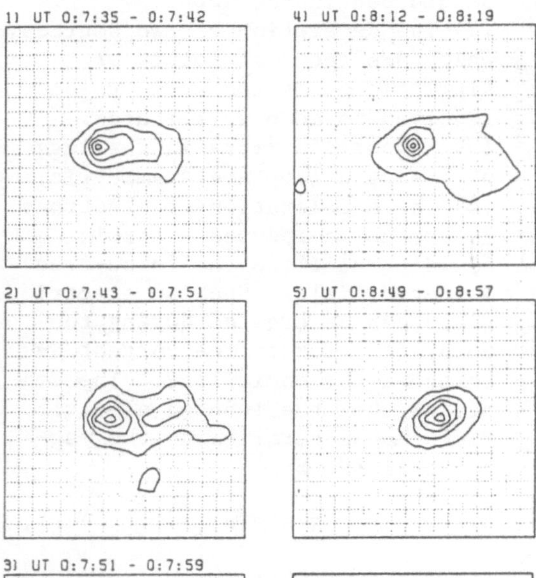

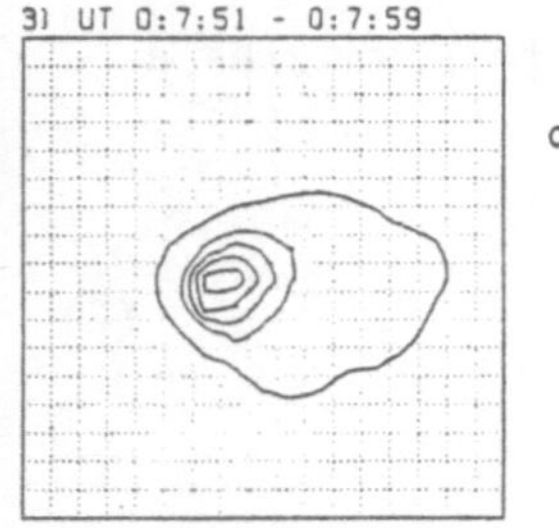

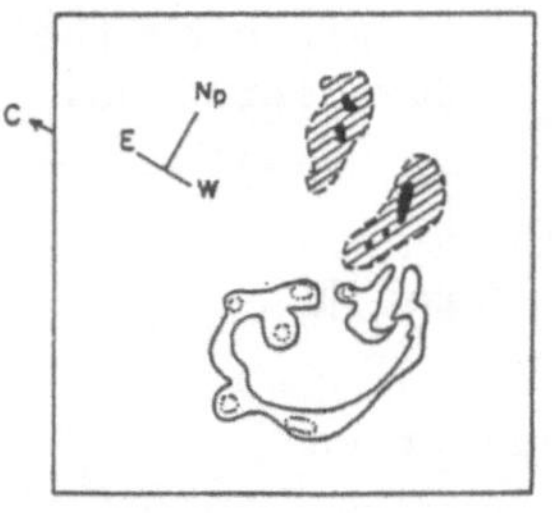

Fig. 5. X-ray images of great impulsive X-ray burst of Sept. 13, 1981. Time profiles are shown in Figure 3(b). A sketch of Hα flare at $0^h07^m0^s$ and sunspots is given in the last panel. The field of view of this map coincides with X-ray maps. Hα flare patches are shown by solid curves and dotted curves show bright portion. Shaded areas represent nearby penumbras. The direction of the solar center (r=9.74' from the center of the map) is shown by an arrow c.

We have not, however, sufficient information yet for choosing the most probable cause of the single source.

In the August 11 event, a clear time delay between the peaks of X-rays and microwaves is observed. Three candidates for the delay can be considered:

(1) Hardening of the electron energy spectrum above 100 keV during the delay, since higher energy electrons contribute more strongly to the radio emissions.
(2) Expansion of sources which are optically thick at radio frequencies as inferred from the radio spectrum which increases with frequency up to 17 GHz.
(3) X-ray and radio sources are not common.

A slight hardening of the X-ray spectrum resulting in an increase with time above 100 keV occurred in the time interval between the peaks of X-rays (30-40 keV) and 17 GHz. However the spectrum became softer thereafter. Accordingly, the delay at 17 GHz can be due to the hardening, but we can rule out (1) for the delays at the lower radio frequencies. Item (3) may not be excluded since we have probably not resolved the fine structures in the X-ray and radio sources. However (2) should be the most probable cause for the delays at least at the lower frequencies, though no increase in the global size of radio source was observed at 17 GHz (H. Nakajima, private communication), and also the X-ray source size was nearly constant as shown in panels (3) and (4) in Figure 6. This is because the radio and X-ray sources may be composed of many unresolved fine filaments and each of them may expand with time due to the dissipation of the force-free current which could be the energy source of the flare. A similar expansion model has been proposed independently by Brown (1982).

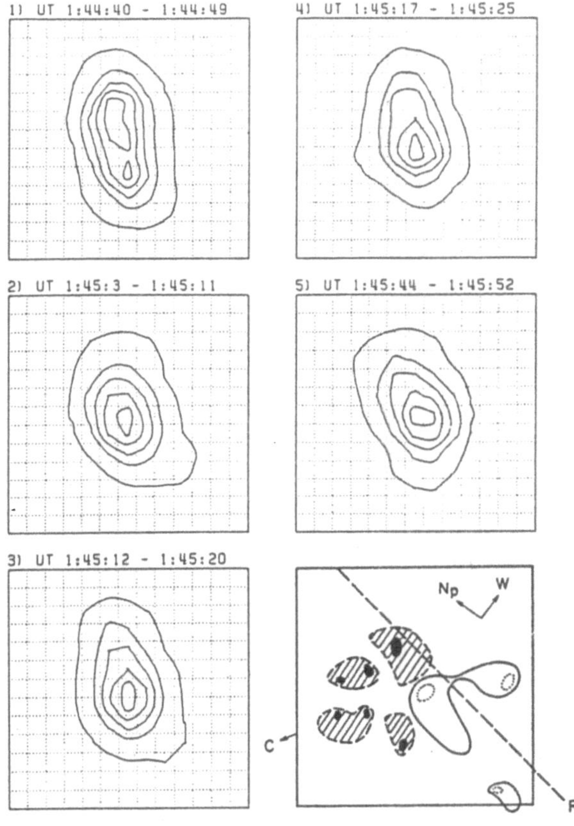

Fig. 6. X-ray images of great impulsive X-ray burst of Aug. 11, 1981. Time profiles are shown in Figure 3(c). A sketch of Hα flare at $01^h55^m45^s$ is given in the last panel (see the caption to Figure 5). r=7.85'. A dashed line R indicates the direction of the peak in the one-dimensional brightness distribution of 17 GHz at $01^h45^m10^s$ UT.

5. CONCLUSIONS

The conclusions of the present morphological study of hard X-ray images of impulsive burst sources are:

1. Spike bursts: Images are single and compact.
2. Impulsive bursts:
   a) The image is almost always single, and extended or compact;
   b) after the main phase, generally the source size becomes smaller and the source position shifts from that of the previous source;
   c) during the increasing and maximum phases of a great impulsive burst, no appreciable variation of the global structure of the X-ray source is observed;
   d) the X-ray source size is much greater than that of the Hα kernel (in two events out of three). It may support the argument by Feldman et al. (1982) that the hard X-ray burst and the soft X-ray burst (which is closely related with the Hα flare) are not casually related to each other.
3. Four possible explanations for the X-ray source to be single are suggested. As one explanation, electric field along the magnetic field is demonstrated to be produced at the decay of force-free current.
4. The time delay in August 11 event between the peaks of X-rays and

microwaves should be attributed to the expansion of optically thick radio sources. However, the delay at the highest frequency (17 GHz) can be ascribed (at least partly) to the hardening of electron spectrum resulting in an increase with time above 100 keV.

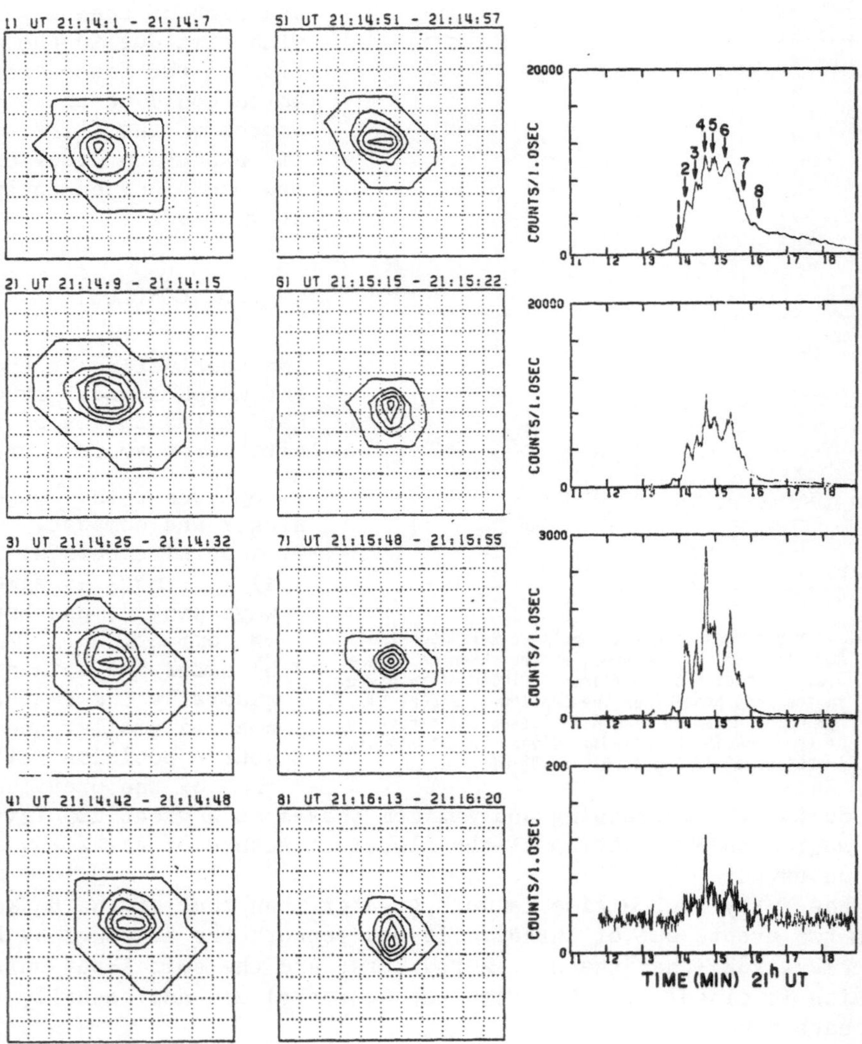

Fig. 7. Time profiles and images of the great impulsive X-ray burst of Sept. 15, 1981. Energy ranges of time profiles are the same as given in the caption to Figure 2.

ACKNOWLEDGEMENT: The deepest thanks of the authors are due to the staff of the Institute of Space and Astronautical Science, especially to the members of the Hinotori team for attention at every stage from the design to the launching and operation of the Hinotori. The authors acknowledge the cooperation of H. Nakajima and T. Kosugi of Nobeyama and S. Enome of Toyokawa for putting the radio data at our disposal. Thanks are also due to H. Zirin and K. Tanaka for the optical information about September 19 event, and to F. Moriyama for the Hα photograph of August 11 event.

REFERENCES

Brown, J.C.: 1982, in these proceedings.

Degaonkar, S.S., Takakura, T., Kaufmann, P., Costa, J.E.R., Ohki, K., and Nitta, N.: 1982, in these proceedings.

Duijveman, A. and Hoyng, P.: 1982, in these proceedings.

Feldman, U., Chung-Chieh Cheng, and Doschek, G.A.: 1982, Astrophys. J. 255, 320.

Hoyng, P., Duijveman, A., Machado, M.E., Rust, D.M., Svestka, Z., Boelee, A., de Jager, C., Frost, K.J., Lafleur, J., Simnett, G.M., van Beek, H.F., and Woodgate, B.E.: 1981, Astrophys. J. Letters 246, L155.

Kanno, M. and Kurokawa, H.: 1982, in these proceedings.

Kurokawa, H. and Kanno, M.: 1982, Hinotori Symposium on Solar Flares (published by ISAS), p.199.

Machado, M.E., Duijveman, A., and Dennis, B.R.: 1982, Solar Phys. 79, 85.

Ohki, K., Takakura, T., Tsuneta, S., and Nitta, N.: 1982, in these proceedings.

Takakura, T., Ohki, K., Tsuneta, S., Nitta, N., Makishima, K., Murakami, T., Ogawara, Y., and Oda, M.: 1982, Hinotori Symposium on Solar Flares (published by ISAS), p.142.

DISCUSSION

KANE: Are the hard X-ray and microwave sources common? If the hard X-ray source is at the footpoint of a loop, can the microwave source be also located near the footpoint of that loop?

TAKAKURA: They are close to each other observationally but not exactly common. Even in the microwave range, the effective source may depend on the frequency if we take into account the non-uniformity of the magnetic field in a single loop. Footpoints having electron density of $10^{12} cm^{-3}$ cannot be radio sources at frequencies as low as 5 GHz.

TIME VARIATIONS OF HARD X-RAY BURSTS OBSERVED WITH THE SOLAR X-RAY TELESCOPE ABOARD HINOTORI (WITH A MOVIE)

Takeo Kosugi* and Saku Tsuneta**
* Tokyo Astronomical Observatory, University of Tokyo, Mitaka, Tokyo 181, Japan
** Department of Astronomy, University of Tokyo, Bunkyo-ku, Tokyo 113, Japan

ABSTRACT. We have developed a new method for synthesizing hard X-ray maps from the raw data of the Solar X-ray Telescope (SXT) aboard Hinotori. Using this method we analyzed five typical SXT events and summarized their images in a movie with a time resolution of about 8 s (half spin period of the satellite). The movie clearly shows that (1) three different classes of bursts, i.e., the gradual thermal burst, the multiple impulsive burst, and the extended outburst, have different structures and show quite different variations from each other, and that (2) the source of the extended outburst is located in the corona above 10^4 km and its shape appears to be a large loop.

1. INTRODUCTION

Since the launch of the Hinotori satellite on 1981 February 21, we have observed more than one hundred hard X-ray solar bursts with the Solar X-ray Telescope (SXT) in the >17 keV range. Many of these bursts have been analyzed and reported at the Hinotori Symposium held in January 1982, at the present seminar, and elsewhere (Tsuneta et al, 1982, 1983; Takakura et al., 1982, 1983; Ohki et al., 1982, 1983; Tsuneta, 1983). The results together reveal a great variety of hard X-ray source structures, such as compact single sources, multiple sources, and large quasi-stationary loop-like structures. Probably the different structures reflect corresponding different physical conditions of hard X-ray emissions. This situation requires us to analyze all the observed events uniformly on the statistical basis.

From this motivation we have developed a new software system which is suitable for analyzing a large mass of data quickly. The present paper describes the essential part of the system. Some physical results have been derived from a preliminary analysis of five typical SXT events made with this system and presented as a movie.

Solar Physics **86** (1983) 333–338. 0038–0938/83/0862–0333 $ 00.90.

2. NEW IMAGING TECHNIQUE OF SXT

SXT is a bi-grid modulation collimator and the transmission bands of SXT scan the solar disk almost linearly according to the satellite spin of about 4 rpm (for details about the hardware design and in-orbit performance of SXT, see Makishima, 1982). The modulated X-ray flux is firstly reduced to one-dimensional scans at various position angles using the precise aspect parameters of the satellite derived from the data of the aspect sensors. This part of the analyzing system was developed by Tsuneta (1983, also see Tsuneta et al., 1982). For the image-synthesis part where two-dimensional images are synthesized from the scans of several directions, we Hinotori group had already possessed two types of techniques: one is the Maximum Entropy Method (MEM; Tsuneta, 1983) and the other is the Arithmetic Reconstruction Technique (ART; Takakura et al., 1982). However, a new method was proposed by one of the authors (TK) for synthesizing a large number of images quickly.

The method is based upon the Fourier synthesis and the so-called CLEAN algorithm. Since observed one-dimensional scans are expressed as

$$g(r,\theta) = \iint f(x,y)p(x\cos\theta + y\sin\theta - r)\, dxdy, \tag{1}$$

where $f(x,y)$ is the brightness distribution of the object and $p(r)$ is the transmission function of the collimator, we have through the Fourier transformation with respect to r (Bracewell and Riddle, 1967; Oda et al., 1976)

$$G(\rho,\theta) = F(\rho\cos\theta, \rho\sin\theta)\, P(\rho). \tag{2}$$

Here $F(u,v)$ is the two-dimensional Fourier transform of $f(x,y)$ in the Cartesian coordinates and $P(\rho)$ is the Fourier transform of $p(r)$. From $G(\rho,\theta)$ we can synthesize a two-dimensional image through the Fourier inversion as

$$\begin{aligned} f_{D,\,obs}(x,y) &= \int d\theta \int G(\rho,\theta)\, \omega(\rho)\, \exp[\, i2\pi\rho(\, x\cos\theta + y\sin\theta\,)]d\rho \\ &= f(x,y) * \mathcal{F}_2^{-1}[P(\rho)\, \omega(\rho)/\rho], \end{aligned} \tag{3}$$

where $*$ denotes the convolution and $\omega(\rho)$ is the so-called reverse filter. Equation (3) means that the image is equal to the original brightness distribution convolved with the synthesized beam. If $\omega(\rho)/\rho$ cancels out $P(\rho)$ completely and accordingly the square bracket of the right side of Equation (3) is unity for all ρ, the synthesized beam is a delta function. However, in practice, $\omega(\rho)$ cannot be increased unlimitedly since the large value of $\omega(\rho)$ increases the noises or measurement errors which are inevitably included in $G(\rho,\theta)$. Moreover, $G(\rho,\theta)$ is given only for discrete points of (ρ,θ) because $p(r)$ is a periodic function and the scan direction θ is already discrete.

In such practical conditions the synthesized beam has a finite width and also spurious features such as a spoke pattern due to the discrete scan directions and the ring-lobes due to the repetition of the transmission function p(r). To remove this dirty beam effect, the CLEAN algorithm (Högbom, 1974) is applied to replace the dirty map by a clean map. The result is as

$$f_{C,obs} = f(x,y) * (\text{ CLEAN beam }). \tag{4}$$

CLEAN is a well-known and well-developed technique in radio astronomy. No essential changes are added to the Högbom's CLEAN. It is to be noted that SXT is similar to radio telescopes of a synthesis type, except that the former measures one-dimensional scans and the latter measures Fourier components directly.

In the actual program, $G(\rho,\theta)$ is set to be zero for ρ where $P(\rho)/P(0) < 0.03$. Three reverse filters, viz., $\omega_1(\rho) = 1/P(\rho)$, $\omega_2(\rho) = [\ 1/P(\rho)\ ]^{1/2}$, and $\omega_3(\rho)$ = const., are prepared to be used selectively. The effective resolution is about 1/2, 2/3, and 1 of the instrumental resolution (FWHM $\sim$ 30 arcsec) for each of these reverse filters. However, the strong reverse filter $\omega_1(\rho)$ increases effects of errors so much, though the transmission function of SXT closely matches its design (Makishima, 1982). Therefore the moderate reverse filter $\omega_2(\rho)$ is used for the present analysis. The computation needs relatively short time (a few seconds of CPU time per map using the FACOM M-200 computer), so that this method provides us a good quick-look system for SXT observations.

3. PRELIMINARY RESULTS

In order to reconfirm some physical results which the previous papers from our group have discussed, we have analyzed five SXT events and summarized their spatial structures and temporal variations in a movie with a time resolution of about 8 s (the half spin period of the satellite). The selected events are listed in Table I. The time

TABLE I: List of the Events Analyzed

			Ref.*
1981 Apr. 2	gradual thermal burst,	compact single source	1,2,5
1981 Oct.15	multiple impulsive burst,	multiple sources	1,4
1981 Apr. 1	extended outburst,	loop-like structure	1,4
1981 Apr.27	extended outburst,	just above the limb	1,3
1981 May 13	extended outburst,	loop-like structure	1,4,5

* Ref. 1. Ohki et al. (1982), 2. Makishima (1982), 3. Tsuneta et al. (1982), 4. Takakura et al. (1982), 5. Tsuneta (1983), for other references see also text.

profiles of these events are given in the literatures cited in the Table. The energy range and the effective resolution are >17 keV and 20 arcsec FWHM respectively. The physical results are summarized as follows:

(1) A gradual burst with a very soft X-ray spectrum (1981 Apr. 2 event) has a compact single source and shows no temporal variations in either position or size (Makishima, 1982).

(2) The multiple impulsive burst on 1981 Oct. 15 shows sudden appearances and disappearances of several sources. A snap-shot map is shown in Figure 1, revealing that three sources exist at this instant. Maps taken at different instants show different features. The existence of physically-connected pairs of sources (Hoyng et al., 1981b) is not obvious from this event alone. However, there is no

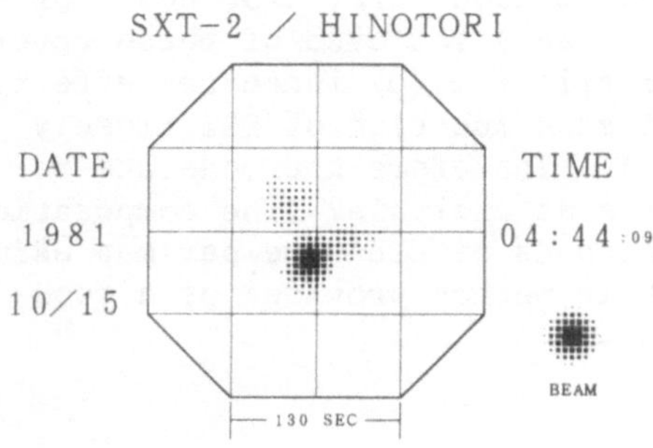

Fig. 1. A snap-shot map of the multiple impulsive burst on 1981 Oct. 15. Top is to the north and right is to the west.

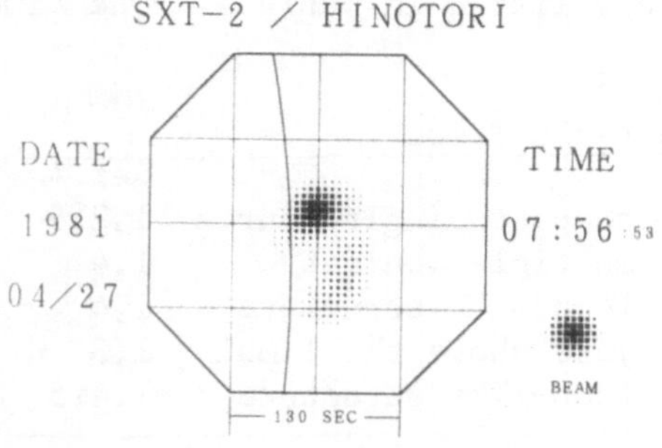

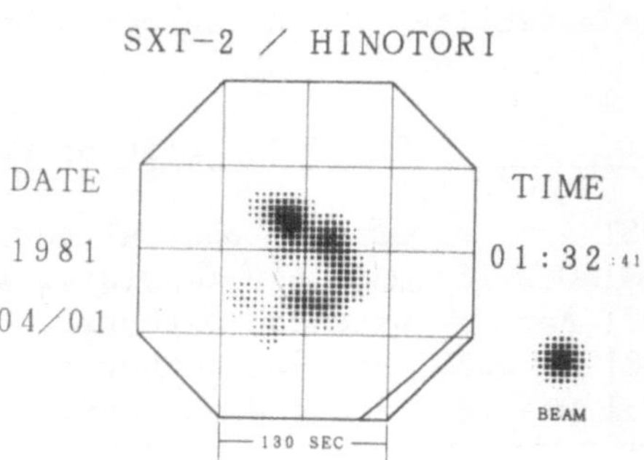

Fig. 2. The extended outbursts on 1981 Apr. 27 (left) and 1981 Apr. 1 (right).

doubt that at least some multiple impulsive bursts are composed of multiple sources, and also that this event is very similar to those reported by the HXIS group (e.g., Hoyng et al., 1981 a, b).

(3) All three extended outbursts studied here (1981 April 1, 1981 April 27, 1981 May 13) have common characteristics: (i) the source is quasi-stationary and located in the corona at more than 10^4 km above the photosphere, (Ohki et al, 1983) and (ii) the source shape appears to be a large loop (Takakura et al., 1982). Since the April 27 event occurred just above the west limb (Figure 2, left), we can measure directly the height to be $>14\times10^3$ km (Tsuneta et al., 1982). In the case of the remaining two events, both of which also occurred near the limb (Figure 2, right), the height is estimated to be several times 10^4 km from projection assuming a vertical structure (Takakura et al., 1982; Tsuneta et al., 1983). Moreover the large source size itself suggests that the source is located at a high altitude. Concerning point (ii), it is important to examine whether the large loop is a large single loop or an apparent loop which is physically the top parts of many loops in an arcade. Though we have not yet had sufficient data to answer decisively, some observational facts imply the arcade interpretation (Sakurai, 1983).

4. CONCLUDING REMARKS

We have developed a new software system for synthesizing maps from the SXT raw data. The method is based upon the CLEAN algorithm and provides us a good quick-look system, since it needs smaller computer time than MEM. On the other hand, MEM has an advantage to the present method in that it gives superresolution if the data require it. Probably these two methods, together with ART and model-fitting method, will be used complementarily.

A preliminary analysis using the new method was introduced in the preceding section. We expect that more SXT data will be promptly analyzed and compared with other data such as the hard X-ray spectrum, Hα observations, and radio observations.

REFERENCES

Bracewell, R.N. and Riddle, A.C.: 1967, Astrophys. J. 150, 427.
Högbom, J.A.:1974, Astron. Astrophys. Suppl. 15, 417.
Hoyng, P., Machado, M.E., Duijveman, A., and 21 co-authors: 1981, Astrophys. J. 244, L153.
Hoyng, P., Duijveman, A., Machado, M.E., Rust, D.M., Svestka, Z., and 7 co-authors: 1981, Astrophys, J. 246, L155.
Makishima, K.: 1982, in Proceedings of Hinotori Symposium on Solar Flares (held at Tokyo, 27-29 January 1982, published by ISAS;

later abbreviated as Hinotori Symposium), p.120.
Oda, M., Muranaka, N., Matsuoka, M., Miyamoto, S., and Ogawara, Y.,: 1976, Space Sci. Instrumentation 2, 141.
Ohki, K., Tsuneta, S., Takakura, T., Nitta, N., Makishima, K., Murakami, T., Ogawara, Y., Oda, M., and Miyamoto, S.: 1982, Hinotori Symposium, p.102.
Ohki, K., Takakura, T., Tsuneta, S., and Nitta, N.: 1983, in this proceedings.
Sakurai, T.: 1983, in these proceedings.
Takakura, T., Ohki, K., Tsuneta, S., Nitta, N., Makishima, K., Murakami, T., Ogawara, Y., and Oda, M.: 1982, Hinotori Symposium, p.142.
Takakura, T., Ohki, K., Tsuneta, S., and Nitta, N.: 1983, in this proceedings.
Tsuneta, S.: 1983, Ph.D. Thesis, University of Tokyo.
Tsuneta, S., Ohki, K., Takakura, T., Nitta, N., Makishima, K., Murakami, T., Ogawara, Y., Oda, M., and Kondo, I.: 1982, Hinotori Symposium, p.130.
Tsuneta, S., Takakura, T., Nitta, N., Ohki, K., Tanaka, K., Ogawara, Y., Murakami, T., Makishima, K., Oda, M., and Kondo, I.: 1983, in this proceedings.

COMPUTED MAGNETIC FIELD STRUCTURE OF THE FLARES OBSERVED BY HINOTORI HARD X-RAY TELESCOPE

Takashi Sakurai
Department of Astronomy, University of Tokyo,
Bunkyo-ku, Tokyo 113, Japan

ABSTRACT. Potential field computations have been carried out to study the location of hard X-ray sources observed by the HINOTORI hard X-ray imaging instrument, SXT. Of the two flares studied, the X-ray source of the 1981 May 13 event, a very unusual gradual flare, appears to lie at the top of an arcade of field lines. In the other event, the 1981 October 15 flare, the observed double source structure is not explained in the present computation, implying the existence of non-negligible electric currents in the flare region.

1. INTRODUCTION

The SXT instrument on board HINOTORI obtained hard X-ray images of dozens of solar flares. Among many simple point-like sources, some of them show loop- or arcade-like structures, or double source structures which might be interpreted as two footpoints of magnetic loops (Takakura et al., 1982; Kosugi and Tsuneta, 1983). In this paper we present the computation of magnetic fields for two flares, in an attempt to study the location of hard X-ray sources in the magnetic field configuration. We hope that this kind of analysis will give important information in modelling the hard X-ray flares.

2. METHOD OF COMPUTATION

We assume that the magnetic field in the corona is current-free. Since current-free (potential) field has no "free" energy, this assumption will be very poor in flare-producing active regions. However there are good reasons to proceed with the potential field computations. (a) If the X-ray images are taken in the late phase of a flare, the magnetic field might have already relaxed to a nearly potential field. (b) The mismatching of computed potential fields with the observations would in any case indicate the existence of sizable electric currents. In this case a more general category of magnetic fields (e.g. the force-free field) is required to fit the observed structure (Sakurai, 1979; 1982).

Solar Physics **86** (1983) 339–344. 0038–0938/83/0862–0339 $ 00.90.

The method adopted here is described in Sakurai and Uchida (1977) and in Sakurai (1981). For weak ($\lesssim 1000$ G) magnetic field regions where magnetographic measurements are available, the present method is essentially the well-known Schmidt method (Schmidt, 1964) with a correction for the sphericity of the solar surface (Sakurai, 1981). Magnetographs tend to saturate on the strong field of sunspots, however, and we utilize the visual measurements of sunspot magnetic field and incorporate them into our computation by making use of our solenoid coil model for sunspot field (Sakurai and Uchida, 1977). In the following examples the magnetograms are from Kitt Peak National Observatory and the sunspot field strengths are taken from Solar Data of the Crimean Astrophysical Observatory.

3. FLARE OF MAY 13, 1981

This event is remarkable in its very long duration of hard X-ray emission. A detailed description and modelling of this gradual event is given by Takakura et al. (1982), Hiei et al. (1982), and Tsuneta et al. (1983). The X-ray source is elongated and slightly bent toward the limb, indicating a gigantic loop. However the overlay of the hard X-ray image and the Hα picture implies that the hard X-ray source lies above and is elongated along the top of the arcade of Hα loop prominence system. The computation also confirms that the X-ray source is located at the top of the arcade above the neutral line running in the north-south direction (Figure 1a). The height of the arcade is $\approx 4 \times 10^4$ km, and the computed field strength of ≈ 50 G is consistent with the value derived from microwave observations (Kawabata et al., 1983).

It is not reasonable to interpret this hard X-ray source as a single magnetic loop which runs orthogonal to the field lines of the arcade. Although there are field lines of this sort (marked by L1 and L2 in Figures 1a, b), their orientation differs somewhat from the direction of the elongation of the hard X-ray source. Moreover the face-on view of the computation (Figure 1b) shows that these field lines are spatially widely separated from the arcade, and their footpoints do not correspond to the location of bright Hα emission. Therefore we conclude that the hard X-ray source lies at the top of the arcade rather than in a single loop. There of course remains a possibility that the field is not current-free, and for example a force-free field computation could give a single loop as the site of the hard X-ray source.

4. FLARE OF OCTOBER 15, 1981

This impulsive flare is studied in detail by Takakura et al. (1982) and by Ohki et al. (1983). The hard X-ray images exhibit a double source structure in the initial phase, which might be regarded as the two footpoints of a flaring loop. In the late phase a third source which itself looks like a large loop appears after the disappearance of the double source.

The potential field computation in this case (Figure 2a) apparently fails to reproduce the loops inferred from the X-ray images. No field lines are found to interconnect the double source in the initial phase. The loop structure seen in the late phase has no counterpart in the computed field lines either. Therefore we conclude that the magnetic field still retains a highly sheared configuration during the period of hard X-ray observation. This might be related to the impulsive nature of this flare. The force-free field computation would be the next step to

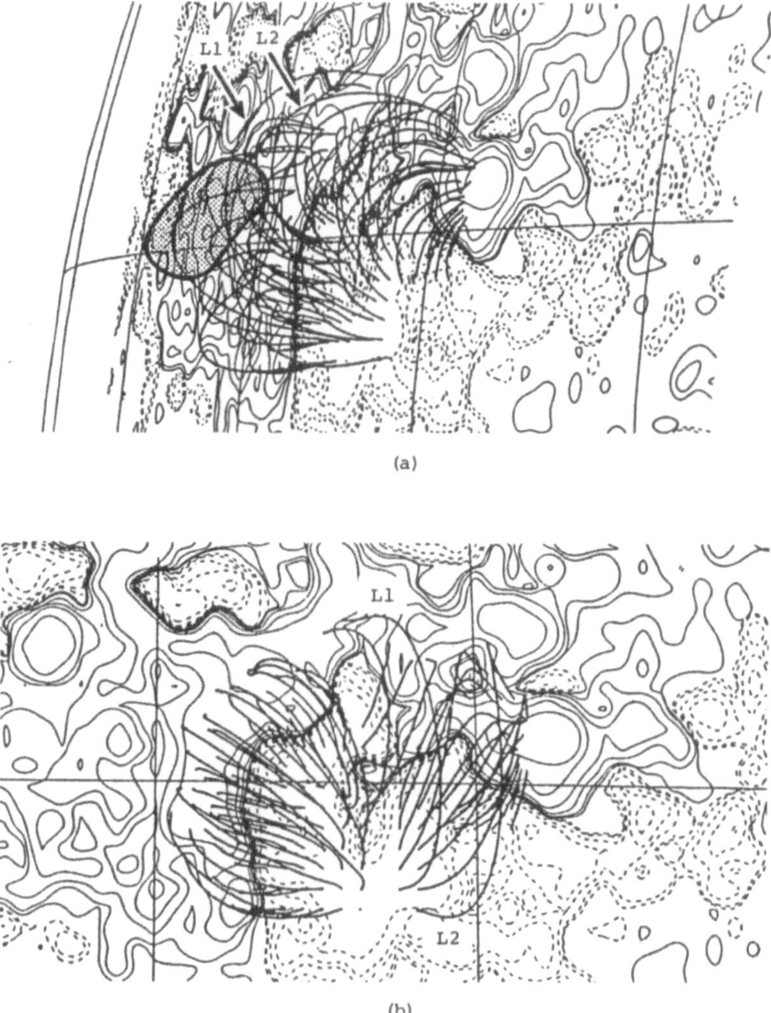

Fig. 1. Computed potential magnetic field lines of May 13, 1981 flare, (a) in proper perspective and (b) face-on view. North is up, west to the right. The hard X-ray source (shaded) lies at the top of the arcade of field lines. Several field lines (marked by L1 and L2) also apparently pass through the X-ray source region but they are widely separated from the arcade.

improve the fitting. The introduction of electric currents in the corona will distort the field and will change the connectivity of field lines. Therefore we may have possiblity to reproduce such field lines that interconnect the double source.

5. SUMMARY

We computed potential magnetic field structures of the flares of May 13, 1981 and of October 15, 1981. The former case is satisfactorily inter-

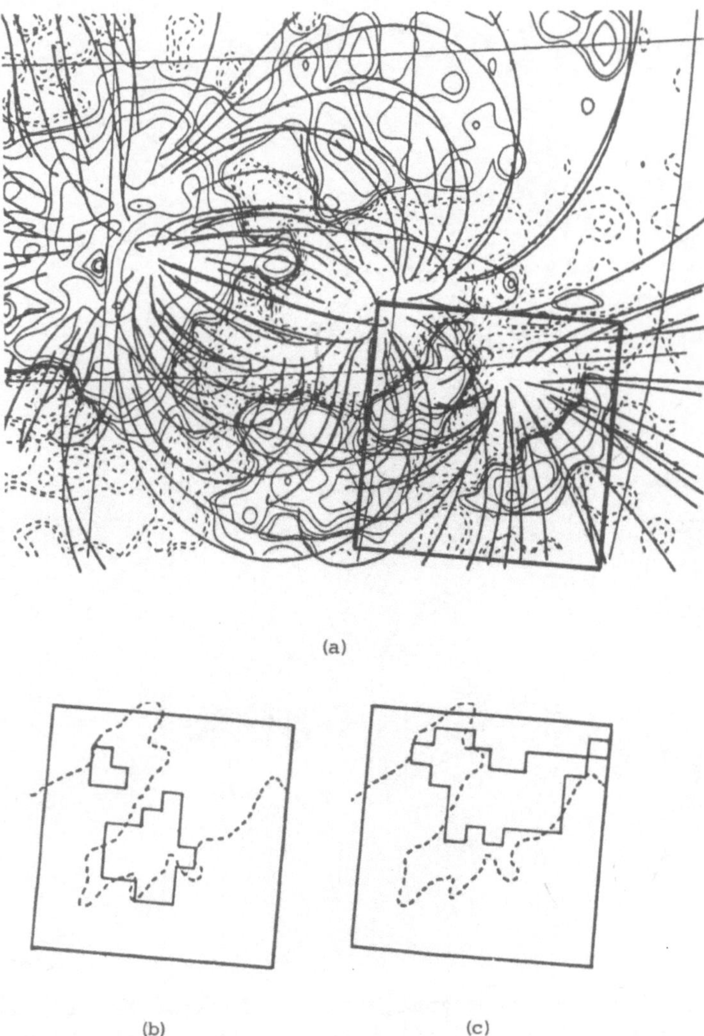

Fig. 2. Computed potential field lines of October 15, 1981 flare (a), together with the hard X-ray images of the flare in the initial (b) and the late (c) phases. North is up, west to the right.

preted within the framework of potential field computations. The latter however shows considerable deviation from the potential field configuration, and a further study, e.g. the force-free field computation, is necessary to give a better reproduction of observed magnetic field structure.

ACKNOWLEDGEMENTS

I would like to thank Dr. J. Harvey for providing me with the magnetograms. Discussions with Drs. T. Takakura, E. Hiei, K. Ohki, and S. Tsuneta are gratefully acknowledged. Thanks are also due to Mr. S. Suzuki for the assistance in data handling.

REFERENCES

Hiei, E., Tanaka, K., Watanabe, T., and Akita, K.: 1982, in 'HINOTORI Symposium on Solar Flares', Institute of Space and Astronautical Science, Tokyo, p.208.
Kawabata, K., Ogawa, H., and Suzuki, I.: 1983, in this issue.
Kosugi, T. and Tsuneta, S.: 1983, in this issue.
Ohki, K., Takakura, T., Tsuneta, S., and Nitta, N.: 1983, in this issue.
Sakurai, T.: 1979, Publ. Astron. Soc. Japan 31, 209.
Sakurai, T.: 1981, Solar Phys. 69, 343.
Sakurai, T.: 1982, Solar Phys. 76, 301.
Sakurai, T. and Uchida, Y.: 1977, Solar Phys. 52, 397.
Schmidt, H.U.: 1964, in W.N. Hess (ed.), Physics of Solar Flares, NASA SP-50, p.107.
Takakura, T., Ohki, K., Tsuneta, S., Nitta, N., Makishima, K., Murakami, T., Ogawara, Y., and Oda, M.: 1982, in 'HINOTORI Symposium on Solar Flares', Institute of Space and Astronautical Science, Tokyo, p.142.
Tsuneta, S., Takakura, T., Nitta, N., Ohki, K., Tanaka, K., Ogawara, Y., Murakami, T., Makishima, K., and Oda, M.: 1983, in this issue.

DISCUSSION

HIRAYAMA: If you have an Hα picture of the flare of October 15, don't you think that you could guess the deviation from the potential field?

SAKURAI: Dr. Ohki informed me that Hα pictures of the October 15 flare were taken at Yunnan Observatory, People's Republic of China. It would be very helpful to compare the computed field lines with the Hα structures in order to infer the deviation from the potential field.

ACTON: Is it possible to infer the geometry and strength of the coronal force-free fields from calculations which use Hα structure to indicate the departure from potential configuration?

SAKURAI: If the 'connectivity' of coronal field lines can be inferred from, e.g., the location of several pairs of Hα patches, we may be able to determine the force-free field by making use of this observational constraint (Sakurai, Pub. Ast. Soc. Japan 31, 209, 1979). The shape of Hα loop prominences can also be a constraint. Another way of determining the force-free field is to measure the vector magnetic field in the photosphere from which the distribution of α ($\nabla \times B = \alpha B$) may be deduced (Sakurai, Solar Phys. 69, 343, 1981).

DYNAMICAL INTERPRETATION OF THE VERY HOT REGION APPEARING AT THE TOP OF THE LOOP

Kazunari Shibata*, Yutaka Uchida**, and Takashi Sakurai***

* Department of Earth Science, Aichi University of Education, Aichi Prefecture
** Tokyo Astronomical Observatory, University of Tokyo
*** Department of Astronomy, University of Tokyo

ABSTRACT. In order to explain the appearance of a hard X-ray source at the top of a loop, we present a model in which the dynamical effects of the dark filament mass infallng along the loop in association with the "disparition brusque" plays an important role. The crash of the infalling mass produces high temperature regions in the low corona above the two footpoints of the loop, and the up-going shocks, created in the crash and strengthened in propagating upwards along the steep density gradient in the tail of the infalling mass, produce a very high temperature (10^8 K) region upon colliding with each other near the top of the loop. Successive occurrence of this process in successively higher loops in magnetic arcade may account for the sources of gradual hard X-ray bursts appearing at the top of the loop-like structure.

1. INTRODUCTION

Observations of limb flares in hard X-rays indicate that, in some events, a hard X-ray source appears high in the corona with certain delay from the low-lying impulsive sources (Takakura et al., 1982; Ohki et al., 1982; Tsuneta et al., 1982; and Tsuneta, 1983, from HINOTORI observations; and Simnett, 1983, from SMM observations). It is suggested from these that the high source is located at the top of a loop-like structure while the low-lying sources are at its footpoints (cf. Sakurai, 1983, for the magnetic field calculation). The low-lying sources are usually attributed to the electron bombardment (Kane, 1973), but the source at the top of a loop is difficult to explain by the same mechanism.

In this paper, we discuss a dynamical model as one of the possible mechanisms which may explain the above-mentioned observations. Our picture is as follows (Figure 1). A rising dark filament may permit a slipping-down of the dense lower part of the dark filament towards the foot-points of the magnetic structure, e.g., Kippenhahn-Schlüter configuration, suspending the dark filament (cf. infall-impact hypothesis by Hyder (1962). The mass may alternatively leak into the loop through some magnetic reconnection around the top of the loop). The potential energy of the suspended mass is converted to the kinetic energy of the infall and then to heat in the crash at the solar surface, and produces

Solar Physics **86** (1983) 345–351. 0038–0938/83/0862–0345 $ 01.05.

hot regions near the footpoints of the loop. Recoiling shocks are created in the crash and rapidly gain strength as they propagate upwards into the tails of the still-infalling gas which have steep density gradient. These strong shocks, ascending from both footpoints, collide with each other at around the top of the loop, and produce a very high temperature region there. In order to check this possibility, we have performed one-dimensional hydrodynamic simulation calculations.

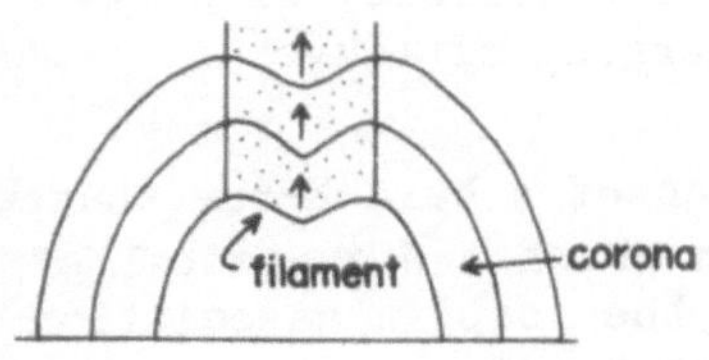

(a) A dark filament prior to the disruption. The initiation of the filament rise follows.

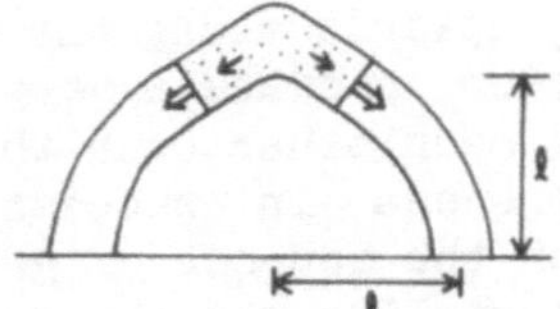

(b) Slipping of gas due to the change of magnetic configuration due to the filament rise. ($t \sim -10^3$ s).

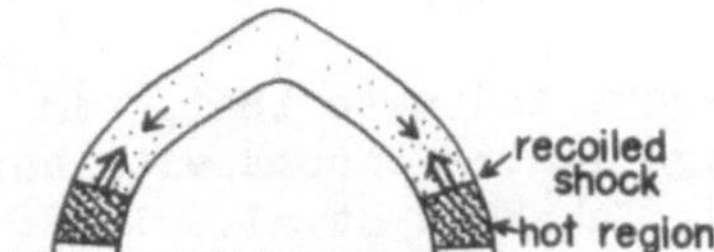

(c) Formation of double hot regions above the footpoints of the loop: generation of recoiling shocks. ($t = 0$ s).

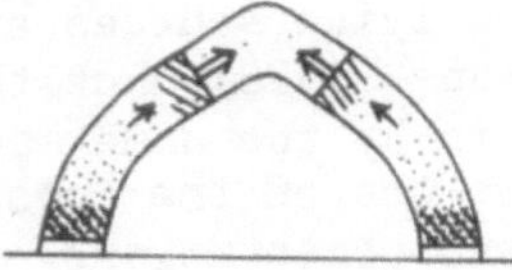

(d) Growth of recoiling shocks in propagating in the tail of infalling mass.

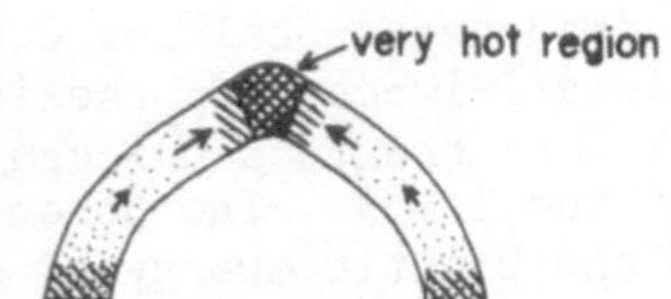

(e) Collision of two very strong shocks and the formation of very hot region at the top of the loop. ($t \sim 10^2$ s).

2. Numerical Simulations

We use the following basic equations with the adiabatic approximation in this first approach:

$$\frac{\partial \rho}{\partial t} + \frac{\partial}{\partial s}(\rho v) + \rho v \frac{\partial \ln A}{\partial s} = 0, \tag{1}$$

$$\frac{\partial}{\partial t}(\rho v) + \frac{\partial}{\partial s}(\rho v^2 + p) + \rho g + \rho v^2 \frac{\partial \ln A}{\partial s} = 0, \tag{2}$$

$$\frac{\partial E}{\partial t} + \frac{\partial}{\partial s}\left[(E + p)v\right] + \rho g v + (E + p)v\frac{\partial \ln A}{\partial s} = 0, \tag{3}$$

where $E = \rho v^2/2 + p/(\gamma - 1)$, $\gamma = 5/3$, $g = g_\odot \cos(s)$, $g_\odot = 2.74 \times 10^4$ cm s^{-1}, $A(s)$ is the cross-section of the loop, and s is the distance along the loop measured from the base level of the photosphere (ie, at $\tau_{5000} = 1$). Other symbols have their usual meanings. These equations are solved numerically by the FCT scheme (Boris and Book 1976). The code used is an extended version of the code used in the study of jet phenomena in the solar atmosphere (Suematsu et al., 1982; Shibata and Suematsu, 1982; Shibata et al., 1982). The configuration of the loop is assumed to be that of Figure 1(b) with the size $\ell = 6\times10^4$ km and assumed to be the same throughout the event. The loop is assumed to be filled with a hydrostatic atmosphere (HSRA + isothermal corona) except for the portion of the dark filament ($T = 10^4$ K) which is initially suspended stably in the dip but now placed at the top of the convex loop. Initially, a pressure equiliblium at the boundary between the dark filament matter and the corona in contact is assumed, but the mass starts to fall due to the unbalanced gravitational force. A lower bound for the temperature is set at 4000 K by considering the radiative input from the surroundings.

Calculation is made only for half of the loop with the top taken as a rigid boundry because of symmetry.

3. RESULTS

Figures 2(a) through (d) show the distributions of ρ, T, v and p as functions of s and t. It is seen in the example shown here ($A(s)$ = const) that, as the dense prominence matter falls down and crashes at the bottom of the loop, it produces a region with a temperature exceeding 10^7 K just above the footpoint as seen in Figure 2(b). As the expansion starts after the compression reaches a peak at about $t \sim 9 \times 10^2$ s, the temperature becomes somewhat lower but remains high. A shock is produced ahead of the expansion, and this up-going shock is strengthened very much as it propagates along the steep density gradient in the tail of the infalling gas as seen in Figure 2(c) and (d).

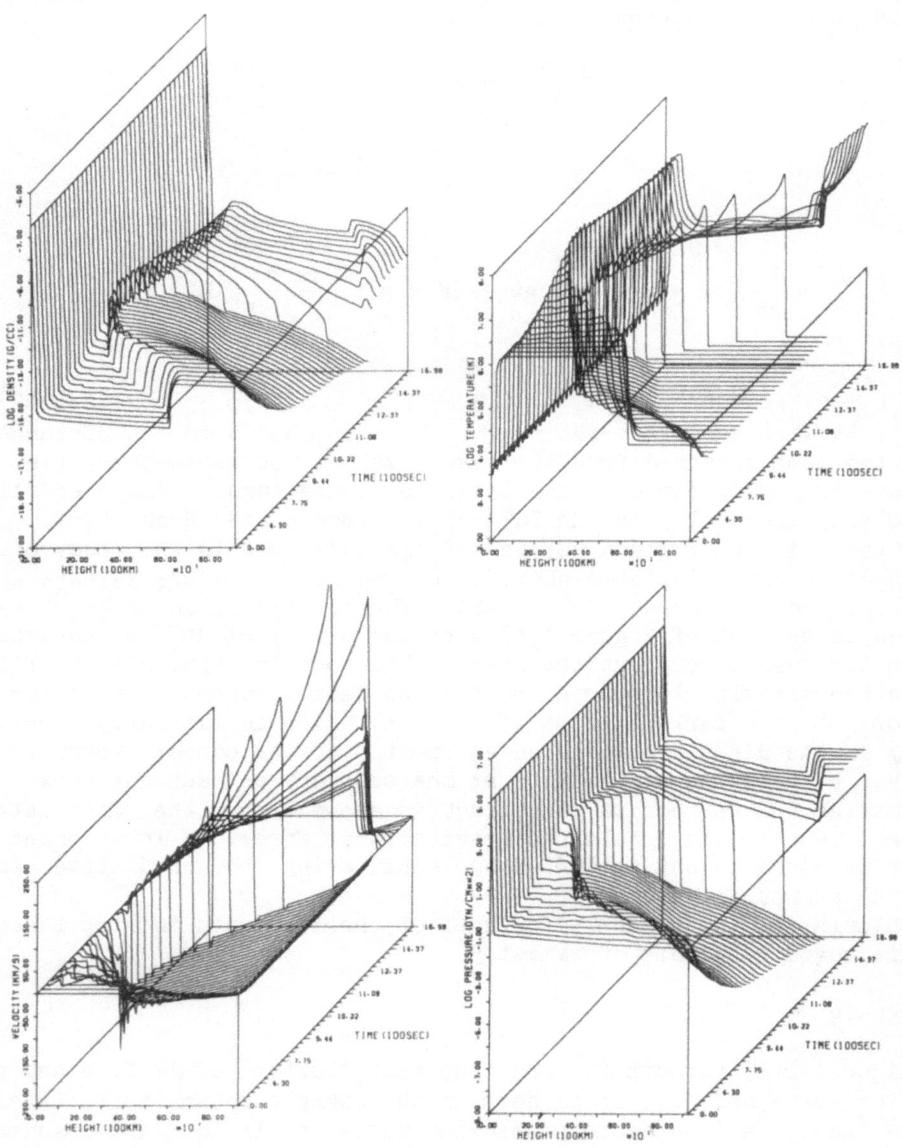

Figure 2. The variation of density(a), temperature(b), velocity(c), and pressure(d) as function of t and s for the case of Figure 1 with $\ell=6\times10^4$ km. Explanations in the text.

Eventually, the shock collides at the top of the loop with another shock coming up from the other half of the loop at about $t \sim 1.1 \times 10^3$ sec, and the temperature rises to more than 10^8 K. The velocity of the shock front and that of the mass flow behind the shock reaches more than 10^3 km s^{-1} before shocks collide as seen from Figure 2(c).

4. DISCUSSION

The model thus gives two hot regions above the footpoints of the loop as well as a very hot region at the top of the loop, and suggests its correspondence to a thermal X-ray source rather than a non-thermal one. However, the highly dynamical processes is expected in the incidence of the shocks produced by the infalling mass and in the collision of two very strong shocks at the top of the loop. These may very well be involved with the acceleration of electrons, either in the strong turbulence, or in the formation of an electrostatic double-layer or two approaching double-layers, and thus the model may well produce non-thermal emissions as well.

The process may take place successively in successively higher-lying loops in the arcade structure, and the source may continue to exist for a considerable length of time. Finally, it may be remarked that the model is not intended to explain the entire sequence of a most energetic flare occurring in the active region. The energy involved in the present model is not sufficient for it since our model is essentially the invisible counterpart of the so-called infall-impact model by Hyder (1962) which suffered the criticism about the insufficiency of the available energy. The model,however, may be suitable for the explanation of the initial phase of spotless double ribbon flares. It is also possible that the hard X-ray producing part of a flare is a biproduct of the more energetic main instability leading to a flare as a whole (Uchida, 1980).

REFERENCES

Boris, J. P. and Book, D. L.: 1976, in Methods of Computational Physics, ed. Adler, B., Vol. 16, p. 85.
Hyder, C.: 1967, Solar Phys. 2, 49.
Kane, S. R.: 1973, in Proc.NASA Symp., High Energy Phenomena on the Sun ed.R. Ramaty and R. G. Stone (NASA Printing), p. 55.
Ohki, K., Tsuneta, S., Takakura, T., Nitta, N., Makishima, K., Murakami, T., Ogawara, Y., Oda, M., and Miyamoto, S.: 1982, in Proc. Hinotori Symposium on Solar Flares (ISAS Printing),p. 102.
Sakurai, T.: 1983, in these Proceedings.
Shibata, K. and Suematsu, Y.: 1982, Solar Phys. 78, 333.
Shibata, K., Nishikawa, T., Kitai, R. and Suematsu, Y.: 1982, Solar Phys. 77, 121.
Simnett, G.: 1983, in these Proceedings.
Suematsu, Y., Shibata, K.,Nishikawa, T., and Kitai, R.: 1982, Solar Phys. 75, 99.

Takakura, T., Ohki, K., Tsuneta, S., Nitta, N., Makishima, K., Murakami, T., Ogawara, Y., and Oda, M.: 1982, in Proc. Hinotori Symposum on Solar Flares (ISAS Printing), p. 142.

Tsuneta, S., Ohki, K., Takakura, T., Nitta, N., Makishima, K. Murakami, T., Ogawara, Y., Oda, M., and Kondo, I.: 1982, in Proc. Hinotori Symposium on Solar Flares (ISAS Printing), p. 130.

Tsuneta, S.: 1983, Thesis, University of Tokyo.

Uchida, Y.: 1980, in Proc. IAU Symp. No 91, Solar and Interplanetary Dynamics, ed. M.Dryer and E. Tandberg-Hanssen (D.Reidel) p. 303.

DISCUSSION

EMSLIE: (1) Your assumed loop length of 10^{10}cm seems somewhat large. (2) The freefall time for the prominence material is on the order of 1000 seconds, much larger than the radiative cooling time for the compressed material. Thus the assumption of adiabatic compression is in some doubt. How does this affect your model calculations?

SHIBATA: (1) The parameters in the calculations can be different. The result does not depend critically on them. The height adopted may not be too unrealistic for the disparition brusque flare filament and in that case, the length of the loop may not be unrealistic either. (2) The time-scales of the steepest part of the compression or the collision of shocks are not very long, and the adiabatic approximation, although not perfect, may not be too bad to be used in the first trial.

PRIEST: (1) In your model, the source of energy is the gravitational potential energy of the prominence material. Is this enough to provide the flare energy or do you need to an additional source? (2) How valid is the adiabatic approximation? (3) In applying the model to a series of loops, one above the other, I would expect the shocks to become MHD ones, with an extra release of energy from the magnetic field.

UCHIDA: (1) You are right. The model is, so to say, the invisible counterpart of the infall-impact model of Hyder, and the amount of energy is not enough to explain the total energy of the most energetic flares in active regions. As seen from the model situation, we intend to apply our model to the relatively quiet double-ribbon flare (<u>disparition brusque</u> flare) especially occurring outside active regions probably related to gradual X-ray sources. (2) The adiabatic approximation may not indeed be fully valid. The steepest part of the compression, however, is pretty short, both at the surface and at the top of the loop, and we think that this approximation is not too bad. (3) It can be shown that β stays smaller than unity and a slow shock along a given tube may represent the situation pretty well.

SVESTKA: We have seen new loops to be formed (or at least excited) many hours after the flare onset (11 hours in the case of the Skylab flare of 29 July 1973). I do not see how your mechanism could work so late in the flare develpment.

SHIBATA: We intend to give you a possible mechanism which produces the appearance of a hot point at around the top of the loop as well as at footpoints in the disparition brusque flare during the impulsive and post-impulsive phase. Other mechanisms may well operate in other phases.

IV: HIGH ENERGY PHOTONS, NUCLEAR PROCESSES, AND PARTICLE ACCELERATION

Chairmen: R. Ramaty and S. R. Kane

SPATIAL STRUCTURE OF HIGH ENERGY PHOTON SOURCES IN SOLAR FLARES

S.R. Kane
Space Sciences Laboratory, University of California,
Berkeley, Calif., U.S.A.

ABSTRACT

Stereoscopic observations of high energy ($\gtrsim$ 100 keV) photon emission from five solar flares have been made with the X-ray spectrometers aboard the ISEE-3 (International Sun Earth Explorer-3) and PVO (Pioneer Venus Orbiter) spacecraft. The observed altitude structure of the photon source and its dependence on the photon energy and time during a flare are compared with the predictions of thermal and non-thermal models of the hard X-ray source. In the case of the *impulsive* source, it is found that (1) the thermal model with adiabatic compression and expansion of a magnetically-confined plasma and the thin target (non-thermal) model are *not* consistent with the observations; (2) the thick target (non-thermal) model and the dissipative thermal model are *partially* in agreement with the observations; (3) the emission probably originates in many individual non-thermal sources distributed in altitude, the lower altitude sources being brighter than those at higher altitude. In the case of the *gradual* source, it is found that (1) models with purely coronal sources are not consistent with the observations; (2) a partial precipitation model with trapped as well as precipitating electrons is consistent with the observations.

1. INTRODUCTION

One of the critical tests for models of solar flares is their prediction regarding the location and altitude dependence of the high energy photon sources during the impulsive and gradual phases. During the impulsive phase, some thermal models, for example, require that the hard X-ray source be located in a low density region at coronal altitudes (cf. Crannell *et al.*, 1978). Non-thermal models may require a high altitude "thin target" source associated with an electron trap (Datlowe and Lin, 1973; Ramaty, 1973; Takakura, 1974), a low altitude "thick target" source due to an electron beam (Brown, 1971; Hudson, 1972) or an extended source due to partial precipitation of electrons from a "leaky" electron trap (Kane, 1974; Melrose and Brown, 1976). The gradual (extended) hard X-ray source, often associated with the second stage of the particle acceleration process, has been assumed to be confined primarily to coronal altitudes (Brown and Hoyng, 1975; Hudson, 1978).

In order to verify the validity of these models, it is important to determine observationally the altitude structure of the impulsive and gradual (extended) hard X-ray sources in solar flares. The X-ray emission to be observed must be >30 keV, and preferably $\gtrsim$100 keV, so that possible ambiguities due to the contamination from the relatively slowly-varying thermal sources can be reduced to a minimum. Moreover, the dependence of the altitude structure on X-ray energy and its variation with time during a flare also need to be measured. No such imaging observations are presently available. An alternative technique consists of viewing the

Solar Physics **86** (1983) 355–365. 0038–0938/83/0862–0355 $ 01.65.

flare from two different directions with the help of hard X-ray spectrometers aboard two or more spacecraft separated in heliographic longitude (Kane, 1981). If the X-ray source is partially occulted by the photosphere from the line of sight of one of the spectrometers, a measure of the altitude dependence of the X-ray source brightness can be obtained. We summarize below such stereoscopic measurements of solar flare X-rays $\gtrsim$100 keV and examine their implications with respect to the models of solar flares.

2. OBSERVATIONS

The stereoscopic observations reported here were made with X-ray spectrometers aboard the ISEE-3 (International Sun Earth Explorer-3) and PVO (Pioneer Venus Orbiter) spacecraft. These instruments have been described elsewhere (Kane *et al.*, 1982). The ISEE-3 hard X-ray spectrometer covers 26 - 3170 keV X-rays in 12 energy channels. The X-ray energy range covered by the PVO instrument is 100 - 2000 keV in 4 energy channels. The time resolution of the two sets of observations is 0.5 - 8 s. The ISEE-3 and PVO spacecraft are located at different distances from the sun. For comparison, the fluxes and arrival times of solar flare photons at the two spacecraft have been converted to equivalent values at the earth's distance from the sun.

Measurements of four well-observed flares, viz., one on 5 October 1978 and three on 5 November 1979, have been partially analyzed earlier (Kane *et al.*, 1979, 1982). Now preliminary measurements for the 14 September 1979 flare are also available. Table I presents the principal characteristics of the five flares. Flares 1 and 2 occurred behind the east limb of the sun. They were in full view of the PVO instrument, but were partially occulted from the ISEE-3 line of sight. On the other hand, flares 3, 4, and 5 were well observed from the earth, their Hα importance being $\gtrsim$1B. All the three flares were in full view of the ISEE-3 instrument, but were partially occulted from the PVO line of sight. For all the five flares, the estimated minimum altitude h_{min} above the photosphere which could be viewed from the occulted spacecraft is also given in Table I.

The time-rate profiles of the hard X-ray bursts associated with flares 1, 3, 4, and 5 have been presented elsewhere (Kane *et al.*, 1979, 1982). Each of the four bursts was primarily impulsive. In the last flare, the impulsive burst was followed by a small gradual burst, which lasted for ~5 min. The X-ray time-rate profile for flare 2 (14 September 1979) is shown in Figure 1. The event started at ~0653 UT. It consists primarily of a gradual hard X-ray burst, on which small fuctuations are superimposed. The observed duration of the burst is $\gtrsim$13 min. and ~5 min. for ~30 keV and ~100 keV X-rays, respectively. The overall time-rate profiles recorded by ISEE-3 and PVO are similar, including the first major maximum at ~0657 UT. It is interesting to note that the small impulsive burst at ~0653:40 UT. and the large impulsive burst at ~0703 UT, which were clearly recorded by the PVO instrument, are not present in the ISEE-3 data. The sources of these impulsive X-ray bursts were almost completely occulted from the ISEE-3 line of sight.

In order to compare the X-ray fluxes from the occulted and unocculted parts of the X-ray source, the following procedure is adopted. Since the ISEE-3 instrument measures 26 - 3170 keV X-rays in 12 energy channels, compared to the four rleatively wide energy channels of the PVO instrument, covering 100 - 2000 keV X-rays, power law X-ray spectra are fit only to the ISEE-3 observations. Fluxes of 150, 350, 750, and 1500 keV X-rays are computed from the observed PVO counting rates, assuming that the shape of the X-ray spectrum incident at PVO is similar to that incident at ISEE-3. If necessary, the observed spectrum at ISEE-3 is extrapolated to cover the PVO energy range. Since the deduced X-ray fluxes for PVO are relatively insensitive to small variations in the assumed incident X-ray spectrum, the results presented in this paper are not expected to be critically dependent on the above assumption.

TABLE I.
Characteristics of the Occulted and Unocculted Hard X-Ray Flares

Flare No.	1	2	3	4		5	
Date	5 Oct. 1978	14 Sept. 1979	5 Nov. 1979	5 Nov. 1979		5 Nov. 1979	
Hα Flare:							
Importance	–	–	1B	2B		1B	
Location	Behind Limb	Behind Limb	S13.4°, E49.0°	S14.2°, E44.6°		S15.4°, E44.0°	
X-ray Burst							
Type	Impulsive	Gradual	Impulsive	Impulsive	Impulsive	Gradual	Total
t_{max} (UT)	0631:41	~0657:11	1741:20	2149:01	2347:22	2347:46	2347:30
Occult. Sp.	ISEE-3	ISEE-3	PVO	PVO	PVO	PVO	PVO
h_{min} (km)	~25000	~30000	0±100	2000±1000	2500±1000	2500±1000	2500±1000
Δt (s)	4	16	8	16	8	8	40
50 keV X-rays (ISEE-3):							
Flux	0.1	2.9	1.3	2.2	16.6	2.2	~8.0
γ	3.4	4.1	3.0	3.3	2.1	3.2	~2.4
150 keV X-rays (ISEE-3):							
Flux	2.45×10^{-3}	1.48×10^{-2}	4.8×10^{-2}	5.9×10^{-2}	1.3	6.4×10^{-2}	0.42
γ	~3.4	5.5	3.0	3.3	~3.0	3.2	3.2
(PVO):							
Flux	0.27±0.01	$(1.11 \pm 0.01) \times 10^{-1}$	$(4.4 \pm 0.3) \times 10^{-2}$	$(2.5 \pm 0.3) \times 10^{-2}$	$(5.7 \pm 0.3) \times 10^{-2}$	$(1.8 \pm 0.2) \times 10^{-2}$	$(3.5 \pm 0.2) \times 10^{-2}$
350 keV X-rays (ISEE-3):							
Flux	$\sim 1.39 \times 10^{-4}$	1.41×10^{-4}	3.6×10^{-3}	4.0×10^{-3}	6.4×10^{-2}	3.0×10^{-3}	3.0×10^{-2}
γ	~3.4	5.5	3.0	3.3	3.6.	4.7	3.2
(PVO):							
Flux	$(4.1 \pm 0.1) \times 10^{-2}$	$(1.99 \pm 0.09) \times 10^{-3}$	$(3.1 \pm 0.2) \times 10^{-3}$	$(1.7 \pm 0.1) \times 10^{-3}$	$(5.1 \pm 0.2) \times 10^{-3}$	$(6.3 \pm 1.1) \times 10^{-4}$	$(2.7 \pm 0.1) \times 10^{-3}$
Flux ratio:							
150 keV	$(9.1 \pm 0.3) \times 10^{-3}$	0.13±0.002	0.92±0.06	0.43±0.04	$(4.4 \pm 0.3) \times 10^{-2}$	0.28±0.04	$(8.3 \pm 0.4) \times 10^{-2}$
350 keV	$(3.4 \pm 0.1) \times 10^{-3}$	$(7.1 \pm 0.3) \times 10^{-2}$	0.86±0.06	0.42±0.03	$(8.0 \pm 0.4) \times 10^{-2}$	0.21±0.04	$(9.0 \pm 0.3) \times 10^{-2}$

Occult. Sp. = Occulted Spacecraft
h_{min} = Occultation height
Δt = Averaging time for spectra
Flux: Photons cm^{-2} sec^{-1} keV^{-1}
Flux ratio: occulted flux / unocculted flux

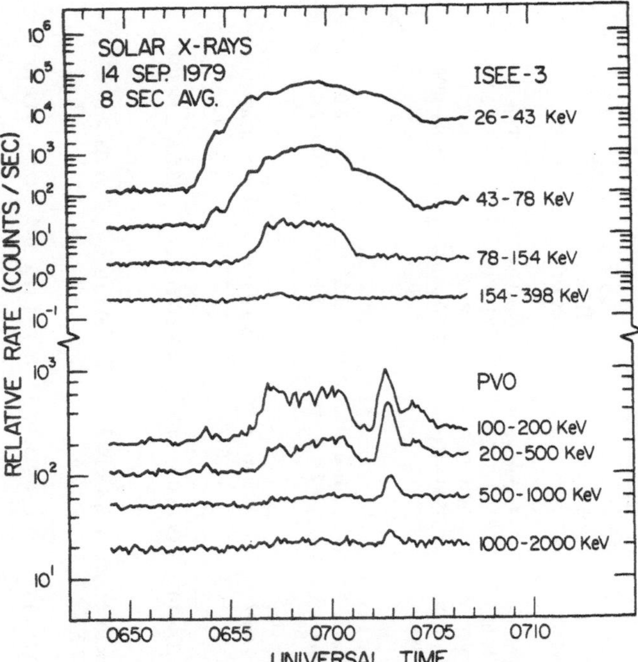

Figure 1. The time-rate profiles of the hard X-ray emission recorded by the ISEE-3 and PVO spacecraft during the 14 September 1979 solar flare. The flare, which apparently occurred behind the East limb of the sun, was in full view of the PVO instrument, but was partially occulted from the ISEE-3 line of sight. Note that only the gradual burst starting at ~0655 UT in the PVO record is present also in the ISEE-3 record. The large impulsive burst at ~0703 UT was almost completely occulted from the ISEE-3 line of sight.

The X-ray spectra for flares 1 (Figure 2), 2 (Figure 3), and 3 through 5 (Figure 4), are presented, following. The spectrum for flare 1 (5 October 1978) has been revised on the basis of the extensive detector calibration described by Kane *et al.* (1982). In the case of flare 2 (14 September 1978), a preliminary X-ray spectrum is shown for the first major maximum. The ratio $r(E, h_{min})$ of the occulted to unocculted X-ray fluxes is computed for photon energies E ~150 keV and ~350 keV from the spectra such as those shown in Figures 2, 3 and 4. The values obtained are listed in Table I.

3. VARIATION OF THE X-RAY SOURCE BRIGHTNESS WITH ALTITUDE AND X-RAY ENERGY

Since the altitude structure of the impulsive and gradual hard X-ray sources could be quite different, it is important that the ratios r of the occulted to unocculted X-ray fluxes be computed separately for impulsive and gradual X-ray components. Also, since the flux ratio could depend on the energy of the observed X-ray emission, the ratios need to be computed separately for different values of the photon energy E, such as 150 keV and 350 keV. In constructing Table I, both of the above requirements have been taken into account.

Figure 5 shows the variation of the occulted to unocculted X-ray flux ratio $r(E, h_{min})$ with the occultation altitude h_{min} and photon energy E. Note the break in the scale for altitude h_{min}. Two values of E are considered in Figure 5, viz., 150 keV and 350 keV. The variation of r is shown separately, left and right, for impulsive and gradual X-ray bursts, respectively.

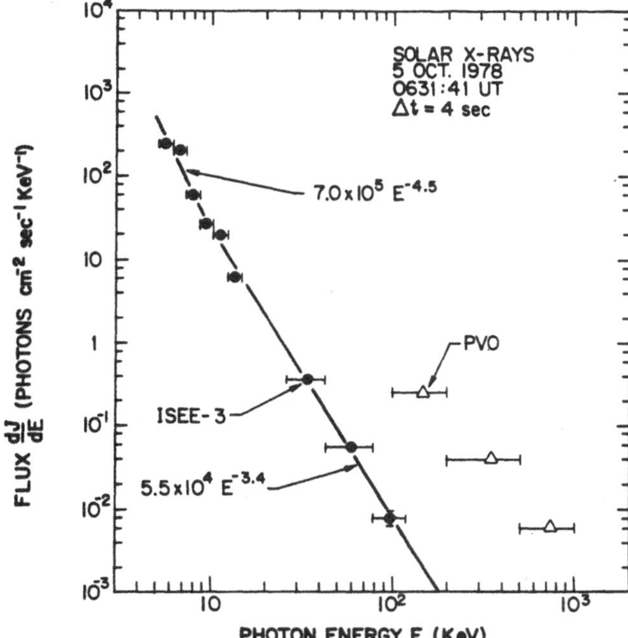

Figure 2. The X-ray spectra observed by the ISEE-3 and PVO instruments at the maximum (~0632 UT) of the impulsive hard X-ray burst on 5 October 1978.

In Figure 5 (left), flares 2, 3 and the impulsive part of flare 4 are included. These three flares occurred in the same active region within a period of six hours, and hence the altitude structure of the X-ray source may be considered essentially similar in the three flares. Flare 1, in which the coronal source was observed, is also shown for comparison. However, this flare occurred almost one year before the flares 2, 3 and 4, and hence may have a somewhat different altitude structure for the X-ray source. Similar remarks apply to Figure 5 (right), where flare 2 and the gradual part of flare 4 are included. In what follows, we discuss the two parts of Figure 5 as if they presented representative observations for a single flare.

(a) Impulsive X-Ray Source

An examination of Figure 5 (left) indicates two principal characteristics of the impulsive X-ray source:

1. The source brightness decreases with increasing altitude h above the photosphere. The decrease in brightness is most rapid as h increases from ~1000 km to ~3000 km. About 95% of the total ~150 keV X-ray emission originates at altitudes $h \lesssim 2500$ km.
2. In the 100 - 500 keV range, the X-ray source brightness at altitudes $h \lesssim 2000$ km does not vary significantly with the photon energy E. At $h \sim 2500$ km, the ~150 keV X-ray emission seems to decrease more rapidly with increasing altitude than the ~350 keV X-rays. This energy dependence is, however, reversed at coronal altitudes.

(b) Gradual X-ray Source

Figure 5 (right) shows the relatively limited observational information available for the gradual X-ray source. A comparison with the altitude dependence of the impulsive source (solid curve) indicates the following:

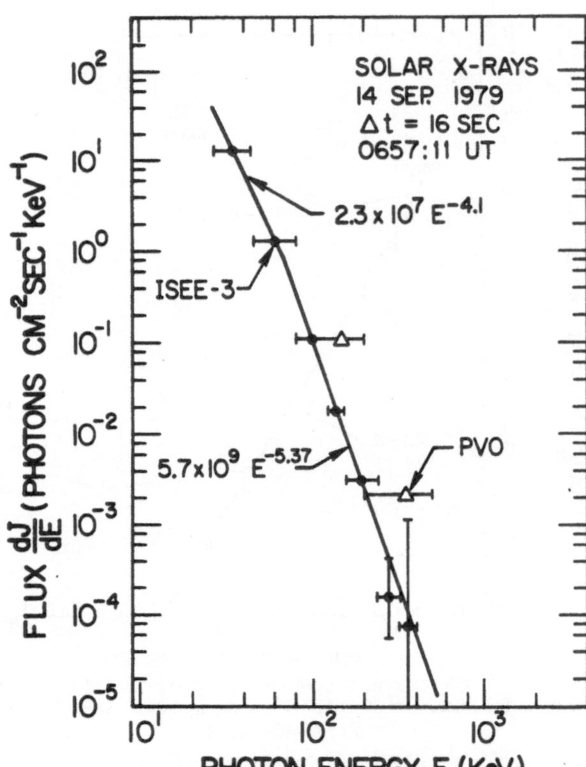

Figure 3. The X-ray spectra observed by the ISEE-3 and PVO instruments at the first maximum (~0657 UT) of the gradual hard X-ray burst on 14 September 1979.

1. For a gradual source, the decrease in brightness with increasing altitude is much less rapid than that for the impulsive source. Hence, the effective height of the gradual source is larger than that for the impulsive source. However, at least in the case of a gradual burst preceded by a large impulsive burst, $\gtrsim$70% of the 100 - 500 keV photon emission originates at altitudes $h \lesssim 2500$ km.
2. In the 100 - 500 keV range, the rate of decrease in the X-ray source brightness with increasing altitude is larger for higher energy photons.

4. EVOLUTION OF THE X-RAY SOURCE WITH TIME

The hard X-ray bursts associated with flares 2 and 5 were relatively large in magnitude as well as duration. It was therefore possible to measure the time variation of the flux ratio $r(E, h_{min})$ for a fixed value of h_{min} and two values of E, viz., ~150 keV and ~350 keV. The results for flare 5 (5 November 1979, ~2347 UT) and flare 2 (14 September 1979) are shown in Figures 6 and 7, respectively. It is important to note that, unlike flare 5 (Figure 6), in which the small gradual burst was preceded by a large, impulsive burst, flare 2 (Figure 7) produced primarily a large gradual burst.

For ~150 keV X-rays, the most important feature in Figure 6 is the increase in the value of r by a factor of ~6 as the character of the X-ray burst changes from impulsive to gradual. This effect can also be seen in ~350 keV X-ray emission, although its magnitude is smaller. The overall variation of r with time is very similar at the two photon energies.

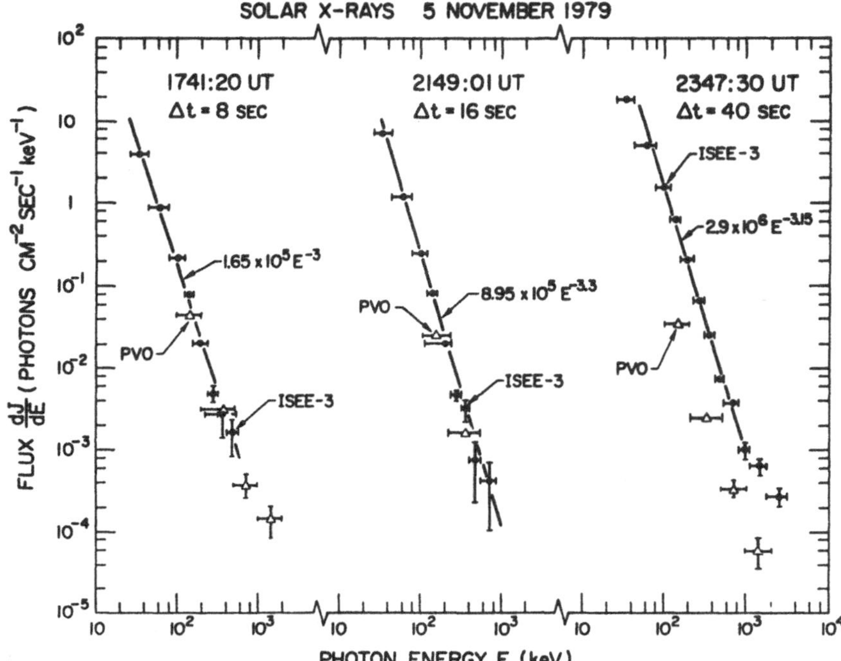

Figure 4. Total event X-ray spectra observed by the ISEE-3 and PVO instruments at the times of three flares on 5 November 1979. After the first flare (~1741 UT), which was in full view of both the ISEE-3 and PVO instruments, within a period of six hours there occurred two additional flares in the same active region which were successively more and more occulted from the PVO line of sight (Kane *et al.*, 1982).

In the gradual burst of flare 2 (Figure 7), the variation of r with time is quite different for ~150 keV and ~350 keV photons. At the time of the first maximum t_{max} (~0656:50 UT), the value of r at ~150 keV is larger than that at ~350 keV only by a factor $\lesssim 2$. As time progresses, the value of r at ~150 keV increases gradually from ~0.1 to ~0.27 and remains at that value for the remainder of the burst. On the other hand, at ~350 keV, r decreases more or less continuously with time from a value of ~0.06 to ~0.01. An examination of the time-rate profile (Figure 1) shows that, whereas the unocculted X-ray spectrum at PVO was gradually hardening with time, the partially occulted spectrum at ISEE-3 was either softening or relatively constant during the interval 0657-0700 UT. Thus, it appears that the energetic electron populations in the coronal and lower altitudes X-ray sources were not well correlated during this flare.

5. COMPARISON WITH THE FLARE MODELS

We now compare the above observations with the characteristics of the hard X-ray source predicted by the different models of the flare emissions (Kane, 1974, 1981). The observed low altitude of the principal part of the *impulsive* hard X-ray source rules out theoretical models requiring a high altitude (low density) X-ray source such as the non-thermal thin target model (Datlowe and Lin, 1973) and the thermal model with adiabatic compression and expansion of a magnetically-confined plasma (Crannell *et al.*, 1978). On the other hand, the non-thermal thick target model (Brown, 1971; Hudson, 1972) and possibly the dissipative thermal model (Brown *et al.*, 1979), which is similar to the thick target model at high photon energies (Emslie, 1980; Brown and Hayward, 1981), are consistent with the observed low altitude of the impulsive X-ray source. The observed variation of the source brightness with altitude above the photosphere is roughly similar to that expected from a thick target model, in which a beam of

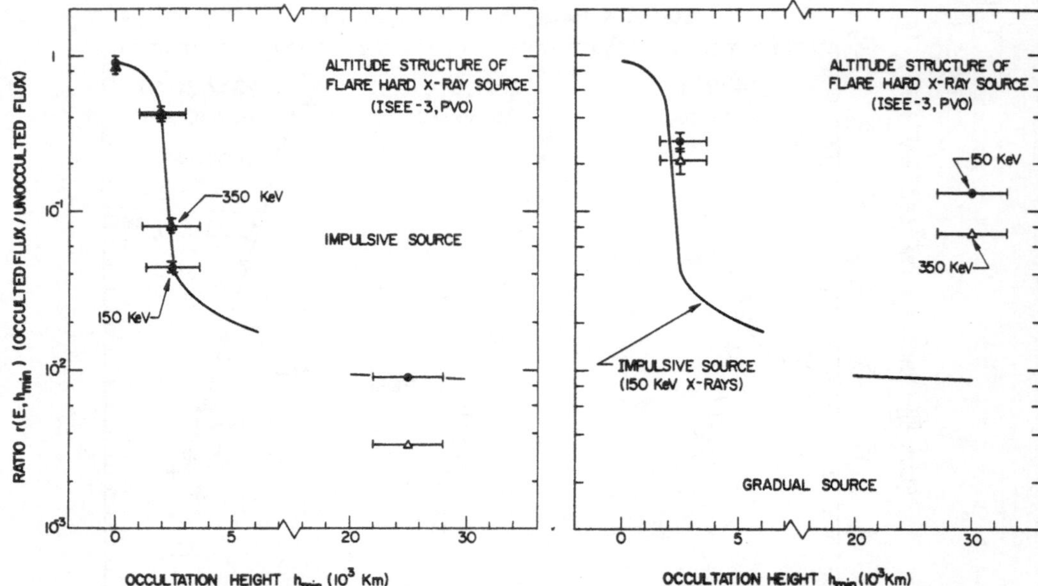

Figure 5. The ratio $r(E, h_{\min})$ of occulted to unocculted X-ray flux plotted against the minimum altitude $h_{\min}$ observable from the occulted spacecraft. Observed values of r for two values of the photon energy E, viz., ~150 keV and ~350 keV, are presented. Two types of sources are considered: (left) impulsive source, (right) gradual source. The solid line at left representing the ~150 keV impulsive source is reproduced at right for comparison.

energetic electrons, accelerated in the corona, moves downwards towards the photosphere (Brown and McClymont, 1975). However, the observations are not in agreement with the predicted $\sim E^{-\alpha} (\alpha = 1.5 - 2)$ dependence of the occulted to unocculted X-ray flux ratio $r(E, h_{\min})$ on the photon energy E. For ~150 keV and ~350 keV X-rays, we expect $R(h_{\min}) = r(150, h_{\min}) / r(350, h_{\min}) \approx 3.6 - 5.4$. From Table I and Figure 5 (left), we find that the observed values of R are much smaller. For $h_{\min} \lesssim 2500$ km, $R \lesssim 1$. Although some of this disagreement could be due to factors, such as the detailed shape of the magnetic field, which have not been taken into account here, it seems more likely that the X-ray source is not homogeneous and continuous in altitude, as assumed in a simple thick target model. The impulsive hard X-ray emission probably originates in many individual sources distributed in altitude, the sources at lower altitudes being, in general, brighter than those located at higher altitudes. Such a distribution could result, for example, from the multi-layer trapping of precipitating electrons if there are inhomogeneities in the magnetic field and electron pitch angle distribution. The individual X-ray sources are not likely to be hot thermal kernels (cf. Brown *et al.*, 1980) because of the very high temperatures required by the high energies of emitted photons. It seems more likely that the X-ray sources are non-thermal and are produced by acceleration and/or trapping of energetic electrons in small "kernels" along the magnetic field. Such a source structure would be consistent with the relatively low observed directivity of the hard X-ray emission (Kane *et al.*, 1980).

The gradual hard X-ray source is often assumed to be located in the corona, where the ambient density is $\lesssim 10^8$ cm^{-3} (cf. Brown and Hoyng, 1975; Stewart and Nelson, 1980). Such an assumption is not consistent with the observations. A substantial fraction of the total gradual hard X-ray emission seems to originate at much lower altitudes. At a given altitude, the relative brightness of the gradual source is greater than or equal to the relative brightness of the impulsive source. This could be due to an enhancement in the ion density caused by "chromospheric evaporation" during the impulsive phase, a plausible explanation, especially for

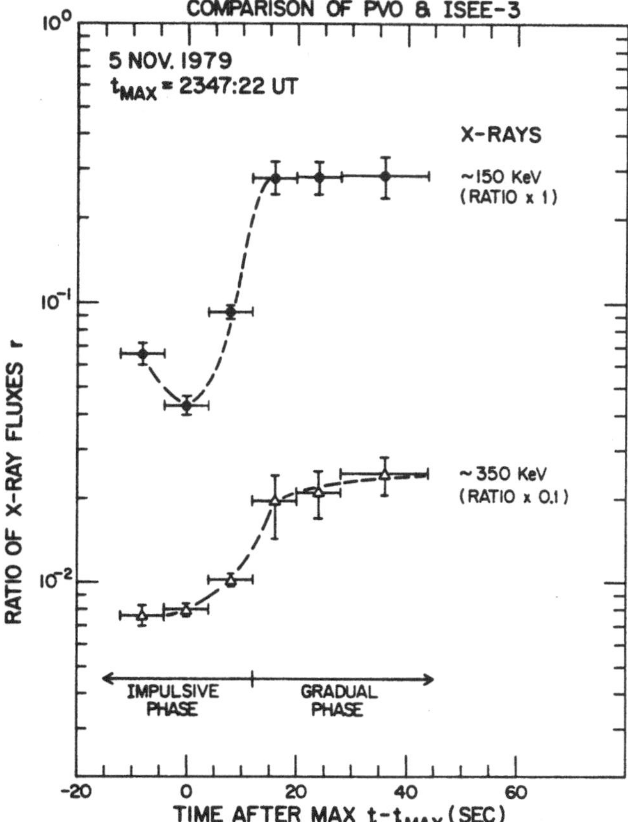

Figure 6. Variation of the ratio r of occulted to unocculted ~150 keV and ~350 keV X-ray fluxes with time during the 5 November 1979 (~2347 UT) flare. Note the large increase in r as the character of the X-ray burst changes from impulsive to gradual.

flares in which a large impulsive X-ray burst can be clearly identified. However, in cases such as flare 2, in which the hard X-ray emission is mostly gradual, it is unlikely that chromospheric evaporation plays a significant role. It appears that, at least during a gradual burst, both chromospheric and coronal hard X-ray sources are present, as would be expected, for example, from a partial precipitation model (Kane, 1974; Melrose and Brown, 1976). In such a model, energetic electrons are continuously injected into a "leaky" magnetic trap from which some electrons precipitate into the chromosphere.

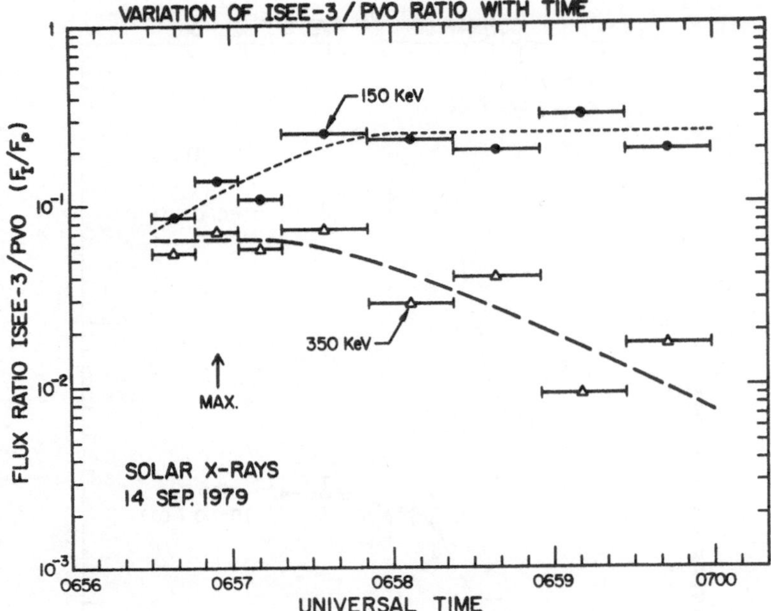

Figure 7. Same as Figure 6, but for the gradual X-ray burst on 14 September 1979. See text for details.

ACKNOWLEDGEMENTS

The author is grateful to R. W. Klebesadel, J. G. Laros, and E. E. Fenimore for providing the PVO data. It is a pleasure to thank Drs. Y. Uchida and K. Tanaka for their hospitality during the joint United States-Japan Seminar on the Recent Advances in the Understanding of Solar Flares held in Tokyo, 5 - 8 October 1982. The research reported here was supported by the National Aeronautics and Space Administration under contract NAS5-25980.

REFERENCES

Brown, J. C.: 1971, *Solar Phys.* **18**, 489.
Brown, J. C., and Hayward, J.: 1981, *Solar Phys.* **73**, 121.
Brown, J. C., and Hoyng, P.: 1975, *Astrophys. J.* **200**, 734.
Brown, J. C., and McClymont, A. N.: 1975, *Solar Phys.* **41**, 135.
Brown, J. C., Melrose, D. B. and Spicer, D. S.: 1979, *Astrophys. J.* **228**, 592.
Crannell, C. J., Frost, K., Mätzler, C., Ohki, K., and Saba, J.: 1978, *Astrophys. J.* **223**, 620.
Datlowe, D. W., and Lin, R. P.: 1973, *Solar Phys.* **32**, 459.
Emslie, A. G.: 1980, Stanford University Institute of Plasma Research Report No. 822.
Hudson, H. S.: 1972, *Solar Phys.* **24**, 414.
Kane, S. R.: 1974, in G. Newkirk (ed.), 'Coronal Disturbances', *IAU Symp.* **57**, 105.
Kane, S. R.: 1981, *Astrophys. Sp. Sci.* **75**, 163.
Kane, S. R., Anderson, K. A., Evans, W. D., Klebesadel, R. W., and Laros, J.: 1979, *Astrophys. J. Lett.* **233**, L151.
Kane, S. R., Anderson, K. A., Evans, W. D., Klebesadel, R. W., and Laros, J.: 1980, *Astrophys. J. Lett.* **239**, L85.
Kane, S. R., Fenimore, E. E., Klebesadel, R. W., and Laros, J.: 1982, *Astrophys. J. Lett.* **254**, L53.
Melrose, D. B., and Brown, J. C.: 1976, *Mon. Not. Roy. Astron. Soc.* **176**, 15.
Ramaty, R: 1973, in R. Ramaty and R. G. Stone (eds.), *High Energy Phenomena on the Sun,*

NASA **SP-342,** p. 188.
Stewart, R. T., and Nelson, G. J.: 1980, *Proc. Astron. Soc. Australia* **3**, 380.
Takakura, T.: 1974, in S. R. Kane (ed.), 'Solar Gamma-, X-, and EUV Radiation', *IAU Symp.* **68**, 299.

DISCUSSION

J. C. Henoux: I don't think the lack of directivity implies that the thick target emission does not explain your observations. Computations by Bai and also by myself show that, even for electron beam X-ray emission, photospheric Compton backscattering greatly reduces the directivity.

S. R. Kane: Your assertion is true only for low energy (<100 keV) photons. Since the relative effect of Compton backscattering tends to decrease with increasing photon energy, stereoscopic measurements of 500 - 1000 keV X-rays can provide a measure of the true directivity of the hard X-ray source.

J. C. Brown: The energy dependence of the flux ratio in the later stages of the Sept. 14, 1979 event is, in a sense, expected in a thick target model, *i.e.*, the 350 keV ratio is lower. I would expect the energy dependence of the ratio in this case to be about E^{-2}, so that the 350 keV ratio should be about 5.1 times smaller than the 150 keV ratio, much as you observe.

S. R. Kane: Yes, it certainly seems that way. I wonder if it is a large density enhancement or some other effect that makes the hard X-ray source look like a thick target source up to heights $\gtrsim$ 30000 km.

H. S. Hudson: It does not seem worthwhile to try to draw general conclusions! You have only three of your excellent stereo observations, and there were only three single-spacecraft occultations reported during the last solar maximum. In particular, the May 12, 1981, HINOTORI event described by Mr. Tsuneta appears to have a much greater extent in height.

S. R. Kane: I agree that we should not draw general conclusions from only three events. We hope to observe some more events and see if these conclusions are still true. As regards the HINOTORI measurements, I do not know what the assigned height corresponds to: is it the geometrical center of the source, or the most intense part, or something else? I hope that we will find a common event so that we may compare the two sets of measurements.

G. A. Doschek: I believe we have soft X-ray spectra for the 5 Nov. 1979 flare you discussed. It may be interesting to compare these data with the ISEE data.

S. R. Kane: Yes! We should certainly compare the two data sets.

OBSERVATIONS OF FINE TIME STRUCTURE IN SOLAR FLARE HARD X-RAY BURSTS

K. Hurly, M. Niel, and R. Talon
Centre d'Etude Spatiale des Rayonnements,
(C.N.R.S.-U.P.S.), B.P. 4346, 31029 Toulouse Cedex, France
and
I. V. Estulin and V. Ch. Dolidze
Institute for Space Research,
Profsojusnaja, 88, 117810 Moscow, U.S.S.R.

ABSTRACT. Hard X-ray (≳100 keV) time histories of solar flares which occurred on 1978 December 4 and 1979 February 18 are presented. The first flare was observed by 3 identical instruments from near-earth orbit (Prognoz 7) and interplanetary space (Venera 11 and 12). Fine time structure is present down to the 55 ms level for the e-folding rise and fall times. These data may be used to localize the emission region by the method of arrival time analysis.

1. INTRODUCTION

Data on the time histories and spatial structures of solar flare hard X-ray emission are important for understanding the processes of energetic electron acceleration and injection. Various methods have been used to date to obtain information on the spatial structure of hard X-ray emission, e.g., multiple spacecraft observations of partially occulted solar flares (Kane et al., 1979,1982), and the use of imaging detectors up to ≃30 keV (Hoyng et al., 1981). Data on the fine time structure of solar X-rays have come from both fortuitous observations (e.g.,Hurley and Duprat, 1977) and dedicated instruments (e.g., Orwig et al., 1981).

In this paper, we present hard X-ray data (≳100 keV) for two solar events which occurred on 1978 December 4 and 1979 February 18, triggering the sensitive, high time resolution data storage modes of the Franco-Soviet SIGNE detectors aboard the Soviet spacecraft Prognoz 7, Venera 11, and Ven-

Solar Physics **86** (1983) 367–373. 0038–0938/83/0862–0367 $ 01.05.

era 12. In the case of the 1978 December 4 observation, this is the first time that identical, widely separated detectors have been used to provide stereoscopic observations of solar flare X-ray emission. We present evidence for what we believe to be the finest time structure yet observed in a solar flare (55 ms) and point out that such data may be used to give information on the spatial structure of the X-ray emission, by the use of arrival time analysis. A more complete report on the 1978 December 4 event and an event which occurred on 1978 December 3 appears elsewhere (Hurley et al., 1983).

2. INSTRUMENTATION

An important feature of the observation of the 1978 December 4 observation is that virtually identical instruments were used on the 3 spacecraft which observed it: 63.6 cm^2 by 3.7 cm thick NaI(Tl) crystals, which faced the sun to within a few degrees. Several high time resolution memories are associated with these detectors for both gamma ray burst and solar X ray objectives; the data presented here are from a memory with 16/1024 s resolution. The absolute times at the spacecraft are derived to 3 ms for the Prognoz and 8 ms for the Venera. More complete descriptions of these experiments have appeared (Barat et al., 1981; Chambon et al., 1979).

At the time of the December 4 observation, the Prognoz satellite was approximately $2x10^5$ km from the earth, and outside the magnetosphere; Venera 11 and 12 were about $48x10^6$ km from earth, but separated by only $2x10^6$ km. For the December 4 observation, the Veneras were about 0.75 A.U. from the sun, while for the February 18 observation, Venera 12 was about 0.87 A.U. from the sun; thus the Venera detectors were equivalent to ≃112 and 84 cm^2 in earth orbit.

3. OBSERVATIONS

The 1978 December 4 event corresponds to an H_α flare of importance SB from McMath Plage region 15694; the 1979 February 18 flare could correspond either to a flare of importance -N (McMath region 15830) or 1B (McMath region 15823). The December 4 event triggered the Prognoz 7, Venera 11, and Venera 12 detectors, which observed emission up to 1 MeV; the February 18 event triggered both the Venera 11 and 12 detectors, which observed emission up to 670 keV. Only the Venera 12 data for the latter event are presented here, since the Venera 11 detector, with a higher threshold (≃250 keV vs. ≃100 keV), did not have a statistically significant time history.

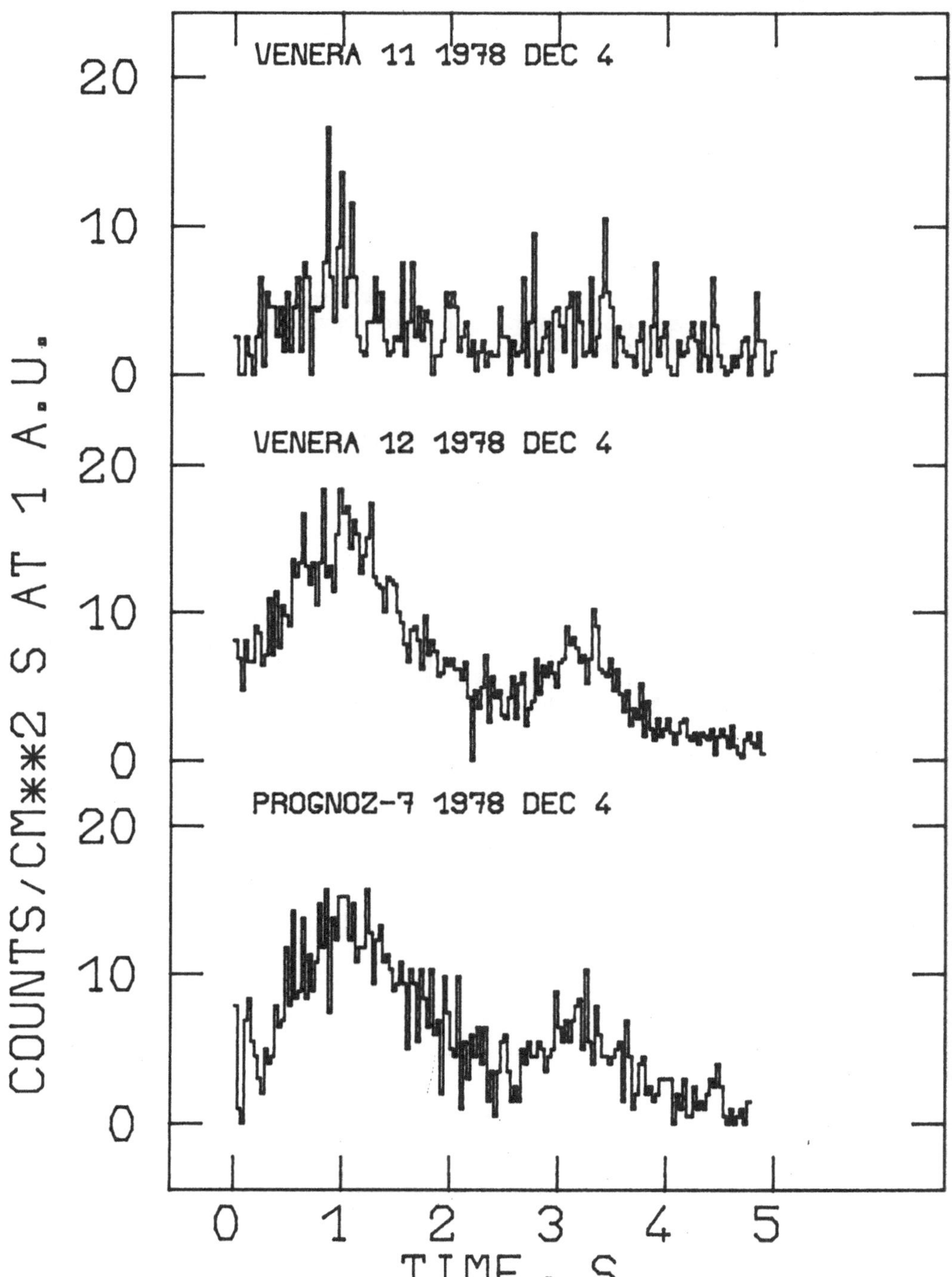

Figure 1. 3 time histories of the 1978 December 4 event.

Figure 1 shows the three time histories for the December event, on a 32/1024 s time scale. The following corrections have been applied to the data. First, the backgrounds have been subtracted. Second, the Venera 12 time history, taken from 99 to 549 keV, and the Venera 11 time history, from 228 to 1082 keV, have been normalized to the Prognoz 7 energy range, 107 to 460 keV, using the observed energy spectrum. Finally, all count rates have been normalized to counts/cm^{2}s at 1 A.U.. The e-folding rise and fall times of the structures in these time histories are in the range 55 ms and above, with corresponding full widths at half maxima of 100 ms and above. The structures are found in the time histories of the three spacecraft. A similar result was obtained for the December 3 event (Hurley et al., 1983).

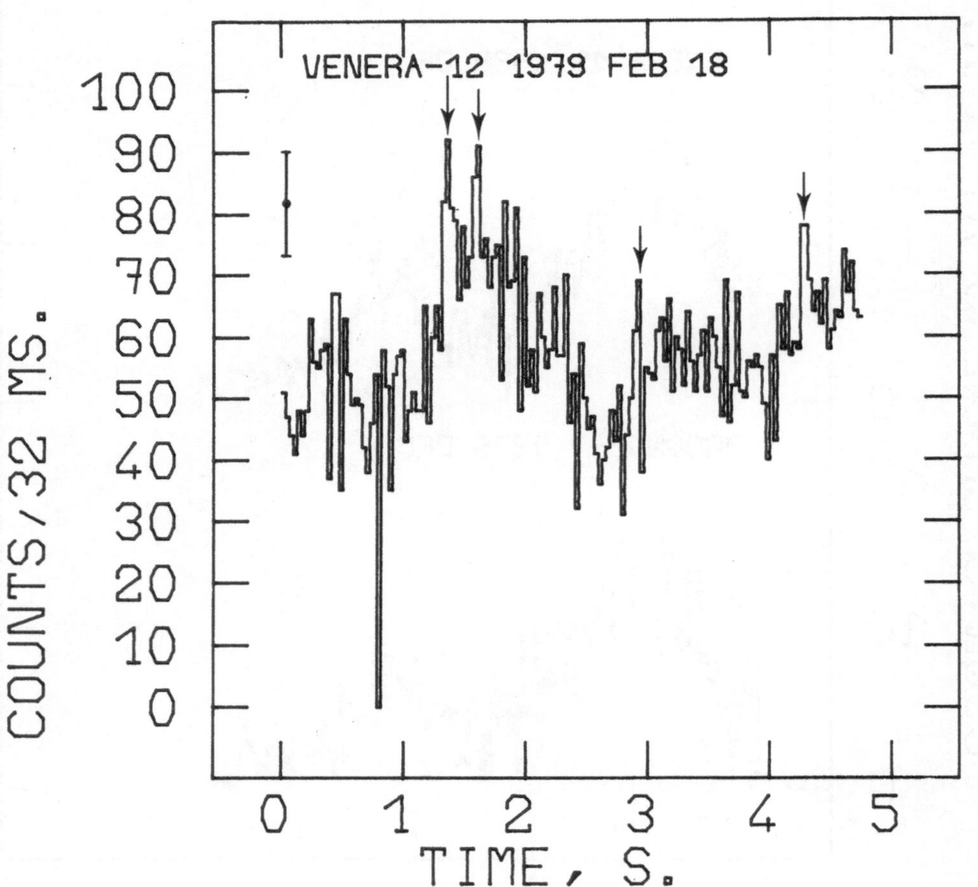

Figure 2. Time history of the 1979 February 18 event. The four structures are significant at the 2 to 4 sigma level with respect to the running average.

The time history of the 1979 February 18 event (Figure 2) also displays fine time structure: four peaks are indicated, which are significant at the 2 to 4 sigma level with respect to the running average count rate. The e-folding times here are in the range 90 ms. and above. In general, comparisons of the Prognoz 7 and Venera 12 data ($\gtrsim$100 keV) with the Venera 11 data ($\gtrsim$250 keV) indicate that faster time structure is present in harder X rays.

4. DETERMINATION OF THE SPATIAL STRUCTURE

The data from two widely separated spacecraft (e.g., Prognoz 7 and Venera 12) may be used to localize the emission region in one dimension (e.g., along a line passing through the center of the sun and the active region). Briefly, the procedure used is the following. The trigger time of one spacecraft, and the distance between that spacecraft and the sun's surface, are used to define an approximate propagation time from the active region to both spacecraft. Spherical wavefronts are assumed. The real-time positions of the sun at the time of the emission, and of the spacecraft at the trigger times, are established to several light-milliseconds. The real-time coordinates of the active region are then established using the heliographic coordinates. The time delay between the two rephased time histories and its associated error are then used to define a locus of permitted points along the line which passes through the active region and the sun's center. The uncertainty in the rephasing of the time histories will introduce an error into the determination of the emitting region. Its magnitude depends on the exact geometry of the observation, but has the value $dH/d\tau$=1300 km/ms for this event (an unfavorable case), where H is the height of the emitting region above the chromosphere, and τ is the lag between the two time histories. In the case considered here, this leads to a large uncertainty: the emitting region can only be said to be between 1.1×10^4 and 1.6×10^5 km above the chromosphere. However, a theoretical minimum uncertainty of 1/4 to 1/10 as much can be achieved, as discussed below.

5. DISCUSSION

Imaging by the arrival time analysis method may be used in the hard X-ray range, at energies well above the limits of imaging detectors currently in use. Assuming that one spacecraft is in near-earth orbit, a favorable geometry for a two spacecraft localization would be with the second spacecraft at 90° to the sun-earth line, along a line passing through the active region. With this geometry, localization could be done both by the arrival time analysis and partial occultation methods (Kane et al., 1979, 1982); the uncertainty due to rephasing would be $dH/d\tau=300$ km/ms. Addition of a third and fourth spacecraft could be used to provide 2 and 3 dimensional localization; the most favorable geometry calls for these spacecraft to be along orthogonal lines passing through the active region. Along one such line, with spacecraft at both ends, $dH/d\tau$ could be reduced to 150 km/ms, resulting in a spatial resolution of about 10" for an uncertainty of 100 ms; this is quite comparable to the results obtained by imaging detectors at lower energies in solar observations.

Such a 4 spacecraft configuration may arise during the International Solar Polar Mission, which will start in 1986. Figure 3 shows ESA's ISPM spacecraft over the solar pole, some four years after launch. A NASA RISE or SIS-type mission spacecraft is shown in a heliocentric orbit, while ESA's DISCO, a near-earth spacecraft, and a Soviet Venera spacecraft provide the other points. If equipped with X-ray detectors, these missions could provide three dimensional images of solar hard X-ray emission.

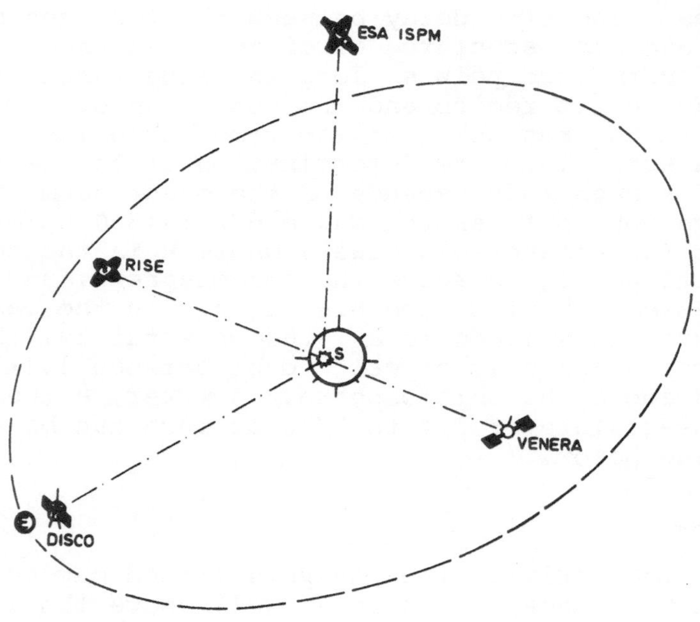

Figure 3. A four spacecraft configuration which might occur in the late 1980's, ideal for imaging hard X-ray emission from the sun by the arrival time analysis method.

ACKNOWLEDGEMENTS

We are indebted to G. Vedrenne for several helpful discussions, and to C. Barat for assistance with the Venera data. This work was supported by CNES Contract 81-212.

REFERENCES

Kane, S.: Anderson, K., Evans, W., Klebesadel, R., and Laros, J., 1979, Astrophys. J. Lett. 233, L151.

Kane, S., Fenimore, E., Klebesadel, R., and Laros, J.: 1982, Astrophys. J. Lett. 254, L53.

Hoyng, P., Machado, M., Duijveman, A., Boelee, A., deJager, C., Fryer, R., Galama, M., Hoekstra, R., Imhof, J., Lafleur, H., Maseland, H., Mels, W., Schadee, A., Schrijver, J., Simnett, G., Svestka, Z., van Beek, H., van Tend, W., van der Laan, J., van Rens, P., Werkhoven, F., Willmore, A., Wilson, J., and Zandee, W.: 1981, Astrophys. J. Lett. 244, L153.

Hurley, K., and Duprat, G.: 1977, Solar Phys. 52, 107.

Orwig, L., Frost, K., and Dennis, B.: 1981, Astrophys. J. Lett. 244, L163.

Hurley, K., Niel,M., Talon, R., Estulin, I., and Dolidze, V.: 1983, Astrophys. J. (accepted).

Barat, C., Chambon, G., Hurley, K., Niel, M., Vedrenne, G., Estulin, I., Kuznetsov, A., and Zenchenko, V.: 1981, Space Sci. Inst. 5, 229.

Chambon, G., Hurley, K., Niel, M., Vedrenne, G., Zenchenko, V., Kuznetsov, A., and Estulin, I.: 1979, Space Sci. Inst. 5, 340.

DISCUSSION

HENOUX: How do you explain the observations where you detect emission peaks from one spacecraft and no emission is seen from the other spacecraft?

HURLEY: There is one case where a spike is lost in a time history due to insertion of the trigger time in place of data. In another case, this is due to poor statistics in the Prognoz time history. A more complete explanation appears in Hurley *et al.* (1983).

GAMMA-RAY OBSERVATIONS FROM HINOTORI

M. Yoshimori*, K. Okudaira*, Y. Hirasima*, and I. Kondo**

* Department of Physics, Rikkyo University, Toshima, Tokyo, Japan

** Institute for Cosmic Ray Research, University of Tokyo, Tanashi, Japan

ABSTRACT. Some interesting results on gamma-ray line emission and its time profiles observed by Hinotori are presented. Possible explanations of gamma-ray line and hard X-ray emissions for the impulsive and gradual flares are discussed. Relationship between the gamma-ray line emission and acceleration and escape of the solar particles is also studied.

1. INTRODUCTION

The acceleration of electrons, protons and ions during solar flares has been conceived to be an essential key problem in the study of high energy solar phenomena. Electrons accelerated to the MeV energy range emit bremmsstrahlung hard X-rays and gamma-ray continua. Protons and ions accelerated to 10 MeV/nuc. produce gamma-ray line emmisions, while at higher energy region produce neutrons and π^o decay gamma-rays. Observations of time profiles and energy spectra of gamma-rays provide a direct probe into the behavior of high energy electrons, protons and ions, especially the timing of the particle acceleration and emission mechanisms.

Over the past years, several observations of solar gamma-ray line emission were made by Chupp et al. (1973), Hudson et al. (1980) and Prince et al. (1982). Also theoretical studies of these phenomena were made by Ramaty et al. (1975, 1979). In the present solar maximum period, two satellites solely dedicated to solar flare observations (SMM and Hinotori) were launched, and numerous gamma-ray emission events have been recorded.

In this paper, some interesting results obtained from the Hinotori observations are presented, and the following problems are discussed; Gamma-ray line intensities and their time profiles, possible explanations of gamma-ray line and hard X-ray emissions for the impulsive and gradual flares, and the correlation between the gamma-ray line emission and escaping solar particles.

Solar Physics **86** (1983) 375–382. 0038–0938/83/0862–0375 $ 01.20.

2. OBSERVATIONS

The gamma-ray spectrometer on board Hinotori is a phoswich type CsI(Tl) scintillator of 8.9 cm diameter and 5.1 cm thickness surrounded by a 0.5 cm thick plastic scintillator for rejection of charged particles. The spectrometer has no collimator and the axis of the spectrometer is parallel to the spin axis of the satellite, which is oriented to the solar direction. The spectrometer covers an energy range of 0.21 - 6.67 MeV in its usual operation mode. The gamma-ray energy spectrum is observed with a 128 channels pulse height analyzer every 2 seconds in the Flare mode and 16 seconds in the Quiet mode. The energy resolution is about 10% (FWHM) at 1 MeV and the detection sensitivity for gamma-ray lines is about 10^{-2} photons cm^{-2} s^{-1}. In-flight energy calibration is provided by three gamma-ray peaks around 0.4, 0.67 and 1.54 MeV resulting from nuclear activation of the CsI(Tl) crystal induced by high-energy protons in the South Atlantic Anomaly. The instrument has been described in detail by Okudaira et al. (1981) and Yoshimori et al. (1983).

Hinotori observed about 40 solar gamma-ray flares with photon emission above 0.3 MeV since its launch on Feb., 21, 1981. For several outstanding flares, significant fluxes of energetic photons above 6 MeV and gamma-ray lines were observed. As examples, gamma-ray energy spectra observed for two large flares are shown in Figures 1 and 2. Figure 1 shows a gamma-ray spectrum for the 1981 Oct. 14 flare, in which the background spectrum is substructed. To display the gamma-ray lines more clearly, the

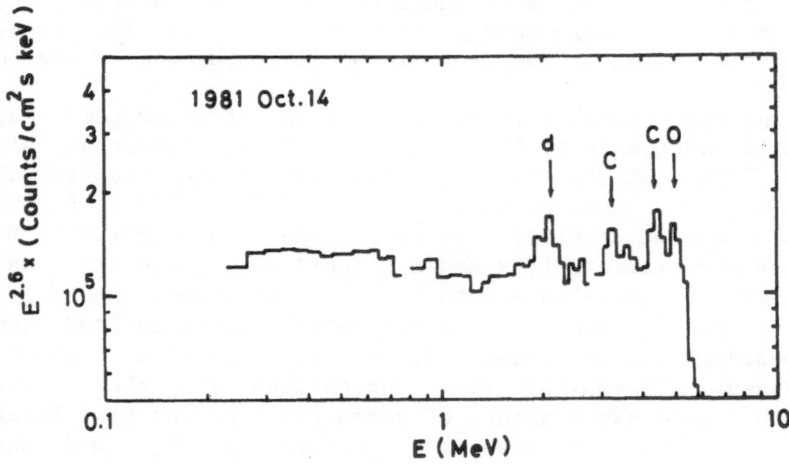

Figure 1. Gamma-ray energy spectrum for 1981 Oct. 14 flare. To display the gamma-ray lines clearly, the observed energy spectrum has been multiplied by $E^{2.6}$ to compensate the gamma-ray continuum.

energy spectrum has been multiplied by $E^{2.6}$ to compensate for the steep gamma-ray continuum. The nuclear deexcitation lines of C at 4.44 MeV and O at 6.14 MeV and the neutron capture line at 2.22 MeV are seen in Figure 1. These gamma-ray lines were first observed by Chupp et al. (1973) and after that time confirmed by many observations. Figure 2 is the gamma-ray spectrum for 1981

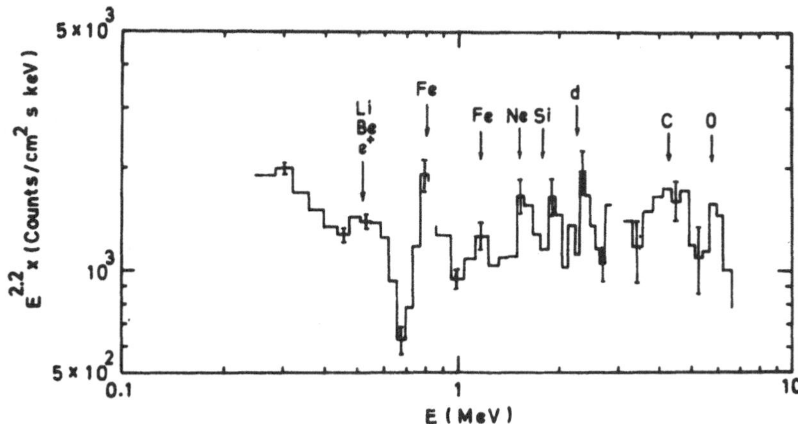

Figure 2. Gamma-ray energy spectrum for 1981 Apr. 27 flare. To display the gamma-ray lines more clearly observed energy spectrum has been multiplied by $E^{2.2}$ to compensate the gamma-ray continuum.

Apr. 27 flare, in which a few more gamma-ray lines such as the nuclear deexcitation lines of Ne at 1.63 MeV, Si at 1.78 MeV and Fe at 0.85 and 1.24 MeV are seen. In addition, a broad peak structure around 0.5 MeV is likely to be a complex of lines of positron annihilation at 0.51 MeV and nuclear deexcitation of Li at 0.48 MeV and Be at 0.43 MeV. The line intensities observed for the 1981 Apr. 27 flare are summarized in Table I. These gamma-ray line intensities are consistent with ones predicted by Ramaty et al. (1979). SMM also observed this flare and similar gamma-ray line emissions were recorded (Chupp 1982). The intensity ratios of neutron capture line at 2.22 MeV to deexcitation line of C at 4.44 MeV are 0.97 $\pm$ 0.97 and 1.31 $\pm$ 0.91 for 1981 Apr. 27 and Oct. 14 flares, respectively. The ratios for these two limb flares are much smaller than those for disk flares. This limb darkening of 2.22 MeV line has also been observed by SMM (Chupp 1982) and provides evidence that the neutron capture takes place in the photosphere.

Obervations of gamma-ray line and hard X-ray time profiles could provide clues on the timing of particle acceletion and the emission of energetic photons during the solar flare. There are

TABLE I: Gamma-ray line intensities for 1981 Apr. 27 flare.

E(MeV)	Origin	Intensity (cm^{-2} s^{-1})
0.51	^{7}Li ^{7}Be e^{+} e^{-}	0.400 ± 0.30
0.84	^{56}Fe	0.230 ± 0.07
1.63	^{20}Ne	0.090 ± 0.04
1.78	^{28}Si	0.048 ± 0.030
2.22	^{2}H	0.029 ± 0.020
4.44	^{12}C	0.030 ± 0.022
6.13	^{16}O	0.028 ± 0.025

at least two types of time profiles for hard X-ray and gamma-ray emissions from solar flares, impulsive and gradual. As an example, time profiles for an impulsive flare (1981 Apr. 4) and a gradual flare (1981 Apr. 27) are shown in Figures 3 and 4. In Figure 3, the upper frame represents the electron bremsstrahlung time profile while middle and lower frames represent predominantly nuclear reaction time profiles. The time resolution is 2 seconds for these time profiles. The time profiles show two peaks, and the flare duration is about 1 minute. As shown in Figure 3, the maximum in hard X-ray time profile is not significantly delayed with respect to the maximum in the profile of gamma-ray line emissions. This implies that for Apr. 4 flare, the nuclear deexcitation reactions occurred simultaneously with electron bremmsstrahlung.

Apr. 27 flare shown in Figure 4 lasted for about 15 minutes, but Hinotori was eclipsed by the earth before the flare is over.

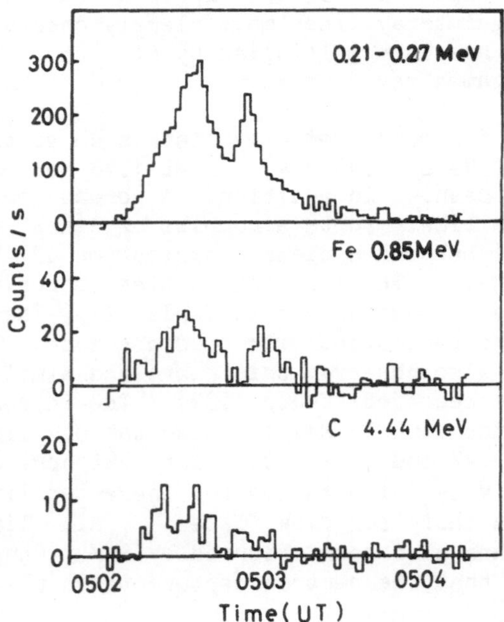

Figure 3. Gamma-ray time profiles for 1981 Apr. 4 flare.

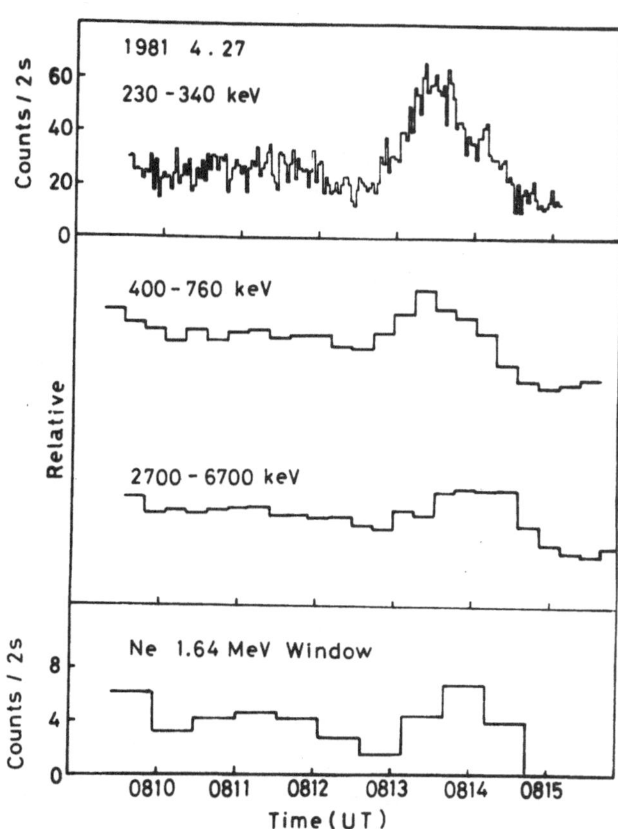

Figure 4. Gamma-ray time profiles for 1981 Apr. 27 flare.

The upper frame represents the hard X-ray time profile with time resolution of 2 seconds. The middle frame represents 2.7 - 6.7 MeV gamma-ray time profile with time resolution of 16 seconds. 2.7 - 6.7 MeV gamma-rays are dominated by gamma-ray lines. The lower frame represents the Ne deexcitation reaction time profile with time resolution of 32 seconds. Figure 4 shows that the 2.7 - 6.7 MeV gamma-ray emission begins to rise nearly at the same time with the onset of hard X-ray increase, but the peak value is reached about 40 seconds later. In addition, the Ne deexicitation line time profile also shows a similar trend to the 2.7 - 6.7 MeV gamma-ray time profile. This implies that for Apr. 27 flare, the mean life time for nuclear reaction was longer than that for electron bremsstrahlung. These different features for impulsive and gradual flares have been found in other flares (Bai, 1982).

Further, the Hinotori hard X-ray imaging observations by SXT revealed different structure for the two types of solar flare. For Apr. 4 impulsive flare, the hard X-ray sources appear to be located at the footpoints of a magnetic loop, while for Apr. 27 gradual flare, the main hard X-ray source is not located at the footpoint but rather in the lower corona (Tsuneta et al., 1982; Takakura et al., 1983). The differences in the time profiles and hard X-ray source locations for the two types of solar flare may provide explanations for the different features in gamma-ray line and hard X-ray emissions.

3. DISCUSSION

Some interesting problems on energetic photon emissions, timing of particle acceleration and correlation between the gamma-ray line emission and escaping solar particles are discussed based on the Hinotori observations.

The observations for two large limb flares on Apr. 27 and Oct. 14, 1981 show a significant decrease in the intensity ratio of 2.22 MeV line to 4.44 MeV line. In order to explain the decrease in this ratio, it is most likely that protons and ions accelerated to energies above 50 MeV/nuc. stream down to the photosphere and produce neutrons and excited nuclei, but that the mean emission depth of the neutron capture line at 2.22 MeV is much larger than that of nuclear deexcitation lines. Therefore, the observations could support a thick-target with protons beamed downward, if this is the case.

For 1981 Apr. 4 impulsive flare, a significant time delay between the hard X-ray and gamma-ray line emission was not observed. In addition, results of the hard X-ray imaging observation suggests that most electrons were beaming down and dissipate their energy at the footpoints of the loop. Therefore, for this flare, we can deduce that electrons, protons and ions are simultaneously accelerated and streamed down to the chromosphere producing hard X-rays as well as the gamma-ray line emission in the thick target region. Then a single-step particle acceleration model seems to be applicable to this event.

On the other hand, for 1981 Apr. 27 gradual flare, a significant time delay was observed between the peaks of hard X-ray and gamma-ray line emissions. Results of the hard X-ray imaging observation suggest that most electrons are trapped in the lower corona and dissipate almost all energy there, with small indication of beaming down to the chromosphere. In contrast, the observation of gamma-ray line emission from both nuclear deexcitation and neutron capture reaction implies that the accelerated protons and ions are beaming down to the photosphere. The observed time delay between the peaks of hard X-ray and gamma-ray line emissions can be explained by either one of the following explanations. Firstly by the concept of second-step acceleration of protons and ions suggested by Bai and Ramaty (1979) and Bai (1982). Secondly by the simultaneous acceleration of electrons and protons in one-step acceleration with a longer lifetime of protons and ions against the energy loss in the ambient medium as suggested by Bai and Ramaty (1979). The fact that the gamma-ray line emmission was observed over a time period of several minutes and long duration of hard X-ray intensity increase (more than 15 min.) requires a mechanism to supply high energy electrons and protons over long time duration.

Although we have discussed about the two distinctively different solar flares, one impulsive and another gradual type, the features described above represent main characteristics for these two types of flares derived from other gamma-ray flares. So at least there are two quite different type of solar flares which caused the acceleration of protons and ions up to 10 MeV/nuc.

As discussed above, the observed gamma-ray line emissions were produced by downward-moving protons and ions in the solar atmosphere. On the other hand, observations of energetic solar protons and ions imply that a part of the accelerated particles escape into the interplanetary space. Therefore, a study of the correlations between the gamma-ray line emission and escaping solar particles will provide clues on the particle propagation in the flare region.

Observation of solar protons have been continued by the Japanese geosynchronous meteorological satellite Himawari since 1979. A significant increase of the solar proton flux with energies between 16 to 500 MeV was observed in association with Apr. 4 flare. The position of Apr. 4 flare determined by Hα observation was reported as S44 W85, and that of Apr. 27 was reported as N16 W90. Thus both flares occured in the so-called preferred heliolongitude region in which the Earth is supposed to be well connected to the Sun by interplanetary magnetic lines of force. After Apr. 27 flare, however, there was very small evidence for the arrival of solar protons to the vicinity of the earth from Himawari observation.

Total number of protons accelerated to energies above 10 MeV in the flare region can be estimated from the fluence of nuclear deexcitation lines observed by Hinotori gamma-ray detector, provided that these protons are isotropically distributed at the source. The fraction of accerelated protons escaped from the flare region into the interplanetary space can be derived by the comparison between the proton flux observed by Himawari and total number of protons derived above. Fraction of escaped protons was found to be 0.1 and <0.01 for Apr. 4 and Apr. 27 flare, respectively. This result confirms those obtained by simultaneous observation by SMM, ISEE-C and HELIOS-3 (von Rosenvinge et al., 1981; Pesses et al., 1981; Evenson et al., 1981). So, the accelerated protons and ions seem to have been well confined within the flare region in the case of Apr. 27 event. This may have close relation to the different features of these two flares in the time profile and hard X-ray source position discussed earlier.

ACKNOWLEDGEMENTS

We wish to acknowledge the contributions and efforts of many people who participated in the Hinotori project. In particular, we thank Professors T. Hayashi and Y. Tanaka for their efforts to realize the project. We are grateful to the members of SXT team for providing hard X-ray imaging data and to the Meteorological Satellite Center for providing the Himawari solar proton data. We also wish to thank Drs. T. Kohno and T. Yanagimachi for analysis of solar proton data.

REFERENCES

Bai,T. and Ramaty,R.: 1979, Astrophys. J. 227, 1072.

Bai,T.: 1982, in R.E.Lingenfelter, H.S.Hudson and D.M.Worrall (eds.), Gamma-Ray Transients and Related Astrophysical Phenomena, AIP, New York, p.409.

Chupp,E.L., Forrest,D.J., Higbie,P.R., Suri,A.N., Tsai,C. and Dunphy,P.P.: 1973, Nature 241, 333.

Chupp,E.L.: 1982, in R.E.Lingenfelter, H.S.Hudson and D.M.Worrall (eds.), Gamma-Ray Transients and Related Astrophysical Phenomena, AIP, New York, p.363.

Evenson,P., Meyer,P. and Yanagita,S.: 1981, 17th Int. Cosmic Ray Conf., Paris, 3, 32.

Hudson,H.S., Bai,T., Gruber,D.E., Matteson,J.L., Nolan,P.L., and Peterson,L.E.: 1980, Astrophys. J. 236, L91.

Okudaira,K., Hirasima,Y., Yoshimori,M. and Kondo,I.: 1981, 17th Int. Cosmic Ray Conf., Paris, 3, 11.

Pesses,M.E., Klecker,B., Gloeckler,G. and Hovestadt,D.: 1981, 17th Int. Cosmic Ray Conf., Paris, 3, 36.

Prince,T.A., Ling,J.C., Mahoney,W.A., Riegler,G.R. and Jacobson,A.S.: 1982, Astrophys. J. 255, L81.

Ramaty,R., Kozlovsky,B. and Lingenfelter,R.E.: 1975, Space Sci. Rev. 18, 341.

Ramaty,R., Kozlovsky,B. and Lingenfelter,R.E.: 1979, Astrophys. J. Suppl. 40, 487.

Takakura,T., Tsuneta,S., Ohki,K., Nitta,N., Makishima,K., Murakami,T., Ogawara,Y., Oda,M. and Miyamoto,S.: 1983, submitted to Astrophys. J.

Tsuneta,S., Ohki,K., Takakura,T., Nitta,N., Makishima,K., Murakami,T., Ogawara,Y., Oda,M. and Kondo,I.: 1982, Proc. Hinotori Symposium on Solar Flares, ISAS, p.130.

von Rosenvinge,T.T., Ramaty,R. and Reames,D.V.: 1981, 17th Int. Cosmic Ray Conf., Paris, 3, 28.

Yoshimori,M., Okudaira,K., Hirasima,Y. and Kondo,I.: 1983, submitted to Nucl. Instr. and Methods.

HIGH ENERGY PARTICLE ACCELERATION IN SOLAR FLARES - OBSERVATIONAL EVIDENCE

E. L. CHUPP*
University of New Hampshire, Durham, NH, USA

ABSTRACT. The recent gamma ray and neutron observations made by the SMM Gamma Ray Spectrometer are reviewed. The implication these observations hold for understanding particle acceleration in solar flares are discussed. The data require that both electrons and ions must be accelerated together to relativistic energies and interact with matter in a time scale of seconds.

1. HISTORICAL PERSPECTIVE AND RECENT OBSERVATIONS

The realization made in 1942, but not reported widely until 1946 (Ehmert, 1948; Berry et al., 1942; Forbush, 1946), that some Hα solar flares have associated impulsive emission of relativistic particles observable at the Earth, immediately raised the question "What is the accelerating mechanism?" In spite of experimental and theoretical efforts over ∿40 years, we have not answered this basic question. Here I refer to the mechanism(s) which produce the impulsive energization or release of ions and electrons of relativistic or sub-relativistic energies in the flaring regions.

Shown schematically in Figure 1 is a chain of events leading to emission of various forms of solar flare radiation observed at the Earth or in space. Observation of γ-ray lines and neutrons provide diagnostics for probing the behavior of sub-relativistic and relativistic ions at the Sun, while the hard X-rays probe the behavior of sub-relativistic and relativistic electrons. It is now well established that the γ-ray lines emitted from a flare region are often

*Presented in behalf of the SMM Gamma Ray Astronomy Team University of New Hampshire, Max Planck Institute for Astrophysics (8046 Garching by Munich, Federal Republic of Germany), and the Naval Research Laboratory, Washington, D.C.

Solar Physics **86** (1983) 383–393. 0038–0938/83/0862–0383 $ 01.65.

sufficiently intense to be detected with modest spectrometers on Earth orbiting satellites such as OSO-7 (Chupp et al., 1973), HEAO-1 (Hudson et al., 1980), HEAO-3 (Prince et al., 1982), SMM (Chupp, 1981), and Hinotori (Yoshimori et al., 1982). Direct neutrons from solar flares have been observed on SMM (Chupp et al., 1982), and at the same time (in one case) with ground level neutron monitors (DeBrunner et al., 1982). In the case of hard X-rays numerous observations have been made from balloons and satellites since 1958 (Peterson et al., 1963), and we have already heard the most recent revelations from Hinotori at this seminar.

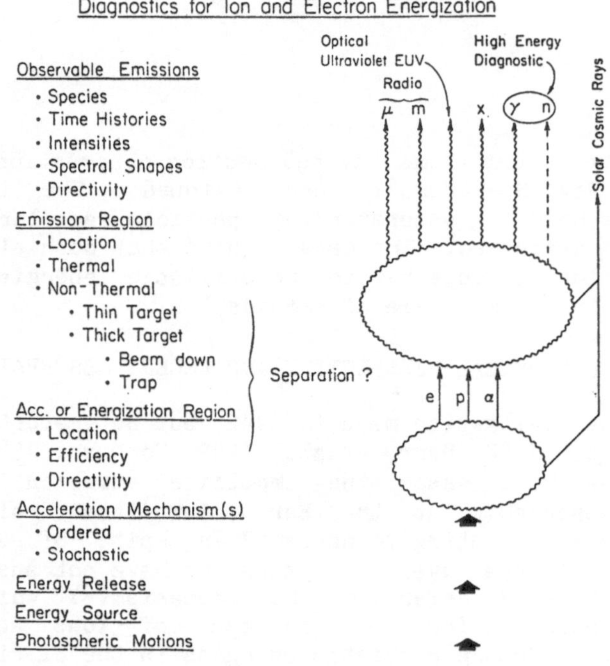

Figure 1. A representation showing how the observed solar flare emissions relate as diagnostic indicators of the energy release/acceleration process. (Conceived by D. Forrest)

In this talk I will concentrate on presenting those observations that our team has made with the <u>SMM Gamma Ray Spectrometer</u> (GRS) that give us insight into the particle acceleration question. Dr. Yoshimori's report on the Hinotori Group's γ-ray observations will, I believe, support what I have to say.

First, I would like to give you some appreciation for the statistics of the types of events we have observed on SMM. (The spectrometer has continued to operate since 17 February 1980 without any apparent degradation.) Until 15 June 1982 we have observed a total of 145 events of extraterrestrial origin yielding photons of energies greater than 300 keV. Of these ∿89 are associated with known solar flares observed at least in Hα and soft X-rays on the GOES satellite, and over 30 of these have excess photon counts in the MeV region indicative of a γ-ray line spectrum with distinct individual γ-ray lines evident in most cases. The remaining events are all apparently of cosmic origin and have photon emission > 1 MeV generally, and in one case to ∿25 MeV. In our present discussion we will consider only solar flare events.

2. OBSERVATIONS RELATING TO PARTICLE ACCELERATION PARAMETERS

The SMM GRS observations provide the following direct information concerning particle acceleration*:

- Impulsive ion and electron acceleration/matter-interaction is virtually simultaneous and can take place on a time scale of seconds or less.
- This process can be repetitive within $\lesssim$10 s and the rise time of particle/matter-interaction can vary from a few to ∿100 s.
- There is strong evidence that the accelerated ion energy spectrum shape is constant from flare to flare although the MeV electron to ion-intensity ratio varies.
- There is evidence that all flares may produce energetic ions.
- The white light continuum emission does not appear to be a signature of ion acceleration.

We will now briefly review the data that support the above assertions. In Figure 2 we show an overlay of the photon time histories from ∿50 keV to 7 MeV for the 0312 UT flare on 7 June 1980 (Forrest et al., 1981), an event with clear γ-ray lines. It is clear that the <u>first</u> impulsive emission at all photon energies begins simultaneously or within less than two seconds. The MeV photons have been shown to be produced solely by ions in this event (Forrest and Chupp,1983; Forrest et al., 1981). Therefore we must conclude that <u>ion and electron</u> acceleration and their interaction with matter occurs simultaneously. The differing apparent rise times (if any) in the separate peaks and the apparent later peaking of the MeV photons <u>does not</u> mean that ion acceleration is delayed relative to electron acceleration since the first γ-ray emission <u>requires</u> that the ions were already accelerated. This figure also illustrates that the acceleration/matter-interaction can be repetitive to $\lesssim$10 s and in this case the total impulsive event lasts only ∿1 minute.

---

*We tacitly assume that particles are <u>only impulsively accelerated</u> and are not merely <u>impulsively</u> released to interact with matter.

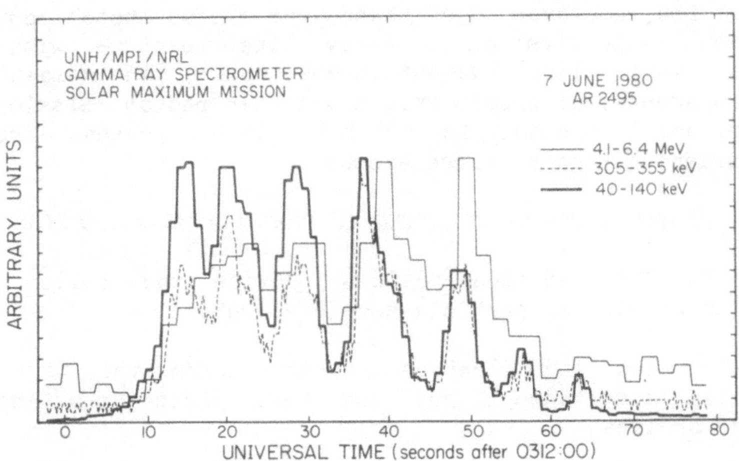

Figure 2. Counting rate versus time for various energy channels of the SMM γ-ray spectrometer during the γ-ray line flare at 0312 UT on 7 June 1980.

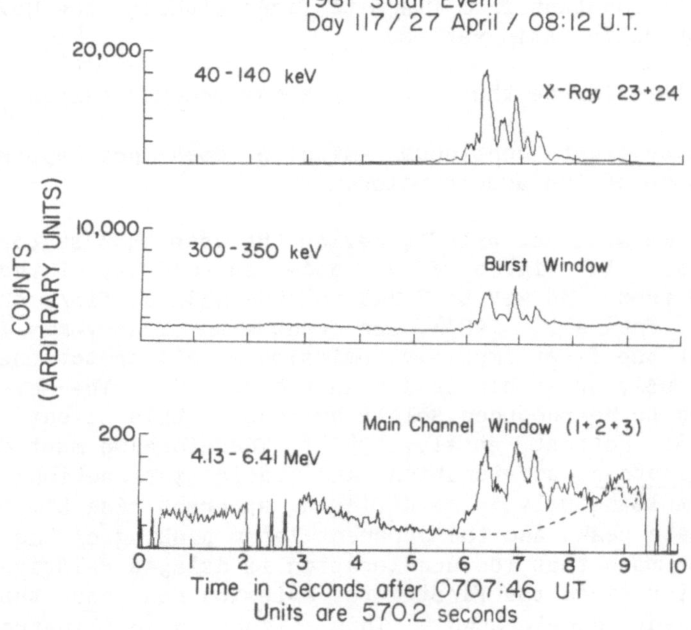

Figure 3. The counting rate versus time for various energy channels of the SMM γ-ray spectrometer during the <u>strong</u> γ-ray line flare at 0812 UT on 27 April 1981.

In Figure 3 we show the time history of the photon emission for an impulsive γ-ray solar flare, 27 April 1981, with a duration of over 10^3 s, which again has a profusion of clear γ-ray lines. An analysis of the rise time (Gardner et al., 1982), of the successive peaks gives ∿50 s, an order of magnitude longer, at least, than for the previous flare discussed. The time history of other γ-ray line events supports our conclusions that electron and ion acceleration are simultaneous; however, there are cases where the γ-ray emission appears suppressed relative to the hard X-ray intensity.

The evidence for accelerated ions with GeV energies comes from direct observations (Chupp et al., 1982), of several 100 MeV neutrons at the satellite. Figure 4 shows the time history for large energy loss events (> 10 MeV) in the SMM GRS during the large limb solar flare of 21 June 1980. During the impulsive phase, lasting ∿1 minute, the instrument responded to high energy photons probably resulting from meson decay (π^0, π^+) and electron bremsstrahlung. After this, for about 17 minutes, the instrument responded to a flux of neutrons arriving at the satellite, giving the characteristic signal shown in the figure. In this case it is believed that the neutrons were produced in a "δ-function" pulse at the Sun at the time of the first impulsive burst shown in the inset. The neutrons arrive at the Earth in accordance with their energy dependent time-of-flight, giving the neutron energy scale shown. It is clear that neutrons of at least

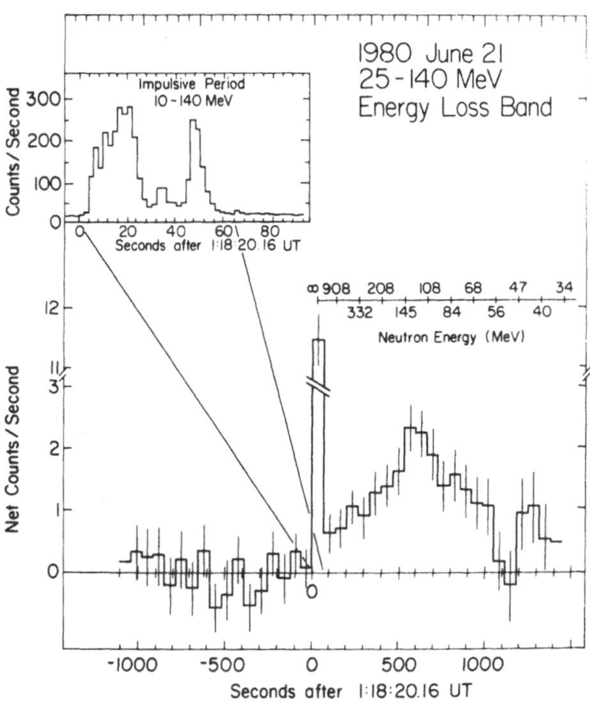

Figure 4.

Lower Panel - Excess counting rate versus time in the SMM GRS is shown for electron equivalent energy loss events between (25-140) MeV before and after start of the impulsive flare at 0118:20 UT. The neutron energy scale, center, assumes the neutrons left the Sun, the light travel time (507 s), before 01180:20 UT.

Inset - The total counting rate versus time after 0118:20 UT for electron equivalent energy loss events between (10-140) MeV during impulsive phase only.

500 MeV kinetic energy are present with a spectrum extending down to ∿50 MeV. Another example of direct solar flare neutrons at the Earth is found in the recent intense flare on 3 June 1982 which occurred at E 72, heliographic longitude. In Figure 5 we show the time history for the large energy losses (> 40 MeV). This event has qualitatively, a similar signature as shown in Figure 4. In this case the impulsive phase γ-rays arrived during the first two major peaks, of duration < 2 minutes, but we have evidence, not shown here, that the first neutrons also arrived during the second major peak. This indicates that neutrons of ∿1 GeV energy were produced at the Sun in the first major impulsive burst. It is clear that in these two cases ions with energies over > 1 GeV were produced within a time scale of seconds, early in the impulsive phase. A ground level enhancement in the Jungfraujoch IGY neutron monitor (DeBrunner, 1982), is also associated with this neutron event.

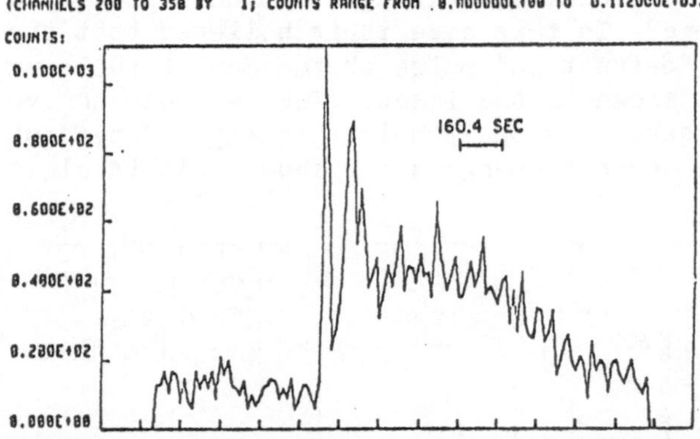

Figure 5.

The raw counts (uncorrected) for large energy loss events (> 40 MeV) is shown during the intense disk flare (S09E72) at 1143 UT on 3 June 1982. A characteristic high energy neutron signal is seen after the impulsive spikes.

The most basic evidence for ion acceleration comes from observation of γ-ray lines during the impulsive phase. Also it has been shown (Forrest et al., 1981; Ramaty et al., 1977), that the excess counts above a bremsstrahlung continuum in the energy band ∿ 4-7 MeV are due to resolved and unresolved nuclear γ-ray lines. Both criteria are applied to all flares with generally consistent results. An example of a rich γ-ray line spectrum is shown in Figure 6 for the large flare on 27 April 1981 discussed above. Since this was a limb flare the normally intense 2.223 MeV line is greatly attenuated because of its photospheric origin; however, the normally prompt nuclear lines are clearly evident and the characteristic sharp fall off in the spectrum at ∿7 MeV (about channel no. 420) is indicative of a pure nuclear origin. We also show in Figure 7 the best example obtained of the 2.223 MeV photospheric γ-ray line which was obtained during the recent intense flare on 3 June 1982. The spectrum shown is confined to the energy band around this line and is compared to the predicted instrument response to a γ-ray line at 2.223 MeV.

From a study of the spectra for many flares such as shown in Figures 6 and 7, we find that the ratio of the prompt γ-ray fluence 4-7 MeV to the 2.223 MeV line fluence is approximately constant. According to theoretical calculations (Ramaty, 1982), this requires that the spectral shape for the accelerated ions be also constant. However, as noted earlier the relative fluence of MeV photons from electron bremsstrahlung to the 4-7 MeV γ-ray fluence is not constant, implying a variable behavior of electrons relative to ions.

An evaluation of the frequency of occurrence of ion acceleration in flares can be made if the spectra from weak γ-ray line flares are summed. We find that features not clearly evident in a single spectrum are enhanced at the channels corresponding to expected γ-ray lines such as the 4.4 MeV and 6.1 MeV lines. While this observation is still preliminary, it suggests that ion acceleration is very common in flares, if not always present. Thus, γ-ray spectrometers with a much greater sensitivity than the SMM GRS could identify lines in individual "weak" events.

As a final point we remark on the question of the origin of the white light flashes often associated with large flares. It is widely believed (Svestka, 1981) that accelerated protons ($\sim$20 MeV) impinging on the photosphere produce the continuum and thus prompt γ-ray lines should appear at the time of the enhancement. That this is not the case in general, is supported by an observation of the γ-ray line time history during the 1 July 1980 solar flare (Ryan et al., 1982). In this case the γ-ray flux peaked $\sim$1 minute before the maximum of the optical continuum emission (Zirin et al., 1981).

3. CONCLUDING STATEMENT

There are several important studies in progress by our team to further define the parameters of the particle acceleration/matter-interaction. These include quantitative determination of estimated absolute proton and electron spectra at the Sun, evidence for beaming and search for a specific signature of ion acceleration by correlating, for example, with the hard X-ray and microwave burst properties. This latter work is not far enough along at this time for presentation. However, the points I have made here are quite firm and it is safe to conclude that the rapidity with which ions and electrons accelerate in the events discussed above forces us to consider that a simple primary process is operative to energize both species.

Figure 6.

The smoothed γ-ray line spectrum from the limb flare on 27 April 1981 which began at 0805 UT. Accumulation time interval (0805:566-0837:03 UT).

Background and flare bremsstrahlung continuum subtracted. Several stronger γ-ray line features from excited nuclear states, which are indicated in the figures, are:

(1) 6.129 MeV - ^{16}O;
(2) 4.438 MeV - ^{12}C;
(3) 2.213 MeV - ^{14}N and 2.223 MeV - $^{1}H(n,\gamma)^{2}H$;
(4) 1.634 MeV - ^{20}Ne;
(5) (1.24 - 1.37) MeV - complex of lines from ^{56}Fe and ^{24}Mg and other nuclides;
(6) (0.8 - 0.9) MeV - ^{56}Fe and other nuclides;
(7) (0.431, 0.478 and 0.511) MeV - ^{7}Be and ^{7}Li resulting from energetic $\alpha \simeq \alpha$ nuclear interactions and positron electron annihilation, respectively.

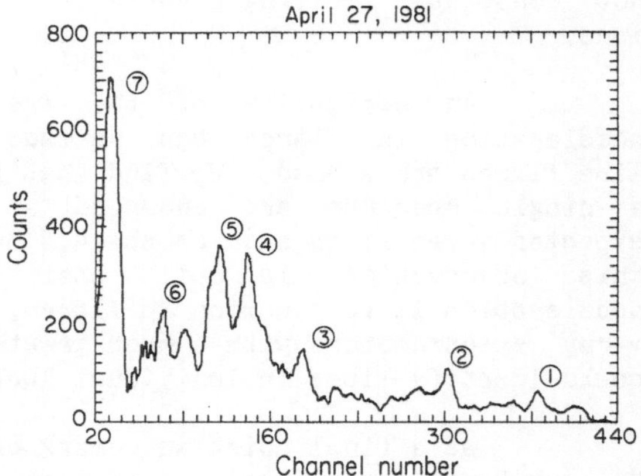

Many line features are strongly broadened due to kinematic effects. The neutron capture line at 2.223 MeV, normally very intense in disk flares, is greatly suppressed in this limb flare because of its photospheric origin.

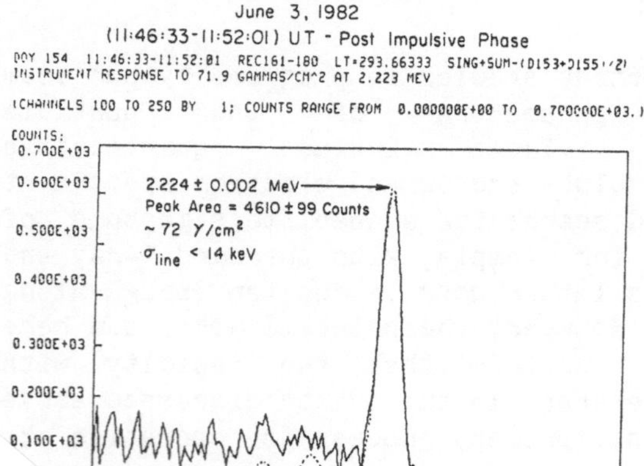

Figure 7.

The γ-ray line spectrum near 2.223 MeV is shown for the intense 3 June 1982 flare during the post-impulsive phase from (1146:33-1152:01) UT. The dotted curve shows the instrument response function fitted to the data yielding the peak parameters shown.

ACKNOWLEDGEMENTS

The author wishes to thank D. J. Forrest, S. Matz and G. Share for helpful discussions and comments. Appreciation goes to Mary Chupp for editing and Diane Belford for typing the manuscript. This work was in part supported by NASA Contract NAS 5-23761.

REFERENCES

Berry, E. V., Hess, V.: 1942, Terr. Mag. 47, 251.

Chupp, E. L., Forrest, D. J., Higbie, P. R., Suri, A. N., Tsai, C., Dunphy, P. P.: 1973, Nature 241, 333.

Chupp, E. L.: 1981, in R. E. Lingenfelter, H. S. Hudson and D. M. Worrall (eds.), Gamma Ray Transients and Related Astrophysics, Amer. Inst. of Physics, New York, p. 363.

Chupp, E. L., Forrest, D. J., Ryan, J. M., Heslin, J., Reppin, C., Pinkau, K., Kanbach, G., Rieger, E., Share, G. H.: Astrophys. J. (Letters) 236, L95.

DeBrunner, H., Flueckiger, E., Golliez, F., Neuenschwander, H., Schubnell, M., Sebor, G.: 1982, Physikalisches Institute, University of Bern, Switzerland, Private Communication.

Ehmert, A.: 1948, Zs. Naturforsch. 3A, 264.

Forbush, S. E.: 1946, Phys. Rev. 70, 711.

Forrest, D. J. and Chupp, E. L.: 1983, In preparation.

Forrest, D. J., Chupp, E. L., Ryan, J. M., Reppin, C., Rieger, E., Kanbach, G., Pinkau, K., Share, G., Kinzer, G.: 1981, 17th Int. Cosmic Ray Conf. 10, 5.

Gardner, B. M., Forrest, D. J., Ryan, J. M., Zolcinski, M. C., Heslin, J. P.: 1982, Bull. AAS 14, 1.

Hudson, H. S., Bai, T., Gruber, D. E., Matteson, J. L., Nolan, P. L., Peterson, L. E.: 1980, Astrophys. J. (Letters) 236, L91.

Peterson, L. E., Winckler, J. R.: 1963, J. Geophys. Res. 64, 697.

Prince, T. A., Ling, J. C., Mahoney, W. A., Riegler, G. R., Jacobson, A. S.: 1982, Astrophys. J. 255, L81.

Ramaty, R., Kozlovsky, B., Suri, A. N.: 1977, Astrophys. J. 214, 617.

Ramaty, R.: 1982, in Hozler, T. E., Mahalas, D., Sturrock, P. A., Ulrich, R. K. (eds.), Physics of the Sun, p. 1.

Ryan, J. M., Chupp, E. L., Forrest, D. J., Matz, S. M., Rieger, E., Reppin, C., Kanbach, G., Share, G. H.: 1982, Astrophys. J., To be published.

Svestka, Z.: 1981, in Priest, E. R., Gordon-Breach (eds.), New York, p. 72.

Yoshimori, M., Okudaira, K., Harasima, Y.: 1982, Hinotori Symposium on SMM Flares, 230, (Report from ISAS - Tokyo).

Zirin, H., Neidig, D. F.: 1981, Astrophys. J. 248, L45.

DISCUSSION

KANE: Do I understand right that your observations require that $\sim$ 100 keV electrons and GeV ions must be accelerated in time $\lesssim$ 1 second?

CHUPP: Yes!

SIMNETT: There is a better test of neutrons at the Earth than the neutron monitor records, which are insensitive to neutrons or protons below $\sim$ 1 GeV. Protons in interplanetary space will be detected by a suitable spacecraft proton detector before any directly accelerated protons can travel along interplanetary field lines. For east limb flares this is an extremely sensitive test.

CHUPP: Your first statement is incorrect: Neutrons and protons > 100 MeV build up a neutron cascade in the atmosphere in which the average energy of the secondary, tertiary, etc , nucleons have a much lower average energy. The neutron monitors are quite sensitive to the lower energy neutrons (see e.g., Alsmiller and Boughner, J. Geophys. Res. , 73, 4935, 1968).

I agree with your second statement that the detection of neutron decay protons is a very sensitive test for neutrons from east longitude disk flares. On the other hand, for flares well connected magnetically to the Earth, the arriving protons from the flare will produce a high background for the later-arriving, low energy, neutron decay protons.

KUNDU: For many years, the Tata Institute in Bombay has been looking for solar neutrons by using balloon-borne experiments -- the results have never been convincing. I am curious to know what is it that made the difference between non-detection of neutrons by them and detection now. Was it because they were not observing at the time of a large flare or was their sensitivity not adequate for detection?

CHUPP: I do not know of any case in which earlier balloon flights were conducted during sufficiently intense solar flares. The observations of solar neutrons reported here were made with a relatively simple neutron detector but under nearly optimal viewing conditions; *i.e.*, intense, short duration (2 minute) gamma ray flares.

GAMMA-RAY LINES AND NEUTRONS FROM SOLAR FLARES*

R. Ramaty[1], R.J. Murphy[1], B.Kozlovsky[2], R.E. Lingenfelter[3]
1-Goddard Space Flight Center, Greenbelt, Maryland 20771
2-Dept. of Physics and Astronomy, Tel Aviv University, Tel Aviv, Israel
3-Center for Astrophysics and Space Sciences, University of California San Diego, La Jolla, California 92093

ABSTRACT

We have derived the energy spectrum of accelerated protons and nuclei at the site of the June 21, 1980 limb flare by a new technique, using observations of the time-dependent flux of high-energy neutrons at the Earth. We find that this energy spectrum is very similar to the energy spectra of 7 disk flares for which the accelerated particle spectra have been previously derived using observations of 4-7MeV-to-2.223 MeV fluence ratios. The implied spectra for all of these flares are too steep ($\alpha T \simeq 0.02$) to produce any significant amount of radiation from π meson decay. We suggest that the observed >10 MeV gamma rays from the June 21, 1980 flare are bremsstrahlung of relativistic electrons.

1. INTRODUCTION

Interactions in the solar atmosphere of flare-accelerated protons, alpha particles and heavier nuclei produce observable gamma-ray lines and neutrons. The detection of this neutral radiation provides unique information on particle acceleration and interaction processes, on the flare mechanism and on properties of the solar atmosphere.

Energetic particle interactions produce a variety of secondary products. For solar flares, the most important of these are neutrons, excited and radioactive nuclei, and π mesons. Solar neutrons are observed both directly at the Earth and in the 2.223 MeV line from neutron capture on hydrogen in the photosphere. Detectable gamma-ray lines are also produced by the deexcitation of nuclear levels and by the annihilation of positrons from the decay of radioactive nuclei and π^+ mesons. The decay of π° mesons could lead to observable >10 MeV gamma-ray emmission, but it appears that for solar flares most of the emission at these energies is due to relativistic electron bremsstrahlung.

Gamma-ray lines from solar flares were first observed by Chupp et al. (1973) with a NaI spectrometer flown on OSO-7. Following

Solar Physics **86** (1983) 395–408. 0038–0938/83/0862–0395 $ 02.10.

these observations, solar gamma-ray lines were seen with the NaI spectrometer on HEAO-1 (Hudson et al., 1980), the NaI spectrometer on SMM (Chupp et al., 1981; Chupp, 1982), the high-resolution Ge spectrometer on HEAO-3 (Prince et al., 1982), and the CsI spectrometer on HINOTORI (Yoshimori et al., 1983). Gamma rays of energies >10 MeV and high-energy solar neutrons near Earth were detected with the gamma-ray spectrometer on SMM (Chupp et al., 1982; Chupp, 1983).

Theoretical investigations of solar neutron and gamma-ray production in flares were carried out in considerable detail by Lingenfelter et al. (1965), Dolan and Fazio (1965), Lingenfelter and Ramaty (1967) and Cheng (1972), prior to the discovery of gamma-ray lines from flares. A number of theoretical studies (e.g., Wang and Ramaty, 1974; Kozlovsky and Ramaty 1974; Ramaty, Kozlovsky and Lingenfelter, 1975, 1979; Crannell et al., 1976, 1979; Ramaty, Kozlovsky and Suri, 1977; Ibragimov and Kocharov, 1977; Ramaty, 1979), performed after the OSO-7 observations, provided consistent interpretations for these observations, developed the details of the basic gamma-ray and neutron production processes and made predictions for further observations. Recent theoretical investigations (Ramaty, Lingenfelter and Kozlovsky, 1982; Ramaty, 1983), using the much more extensive SMM and HEAO data, showed that gamma-ray production in flares takes place predominantly in thick-target interactions by accelerated particles whose energy spectra do not vary much from flare to flare. The short acceleration time of nuclei in flares implied by these data was investigated by Bai (1982), and the problem of magnetic mirroring and its effect on gamma-ray production was studied by Zweibel and Haber (1983).

In the present paper we briefly review the theory of gamma-ray and neutron production in solar flares and present a theoretical analysis of the recent neutron and >10 MeV gamma-ray observations. In the discussion that follows we consider only thick-target interactions.

2. NEUTRONS

The 2.223 MeV line was expected (Lingenfelter and Ramaty, 1967) theoretically to be the strongest line from flares and this prediction was confirmed by observations (Chupp et al., 1973; Hudson et al., 1980; Chupp, 1982). The limb darkening of the 2.223 MeV line caused by Compton scattering in the photosphere, predicted by Wang and Ramaty (1974), has also been recently observed (Chupp, 1982; Yoshimori et al., 1983). This effect, together with the observed time delay of the line flux (Chupp et al., 1981; Prince et al., 1982) and the precisely determined line energy (Prince et al., 1982; Chupp, 1983), provides clear evidence that the observed emission at 2.223 MeV is indeed due to neutron capture and that this capture takes place in the photosphere.

For solar flares whose duration is much shorter than the typical neutron transit time from the Sun to the Earth, the time dependence

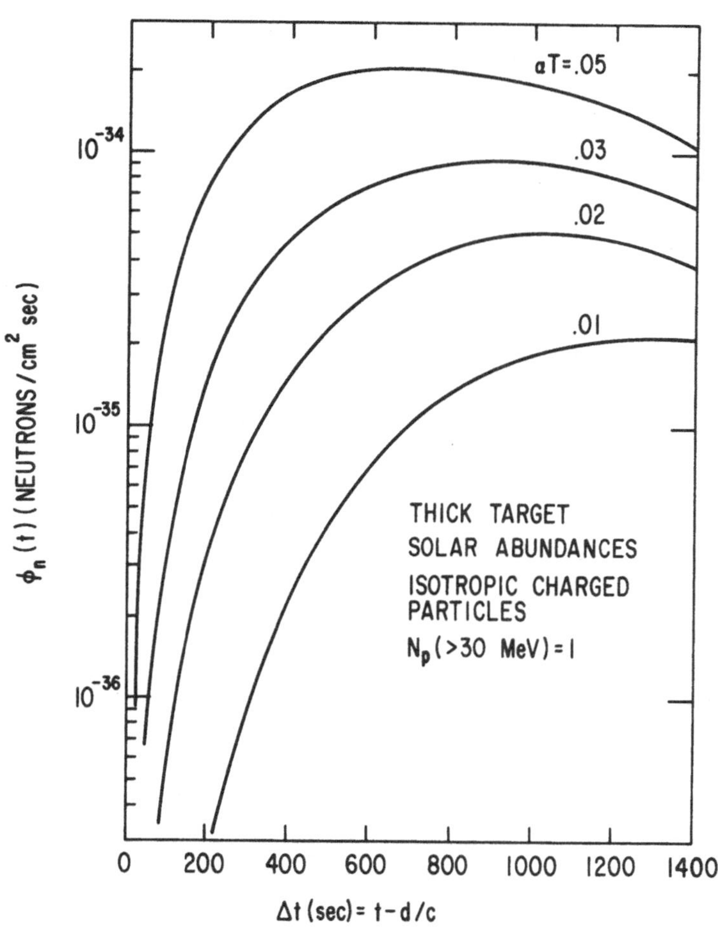

Figure 1. Time-dependent neutron fluxes at Earth from neutron production at the Sun.

of the high-energy neutron flux at Earth is a direct measure of the energy spectrum of the neutrons released from the Sun. The recent high-energy solar neutron observations (Chupp et al., 1982) confirm the time dependence of the neutron flux at Earth predicted earlier (Lingenfelter et al., 1965; Lingenfelter and Ramaty, 1967). Extending these previous studies, we have carried out a detailed investigation of the energy and angular distribution of neutron production in flares.We show in Figure 1 calculated time-dependent neutron fluxes at Earth resulting from various energetic particle spectra at the Sun.

For these calculations we have assumed an isotropic distribution of energetic particles incident on the solar atmosphere, producing a burst of neutrons at t =.0 in a thick-target model. We have normalized the number of energetic particles to 1 proton above 30 MeV and we have assumed that the composition of these particles is the same as that of the photosphere. We have further assumed that for all particle species, the energy spectrum incident on the atmosphere is given by the Bessel function appropriate for stochastic Fermi acceleration to nonrelativistic energies, $N(E) \propto K_2[2(3p/(mc\alpha T))^{1/2}]$ (Ramaty, 1979; Forman, Ramaty and Zweibel, 1983), where K_2 is the modified Bessel function of order 2, E and p are particle energy and momentum per nucleon, m is the proton mass, and α and T are, respectively, the stochastic acceleration efficiency and the particle residence time in the acceleration region. We have also assumed that after their production, the neutrons escape freely from the Sun. But because the neutrons have mean free paths comparable to the stopping range of the protons which produce them, this assumption requires that the charged particles be trapped magnetically in the thick-target interaction region at column depths significantly less than their ranges.

As can be seen from Figure 1, the rise time to maximum of the neutron flux becomes shorter as αT increases. This follows because larger values of αT imply more high-energy particles, which in turn produce more high-energy neutrons. We have compared the time-dependent neutron fluxes of Figure 1 with observations of the June 21, 1980 limb flare (Chupp et al., 1982; D. Forrest, private communication, 1982). We find that the calculated curve for αT = 0.02 fits the observed time-dependent neutron flux very well. Larger values of αT produce curves which rise to maximum too fast, while for αT less than about 0.02 the rise is slower than observed. As we shall see in Section 4, this value of αT, deduced from neutron observations of a limb flare, is essentially the same as the αT deduced from deexcitation-to-neutron capture line ratios for 7 disk flares.

By normalizing the αT = 0.02 curve in Figure 1 to the absolute value of the time-dependent neutron flux from the June 21, 1980 flare (D. Forrest, private communication, 1982), we find that the number of protons above 30 MeV was 1.0×10^{33} and that the total neutron production was 2.3×10^{30}. For a disk-centered flare, this neutron production would imply a 2.223 MeV fluence of ~190 photons/cm^2, based on a neutron-to-2.223 MeV photon conversion factor of 0.23 (Ramaty, 1983). The observed (Chupp, 1982; D. Forest, private communication, 1982) 2.223 MeV fluence of ~6 photons/cm^2 for the June 21, 1980 limb flare provides clear evidence for the limb darkening of this line.

3. NUCLEAR DEEXCITATION LINES

A variety of gamma-ray lines are produced in solar flares from the deexcitation of nuclear levels. These levels are populated by inelastic collisions (e.g., $^{12}C(p,p')^{12}C^{*4.44}$), spallation reactions (e.g. $^{20}Ne(p,p\alpha)^{16}O^{*6.13}$), nonthermal fusion reactions (e.g., $^{4}He(\alpha,p)^{7}Li^{*0.478}$) and the decay of radionuclei produced by spallation reactions (e.g., $^{16}O(p,p2n)^{14}O(e^{+})^{14}N^{*2.31}$). Using laboratory measurements of the excitation functions of a large number of such reactions, calculations have been made (Ramaty, Kozlovsky and Lingenfelter, 1979) of the resultant gamma-ray spectra. An example is shown in Figure 2. Here the energies of 1.5×10^5 photons were derived from a Monte-Carlo calculation for thick-target interactions of isotropic energetic particles having solar abundances and αT = 0.02. These photons were then binned in 100 keV intervals, appropriate for a NaI detector such as that flown on SMM. The upper

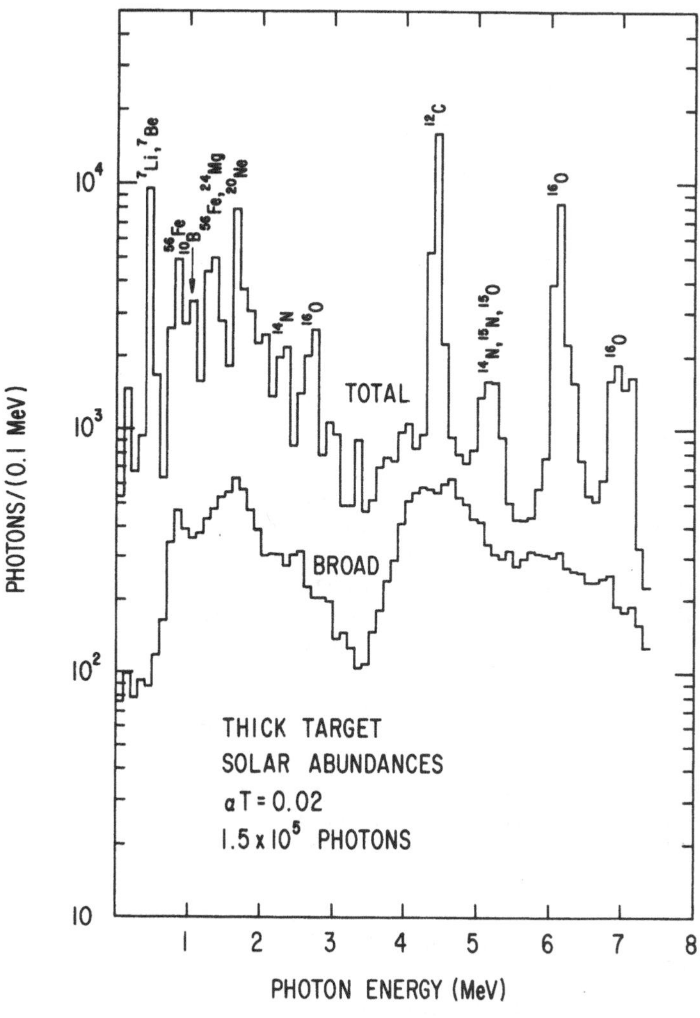

Figure 2. Energy spectrum of prompt nuclear deexcitation radiation.

histogram shows the total gamma-ray spectrum, while the lower one shows the broad component, i.e., gamma rays from the interactions of energetic nuclei heavier than He.

The total spectrum in Figure 2 contains a variety of narrow lines which result mostly from the deexcitation of ambient nuclei excited by fast protons and alpha particles. As can be seen, the strongest such lines are at 6.129 MeV from ^{16}O, at 4.438 MeV from ^{12}C, at 1.634 MeV from ^{20}Ne, at 0.847 MeV from ^{56}Fe, at ~1.3 MeV from ^{24}Mg and ^{56}Fe, and at ~ 0.45 MeV from ^{7}Li and ^{7}Be. The excited states $^{7}Li^{*0.478}$ and $^{7}Be^{*0.431}$ are formed in flares by nonthermal fusion reactions between alpha particles (Kozlovsky and Ramaty, 1974).

While the relative widths of these narrow lines, broadened by the recoil velocities of the heavy target nuclei, are only on the order of 1 to 2 percent, the widths of the broad lines, reflecting the velocities of the projectiles themselves, are much larger. Consequently, only a few discrete features can be discerned in the broad component. As can be seen, there is a broad feature between 4 and 5 MeV, mostly from ^{12}C, another one between 1 and 2 MeV from ^{20}Ne, ^{24}Mg, ^{28}Si and ^{56}Fe, and a broad line at ~ 0.85 MeV from ^{56}Fe.

But the contribution of the broad component to the total emission in Figure 2 is quite small. In thick-target interactions, this is caused by suppression of the contribution of heavy nuclei in comparison with that of protons and alpha particles resulting from their larger energy loss rates. However, the contribution of the broad component can be much larger if the abundance of accelerated heavy nuclei relative to accelerated protons significantly exceeds that in the photosphere. In particular, a higher Fe abundance in the accelerated particles could produce a more intense 0.847 MeV line of FWHM less than 200 keV.

The 4.438 MeV and 6.129 MeV lines have been observed from several flares (e.g. Chupp, 1982). These two lines, as well as several of the other strong lines shown in Figure 2, including the ^{7}Li-^{7}Be feature at ~0.45 MeV, have been seen from the April 27, 1981 limb flare (Chupp, 1983; Yoshimori et al., 1983).

In addition to the nuclear lines shown in Figure 2, gamma-ray emission from flares should also contain a significant contribution from electron bremsstrahlung (see also Section 6). In the 4-7MeV band, however, most of the emission appears to be nuclear (Ramaty, Kozlovsky and Suri, 1977; Ibragimov and Kocharov, 1977). This result is supported by recent data (Chupp, 1982) which indicate that for all disk flares from which gamma-ray lines were seen, the ratio of the observed fluence in the 4-7MeV band to the fluence in the 2.223 MeV line does not vary much from one flare to another. This approximate constancy indicates that both radiations are produced by the same population of energetic particles. If one of them were produced by electrons and the other by nuclei, one would expect a much more variable ratio than observed. Indeed, the ratio of the continuum fluence at ~ 1 MeV (which is mostly electron bremsstrahlung) to the 4-7MeV fluence varies significantly from flare to flare (Chupp, 1982).

4. COMPARISONS OF THE NEUTRON AND NUCLEAR DEEXCITATION LINE CALCULATIONS WITH DATA

Ratios of photon production in the 4-7MeV band to the total neutron production as functions of αT, in both the thick and thin-target models, are shown in Figure 3 (from Ramaty, 1983). These calculations are based on our detailed studies of nuclear deexcitation lines and neutron production in energetic particle reactions. The curves labelled EP1 are for a set of energetic particles enhanced in He and heavier nuclei (see Ramaty, 1983), while those labelled EP2 are for energetic particles with solar abundances.

From the comparison of the curves of Figure 3 with the observed 4-7MeV-to-2.223 MeV fluence ratio for the June 7, 1980 flare, and from the assumption that the αT of the particles from this flare observed (R. McGuire, private communication, 1981; see also Ramaty, 1983) in interplanetary space is similar to the αT of the particles interacting at the Sun, it was shown (Ramaty, Lingenfelter and Kozlovsky, 1982; Ramaty, 1983) that gamma-ray production in the June

7 flare proceeded predominently via thick-target interactions. This conclusion also followed from the comparison of the total number of protons seen in interplanetary space with the number required to produce the gamma rays at the Sun (Von Rosenvinge, Ramaty and Reames, 1981), and from the general absence of spallation products (e.g. ^{2}H, Li, Be, B) in the observed (McGuire, Von Rosenvinge and McDonald, 1979) interplanetary particle fluxes (see also Ramaty, 1983).

By comparing the observed 4-7MeV-to-2.223 MeV fluence ratios for other flares with the thick-target calculations (Figure 3), and by taking into account the flare positions on the solar disk, values of αT were deduced (Ramaty, 1983) for the flares listed in Table I. As can be seen, the implied αT varies little from flare to flare. This result also applies to limb flares for which αT cannot be easily deduced from observations of the 4-7MeV-to-2.223 MeV fluence ratios because of the strong attenuation

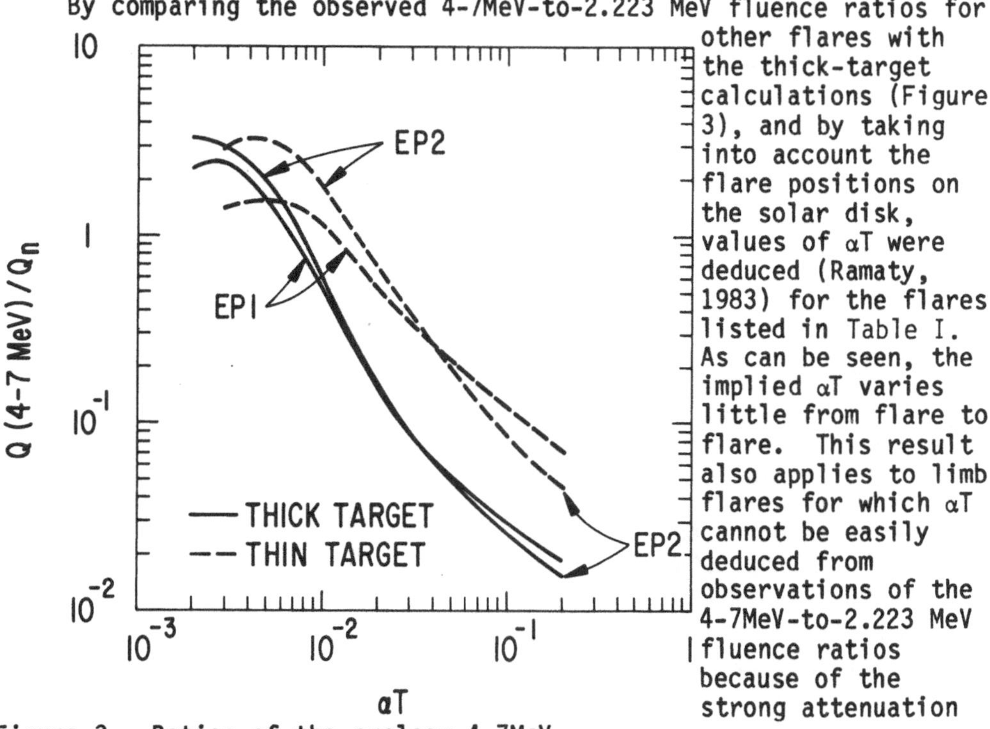

Figure 3. Ratios of the nuclear 4-7MeV photon to neutron production.

of the 2.223 MeV line in the photosphere. But as we have seen in Section 2, αT can be obtained for such flares from high-energy neutron observations. For the June 21, 1980 limb flare, we found that $\alpha T \simeq 0.02$, in excellent agreement with the disk flare results given in Table I. This finding demonstrates, first of all, the consistency of the gamma-ray line and neutron production theory in flares, since the two determinations of the energetic-particle spectrum rely on different physical proceses and utilize essentially independent sets of nuclear data. In addition, the small variability of αT implies that energetic protons and nuclei are probably accelerated in all gamma-ray flares by the same mechanism and under rather similar conditions.

The range of αT deduced from both gamma-ray and neutron observations is very similar to the αT range of 0.014 to 0.036 obtained from fitting Bessel function spectra to energetic proton measurements in interplanetary space (McGuire, Von Rosenvinge and McDonald, 1981). Therefore, the same mechanism probably accelerates both the energetic particles which produce gamma rays at the Sun and those observed in interplanetary space.

The last column in Table I gives the number of protons greater than 30 MeV required to produce the observed gamma-ray emission by thick-target interactions. These numbers were obtained from our calculation of the total 4-7MeV photon production, the observed

TABLE I: Energetic particle parameters deduced from gamma ray data of disk flares.

Flare	Location	4-7MeV-to-2.223 MeV Fluence Ratios	αT	N_p(>30 MeV)
1972, Aug. 4	E08N14	0.68±0.09	0.019	1.6×10^{33}
1978, July 11	E43N18	~ 0.71	0.020	1.3×10^{33}
1979, Nov. 9	E00N16	1.32±0.33	0.014	2.6×10^{32}
1980, June 7	W74N12	1.74±0.27	0.015	6.6×10^{31}
1980, July 1	W37S12	0.94±0.19	0.016	1.9×10^{31}
1980, Nov 6	E74S12	1.44±0.2	0.017	1.0×10^{32}
1981, Apr. 10	W37N09	1.38±0.16	0.014	9.7×10^{31}

4-7MeV photon fluences and the deduced αT's (see Ramaty, 1983, for more details). The total number of $\sim 10^{33}$ protons >30 MeV required in the large flares of August 4, 1972 and July 11, 1978 is quite comparable to the total number inferred above from high-energy neutron observations of the June 21, 1980 flare.

Using a >30 MeV proton number of 10^{33} and an αT=0.02 for the June 21, 1980 flare, we predict that the 4-7MeV fluence from nuclear deexcitation to be ~ 130 photons/cm^2 for this flare. This value constitutes about 2/3 of the observed (D. Forest, private communication, 1982) 4-7MeV fluence of ~200 photons/cm^2, a result which confirms the finding that this energy band is predominantly nuclear.

5. POSITRONS

The 0.511 MeV line due to positron annihilation has been observed from several solar flares (Chupp et al., 1973; Share et al., 1980; Chupp, 1982). Calculations of positron production from the decay of π^+ mesons and radioactive nuclei were carried out previously (e.g., Ramaty, Kozlovsky and Lingenfelter, 1975). Here we give results of new calculations of positron production based on a large number of positron emitters produced in nuclear reactions involving all reasonably abundant isotopes up to those of Ni. In Figure 4 we show the ratio of the total, time-integrated positron production to the 4-7MeV nuclear gamma-ray production as a function of αT for thick-target interactions and solar composition for the energetic particles.

The initial energies of positrons from radioactive nuclei are of the order of several hundred keV while those from π^+ decay are from about 10 to 100 MeV. Because only about 10^{-3} of the total positron production is from π^+ decay for an $\alpha T = 0.02$, the initial energies of the bulk of the positrons are expected to be less than an MeV.

The slowing down of positrons from the energies at which they are produced to energies comparable with those of the ambient electrons where they annihilate, and the subsequent annihilation process have been studied in considerable detail (Crannell et al., 1976; Ramaty, 1983). Positrons with an initial energy of ~ 0.5 MeV slow down and annihilate in about$(2 \times 10^{12}/n_H)$ sec, where n_H is the density of the ambient hydrogen (Ramaty, 1983). If n_H is high enough so that the slowing down and annihilation time is much shorter than the half-lives of the dominant positron emitters, the time dependence of the 0.511 MeV line flux is determined only by the decay rates of the positron emitters and their relative contributions to the total positron production.

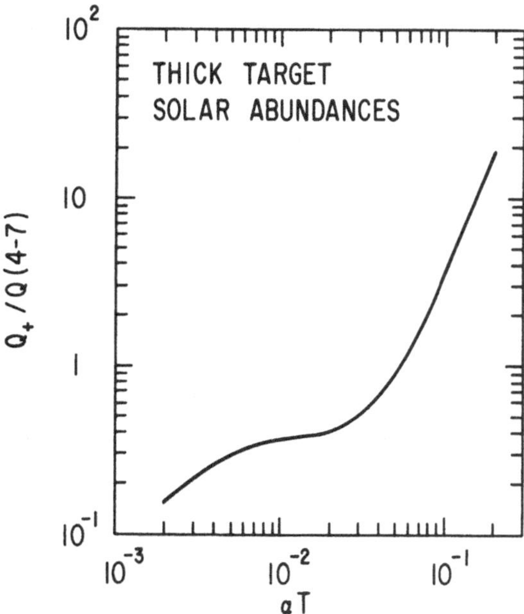

Figure 4. Ratios of the total positron to nuclear 4-7MeV production.

In Figure 5 we show the time-dependent 0.511 MeV line flux at Earth for t>1 sec from a burst of positron-emitter production at t = 0 for thick-target interactions, $\alpha T = 0.02$ and solar abundances for the energetic particles. The quantity $\bar{f}_{0.511}$, which takes into account the formation and breakup of positronium, is the average number of 0.511 MeV photons produced per annihilating positron. In a low density medium ($n_H < 10^{14} cm^{-3}$) where triplet positronium annihilates before it can be broken up, $\bar{f}_{0.511} \simeq 0.65$ (Bussard, Ramaty and Drachman, 1979). For higher density solar plasmas, however, we expect that $\bar{f}_{0.511} \simeq 1$.

The 0.511 MeV annihilation line has so far been detected from only a few flares. For the August 4, 1972 disk flare (Chupp et al., 1973), the ratio of the 0.511 MeV line fluence to the 4-7MeV fluence, measured during the first 500 seconds of the flare, was 0.33±0.11 (Chupp et al., 1973; see also Ramaty, 1983). As shown in Table 1, the energetic particle spectrum of this flare had an αT close to 0.02. From the time dependence calculated for such an αT (Figure 5), we expect that 60 percent of all the positron emitters should have decayed in 500 seconds. Therefore, the ratio of the total positron production to the 4-7MeV production, for $\bar{f}_{0.511} = 1$, should have been 0.55±0.17. This value is in good agreement with the expected value of ~0.4 (Figure 4) for that αT.

For the June 21, 1980 limb flare the time-integrated 0.511 MeV fluence for a period of 700 seconds was ~ 20 photons/cm² (G. Share, private communication, 1982) and the nuclear 4-7MeV fluence was estimated to be ~ 130 photons/cm². Using $\alpha T = 0.02$, deduced from the high-energy neutron time dependence, we expect that about 70 percent of all the positron emitters should have decayed in 700 seconds (Figure 5), implying a positron emitter-to-4-7MeV production ratio of ~ 0.22. This value is smaller than the calculated ratio of ~ 0.4 for $\alpha T = 0.02$. This suggests that the 0.511 MeV line from limb flares could be attenuated by Compton scattering in the solar atmosphere.

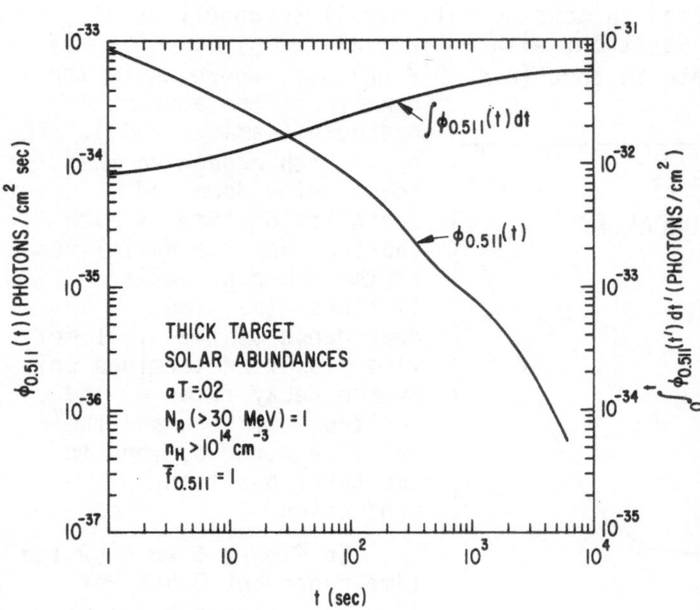

Figure 5. Time dependent 0.511 MeV line.

6. HIGH-ENERGY GAMMA-RAY EMISSION

At energies greater than ~ 10 MeV, most of the gamma rays in solar flares are expected to be produced by electron bremsstrahlung and π° decay (Ramaty, Kozlovsky and Lingenfelter, 1975; Crannell et al., 1979). The possibility of producing nuclear line emission in this energy region was studied by Crannell, Crannell and Ramaty (1979) who found only one important line at 15.1 MeV. However, the contribution of this line to the total >10 MeV emission is negligible.

In Figure 6 we present results of calculations of ratios of the

π^+-to-π° meson production and the π°meson-to-4-7MeV nuclear photon production for thick-target interactions and isotropic energetic particles with solar abundances. Using these ratios, we have evaluated the >10 MeV photon production from pion decay, by combining the contributions of the gamma rays from π° decay with the bremsstrahlung of positrons and electrons from $\pi^\pm$ decay. We also show in Figure 6 the ratio of this >10 MeV photon production to the 4-7MeV nuclear radiation production.

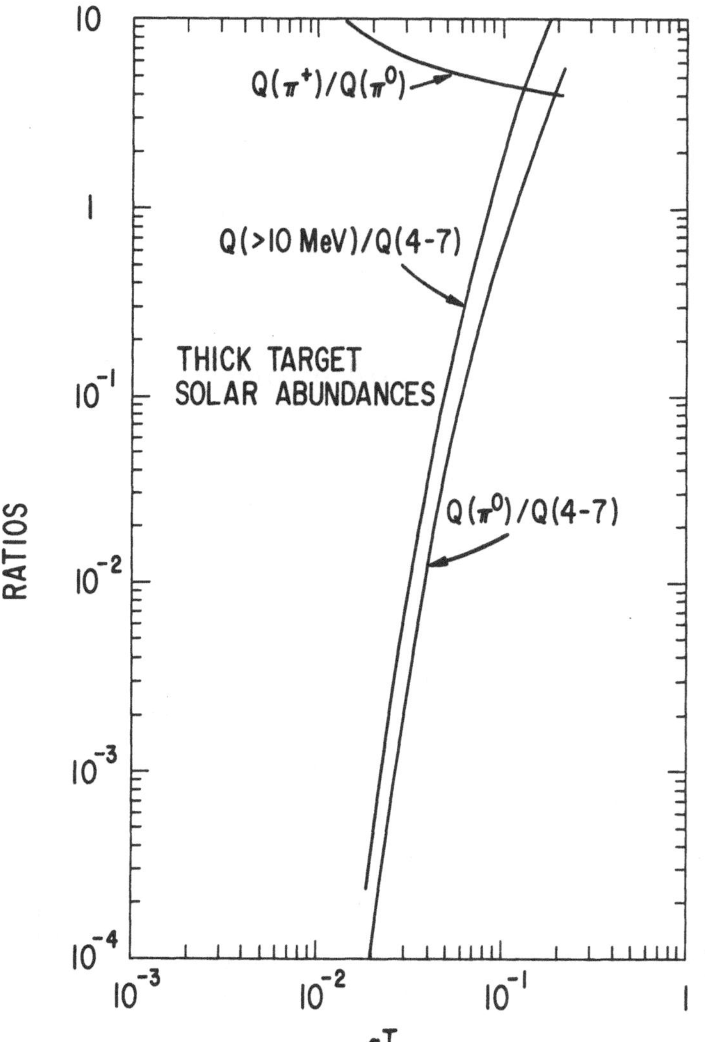

Figure 6. π meson and >10 Mev photon production from πo decay and $e^\pm$ bremmsstrahlung from $\pi^\pm$ decay.

In addition, we have evaluated the >10 MeV photon production from directly-accelerated electron bremsstrahlung. We find that if the ratio of electrons-to-protons at energies greater than 30 MeV is $\sim 10^{-3}$, then at photon energies >10 MeV electron bremsstrahlung exceeds photon production from pion decay by factors of $\sim 10^3$ for $\alpha T=0.02$, ~ 70 for $\alpha T=0.03$ and ~ 5 for $\alpha T=0.05$. These ratios are only weakly dependent on the spectrum of the electrons.

The >10 MeV fluence observed (Chupp et al., 1982; D. Forrest, private communication, 1982) from the June 21, 1980 flare was $\sim$ 30 photons/cm^2. This fluence could not have resulted from pion decay if αT = 0.02, because with a nuclear 4-7MeV fluence of ~130 photons/cm^2 for this flare (Section 4), the resultant >10 MeV fluence from pions would be only ~0.06 photons/cm^2 (Figure 6). Bremsstrahlung of directly-accelerated electrons, on the other hand, could have produced the observed >10 MeV gamma rays if the number of electrons of energies >30 MeV was ~ 5 x 10^{29}, i.e. about 5 x 10^{-4} of the number of protons of similar energies. This interpretation of the >10 MeV continuum is quite consistent with a simple extrapolation of the continuum measured (Chupp, 1982) between 0.29 and 1MeV. The extrapolation would also imply a bremsstrahlung contribution in the 4-7MeV band of about ½ of the estimated nuclear contribution.

7. SUMMARY AND CONCLUSIONS

We have reviewed the theory of gamma-ray and neutron production in solar flares. We have discussed the production of neutrons, the attenuation of the 2.223 MeV line by Compton scattering in the photosphere, the production of a variety of observable prompt nuclear deexcitation lines, methods for determining the number and energy-spectrum of the charged particles accelerated in flares, the production of positrons and the time dependence of the 0.511 MeV line, and photon production at energies >10 MeV.

We have presented new calculations of time-dependent high-energy neutron fluxes at Earth resulting from energetic-particle reactions at the Sun. These calculations provide a new technique for determining the accelerated-particle energy spectrum at the Sun by comparing the time profiles at the Earth, calculated for isotropic charged-particle distributions incident on the solar atmosphere, with the time-dependent high-energy neutron flux observed from the June 21, 1980 flare. We found that the energy spectrum of the accelerated charged particles in this limb flare was essentially the same as in the 7 disk flares for which the spectrum was determined from the 4-7MeV-to-2.223 MeV fluence ratios. This result strongly suggests that a single mechanism, operating under similar conditions, accelerates the protons and nuclei in all gamma-ray producing flares.

We have also calculated the production of neutral and charged π mesons and the gamma-ray production >10 MeV resulting from their decay. We find that for accelerated-particle spectra with $\alpha T \simeq 0.02$, deduced from observations of both the high-energy neutron time profile for a limb flare and 4-7MeV-to-2.223 MeV fluence ratios for disk flares, the >10 MeV photon fluences from pion decay are <10^{-3} of the 4-7MeV fluences of these flares. This is much less than the >10 MeV emission observed from the June 21, 1980 limb flare. This emission, however, could have been bremsstrahlung of directly-accelerated electrons provided that the electron-to-proton ratio at energies >30 MeV was ~5 x 10^{-4}.

ACKNOWLEDGEMENTS

We wish to acknowledge E. L. Chupp, D. J. Forrest and G. H. Share for providing data prior to publication. Financial support from NASA through the Solar Terrestrial Theory Program and from the NSF through the Solar Terrestrial Program is also acknowledged.

REFERENCES

Bai, T.: 1982, in R. E. Lingenfelter et al. (eds.) Gamma-Ray Transients and Related Astrophysical Phenomena, AIP, N.Y., p. 409.

Bussard, R.W., Ramaty, R. and Drachman, R.J.: 1979, Ap.J. 228, 928.

Cheng, C., C.: 1972, Space Sci. Rev. 13, 3.

Crannell, C. J., Crannell, H. and Ramaty, R.: 1979, Ap. J. 229, 762.

Crannell, C. J., Joyce, G., Ramaty, R. and Werntz, C.: 1976, Astrophys. J. 210, 582.

Chupp, E. L.: 1982, in R. E. Lingenfelter et al. (eds.) Gamma-Ray Transients and Related Astrophysical Phenomena, AIP, N.Y., p. 363.

Chupp, E. L.: 1983, this volume.

Chupp, E. L. et al.: 1973, Nature 241, 333.

Chupp, E. L. et al.: 1981, Astrophys. J. Lett. 244, L171.

Chupp, E. L. et al.: 1982, Astrophys. J. Lett. 263, L95.

Dolan, J. F. and Fazio, G. G.: 1965, Rev. Geophys. 3, 319.

Forman, M. A., Ramaty, R. and Zweibel, E. G.: 1983, in P. A. Sturrock et al. (eds.), The Physics of the Sun, in press.

Hudson, H. S. et al.: 1980, Astrophys. J. Lett. 236, L91.

Ibragimov, I. A. and Kocharov, G. E.: 1977, Sov. Astron. Lett. 3 (5), 221.

Kozlovsky, B. and Ramaty, R.: 1974, Astrophys. J. Lett. 191, L43.

Lingenfelter, R. E., Flamm, E. J., Canfield, E. H., and Kellman, S.: 1965, J. Geophys. Res. 70, 4077, 4087.

Lingenfelter, R. E. and Ramaty, R.: 1967, in B. S. P. Shen (ed.), High Energy Nuclear Reactions in Astrophysics, W. A. Benjamin, New York, p. 99.

McGuire, R. E., Von Rosenvinge, T. T., and McDonald, F. B.: 1981, 17th Internat. Cosmic Ray Conference Papers 3, 65.

Prince, T. A. et al.: 1982, Astrophys. J. Lett. 255, L81.

Ramaty, R.: 1979, in J. Arons et al. (eds.) Particle Acceleration Mechanisms in Astrophysics, AIP, New York, p. 135.

Ramaty, R.: 1983, in P. A. Sturrock et al. (eds.) The Physics of the Sun, in press.

Ramaty, R., Kozlovsky, B., and Lingenfelter, R. E.: 1975, Space Sci. Rev. 18, 341.

Ramaty, R., Kozlovsky, B., and Lingenfelter, R. E.: 1979, Astrophys. J. Suppl. 40, 487.

Ramaty, R., Kozlovsky, B., and Suri, A. N.: 1977, Ap. J. 214, 617.

Ramaty, R., Lingenfelter, R. E., and Kozlovsky, B.: 1982, in R. E. Lingenfelter et al. (eds) Gamma-Ray Transients and Related Astrophysical Phenomena, AIP, New York, p. 211.

Share, G. H. et al.: 1980, Bull. Amer. Astr. Soc. 12, 891.

Von Rosenvinge, T. T., Ramaty, R., and Reames, D. V.: 1981, 17th Internat. Cosmic Ray Conference Papers, Paris 3, 28.

Yoshimori, M. et al.: 1983, this volume.

Wang, H. T., and Ramaty, R.: 1974, Solar Phys. 36, 129.

Zweibel, E. G. and Haber, D.: 1983, Astrophys J., (in press).

DISCUSSION

PRIEST: Do you really think the high energy particles are produced by stochastic Fermi acceleration or would you put your money elsewhere?

RAMATY: The gamma ray observations cannot, at present, prove or disprove whether solar flare ion acceleration is due to the stochastic Fermi acceleration or not. However, all of the data is consistent with this mechanism, including the very rapid examples of gamma ray emission. Rapid stochastic acceleration requires a sufficiently rapid velocity of scatterers and a sufficiently short scattering mean free path.

KANE: Can your stochastic process accelerate both GeV ions and electrons?

RAMATY: Stochastic Fermi acceleration is probably too slow to accelerate relativistic (10 MeV) electrons in solar flares. Therefore, I think that, in the June 21, 1980 flare, the gamma ray continuum > 10 MeV was due to nuclear reactions via π° and $\pi^\pm$ decay. Stochastic acceleration, however, may be important for electrons $\lesssim$ 1 MeV.

Note added in proof: (R. Ramaty): The above conclusion on the origin of the >10 MeV photons from the June 21, 1980 flare is different from that arrived at in the paper by Ramaty et al. in this volume. The conclusion of the paper is based on a much more detailed analysis than the comment made orally at the time of the Seminar.

CHARACTERISTICS OF GAMMA-RAY LINE FLARES AS OBSERVED IN HARD X-RAY EMISSIONS AND OTHER PHENOMENA

Taeil Bai
Institute for Plasma Research, Stanford University,
Stanford, Calif., U.S.A.
and
B. R. Dennis, A. L. Kiplinger, L. E. Orwig, and K. J. Frost
Goddard Space Flight Center, Greenbelt, Code 684, MD 20771

ABSTRACT. **Observations of gamma-ray lines from solar flares by SMM demonstrated that energetic protons and heavy ions are accelerated during the impulsive phase. In order to understand the acceleration mechanism for gamma-ray producing protons and heavy ions, we have studied the characteristics of the flares from which gamma-ray lines were observed by SMM. In order to identify the characteristics unique to the gamma-ray line flares, we have also studied intense hard X-ray flares with no gamma-ray line emissions. We have found the following characteristics: 1) Most of the gamma-ray line flares produced intense radio bursts of types II and IV. 2) For most of the gamma-ray line flares, the time profiles of high-energy ($\gtrsim$ 300 keV) hard X-rays are delayed by order of several seconds with respect to those of low-energy hard X-rays. The delay times seem to be correlated with the spatial sizes of the flares. 3) In Hα importance, the gamma-ray line flares range from sub-flares to importance-3 flares. 4) The hard X-ray spectra of the gamma-ray line flares are generally flatter (harder) than those of flares with no gamma-ray line emission.**

From these characteristics, we conclude that the first-order Fermi acceleration operating in a flare loop is likely to be the acceleration mechanism for energetic protons and heavy ions as well as relativistic electrons.

1. INTRODUCTION

Energetic protons and heavy ions accelerated in solar flares produce nuclear gamma-ray lines by interacting in the solar atmosphere (Lingenfelter and Ramaty, 1967; Ramaty, Kozlovsky, and Lingenfelter, 1975). Therefore, by observing solar gamma rays we can get information on the acceleration of these particles in solar flares. The gamma-ray detector was installed on the Solar Maximum Mission (SMM) just to do that. SMM has so far observed gamma-ray lines from more than ten flares (Rieger, 1982), exceeding our expectation.

Solar Physics **86** (1983) 409–419. 0038–0938/83/0862–0409 $ 01.65.

One of the important results emerging from these observations is that protons and heavy ions are accelerated rapidly during the impulsive phase (Forrest et al., 1981; Chupp, 1982; Chupp et al., 1982). The fact that gamma-ray lines are produced during the impulsive phase was known from earlier gamma-ray observations (Chupp et al., 1973; Hudson et al., 1980), but the rapidity of the acceleration is a discovery made by SMM. At least in some flares, the characteristic ion acceleration time seems to be as short as a few seconds (e.g., Forrest et al., 1981; Chupp et al., 1982).

Until recently, a widely accepted view on ion acceleration in solar flares is that protons and heavy ions as well as relativistic electrons are accelerated during the second phase of a flare, which is manifested by radio bursts of types II and IV observed several minutes later than the impulsive phase (Wild, Smerd, and Weiss, 1963; de Jager, 1969; Svestka, 1976). This view was obtained from associations of energetic protons escaped into the interplanetary medium and radio bursts of types II and IV. However, recent gamma-ray observations require development of new ideas and interpretations. Based on these new observations, Bai (1982) proposed that the first-order Fermi acceleration by shocks passing through a closed flare loop is a likely mechanism for the rapid ion acceleration. However, Chupp et al. (1982) proposed that a single acceleration mechanism accelerates everything--nonrelativistic electrons producing impulsive hard X-rays, reltivistic electrons, and gamma-ray producing protons and heavy ions.

The purpose of this paper is to study the characteristics of the SMM gamma-ray line flares as observed in hard X-ray emissions and other phenomena and to deduce a likely acceleration mechanism.

2. CHARACTERISTICS OF GAMMA-RAY LINE FLARES

A. Identification of Gamma-Ray Line Flares

Some of the flares from which SMM observed gamma-ray lines are discussed in several papers (e.g., Forrest et al., 1981; Chupp, 1982; Chupp et al., 1981; Chupp et al., 1982). However, a complete list of gamma-ray line flares observed by SMM until the end of October 1981 is found in Rieger (1982). In Figure 3 of Rieger (1982), the detection of gamma-ray lines is indicated by horizontal arrows. In addition to the ten gamma-ray line flares found here, we add two flares that occurred on October 7 and 14, 1981, as gamma-ray line flares. These two flares produced gamma-ray continua extending to 10 and 30 MeV, respectively, and showed other characteristics of gamma-ray line flares. We reasoned that the 2.2 MeV line from these flares could have been attenuated below the detectable level because of their locations at the limb. We learned at this very meeting (the U.S.-Japan joint seminar) that gamma-ray lines had indeed been detected from these flares by Hinotori (Yoshimori et al., 1983).

In this study we include only the gamma-ray line flares observed by SMM, because the characteristics appearing in hard X-ray emission are important as will be shown later, and SMM provides good quality hard X-ray data easily available to us. In this study we do not regard as gamma-ray line flares the flares from which SMM did not detect gamma-ray lines but detected the 4 ~ 7 meV continuum excess, which is interpreted to be of nuclear origin (see Chupp, 1982; Ramaty, Lingenfelter, and Kozlovsky, 1982). The reason is as follows. Protons and heavy ions with relatively low energy (4 ~ 10 MeV/nucleon) can produce gamma rays in the 4 - 7 MeV range; but in order to produce the 2.2 MeV line, relatively high-energy (several tens of MeV/nucleon) particles are necessary. Therefore, the fluence ratio between the 2.2-MeV line and the 4 ~ 7 MeV continuum is a measure of the hardness of the energy spectrum of protons or heavy ions. For most gamma-ray line flares occurring in the disk, this ratio is about unity (Chupp, 1982; Ramaty, 1982), which indicates rather similar proton spectra for these flares. Thus, the flares which produced detectable fluxes in the 4 ~ 7 MeV range but not in the 2.2 MeV line must have accelerated protons and heavy ions only to low energies (mostly up to 10 MeV/nucleon). For this reason, we do not regard these flares as gamma-ray line flares in our study.

B. Observed Characteristics

First of all, emphasis of our study was laid upon the characteristics as appeared in hard X-ray emissions, for the following reason. A large fraction of flare energy seems to go into the electron acceleration, and according to many models the acceleration of protons is a secondary phenomenon resulting from the energy deposition (or the acceleration itself) of the energetic electrons. Therefore, the characteristics appearing in hard X-ray emission may provide important clues to the proton acceleration mechanism. We first tried to find the delay of high-energy hard X-rays, the observation of which from the August 1972 flares was the basis for the original proposal for the "second-step" acceleration (Bai and Ramaty, 1979). We then studied the hard X-ray spectra.

We studied the association of radio bursts of types II and IV. Earlier studies (e.g., Wild, Smerd, and Weiss, 1963; Svestka and Fritzova-Svestkova, 1974) showed good association of proton flares with these two types of radio bursts.

We also studied Hα importance and the maximum flux density of microwaves observed at about 10 GHz, where the coverage is reasonably good.

Figure 1 shows the number of hard X-ray events versus the HXRBS peak rate. The HXRBS peak rate is the peak total counting rate observed by HXRBS in all the energy channels from 30 ~ 500 keV (Dennis et al., 1982). The shaded portion indicates the gamma-ray line flares.

As can be seen, intense hard X-ray flares are more likely to be gamma-ray line flares. Conversely, gamma-ray line flares are generally intense hard X-ray flares. Therefore, we studied intense hard X-ray flares in order to find out the characteristics unique to the gamma-ray line flares.

The results of our study are tabulated in the tables: results from gamma-ray line flares in Table I; non-gamma-ray line flares, Table II. The following characteristics of gamma-ray line flares are self-evident from these tables.

1. Emission of radio bursts of types II and IV is one of the characteristics common to all the gamma-ray line flares with few exceptions, whereas only one among non-gamma-ray line flares produced types II and IV radio bursts.

2. The delay of high-energy hard X-rays is another characteristic of gamma-ray line flares. The delay is found from almost all the gamma-ray line flares, but it is found from only one non-gamma-ray line flare. Note here the finding by Gardener et al. (1981) that for most gamma-ray line flares the 4 ~ 7 MeV gamma-ray time profile is delayed with respect to the hard X-ray (~ 80 keV) time profile.

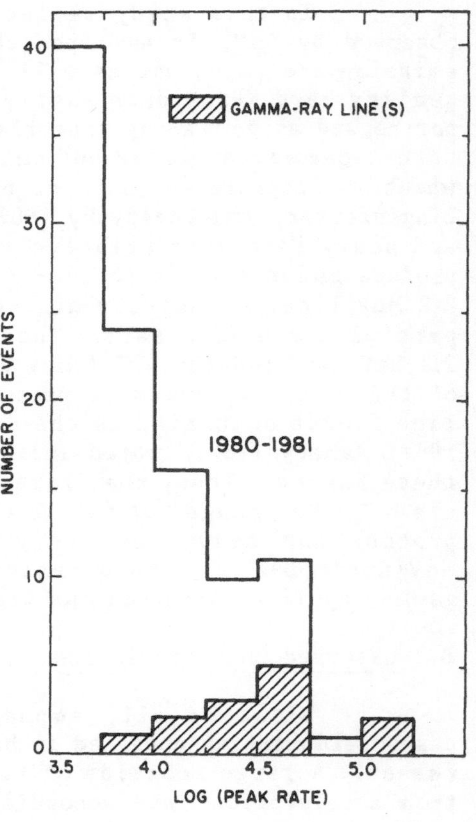

Figure 1. Number of hard X-ray events versus HXRBS count rate. The horizontal axis represents the peak count rate measured in the 30 ~ 500 keV range by HXRBS. From the shaded area, which represents the gamma-ray line flares, we can see that gamma-ray line flares are very intense hard X-ray flares.

3. Gamma-ray line flares come in all sizes of Hα importance. The Hα importance of gamma-ray line flares, however, seems to be smaller, on the average, than the Hα importance of proton flares (flares associated with interplanetary energetic protons). Less than 10 percent of proton flares (4 out of 52) are subflares (Kahler, 1982), whereas 3 out of 12 gamma-ray line flares are subflares.

4. Gamma-ray line flares show flatter (harder) hard X-ray spectra than non-gamma-ray line flares. For gamma-ray line flares,

power-law spectra extend to the highest energy channel (to 500 keV), with occasional high-energy flattening. On the other hand, for non-gamma-ray lines flares, hard X-ray spectra generally show high-energy steepening, and even in cases when a single power-law spectrum fits well to high-energy data, the spectrum is steeper than those of gamma-ray line flares. This finding is evident in Figure 2.

Upon inspecting Table I closely, we find that there is some correlation between the spatial size of the flare and the delay time. Sub-flares (1982 June 7, 1980 July 1, 1980 Feb. 26) exhibit either small delays (< 3 s) or no delay. On the other hand, importance-3 flares (1981 Apr. 1, 10) exhibit relatively large delays (~ 10 s).

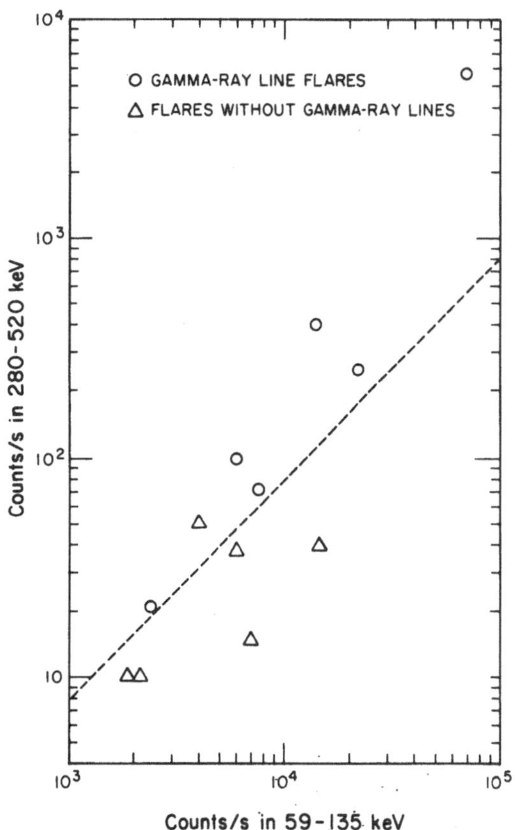

Figure 2a. Hardness of hard X-ray spectra of very intense X-ray flares. Gamma-ray line flares (O) show harder spectra than flares without gamma-ray lines (X).

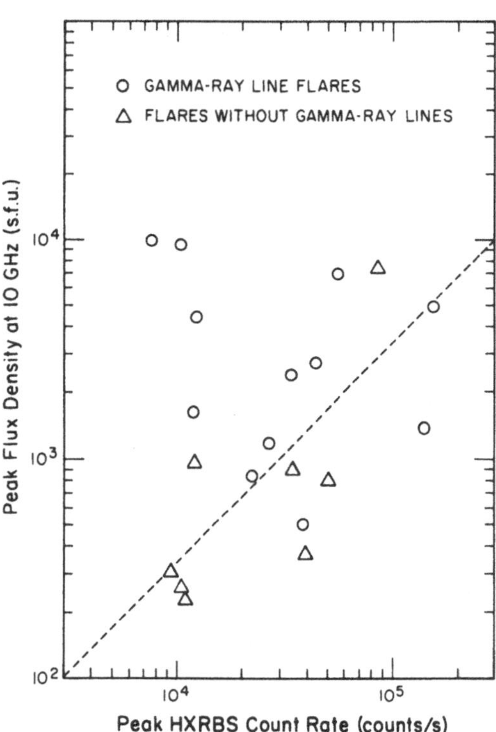

Figure 2b. Microwave flux density at 10 GHz versus peak HXRBS rate. Since 10 GHz microwaves are produced by relativistic electrons, this figure also shows that the electron spectra of gamma-ray line flares are harder.

TABLE I: SMM γ-Ray Line Flares

Date	HXST	T	Hα	Location	Radio Bursts II	Radio Bursts IV	Delay (s)	Hard X-Ray Spectral Index
6/07/80	0312:10	60	SN	N14W70	2	1	x	2.8
6/21/80	0118:20	60	1B	N17W91	2	3	x	2.0
7/01/80	1626:50	180	SB	S12W38	3	2	2.5	3.4
11/06/80	0344:40	>420	2B	S12E72	3	3	2.0	3.9
2/26/81	1424:40	120	SB	S13E53	-	-	0.5	3.6
4/01/81	0134:00	1200	3B	S43W52	3	2	10	3.6
4/10/81	1626:20	>600	3B	N03W38	3	2	8	3.7
4/26/81	1144:20	>660	2B	N12W74	-	3	120	3.7
4/27/81	0805:00	1320	2B	N16W90	2	2	12	3.9 (H)
7/19/81	0533:20	280	1B	S29W56	2	1	7	4.4 (H)
10/07/81	2255:40	>480	1B	S13E90	2	-	9	3.5 (H)
10/14/81	1705:30	>130	1B	S06E90	2	2	2.5	3.1

TABLE II: Big Hard X-Ray Flares with no Gamma-Ray Lines

Date	HXST	T	Hα	Location	Radio Bursts II	Radio Bursts IV	Delay (s)	Hard X-Ray Spectral Index
4/28/80	2039:54	10	-	-	-	-	x	4.5 (S)
5/09/80	0712:00	210	1B	S20W35	-	-	x	4.0 (S)
6/04/80	0654:10	60	SB	S14E59	-	-	x	4.2
6/29/80	1041:30	70	-	-	-	-	x	4.0
7/21/80	0255:50	120	SN*	S15W60	-	-	2	3.5
10/14/80	0605:20	480	3B	S09W07	1	1	x	6.5 (S)
11/07/80	0203:20	240	2B	N07W11	-	-	x	4.5
11/12/80	0448:20	360	3B	N10W72	-	-	x	4.5 (H)

*Late observation at 0307 UT.

For limb flares Hα importance is not a good indication of the flare size, but *Hinotori* hard X-ray images are available for two of the limb flares (1981 Apr. 27, 1981 Oct. 7) in Table I. From these two flares *Hinotori* found very high (> 10^9 cm) hard X-ray sources, which probably are parts of high loops (Ohki *et al*. 1983). These two flares of large spatial size also exhibit large delays (~ 10 s). The 1981 April 26 flare exhibits a very large delay (~ 120 s). Even though this was only a 2B flare, it showed characteristics of high coronal X-ray sources. First, its hard X-ray time profiles show gradual rise and gradual fall with no apparent rapid variations, and the hard X-ray emission lasted for ~ 20 minutes. These are exactly the characteristics of hard X-ray emission from the high corona (Frost and Dennis 1971; Hudson 1978; Hudson, Lin, and Stewart 1982). Second, the large delay of high-energy hard X-rays, whether it is due to collisional losses or due to the second-step acceleration, is compatible only with low ambient densities ($\lesssim 10^9$ cm^{-3}) (cf. Bai and Ramaty 1979). Therefore, the 1981 April 26 flare seems to fit in the picture. Anyway, this flare deserves a more detailed study.

Notes to Tables I and II:

HXST: Hard X-ray start time. Time when the flux in the 135-218 keV energy band began to exceed one-tenth of the maximum flux.

T: Duration between the hard X-ray start time and the hard X-ray end time.

For Hα importance, location, and types II and IV radio bursts, *Solar-Geophysical Data* and Rust *et al*. (1982) have been used. The numbers in the radio-bursts column indicate intensity classes following *SGD*:

1: <50 solar flux unit.
2: 50-500 sfu.
3: >500 sfu (20^{-22} W m^{-2} Hz^{-1}).
Short horizontal bars in this column indicate no report in SGD.

Delay: Delay of the hard X-ray time profile of the highest energy band (317-531 keV) with respect to the time profile of the lowest energy band (30-59 keV). In some cases the time profiles of the low-energy bands are flooded by a slowly varying low-energy component; in such cases the delay is measured with respect to the lowest energy band that is not flooded by the low-energy component. In most flares hard X-ray time profiles show multiple spikes (or bursts), and the delay time for one spike is not necessarily the same as the delay time for another. The delay times in these tables are "typical" delay times. The x means no noticeable delay.

Spectrum: (H) = high-energy hardening; (S) = high-energy softening.

3. DISCUSSION AND CONCLUSION

From the first characteristic, we can conclude that the acceleration of ions must be due to shocks, either directly or indirectly. Radio bursts of types II and IV are believed to be produced in the high corona by shocks (Wild, Smerd, and Weiss 1963; Svestka 1976). From small delays of gamma rays, we can conclude that ion acceleration must be carried out by shocks while they are still in the flare loop(s). From the second and fourth characteristics (delay of high-energy hard X-rays; high-energy spectral flattening), we can conclude that the same acceleration mechanism might accelerate both relativistic electrons and protons. It is well known that Fermi acceleration is efficient for accelerating energetic ions and relativistic electrons but inefficient for accelerating low-energy particles (e.g., Ginzburg and Syrovatskii 1964).

The second-order Fermi acceleration is due to random scatterings, and the first-order Fermi acceleration is due to systematic scatterings (Fermi 1949; 1954). Because a shock front propagating in a flare loop provides systematic scatterings, we arrive at the logical conclusion that the first-order Fermi acceleration in closed flare loops is a likely acceleration mechanism for ions and relativistic electrons. The acceleration rate of this mechanism is given by (Bai *et al.* 1983)

$$\frac{dE}{dt} \simeq 4\langle\mu^2\rangle \left(\frac{V_s}{d}\right) E \quad , \tag{1}$$

where $\langle\mu^2\rangle^{1/2}$ is the rms pitch angle, E energy, V_s the shock speeed, and d the half-loop length. Bai *et al.* (1983) showed that for reasonable parameters this acceleration mechanism can accelerate protons and relativistic electrons rapidly enough to meet the observed constraints.

Bai *et al.* (1983) proposed that acceleration was due to scattering at the shock front (see Wentzel 1963, 1964; Jokipii 1966). However, particles passing through the shock front can be accelerated due to scatterings with irregularities in the upstream and downstream (e.g., Blandford and Ostriker 1978; Lee and Fisk 1982 and references therein). This is also a first-order Fermi acceleration. The relative importance of these two mechanisms is a subject of a future study. Another acceleration mechanism associated with shocks is $\vec{V} \times \vec{B}$ acceleration (Pesses *et al.* 1981), which is effective in perpendicular shocks.

As mentioned earlier, the Hα importance of gamma-ray line flares is smaller on the average than the Hα importance of proton flares. Furthermore, only a small fraction ($10^{-3} \sim 10^{-2}$) of protons accelerated in gamma-ray line flares escape into the interplanetary medium (von Rosenvinge, Ramaty, and Reames 1981; Pesses et al. 1981). Therefore, it seems that protons found in the interplanetary medium are accelerated by a diffferent mechanism than gamma-ray producing protons. Interplanetary protons are probably accelerated in the open-field configuration in the high corona.

Research at Stanford was supported by NASA Grant NGL 05-020-272. Travel grants for attending this meeting were provided by NSF. We appreciate discussions with Dr. D. J. Forrest.

REFERENCES

Bai, T.: 1982, in Gamma-Ray Transients and Related Astrophysical Phenomena, R. E. Lingenfelter, H. S. Hudson, and D. M. Worrall (ed.), (New York: AIP), p. 409.

Bai, T., Hudson, H. S., Pelling, R. M., Lin, R. P., Schwartz, R. A., and von Rosenvinge, T. T.: 1983, Ap. J. (in press).

Bai, T., and Ramaty, R.: 1979, Ap. J. **227**, 1072.

Blandford, R. D., and Ostriker, J. P.: 1978, Ap. J. (Letters) **221**, L29.

Chupp, E. L.: 1982, in Gamma-Ray Transients and Related Astrophysical Phenomena, R. E. Lingenfelter, H. S. Hudson, and D. M. Worrall (ed.), (New York: AIP), p. 363.

Chupp, E. L. et al.: 1981, Ap. J. (Letters) **244**, L167.

Chupp, E. L, Forrest, D. J., Higbie, P. R., Suri, A. N., Tsai, C., and Dunphy, P. P.: 1973, Nature **241**, 333.

Chupp, E. L. et al.: 1982, Ap. J. (Letters) (in press).

de Jager, C.: 1969, in COSPAR Symp. on Solar Flares and Space Sci. Res., C. de Jager and Z. Svestka (ed.), (Amsterdam: North Holland), p. 1.

Dennis, B. R. et al.: 1982, NASA TM 83925.

Fermi, E.: 1949, Phys. Rev. **75**, 1169.

Fermi, E.: 1954, Ap. J. **119**, 1.

Forrest, D. J. et al.: 1981, in Proc. 17th Intl. Cosmic Ray Conf. (Paris), Vol. **10**, p. 5.

Frost, K. J., and Dennis, B. R.: 1971, Ap. J. **165**, 655.

Gardener, B. M. et al.: 1981, Bull. AAS **13**, 903.

Ginzburg, V. L, and Syrovatskii, S. I.: 1964, The Origin of Cosmic Rays (New York: MacMillan).

Hudson, H. S.: 1978, Ap. J. **224**, 235.

Hudson, H. S., Bai, T., Gruber, D. E., Matteson, J. L., Nolan, P. L., and Peterson, L. E.: 1980, Ap. J. (Letters) **236**, L91.

Hudson, H. S., Lin, R. P., and Stewart, R. T.: 1982, Solar Phys. **75**, 245.

Jokipii, J. R.: 1966, Ap. J. **143**, 961.

Kahler, S. W.: 1982, Ap. J. **261**, 710.

Lee, M. A., and Fisk, L. A.: 1981, in Proc. 17th Intl. Cosmic Ray Conf. (Paris), Vol. **3**, p. 405.

Lingenfelter, R. E., and Ramaty, R.: 1967, in High Energy Nuclear Reactions in Astrophysics, B. S. P. Shen (ed.), (New York: Benjamin), p. 99.

Ohki, K. et al.:1983, in this volume.

Pesses, M. E., Tsuratani, B. T., van Allen, J. A., and Smith, E. J.: 1979, J. Geophys. Res. **84**, 7297.

Pesses, M. E., Klecker, B., Gloeckler, G., and Hovestadt, D.:1981, in Proc. 17th Intl. Cosmic Ray Conf. (Paris), Vol. **3**, p. 36.

Ramaty, R.:1983, in The Physics of the Sun, T. E. Holzer, D. Mihalas, P. A. Sturrock, and R. K. Ulrich (ed.), to be published.

Ramaty, R., Kozlovsky, B., and Lingenfelter, R. E.:1975, Space Sci. Rev. **18**, 341.

Ramaty, R., Lingenfelter, R. E., and Kozlovsky, B.:1982, in Gamma-Ray Transients and Related Astrophysical Phenomena, R. E. Lingenfelter, H. S. Hudson, and D. M. Worrall (ed.), (New York: AIP), p. 211.

Rieger, E.: 1982, in Hinotori Symp. on Solar Flares (Tokyo: Inst. Spac. Astronautical Sci.), p. 246.

Rust, D. M., Nelson, J. J., Pryor, L. H., Frank, Z. A., and Boggess, A. M.:1982, preprint.

Svestka, Z.: 1976, Solar Flares (Dordrecht: Reidel), Chap. VI.

Svestka, Z., and Fritzova-Svestkova, L.:1974, Solar Phys. **36**, 417.

von Rosenvinge, T. T., Ramaty, R., and Reames, D. V.:1981, in Proc. 17th Intl. Cosmic Ray Conf. (Paris), Vol. **3**, p. 28.

Wentzel, D. G.:1963, Ap. J. **137**, 135.

Wentzel, D. G.:1964, Ap. J. **140**, 1613.

Wild, J. P., Smerd, S. F., and Weiss, A. A.:1963, Ann. Rev. Astron. Astrophys. **1**, 291.

Yoshimori, M. et al.:1983, in this volume.

DISCUSSION

KAI: You have obtained a good correlation between the X-ray production and type II bursts. It seems to me rather surprising. For example, a type II burst starting after the impulsive burst had completely faded for the 7 June 1980 event.

BAI: Type II bursts typically occur several minutes later than the impulsive phase. But the shocks which we believe are responsible for type II bursts must originate at the time of impulsive energy release. I think that the shocks accelerate protons and relativistic electrons while they are still in the impulsive flare loop. When these shocks propagate to the high corona in a few minutes, they produce type II bursts.

BRUECKNER: How is the Fermi accelerating mechanism dependent on the density and velocity in the shock front? Type II bursts, according to my estimates, are connected with initial densities as high as 10^{10} cm^{-3}; this is before the type II burst becomes visible at the 100 MHz level. However, according to my estimates, supersonic shocks are expected already 20 seconds after flare onset (see my paper in this volume).

BAI: The density of the medium has no direct influence on the Fermi acceleration, but the collisional energy loss rate is proportional to the density. Therefore, the higher the density, the higher the "injection energy" for the Fermi acceleration. Your interpretation that shocks producing type II burst start early is consistent with what I have discussed.

YOSHIMORI: What are the hard X-ray energy exponents for the non-delayed class of events? Is there any difference between the energy exponents for non-delayed and delayed classes of events?

BAI: Please refer to Tables I and II.

EMSLIE: How does your proposed Fermi acceleration model explain the correlation you found between high-energy hard X-ray delay and hard X-ray burst duration?

BAI: The correlation between the hard X-ray delay and the burst duration was discussed in my talk but is omitted in the text due to lack of space. I think the common link is the size of the loop. As discussed in the text, the larger the loop, the larger the delay. Then, the implied correlation between the size of the loop and the burst duration might be because it takes a longer time for a larger loop to release energy.

NARROW-BAND DECIMETER BURSTS AND X-RAY EMISSIONS - POSSIBLE EVIDENCE OF NEGATIVE ABSORPTION OR MASER EFFECT

Shinzo Enome
Toyokawa Observatory, Research Institute of Atmospherics, Nagoya University, Toyokawa 442, Japan

ABSTRACT. Bursts with peculiar time evolution and narrow-band microwave spectra are studied in detail based on X-ray data of HINOTORI and microwave data of Toyokawa Observatory. Discussions are given for emission mechanism and burst scenario.

1. INTRODUCTION

During the active period of HINOTORI, a Japanese satellite for solar studies, more than 600 solar flares were observed by HXM (Hard X-ray Monitor) and FLM (FLare Monitor) of that satellite (Enome, 1983; Tanaka, 1983). Most of these events, though we can not give an exact number of percentage here, are impulsive, and gradual and intense extended events. A good coverage of these intense extended events by HINOTORI can be seen in preliminary results of analyses in the proceedings of HINOTORI Symposium held in Tokyo in January, 1982, and in the present proceedings of US-Japan Seminar on Recent Advances of the Understanding in Solar Flares, where variety of developments and types of flares is evidenced.

Impulsive bursts, which are most frequently observed and extensively studied, are always associated with hard X-ray spikes and microwave spikes with a smooth soft X-ray emission delayed from the impulsive components by a few minutes. Besides these typical events there are small number of events which have peculiar scheme of time evolution and are sometimes associated with narrow-band decimeter bursts.

In this paper some of these peculiar events are studied in detail and physical processes involved are discussed and suggested.

2. OBSERVATIONS

To illustrate time histories of hard X-rays, soft X-rays, and

Solar Physics **86** (1983) 421–426. 0038–0938/83/0862–0421 $ 00.90.

microwaves, HINOTORI quick-look profiles in two channels in hard X-rays and two channels in soft X-rays, and also radio profiles of Toyokawa Observatory at four frequencies from 1 GHz to 9.4 GHz with time axis adjusted to HINOTORI quick-look profile are used.

The first example of simple impulsive event is shown in Figure 1, which occurred on October 15, 1981 and started at 0443 UT and peaked at 0444 UT. The microwave spectrum of this burst has positive slope as

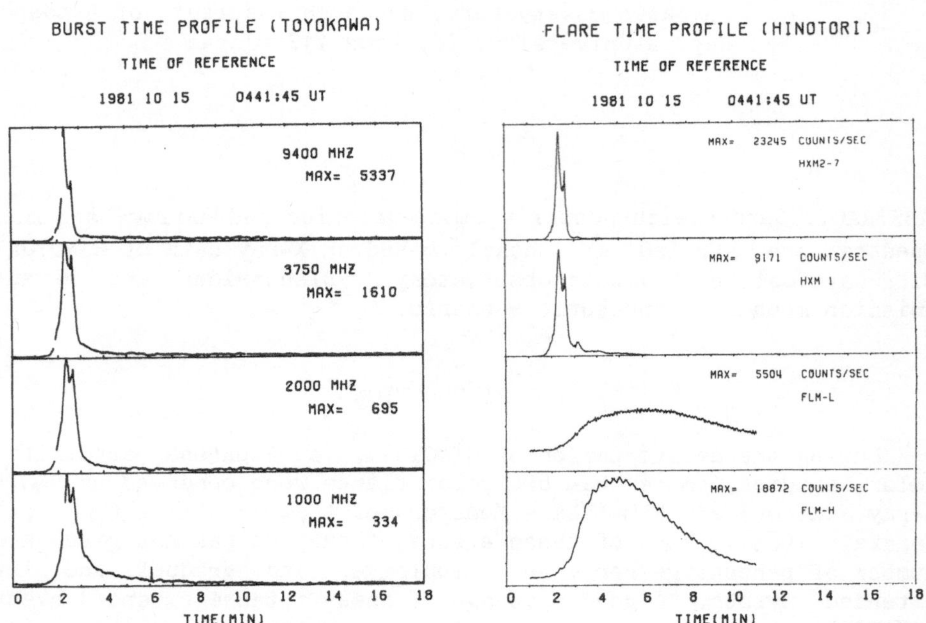

Figure 1. Time profiles of the burst of 0444 UT on October 15, 1981 are shown in microwaves at 9.4, 3.75, 2, and 1 GHz from the top to the bottom of the left-hand side sub-frames, and in hard X-rays at energies above 40 keV for HXM channel 2 - 7, at energies between 17 to 40 keV, and in soft X-rays at energies between 2 and 5 keV for FLM low channel and at energies between 5 and 15 keV for FLM high channel from the top to the bottom of the right-hand side sub-frames.

a function of frequency higher than 9.4 GHz. The time profile of the HXM of channel 2 through channel 7, which cover photon energies above 40 keV, resembles that of 9.4 GHz very much, which suggests common origin of radiation. The HXM channel 1 profile, which corresponds to photons of 17 to 40 keV, is similar to 3.75 GHz profile with respect to the main peak, secondary peak and small enhancement in the late decay phase. Soft X-ray emissions of FLM in low and high energy channel are shown on the third and the lowest sub-frame of the HINOTORI quick-look plot. Both time profiles show gradual and smooth evolution and development of soft X-ray component, from which one can derive emission measure and temperature under isothermal assumption (Watanabe et al., 1983). Another example is shown in Figure 2 of an impulsive burst with more complex time structure which is composed of many spikes. General correspondence between hard X-rays and microwaves is also very good in this case except for 1 GHz time profile. This burst occurred on October 15, 1981 at 0553 UT. Soft X-ray emissions

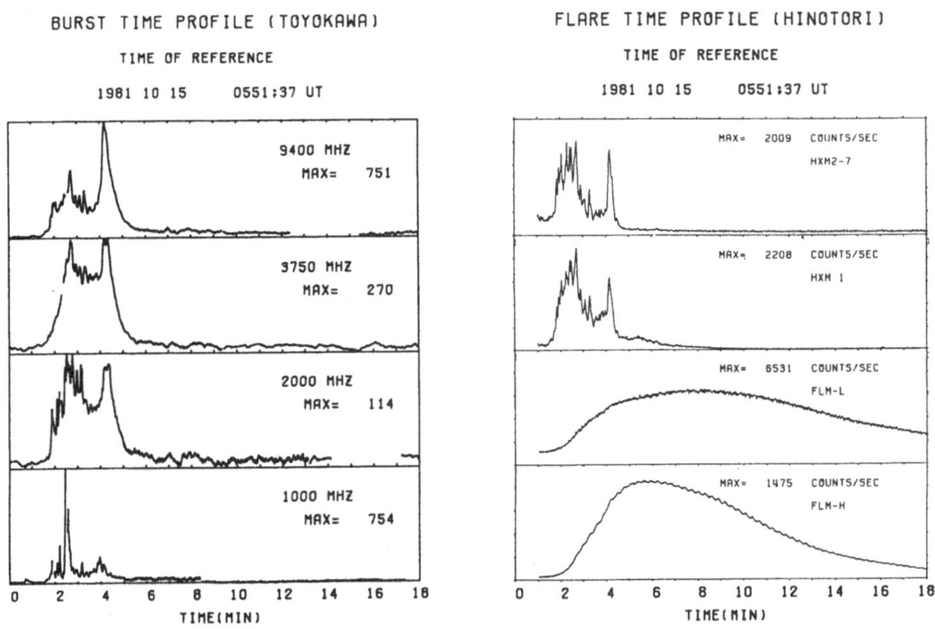

Figure 2. Time profiles of the burst of 0553 UT on October 15, 1981 are shown, where configuration of sub-frames and their content are the same as in Figure 1.

are again very gradual and smooth. These two examples are typical impulsive bursts with impulsive spike of hard X-rays and microwaves which corresponds to non-thermal component of flare plasma, and also delayed gradual and smooth soft X-ray emission which corresponds to thermal component of flare plasma.

In Figures 3 and 4 are shown two examples of unusual events. The third event also occurred on October 15, 1981. This burst is composed of two parts. The first spike started at 0638 UT with duration of about 1 minute in hard X-rays. The second spike started at 0641.5, whose intensity is about one tenth of the first spike. Microwave spectrum of the first spike is broad band with its peak around 4 GHz. The spectrum of the second spike, on the other hand, is extremely strong only at 1 GHz, which is quite different from the first spike. The time profile of the 1 GHz second peak is composed of many short duration sub-spikes, whose duration is of the order of 10 seconds. These spectral and temporal structures are characteristics of narrow-band decimeter bursts. Soft X-ray emissions have also two

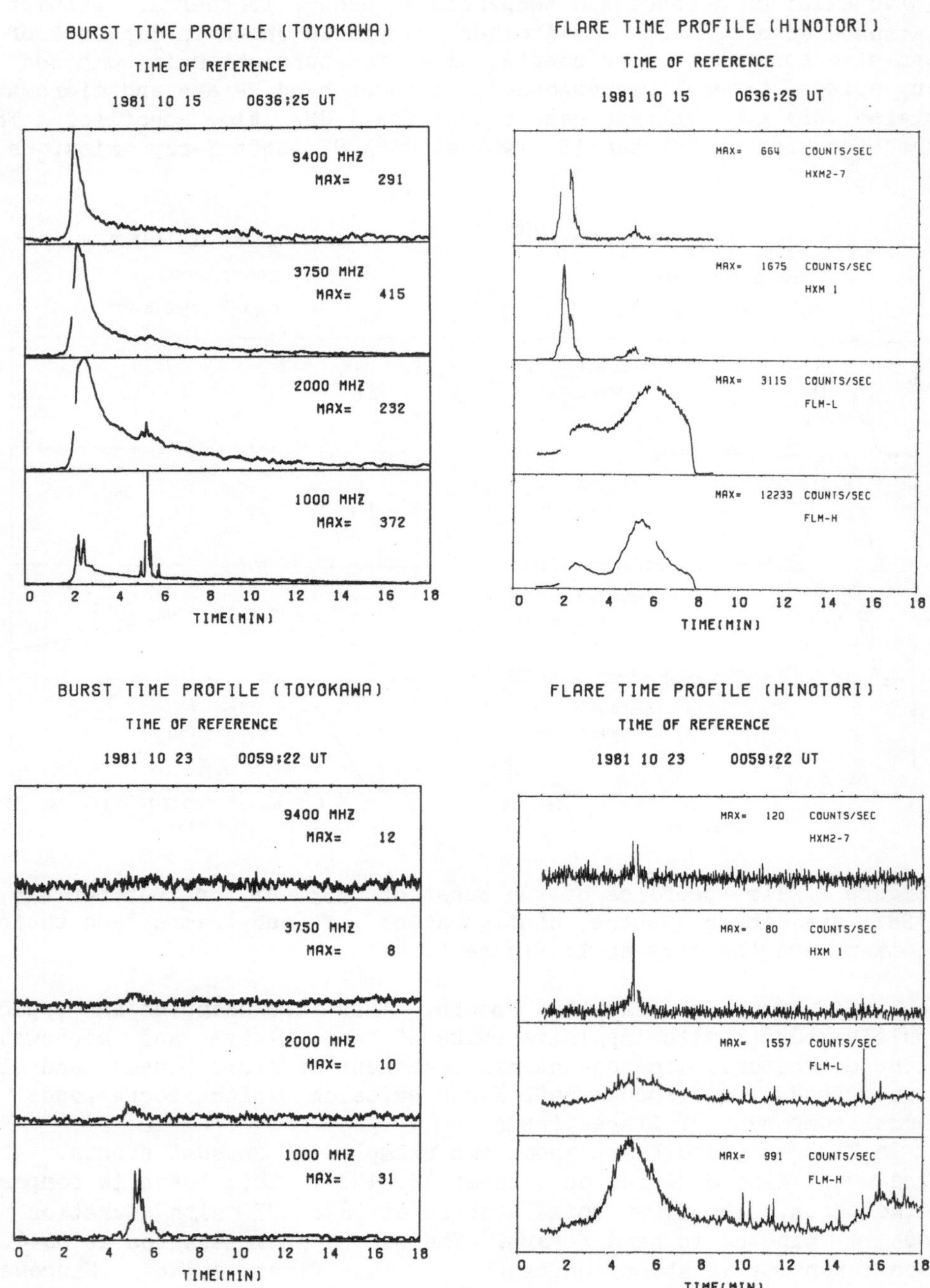

Figures 3 and 4. Time profiles of the bursts of 0603 UT on October 15 and of 0104 UT on October 23, 1981 are shown as in Figures 1 and 2.

gradual enhancements, however, the second soft X-ray enhancement starts much earlier than the second spike of X-ray, and which is not the standard relation of hard X-ray and microwave spikes to gradual soft X-ray emissions that appreciable increase of soft X-ray emission is just after the onset of the impulsive spike. In this case, the second spike started when the soft X-ray emission was more than half the maximum value. The fourth burst has similar relation between soft X-ray enhancement and spike of hard X-ray and microwaves. The latter spikes, in this case, corresponded in time to the peak of the soft X-ray emission. The microwave spectrum is also narrow-band around 1 GHz.

3. INTERPRETATION

A possible interpretation of these events is given below in terms of maser effect or negative absorption (Melrose and Dulk, 1982b). They have shown that even in non-relativistic case a maser effect may be driven by loss cone anisotropy, and growth rate of the maser amplification is given explicitly. Based on this maser mechanism Melrose and Dulk (1982a) proposed a new explanation of evolutional sequence of X-ray, EUV, and decimeter bursts. Here we propose somewhat different scenario to account for these peculiar events as explained above, in which suggestion is obtained from another model by Sprangle and Vlahos (1982) to give explanation for secondary acceleration of electrons outside a flaring loop.

Our description of the events is as follows:

i) Electrons and probably protons are accelerated by a very rapid and effective plasma process such as coalescence effect (Tajima et al, 1982) to energies of a few hundred keV and MeV respectively; When protons are accelerated above tens of MeV, there will be a prompt line gamma rays to be detected in impulsive phase (Forrest et al, 1981); It should be emphasized that during impulsive acceleration process strong turbulence of 250 km/s is observed by SOX onboard HINOTORI (Tanaka et al. 1982), which has to be considered in constructing an acceleration model;

ii) These electrons will emit hard X-rays and broad-band microwaves by bremsstrahlung and gyro-synchrotron radiation as observed in impulsive phase;

iii) Non-thermal electrons in impulsive phase will be trapped and loose energy by electron-electron collisions and thermalized to emit the first gradual soft X-ray radiation;

iv) MeV-order protons will also be trapped in a magnetic loop and thermalized by protons-electron collisions, whose relaxation time is 43 times or square root of proton-electron mass ratio times; This thermal component of flare plasma is seen in secondary enhancement of soft X-rays, delayed from the impulsive phase by 5 minutes in the case of the third event;

v) The thermal electons, possibly heated by non-thermal protons as explained by longer relaxation time, will have loss-cone instability, which can excite by maser effect narrow-band decimeter radiations as

observed at the maximum peak of the secondary soft X-ray enhancement;
vi) This narrow-band radiation then can accelerate electrons outside a flaring loop (Spranger and Vlahos, 1982), which are evidenced by small enhancement of hard X-ray radiation synchronized with narrow-band decimeter emission;
vii) In the case of the fourth flare where no impulsive phase is observed, a possible explanation is that no electrons are accelerated in impulsive phase;
viii) If narrow-band radiation is generated within the framework as explained in v), 10-second life time of the decimeter radiation is interpreted that loss-cone instability is maintained during the corresponding life time, from which density of the ambient gas within a magnetic loop is deduced to be 6 X 10^8 cm^{-3}.

4. CONCLUDING REMARKS

Relations between narrow-band decimeter bursts and X-ray emissions were first examined in this study. Although the present investigation is limited within a qualitative regime, a possible evidence of maser effect or negative absorption is proposed to explain the observed sequence of radio-X-ray phenomena.

Finally it is somewhat serious difficulty to be considered that narrow-band decimeter radiations observed in the third and fourth flares are not polarized. If maser effect is in operation in the relevant events, microwave radiation has to be nearly 100 % polarized. The only existing mechanism, which can depolarized circular radiation, is mode coupling (Cohen, 1958).

The author would like to express his sincere thanks to each member of HINOTORI team and PI's of HINOTORI for making the quick-look data available to him.

REFERENCES

Cohen, M.:1960, **Ap. J. 131**, 664.

Enome, S.:1983, **Proc. SMM Symposium**, COSPAR XXIV (Ottawa) in press.

Forrest, D. J., Chupp, E. L., Ryan, J. M., Rappin, C., Rieger, E., Kanbach, G., Pinkau, K., Share, G., and Kinzer, R. L. 1981, **Proc. 17th Int. Cosmic Ray Conf.** (Paris) **10**, 5.

Melrose, D. and Dulk, G.:1982a, **Ap. J. 259**, 884.

Melrose, D. and Dulk, G.:1982b, **Ap. J. 259**, L41.

Sprangle, P. and Vlahos, L.:1982, preprint.

Tajima, T., Brunel, F., and Sakai, J.:1982, **Ap. J. 258**, L45.

Tanaka, K., Watanabe, T., Nishi, K., and Akita, K.:1982, **Ap. J. 254**, L59.

Tanaka, K.:1983, **Proc. IAU Colloquium 71**, "Activities of Red Dwarf" (Catania) in press.

Watanabe, T., Tanaka, K., Akita, K., and Nitta, N.:1983, **Proc. US-Japan Seminar** (Tokyo), in press.

PARTICLE ACCELERATION IN THE 1981 APRIL 1 FLARE

Hiroshi Nakajima
Nobeyama Solar Radio Observatory of the Tokyo Astronomical Observatory, University of Tokyo, Nobeyama, Nagano-Pref. 384-13, Japan.

ABSTRACT. The large microwave burst of 1981 April 1, which was accompanied by both hard X-ray and γ-ray emissions, was analyzed to study the acceleration of particles in the impulsive phase. The analysis suggests the following results. (1) Electrons were accelerated up to energies of several hundred keV in a low loop. On the other hand, electrons were accelerated to relativistic energy without injection of pre-accelerated electrons near the top of a large loop where energetic ions were also probably accelerated. (2) The mechanism for accelerating electrons to relativistic energy and also ions was different from that for accelerating electrons up to energies of several hundred KeV and was closely related with upward motion of a flare loop.

1. INTRODUCTION

The acceleration of particles is regarded to be one of the essential problems in the study of solar flares. It has found general acceptance that acceleration of particles in large solar flares occurs in two stages or phases (Wild et al., 1963; de Jager, 1969; Frost and Dennis, 1969; Bai and Ramaty, 1976; Hudson et al., 1982). In an impulsive phase characterized by impulsive microwave and hard X-ray bursts, electrons are accelerated to energies of several hundred KeV. In the second stage, the electrons accelerated in the impulsive phase would be further accelerated to relativistic energies and ions also accelerated to energies >10 MeV. This stage is characterized by gradual microwave and hard X-ray bursts and by type II and Type IV bursts.

However, the γ-ray spectrometers on board the Solar Maximum Mission (SMM) and the Hinotori have recently given the quite important result that both energetic ions and relativistic electrons may be accelerated almost simultaneously with electrons at energies <100 KeV during the impulsive phase in large solar flares (Forrest et al., 1981; Chupp, 1981; Rieger, 1982; Yoshimori et al., 1982). The time profiles of the γ-ray lines are delayed with respect to the hard X-ray time profiles, with delay times ranging up to 1 min (Gardner et al., 1981).

Solar Physics **86** (1983) 427–432. 0038–0938/83/0862–0427 $ 00.90.

This observational fact has made it necessary to revise our ideas about two-stage acceleration.

Recently, a second-step acceleration model has been proposed by Bai and Ramaty (1979), and Bai (1982) in order to explain the acceleration of energetic particles in the impulsive phase. According to this model, energetic ions as well as relativistic electrons are accelerated by the first-order Fermi acceleration due to the passage of a shock wave through a closed flare loop. The second-step acceleration is impulsive-like and takes place in the same flare loop as the impulsive acceleration.

In this paper, I present observations with the 17-GHz interferometer at Nobeyama (Nakajima et al., 1980) of the large solar flare of 1981 April 1, which was also accompanied by both hard X-ray and γ-ray emissions. This flare provided a rare opportunity to study the acceleration of relativistic electrons and energetic ions in the impulsive phase.

2. OBSERVATIONS

The microwave burst was associated with a class 3 B Hα flare at S43, W52 (NOAA Solar-Geophysical Data).

The microwave time profile at 17 GHz was composed of three major peaks with similar time profiles at time intervals of about 10 minutes. Each of the three peaks consisted of both an impulsive component and a following gradual component. At meter- wavelengths, a type IV burst was observed with the dynamic spectrograph at Nobeyama. The type IV burst showed a time profile similar to that of the microwave burst. A Type II burst also occurred lasting from 01:37.5 to 01:57.5 UT (NOAA Solar-Geophysical Data).

Figure 1 shows time variation of east-west one-dimensional profiles observed with the 17-GHz interferometer at Nobeyama. Figure 1 clearly shows that the microwave source was composed of two kinds of sources. One was strongly polarized ($\sim$40%) and had time profiles with rapid rise and fall. These sources stayed almost at the same location during the course of evolution; hereafter, these sources are referred to as S-sources. The other sources were weakly polarized (<10%) with the same polarization sense as the S-sources and had gradual time variations. These sources (referred to as M-sources) started at almost the same location as the S-sources, but moved rapidly in the early phase of the burst and then slower at the second and third peaks.

The sizes of both S- and M-sources are estimated to be nearly constant at $\sim$30" during the course of evolution of the event.

Figure 2 compares the peak brightness of S- and M-sources at 17 GHz, their source locations, and the hard X-ray and γ-ray emissions. The hard X-ray and γ-ray emissions were observed with the Hard X-ray Monitor Spectrometer and the Solar γ-Ray Detector on board the Hinotori (Ohki et al., 1982; Yoshimori et al., 1982). As the S- and M-sources were resolved by the 17-GHz interferometer during the second and third peaks, time profiles of peak brightness of the two kinds of sources could be obtained independently. The time profiles of peak brightness can be considered to be roughly proportional to the time profiles of flux density, because the source sizes of the S- and M-sources were nearly constant during the course of evolution. During the first peak, where the S- and M-sources were not resolved, the source location was derived from the total intensity profile (I-profile) on the assumption that the microwave emission from the M-source was much larger than that of the S-source; additional information was derived from the polarization profile (V-profile; the difference between two circular polarization components) by taking advantage of the fact that the S-source was highly polarized.

It is clearly seen that the time profiles of both the microwave emission from the S-source and the hard X-ray emission showed rapid time variations, and the temporal coincidence between them was nearly perfect. On the other hand, the time profiles of the microwave emission from the M-source and the γ-ray emission showed gradual time variations and the temporal coincidence between them was also good. It is also seen that the peaks of the γ-ray emission lagged behind the

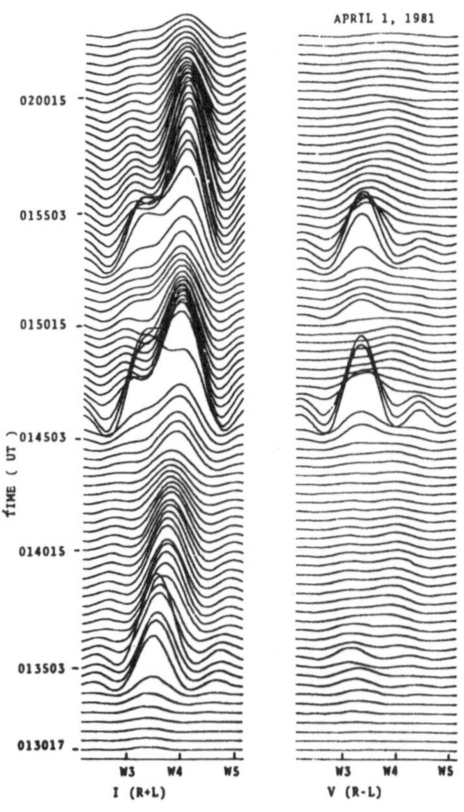

Fig. 1. Time variation of east-west one-dimensional profiles observed with the 17-GHz interferometer at Nobeyama. The vertical scale of the V(=R-L)-profiles is two times that of the I(=R+L)-profiles. The profiles are synthesized with a flat data window, so the sidelobe levels are rather large. The spatial resolution is ∿34".

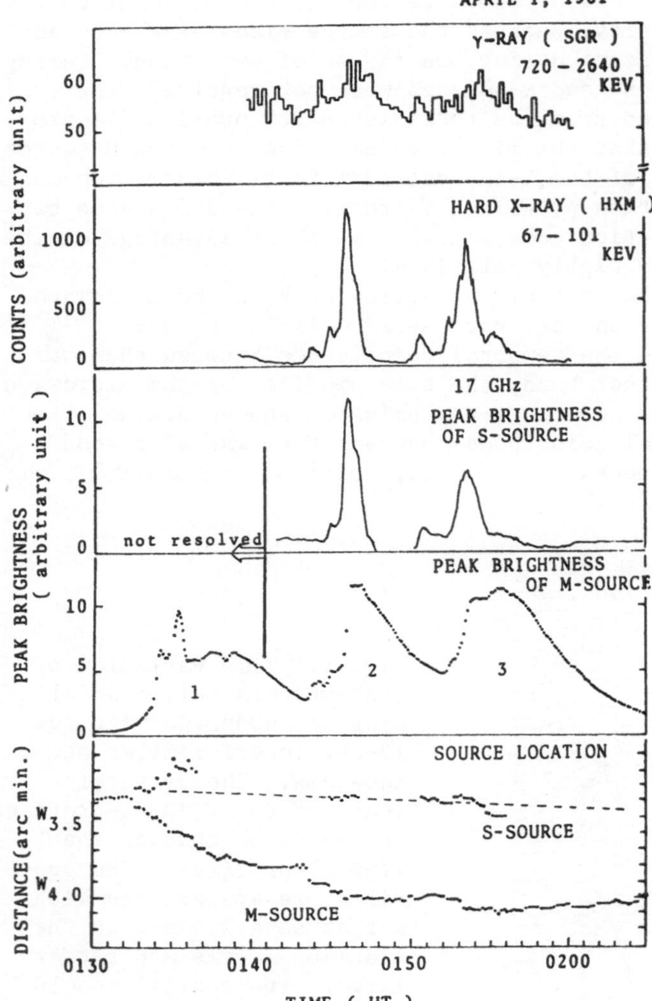

Fig. 2. Time profiles of peak brightnesses of the S- and M- sources at 17 GHz, and those of the hard X-ray (67 - 101 KeV) and γ-ray (720 - 2640 keV) emissions. Also source locations measured by the 17 GHz interferometer are plotted.

peaks of the hard X-ray emission by ∿ 1 min, though the hard X-ray emission showed no systematic delay at energies of from 15 keV to 312 keV (Ohki et al., 1982). Kawabata et al. (1982) pointed out by comparison of the γ-ray and millimeter-wave (35 GHz) observations that the millimeter-wave sources were also double — one source corresponding to γ-ray emissions at energies < 360 keV and the other to γ-ray emissions at energies > 360 keV. Our analysis at 17 GHz agrees well with their observations. The facts strongly suggest that the emission from the S-source was due to electrons at energies < several hundred keV and the emission from the M-source was due to relativistic electrons.

A close comparison between the time profiles of the microwave emissions from the S- and M-sources shows that they started almost simultaneously, within 20 sec. The near-simultaneity of the onset times of the S- and M-sources suggests that these two sources were closely related to each other.

The lowest panel in Figure 2 clearly shows that the M-source moved toward the west limb discontinuously, and then reached 0.6' far from the S-source. If we interprete the westward motion of the M-source as the projected effect of upward motion, the M-source reached the height of $\sim 1.2\times 10^5$ km above the photosphere.

The microwave to millimeter-wave spectra shows for each of the three major peaks that the turnover frequencies are lower at times when the microwave emission from the M-source was prominent than the peak times of the microwave emission from the S-source. From the observation that the high frequency part of the spectrum at the former times is due to electrons at higher energies than that at the latter times, we can deduce that the magnetic field in the M-source is much weaker than that in the S-source. This result is consistent with the observation that the polarization degree at 17 GHz of the M-source is much weaker than that of the S-source.

3. SUMMARY AND CONCLUSIONS

The microwave burst of 1981 April 1 was composed of two kinds of sources which began to flare up repeatedly almost simultaneously during impulsive phases. The emission from one kind of source was due to electrons at energies < several hundred KeV, and that from the other kind of source was due to relativistic electrons. The former source was located in a strong magnetic field region and showed no motion. On the other hand, the latter was located in a weaker magnetic field region and reached the height of 1.2×10^5 km above the photosphere, showing discontinuous upward motion.

The analysis of the event strongly suggests that, although relativistic electrons as well as energetic ions are accelerated during the impulsive phase, they are accelerated without injection of pre-accelerated electrons near the top of a large loop different from that where electrons are accelerated up to energies of several hundred KeV, possibly by a different mechanism. The acceleration mechanism of relativistic electrons and also energetic ions are closely related to upward motion of a flare loop in the corona.

The present observation does not seem to support the second-step acceleration model proposed by Bai (1982) in the following respects:

1. The observation suggests that the 1981 April 1 burst took place in two different loops rather than in a large single loop.

2. If the 1981 April 1 burst took place in a single large loop and electrons were accelerated by the first-order Fermi mechanism due to the passage of a shock wave through the large loop, large variations of both the location and size of the microwave source should have been observed around the second and third microwave peaks. However, such large variations were not observed.

ACKNOWLEDGEMENTS

The author would like to thank Dr. M. Yoshimori, Mr. N. Nitta, Prof. K. Kawabata, and Dr. S. Enome for providing the γ-ray, hard X-ray, millimeter-wave, and microwave data, respectively.

REFERENCES

Bai, T.: 1982, in Gamma Ray Transients and Related Astrophysical Phenomena, eds. R.E. Lingenfelter, H.S. Hudson, and D.M. Worrall (A.I.P., New York), p.409.

Bai, T. and Ramaty, R.: 1976, Solar Phys. 49, 343.

Bai, T. and Ramaty, R.: 1979, Astrophys. J. 227, 1072.

Chupp, E.L.: 1982, in Gamma Ray Transients and Related Astrophysical Phenomena, eds. R.E. Lingenfelter, H.S. Hudson, and D.M. Worrall (A.I.P., New York), p.363.

De Jager, C.: 1969, in COSPAR Symp. on Solar Flares and Space Sci. Res., eds. C. de Jager and Z. Svestka (North Holland, Amsterdam), p.1.

Forrest, D.J., Chupp, E.L., Ryan, J.M., Reppin, C., Rieger, E., Kanbach, G., Pinkau, K., Share, G., and Kinzer, G.: 1981, in Proceedings of the 17 th International Cosmic Ray Conference (Paris) 10, 5.

Frost, K.J., and Dennis, B.R.: 1971, Astrophys. J. 165, 655.

Gardner, B.M. et al.: 1981, Bull. Am. Astron. Soc. 13, 903.

Hudson, H.S., Lin, R.P., and Stewart, R.T.: 1982, Solar Phys. 75, 245.

Kawabata, K., Ogawa, H., Takakura, T., Tsuneta, S., Ohki, K., Yoshimori, M., Okudaira, K., Hirashima, Y., Kondo, I.: 1982, in Hinotori Symposium on Solar Flares, (Tokyo, Jan. 27-29, 1982), p.168.

Nakajima, H., Sekiguchi, H., Aiba, S., Shiomi, Y., Kuwabara, T., Sawa, M., Hirabayashi, H., Kosugi, T., and Kai, K.: 1980, Publ. Astr. Soc. Japan 32, 639.

Ohki, K., Nitta, N., Tsuneta, S., and Takakura, T.: 1982, in Hinotori Symposium on Solar Flares, (Tokyo, Jan. 27-29, 1982), p.69.

Rieger, E. 1982, in Hinotori Symposium on Solar Flares, (Tokyo, Jan. 27-29, 1982), p.246.

Wild, J.P., Smerd, S.F., and Weiss, A.A.: 1963, Ann. Rev. Astr. and Astrophys. 1, 291.

Yoshimori, M., Okudaira, K., Hirashima, Y., and Kondo, I.: 1982, in Hinotori Symposium on Solar Flares, (Tokyo, Jan. 27-29, 1982), p.230.

VII. OPEN DISCUSSION OF CONTROVERSIAL POINTS

Chairmen: Y. Uchida and C. de Jager

INTRODUCTION

At the conclusion of the Seminar, an entire morning session was devoted to open discussion centered on four topics with controversial issues. The overall chairman of the session was Y. Uchida, and a discussion leader introduced each topic and guided that part of the discussion. The topics and discussion leaders were the following:

"The stages of particle acceleration"	E. Chupp
"Location of X-ray and microwave sources"	H. Hudson
"Is the electron distribution thermal or non-thermal?"	G. Dulk
"Mass transport and interaction of loops"	G. Doschek

We adopted the guideline that the discussions should center as much as possible on deriving physical picture rather than on the further presentation of individual data.

The transcription of the discussion from magnetic tapes has been heavily edited, with numerous alterations or deletions of the verbatim comments. We hope sincerely that the editing did not make any substantial changes in the essential content of the interesting discussions.

(H.S.Hudson and Y.Uchida)

DISCUSSION

UCHIDA: For this free discussion session, we have chosen four topics on the basis of interest and controversy, and four people have been asked to undertake the role of the discussion leaders. The first topic is "The stages of particle acceleration" led by Dr. Chupp, the second "The location of X-ray and microwave sources" led by Dr. Hudson, the third "Is the electron distribution thermal or non-thermal" led by Dr. Dulk, and the fourth "The mass transport and the interaction of loops" led by Dr. Doschek. In presenting your opinion or in discussing the topic, you are asked to identify yourself clearly. We are recording the discussion for the later use in editing the Proceedings, and the

Solar Physics **86** (1983) 435–461. 0038–0938/83/0862–0435 $ 04.05.

tape-recorder can not recognize your face. Secondly, the native speakers of English please speak distinctly and slower than usual so that Japanese colleagues in the audience will not be lost. Just try to imagine that you are in the audience of a discussion in French, which you may be able to read but have not been exposed to so frequently! I will remind the speaker to speak slowly if he goes too fast, and if he still goes too fast, I may switch the language of the session to Japanese by the prerogative of the chairman, probably up to 50% of the time legitimately(?!). Now I call the first discussion leader, Dr. Chupp, for the first topic.

I. THE STAGES OF PARTICLE ACCELERATION

CHUPP: Thank you, Dr. Uchida. Since I have been asked to lead a general discussion on this topic I will assume that we can start at a fairly advanced level. I will say at the outset that I don't know what the answer to the particle acceleration question is, but we have certainly learned some important things from our recent observations.

First of all I would like to remind you of some of the inferences from the gamma-ray and hard X-ray observations. A summary of these inferences appears in Table I. The last item may be more a question of sensitivity, and we may not know for sure yet whether any given flare will accelerate particles to high energies.

TABLE I

Inferences on Particle Acceleration in Flares

1) Ion and electron acceleration is impulsive. The particle/matter interaction is simultaneous to within a few seconds.
2) GeV energies are attained within seconds.
3) The acceleration process may be repetitive within ten seconds or less.
4) Rise times of the particle/matter interactions range from less than one second to about 100 seconds.
5) There is strong evidence for a constant ion energy spectrum, but the MeV electron/ion ratio varies.
6) There is evidence that all flares accelerate energetic ions.

I would like to emphasize that one observes directly the interaction rate of the energetic particles. One measures the flux at the Earth and then corrects for the distance to get the interaction rate; these are prompt emissions. There are two possible limiting geometries. Suppose one has a trap at the Sun; the reaction rate is given by the product of the density of fast particles with the density of the ambient particles in the reaction, together with the cross-section, to get a total rate. The acceleration mechanism injects the fast particles at a certain rate into the trap.

The other possible type of geometry is the beam-type geometry, where particles flow down the field lines and produce a thick-target type of interaction. In this kind of geometry the time for ionization losses by the particles will be much shorter than the time scale of variation observed in the emission. The decay times of the individual bursts may be very short, perhaps as short as the rise time. This is implied by the symmetry of the time profiles, that is the rise and decay times of an individual burst appear to be the same.

I would like to start the discussion off with two questions. First, is it possible to explain the X-ray, gamma-ray, and microwave observations without invoking only a single acceleration process? We must simultaneously accelerate protons and electrons on the time scales that we observe. Is there any way to avoid the use of a single process? Secondly, once the audience has dealt with this question, what is the role of the secondary acceleration process - second stage, second step, or second phase or whatever you may wish to call it. With these remarks I will now open the floor for discussion.

RAMATY: It is necessary to qualify a couple of points in your table, I think. I would prefer "hundreds of MeV in tens of seconds" for the acceleration speed. For the highest energies you really rely upon the neutrons, for which the highest energy is about 500 MeV. Another issue is the spectral distribution. I think that the gamma-ray data now show the existence of more than one spectral component.

CHUPP: This is true. We saw yesterday from your calculations that some variety in the spectrum exists. However the observational results that I have listed are the predominant characteristics of some particular acceleration mechanism. I believe that the theorists should try to come up with a mechanism that satisfies these properties.

BAI: About acceleration times, we observe time scales of seconds in the June 7 event, and tens of seconds in other events. As I pointed out yesterday, a time of a few seconds can be achieved with a first-order Fermi mechanism, in which the energy increases linearly with time. One can associate the acceleration time with the length of the loop, as it is traversed by approaching shock fronts. It is interesting that the June 7 and June 21 events were very impulsive. This means according to my model that the length of the loop was very small, resulting in prompt acceleration. On the other hand the April 27 flare and October 7 and 14 events were very gradual and had longer time scales, corresponding to larger size scales.

Secondly I am against the idea of having only a single acceleration mechanism. There is an immediate number problem; there are more protons than electrons at high energies, and more electrons than protons at low energies. We don't have direct evidence on the protons at low energies, but we would have trouble with energy content without a low-energy cutoff. Another thing is that the energy loss time is so different for protons and electrons. The high-energy electrons have a very long energy-loss time. At low energies the acceleration must compete with particle energy loss; this is a basic reason for having two acceleration processes even in the impulsive phase.

CHUPP: You point out that there are very few cases of very rapid acceleration of ions. This was a point that I have also emphasized, namely that the acceleration may be very rapid. However don't forget that as soon as you see the first gamma-ray in an event, you know that the particles have been already accelerated to a sufficiently high energy to produce it. The hard X-rays and the gamma rays may start simultaneously; even if the time of maximum may differ.

BAI: It may be difficult to pin down the time of acceleration because of the background problem. In addition, for the few-MeV gamma-rays, there may also be a bremsstrahlung component, so those initial few gamma-rays may be produced in this way. We don't know that they are of nuclear origin.

CHUPP: That's so, and so we should strive for better sensitivity and time resolution.

ZIRIN: What about electron acceleration? We have done a little work on the cross-correlation of spikes in the radio and X-ray emission, and find that there is a small delay. In general those who have done such comparisons find a delay on the order of a second. Since the radio emission comes from energies on the order of 100 keV, while the X-ray emission comes from about 40 keV electrons, this would give you about a one-second rise time for the acceleration to 100 keV. Has anyone studied this matter? There must be a lot of observational material.

ACTON: I have been impressed by the high efficiency for the acceleration of particles by shocks. Pesses has pointed out that if the geometry is favorable, extremely high energies may readily be obtained. Eric Priest showed very interesting models of fields that contain islands where conditions would be favorable for acceleration; little kinks in the field lines that represent shocks going out from the disturbed region. This seems to be a very good example of a two-stage or two-step system, and if one has just the right conditions then one may get the right acceleration. Well, it seems to work in Eric (Priest)'s computer. But I don't know how to address the issue of energetics. What is the energy source for the shock? Is it particles streaming down, or is it some sort a flow energy? Could someone address this question?

PRIEST: The energy is the magnetic free energy that is in the field. In the shock wave, this energy is converted into heat and kinetic energy. I think it is also important to notice that the repetition rate - perhaps ten seconds - could be explained by the repetitive formation of the pairs of islands by the tearing-mode instability. I did some back-of-the-bedclothes calculations last night, and found that the neutral point forms relatively slowly, on the tearing-mode time scale, and that - if you take typical values of the length scale and the Alfven speed - can easily give you the ten-second time scale. Once the islands form, they then are destroyed very rapidly by the coalescence instability, and that takes place on an Alfven time scale, which can provide an 0.1 s time scale quite easily. This coalescence instability is not new; we are finding it in our simulations but it has previously been discussed by Tajima and by Sakai and by several others. I think it's a well-established instability.

UCHIDA: I think the discussion could proceed in several directions. This picture of the acceleration, and the details of the mechanisms, might be a subject to return to after we dispose of questions of the type that Zirin asked.

CHUPP: But there are some important issues here. I would like to ask Eric Priest about shocks - they are currently in vogue, but are there other theoretical possibilities that we could turn to?

PRIEST: Heyvaerts has just made a very extensive survey of possible acceleration mechanisms, concluding that they are very difficult. Shock waves may have been studied the least, but they appear to be the most potent. I should point out that there are several places where you may get shock waves - slow shocks in the area Loren Acton mentioned - and fast shocks at the reconnection site. In addition fast shock waves will move away from the initial site of the flare as a result of plasmoid ejection. Also this is a very complicated system we are dealing with - shocks may be produced in other ways also. I think there are many ways in which shock waves can occur, and I think that this area of shock wave acceleration is one of the very important topics for the future.

KANE: I think the question of two stages of acceleration, or many more, depends upon two observational quantities: The rate of acceleration and the energy achieved. The earlier observations showed us an impulsive phase for which the energy attained appeared to be low, electrons up to a few hundred keV. This could easily be explained by an electric field, it seems to me. However we are in a different situation now. We have a problem not only to accelerate electrons and protons simultaneously in a few seconds, but also to get ions to hundreds of MeV at least. Can it be done by shocks, or can we still use direct electric-field acceleration? We must face this problem theoretically.

SIMNETT: I would like to comment on something that Reuven Ramaty brought up, that the neutrons may need a few hundred MeV; if we think of the particles in terms of the velocities and consider type III bursts, we will note velocities there of a few tenths of the speed of light. Protons at similar velocities would have energies in the right range to produce the neutrons. Another quantum of information that we need to remember, reflecting on something that Sharad Kane mentioned, is that the highly-relativistic electrons, at energies a factor of ten times their rest energy, always have a delay on the order of 15 minutes beyond the time of the type-III-burst electrons. I think this definitely implies a second acceleration stage, and it might point us towards a velocity-dependent acceleration mechanism.

RAMATY: Shock waves are known to be important in the bow shock, in the interplanetary medium, and for cosmic rays, so I don't think that we can avoid their application in solar flares. I would like to mention another observation that has not come up before, even though one of the authors (Dr. Yanagita) is in the audience. That is the correlation of relativistic electron events with gamma-ray flares. As Simnett said, in most cases the relativistic electron flux is far below the proton flux at the same energy, say 40-50 MeV, and these electrons will not affect the gamma-ray line observations because their radiation is by Bremsstrahlung. Relativistic electrons which seemed to be quite prompt were seen. These are quite a mystery to me. Although we have in one flare (1980 June 21) good evidence that the high energy continuum is due to protons, the electrons are also accelerated quite promptly. As far as I understand shocks - I am a theorist and don't really understand shocks well enough - it is difficult to accelerate relativistic electrons in shocks (shocked laughter). If it does turn out to be impossible to accelerate relativistic electrons in shocks, then the observations will really require another mechanism.

CHUPP: We must be careful, since the particles seen in space may indeed come from a different process than those responsible for the gamma rays. One historical reason for our confusion about different acceleration stages was that people used to think that the highest energies could only be attained in a delayed acceleration. Now it seems that conditions appropriate for shock-wave or turbulent acceleration may occur in different phases of the flare.

UCHIDA: Rather than jumping into the more difficult problem of simultaneously accelerating protons and relativistic electrons, are we in the situation where we understand really thoroughly the problem of accelerating the lower-energy electrons that produce the hard X radiation? Do we understand where the source is located and how the acceleration is carried out?

BROWN: I think the point here, as discussed by Andre Duijveman yesterday, is that it is relatively easy to get 60 - 100 keV for an individual energy, the problem being the total number of electrons. If the electrons are accelerated rather than heated, the fraction of energy going into these electrons is embarrassingly high. Andre (Duijveman) gave us numbers like at least 20%, and we simply don't have any way of doing that. I don't think that we have the slightest idea of how to get this large a number of electrons. I think that the protons are in a somewhat different category since we may need only a small fraction to be accelerated.

KAI: I would like to show one example which contains a very crucial point about whether the acceleration occurs in one step or two steps. These data have already been shown by Dr. Chupp, and reveal a two-second delay. Nakajima has found that the microwave time profile has two sets of peaks. The first correspond very well to the hard X-ray peak, and the second to the gamma rays; we think that this implies two kinds of accelerations although we do not know how to identify them yet. There is only very weak hard X-ray emission corresponding to the second microwave peaks.

KANE: Can the theorists explain the one-second delay? Nakajima is saying that it is the result of a second acceleration mechanism, but I think that we have to be careful here. If we compare hard X-rays with microwave emission, we seem to see larger delays at the higher energies. This implies a partial trapping and would produce a dispersion in peak times. Takakura mentioned yesterday that the microwave source might be expanding, which could have the same effect.

ACTON: Back to Dr. Uchida's important question about the acceleration of the primary electrons, to my understanding the only population of electrons that can be accelerated in any of these processes is a sort of runaway tail. I wonder if anyone has a new process that we haven't heard about that can accelerate a significant population, for example in one of these magnetic field-annihilation islands that seem to be appearing out of the simulations?

UCHIDA: Regarding Eric Priest's calculations on this point, I would like to ask whether the boundary conditions reflecting Alfven waves back and forth have something to do with the locations where the X-type neutral point is created; where are the second and third neutral points created?

PRIEST: We have done this experiment in a variety of numerical boxes, and in particular we have tried to make the box as large as possible, so that the boundaries are as far as possible away from the reconnection site. The creation and annihilation of the neutral points, and their period, are not functions of the boundary conditions.

UCHIDA: Then what determines where and how the second, third, and fourth X-type neutral points are created? What determines the location physically?

PRIEST: I think that is something we have to look into in more detail. These have been preliminary results; but the wavelength of the production of the islands is determined by the maximum wavelength which is the maximum growth rate. The wavelength for maximum growth rate can be gotten from Furth, Killeen and Rosenbluth...

UCHIDA: Modified by the line-tying effect at the bottom of the loop?

PRIEST: Yes.

UCHIDA: If such a situation is created and the crest of the disturbance propagates very quickly, the acceleration related to shocks may be there, but there are also several other possibilities. For example in the situation that Shibata and I proposed, the recoiling shocks collide around the top of the loop, causing first-order Fermi acceleration. In this connection, however, we must deal with two-fluid calculations in which the electrons and ions can move differentially. Probably Washimi can talk here about the creation of an electrostatic double layer by a stream in a magnetic-field gradient, with the resulting buildup of electric potential there.

WASHIMI: We are making a theory for energetic electrons, not energetic protons. The MHD theory is a powerful one for dealing with large-scale energy conversion, but for electron acceleration it is necessary to calculate an electric field, which cannot be dealt with by MHD theory. We assume that a fast plasma flow is created by reconnection, as for example shown by Sato and Hayashi with MHD computer simulations. We thus assume a plasma flow. In this case the ions have the main kinetic energy. If this initial flow comes to a strong magnetic field region, a potential will be created by the mirror effect, because the ions are anisotropic and can penetrate, whereas the electron distribution is still isotropic. The ion density becomes locally larger than the electron density. However the charge neutrality condition is a very strong one, so a large potential is created to restore neutrality. This large potential drop accelerates the electrons and decelerates the ions, converting their kinetic energy. If, as the simulations show, a large flow velocity (on the order of the Alfvén velocity) occurs at the reconnection site, then the energy of the accelerated electrons will be several tens of keV, which will be enough to explain the fast electrons and hard X-ray emission observed.

TAKAKURA: I am wondering why the theorists have given up on force-free current systems for the particle acceleration. If the current decays due to some large anomalous resistivity, a parallel electric field will be produced. This will efficiently accelerate electrons. The protons have larger mass and thus require a longer acceleration time. The magnitude of the potential drop could be 10 GV, so some protons could be accelerated to 100 MeV in the impulsive phase by this means. This of course requires an anomalous resistivity, and the theory does not tell us how this comes about. Observationally, we need a quick dissipation of a current system to provide the flare energy. If we suppose that current flows in a force-free magnetic configuration, we will need to have parallel electric fields because of the observed rapid decay.

On a separate subject, we have to be careful about the terminology in the subject of second-phase acceleration. If the acceleration efficiency is low, the acceleration will take time. Even if the starting times for the two species are the same, the times of maximum will still differ. This would not in this case imply a separate second-phase acceleration mechanism.

HUDSON: I don't see how Washimi's mechanism could work in the impulsive phase. Would it be fast enough?

UCHIDA: Suppose that a neutral plasma flows into a field gradient. The protons can penetrate into a stronger field than the electrons. Thus an electric double layer forms, and its magnitude is proportional to the initial down-flow energy. So, in his picture, the electron acceleration in the impulsive phase is a biproduct of other dynamical process, rather than being the very first creature in a flare. Actually, a dynamical phenomenon like the dark filament rise precedes the impulsive phase by 20 min.

ZIRIN: I have been starting to worry about energetics from another direction. From the optical and the X-ray evidence it looks like you have a real difference betweeen the impulsive phase and the main phase of a flare, at least in the two-ribbon situation. There is some evidence that most of the energy is emitted fairly late in such cases. Yet I understand now that the protons must be accelerated in the impulsive phase. It seems confusing since we have evidence that the impulsive phase is not so dominant energetically, at least in the big two-strand flares.

CHUPP: Well, the impulsive-phase acceleration of ions is an observational fact, and it does seem to make the energetics more difficult.

RAMATY: There is still less energy carried by the protons than the electrons in the impulsive phase. The question of total energetics has not changed with these new gamma-ray observations. For the 1980 June 7 flare, the total proton energy above 20 MeV is at most one tenth, if that, of the electron energy above 20 keV.

ENOME: I would like to make a comment on the subject of acceleration mechanisms. It seems to me that the theorists may have a louder voice than observers here. They are routinely thinking about these complicated matters. The question of turbulence or shock waves for high-energy particle acceleration should also be asked of the observers. We have many optical and X-ray observers here. Can we ask them if there is a chance to observe the turbulence or shocks in any of the wide range of observations they have made? The HINOTORI observations, for example, show turbulence in the initial phase of a flare. The velocity of the turbulence - some 200 km/s - may not be enough to satisfy Ramaty's requirements. My comment is just that we should look more carefully to extract clues from the observations.

KUNDU: I have a comment with regard to Dr. Kai's data, showing two peaks in the 1980 June 7 event. One corresponded to the hard X-rays, and one to the gamma-rays. I have seen several of these events, in which in the microwaves you have several peaks which do not necessarily correspond to the integrated spectrum all the time; whenever the spectrum hardens you have got another peak. I would like to point out that it is probably the same acceleration process and that it is related to the spectral hardening phenomenon. That is why you see a second peak in the microwaves, since the higher energy electrons will efficiently produce microwaves. I don't personally think that it is a two-step process. The spectrum hardens and you see more microwave radiation, and I don't think that this requires a two-stage acceleration process, at least not within the short time difference between the hard X-rays and gamma rays.

UCHIDA: Thank you Dr. Kundu. We probably have to proceed to the next topic, since we have three more to go. Dr. Hudson please.

II. LOCATION OF X-RAY AND MICROWAVE SOURCES

HUDSON: I would like to take about three minutes regarding sources structures, before inviting general discussion. In addition to the gamma rays, another highlight of the present solar maximum has been the achievement of imaging observations of hard X-rays and of microwaves. In other words during this maximum we have really for the first time begun to see the detailed structure of the non-thermal sources in solar flares. Of course, the discussion of the term 'non-thermal' should be reserved until later. If we stretch a bit and consider 1973 to be a part of the present solar cycle, we can include the Skylab results. These beautiful observations of the spatial structure of the thermal sources showed them to consist almost entirely of loops. Now here is a little test for the non-native-English speakers: The loops were ubiquitous.

I would like to suggest a classification of the X-ray observations, both hard and soft, as summarized in Table II. Among the hard X-ray phenomena, we have observations of the impulsive sources; I think that there are perhaps three kinds of impulsive sources. First there are the very fast spikes, on time scales below one second; second, the smoother impulsive time profiles for which the prototype is the large spikes of the August 1972 flares; and finally the second-step kind of phenomenon. In the gradual part of the development of X-ray emission, one has a few observations of gradual hard sources that may be associated with meter-wave flare continuum. We have a nice observation now of the 1981 May 13 event by HINOTORI. There is also a gradual softer spectrum of the sort observed in the stereoscopic data from PVO and ISEE-3, I believe on November 5, 1979. Finally there is a new component recorded by the Berkeley-San Diego-Toulouse balloon collaboration of a thermal X-ray spectrum in the hard spectral range. This is probably associated with loops of the type that we have known for some time, except that the high temperatures raise some difficult theoretical problems. We also have as a separate category the post-flare loops, prominent in the large two-ribbon flares and leading to the loop prominence systems seen in H-alpha. Finally we have 'Svestka's giant arches.' These may have been observed before by Webb and Kundu, but the SMM observations have greatly clarified their nature. They are particularly interesting because of their association with Type I noise storms. In the future I anticipate much interesting research in the associations between coronal X-ray and radio emissions of various types.

Table II

Classification of Flare X-ray Sources

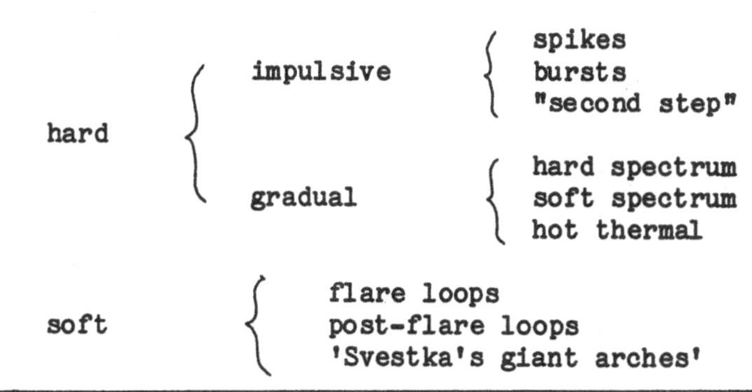

hard	impulsive	spikes
		bursts
		"second step"
	gradual	hard spectrum
		soft spectrum
		hot thermal
soft	flare loops	
	post-flare loops	
	'Svestka's giant arches'	

Here are some observational problems, and I believe that even on the observational level there are differences of opinion. It is clear that we lack sufficient spatial resolution in X-radiation, and higher spectral resolution would also be very helpful. For example, HINOTORI seems to see patchy impulsive emission, whereas the SMM hard X-ray imaging revealed several events with an almost symmetrical double foot-point structures. Are these observations reconcilable? Another apparently unexplainable observation is the apparent compactness of the microwave sources as viewed by the VLA. These are certainly small compared with the size scales of the X-ray sources.

Our main goal in the physics of solar flares seems to be to describe and understand the energy release and particle acceleration mechanisms. We want to do this not only to understand solar flares, but also because many of the mechanisms involved will be applied elsewhere in the Universe, where the solar phenomena often provide the basic tools for working on problems in stars, stellar systems, galaxies, or other objects. Some of the intermediate steps are the following: (A) We are puzzled by the energization of the thermal sources. This problem is especially difficult for the very hot sources discovered recently, for which the energy input requirement is very great. (B) The impulsive-phase acceleration remains unknown, although we are beginning to learn something about the plasma properties in the flaring loops. (C) We are baffled by the relationship between the hard and soft X-ray sources. I thought in this context that it was interesting for Dr. Takakura to have suggested that there may observationally be different structures involved in the impulsive phase. (D) We have little physical knowledge of the natures of many of the transient sources in the corona. With these new X-ray imaging systems we are getting our first look at many of these objects. Finally, (E) There is the problem of cold matter in dynamic solar phenomena. As pointed out by Dr. Brueckner, the energy involved in the mass motions greatly exceeds that observed in the non-thermal high-energy phenomena, and the relationships over this wide range of plasma parameters have hardly begun to submit to explanation.

With these introductory remarks about X-ray structures and solar flares, I will open the floor for discussion.

DOSCHEK: The UVSP experiment on the Solar Maximum Mission has observed EUV flashes in coincidence with hard X-ray spikes. This provides a clear impression that the footpoints of multiple loops are being excited by non-thermal electrons.

HUDSON: This observation goes back to the initial observations by Kane and Donnelly of simultaneous EUV and hard X-ray bursts, using the SFD ionospheric disturbance to measure the EUV flux. In addition the Big Bear observations of the Helium D3 line seemed to show numerous flashes at discrete points in the H-alpha flare area. Perhaps these were not paired and were not the footpoints of individual magnetic flux tubes.

BROWN: Isn't George Doschek talking about synchronism at about the one-second level? If so, there is no real problem, because the electrons can travel more than 10^{10} cm; thus simultaneity is no guarantee of co-spatiality.

HUDSON: I think the synchronism is much better than one second on a statistical basis from the Donnelly and Kane results.

FROST: We don't have to rely upon the SFD observations. The geometry of the UVSP transition-region emission lines matched that of the HXIS sources, and both agreed in temporal profile exactly with the hard X-ray spikes, so I agree with George Doschek's conclusion that the SMM data have really found the sources in the chromosphere of the EUV bursts, and they are cospatial with the hard X-ray sources. Understanding the target will be difficult - I don't know if there is any substitute for hard X-ray images - but there is no doubt about the agreement of the positions and times to within the instrumental resolution.

BRUECKNER: I had concentrated in my talk on the jets which provide upward material acceleration. But there is a second class of events that we see in the higher-resolution observations of the transition zone, which we call the turbulent events. All over the Sun we see small kernels, usually unresolved at the instrumental resolution of one arc sec but occasionally as large as three arc sec. They show turbulent velocities of up to 400 km/s, developing to that level in less than the 10-s resolution time of the instrument. If you observed these in the quiet Sun, with magnetic fields of no more than 10 - 50 gauss, you could imagine scaling them up to active-region field strengths. These are non-thermal accelerations, and can easily be recognized because ions with different mass-to-charge ratios attain the same velocities. And they are all over then Sun.

HUDSON: Do the jets produce electrostatic fields, fast electrons, and therefore hard X-rays?

BRUECKNER: Not likely because of the small energy involved in each event.

BAI: What is the mass?

BRUECKNER: The energy is on the order of 3×10^{26} ergs, the mass 6×10^{10} grams. But the integrated mass and energy may be quite large because they occur all over the Sun, where we estimate about 750 events per second in total.

ANTONUCCI: I would like to return to the SMM ultraviolet spectrometer observations, because they are very important in this respect. There is one flare in which the OV line shows seven sites which flare in sequence, on a three-arc-sec size scale, and for which the team found exact agreement with individual spikes observed by the hard X-ray burst spectrometer. This is the point made by both Doschek and Frost. Regarding the HXIS observations, there were multiple bright points in several flares. I think the main problem right now is spatial resolution.

I would like to comment on turbulence in response to Dr. Enome's remarks. Certainly in soft X-rays we see turbulence before the hard X-ray peak, before the real input of energy, and before the blue shift begins. This is quite well established observationally. With the bent-crystal spectrometer we can determine these relationships with quite good time resolution.

DE JAGER: The HXIS observations have been mentioned. I hope that Andre Duijveman will make some comments, since he has been studying the footpoints in some detail. I agree with Ester Antonucci that the situation is not so straightforward as we would like it to be. We find three footpoints in some cases. We have been trying to deconvolve the response knowing the instrumental profile in detail. The footpoints appear to be very patchy, and we guess that if one could see them at high resolution one would find a very detailed structure. The spiky first phase of a flare tends to show the patchy structure. You find peaks mostly seen in hard X-rays, but also in soft X-rays - note that in some flares you do not see hard X-rays at all - and the general characteristic of the emission from the patches is spikiness. This is also associated with the strong turbulence noted by Ester Antonucci.

HUDSON: The concensus seems to be that there is no observational discrepancy between the two hard X-ray imaging systems. We seem to be up against the limit of resolution, however.

DUIJVEMAN: To answer Doschek's comment about the correspondence of the OV brightenings with hard X-ray bursts. We have one example in which we had three brightenings of hard X-rays, one of them quite distant, and the UVSP executed a small raster containing one of the footpoints.

They observed a burst exactly simultaneous with the footpoint burst. This is as far as I know the only good example of simultaneity in this sense.

I also wanted to remark on the frequency of occurrence of footpoints. We do not have a lot of examples, and there are several reasons for this. First the flare must be rather large, with a footpoint separation larger than the 8 arc sec resolution of the instrument. Second, the flux must be large in order to image the hard X-rays with good enough statistics. Also this structure is only seen during the impulsive spikes - after the impulsive phase the hard and soft images seem to coincide. As a final comment, HXIS did not observe a good hard X-ray burst on the limb, and in this respect HINOTORI has an advantage.

DOSCHEK: Perhaps we have enough information to put an upper limit on the sizes of these little kernels. The UVSP has, I think, 4 x 4 arc sec resolution at best. If the OVII data from the Aerospace experiment has been interpreted correctly, then we have densities of around $2x10^{12}$ and from our emission measure we can calculate the characteristic size of the kernels. This turns out to be on the order of 1000 km, or 1.3 arc sec, so it might be very hard to resolve them fully.

ZIRIN: I am sure that we could resolve these in the optical region. I doubt if an EUV kernel would occur without an optical one, so we know that they have a lot of fine structure. It is more important to try to figure out how they get there. The footpoints of the thermal events seem to provide an interesting question, however. Initially they seem to be large, especially in these three-strand events. I wonder if this is due to a change in the ratio of parallel to perpendicular conductivity? Is there a mechanism for the diffusion of energy across field lines that would account for this increase of size, or is there another mechanism, such as radiative excitation, that explains these really big, bright footpoints?

HUDSON: I wanted to make one comment about the identity of the hard X-ray and EUV flashes. The solar atmosphere has a lot of structure, especially at interesting moments during activity. There is certain to be considerable horizontal inhomogeneity. The SMM observations of the 1980 November 6 event showed coronal 'flaring' without a strong response in the chromosphere. In such cases the hard X-ray and EUV responses should be quite different.

ACTON: I have difficulty thinking of things that I can't draw pictures of, and Dr. Zirin's comment reminds me of something that has puzzled me for a long time. If one looks at the H-alpha image of a large solar flare, one sees an enormously complicated and convoluted object in the chromosphere, extending over a very large area. We now think that this brightening results from heat conducted from above. This says that in the corona the hot volumes must be interconnected in a most complex topology. The means by which this complex topology is established might be a key to understanding the whole flare process. I have become convinced that loops are physically interacting. But if I try to draw a picture of interacting loops, I find that the interaction can only take place on a surface. How can appreciable magnetic flux be annihilated there? The result in any case is that substantial volumes are filled with hot plasma. How does it get there? It seems to me that there are things happening to affect the transport of energy transverse to the field lines, and in a very complicated topology. I wonder if there is anybody here smart enough to explain how this happens?

GROUP: (Hollow laughter.)

HUDSON: On a wholly separate subject, I would like to remind the audience that although we are seeing wonderful observational results from this present solar maximum, the next one is only a short time away. So far as I know, there are no firm plans anywhere for a new program of observations that would help Loren Acton answer his question. I think that in the next year or so we have to make up our minds about what new observations we need worst.

TANDBERG-HANSSEN: The UVSP instrument on SMM has been mentioned several times, and I would like to clarify the technical point raised about its resolution. One of the good contributions of UVSP has been to give some information about the location of the bright points, as we call them, that seem to coincide with the hard X-ray emission. The way in which we do this is to use the high resolution, as we have a capability for one arc sec resolution, but as George (Doschek) pointed out, we have mainly used a larger aperture of 3 x 3 arc sec, for these observations. From the timing we can locate the emission very precisely, and it does occur exactly in the transition region at the footpoints of the loops. We have also examined these kernels that Zirin can see. Away from the active regions, in the quiet Sun, we see small brightenings in OV and CIV possibly related to the phenomena Guenter Brueckner mentioned. These occur quite frequently. These 'microflares' are a step down in brightness from the Skylab observation of bright points. The observations at longer wavelengths thus seem to be very consistent, and we hope to continue using the data to help with the problems associated with the X-ray emission.

BRUECKNER: To continue briefly about bright points, I would like to emphasize that our rocket observations with arc-sec resolution show bright points without turbulence. The turbulent events are different from the jets. To really straighten things out, we need simultaneously high spectral, spatial, and temporal resolution.

PRIEST: May I ask in a summary way whether there is in fact good agreement between the SMM and HINOTORI observations? As far as the hard X-ray observations are concerned, are there any discrepancies that are unresolved?

DE JAGER: We do not have simultaneous observations, so I think that this question cannot be answered exactly. The HINOTORI data do look different to me, with more fluctuations compared with the rather stable sources observed by our instrument. We often see multiple structures, and in my opinion with high resolution we would see very fine structure. So the difference is perhaps a matter of resolution and signal-to-noise ratio. To add to Guenter Brueckner's list of items, I would like to have detailed knowledge of the magnetic field above the flaring zone. When we see the patches in brightness and call them footpoints, we may be inferring too much - all we see is patches. We are not in all cases sure that they are footpoints. The gradual phase is to me less interesting right now because it is the after-effect of the flare. For the primary acceleration the 'footpoints' are more important, and we would very much like to see them at the limb. Unfortunately HXIS really observed only one good flare at the limb, and that was the 1980 April 30 event of which I am always speaking. In that flare there is some elevated structure. During the patchy phase in this flare you see the loop being filled, in my opinion by convection. To use data of this kind properly, you absolutely must have knowledge of the magnetic field. It is essential to our understanding.

PRIEST: Are those observations consistent with the HINOTORI observations that show hard X-ray emission from well above the photosphere? Perhaps we could have a comment from the HINOTORI people.

TAKAKURA: I don't think we have a serious discrepancy between us and HXIS observationally. From the point of view of interpretations, I am not so sure! One item concerns the purely coronal hard X-ray sources. We have observed four or five of them, but I don't think HXIS observed any. This is because we had many X class flares, while HXIS observed only one X class flare. In that single case their observations of the extended source may be confused with background problems. Some discrepancy might be due to the relative contrast, for example in our 1981 May 13 event or even in some impulsive events. Double sources may appear if we set a high threshold level, say 89% of the peak brightness.

KIPLINGER: I have looked at very many events, and I am not sure that there is such a thing as a gradual flare. I think that if you look with high enough time resolution and sensitivity, you will almost always see an impulsive component. In our data it may occur in 95% of the flares. The most convincing case for a real gradual event seems to me to be this 1981 May 13 event that Tsuneta discussed. But in that case can we be sure that there was not some tiny impulsive phase that another instrument could have detected? The impulsive phase implies to me a rapid reconnection of magnetic field, while a slower reconnection in the gradual phase may provide the later optical and soft X-ray energy release.

I have done some statistical analysis that I think you may be interested in. I calculated the ratio of the amplitude of the impulsive component to the amplitude of the gradual component above the background levels of our HXRBS data. This is just the contrast of the impulsive component. It was a lot of work to put this together, so I hope you appreciate this! I find that the bigger events have less contrast for the impulsive emission, in the sense that bigger events tend to have stronger gradual emission. Now I have divided the SMM data into two halves, with the second half (1981) corresponding to the period of HINOTORI observations. The same tendency occurs in both halves, but the relative amount of impulsive emission seems definitely to have diminished for the HINOTORI period. This may have a bearing on why the classes of events observed may have differed, and I suggest that it may be related to the evolution of the Sun through the activity cycle.

UCHIDA: Thank you very much, Dr. Kiplinger, for pointing out an interesting aspect of the solar-cycle phase dependence. We now have to call Dr. Dulk for the third topic since the time is pressing. Dr. Dulk, please.

III. IS THE ELECTRON DISTRIBUTION THERMAL OR NON-THERMAL?

DULK: To discuss this basic question I would like to confine us to the impulsive phase. As a further restriction, I would also like to discuss only flares that occur in magnetic loops, rather than consider the possibility of unstable neutral sheets. People have talked about thermal and nonthermal models for a long time, and I would like to summarize the situation. Various people speak of a nonthermal model if (i) the hard X-ray spectrum has a power-law form, (ii) if there is evidence for foot-point brightening, (iii) if there are high-energy electrons present, say 100 keV, (iv) if one thinks there is a beam present, or (v) if there is evidence for two populations, hot and cold. I don't know if the terms 'thermal' and 'nonthermal' really mean anything. It would seem to be more reasonable physically if one talks about a trap plus precipitation in these flux-tube flares. The discussion should then center on the radiation process, whether thick-target or thin-target. Then one could ask whether either the trap emission or the footpoint emission might be negligible. One should try to determine whether the distribution function of the high-energy electrons is thermal or nonthermal; this is where the terms really mean something. If the distribution is Maxwellian, one can use the term thermal, and if it is non-Maxwellian, nonthermal. This question can be asked for both the trapped and precipitating populations if they occur. If there is a background population, is it important or not? If it is dense enough, one may have thick-target emission even from the top of the loop.

With this kind of separation of the problem, we can proceed to more substantial questions. Does the energy release produce an essentially thermal (Maxwellian) population? Does it produce a directed beam or an isotropic distribution? These are kinds of questions that we should really put to the theorists.

One aspect of the physics involved in this question that I find to be particularly interesting is the role of maser action for the microwave source. It could be involved in precipitating the electrons and in forming the soft x-ray plasma. These ideas are not very familiar to many of you and I would therefore like to review briefly and indicate how masers may operate in the flare context. Of course the term 'maser' is an acronym for 'microwave amplification by the stimulated emission of radiation.' The important point is that the stimulated emission competes against absorption - if it wins, the result is sometimes called 'negative absorption.' In general it requires a population inversion relative to the equilibrium distribution. That could occur in several forms. The most familiar is the laser case of an overpopulation of a higher-energy level in an atom. It is also quite familiar that the 'bump on the tail' distribution may lead to a high level of waves. A third example is one Ramaty has recently been working on, in the gamma-ray spectral region. Finally we come to the case that may be important in the present context, namely the existence of an anisotropy in the particle distribution function.

An anisotropic distribution function results in a 'vacuum' in parts of velocity space. Since Nature likes to fill in a vacuum, the anisotropy will drive waves that can accomplish this. The easiest way to view this is in the space of velocity components parallel and perpendicular to the magnetic field. Contours in this space are circles for an isotropic distribution. On the other hand a mono-energetic beam exists as a single point, and is a strongly non-equilibrium distribution that could give rise to waves. A loss-cone distribution results if electrons with small pitch angles are depleted, and we could have either one-sided or two-sided loss-cone distributions.

Could loss-cone distributions exist in solar flares? They surely will if we consider a loop geometry with some convergence towards the footpoints and an electron source at the top. The downward-moving electrons will mirror and go back up unless they have small pitch angles. These are lost to collisions, and the result is a classical loss-cone distribution. This provides a source of free energy to drive a maser. Waves that can feed on the perpendicular energy can interact resonantly and grow unstably. Electromagnetic waves at the electron cyclotron frequency are a particular example, and they can put the electrons into the loss cone. There are two important consequences. First, the waves cause electrons to diffuse into the loss cone very rapidly; the electrons precipitate and simulate the behavior of a distinctive acceleration in making thick-target hard X-ray emission. Second, approximately half of the energy of the accelerated electrons goes into the maser, roughly the same as into the precipitating electrons. We are familiar with the large magnitude of this impulsive-phase energy. This energy is available to heat a larger volume rapidly. I see this as an answer to Loren Acton's requirement for cross-field energy transport, and it could be an enormous effect.

Discussion !

UCHIDA: I feel that this is a rather restricted side of the problem which we wanted to discuss here.

DULK: But a very relevant one, I think, to my original five questions. The maser interaction can produce a thick-target emission. It obviates the need for a beam in the classical sense.

RAMATY: What you are talking about is merely an anisotropy. From an initial isotropic distribution, those electrons with small pitch angles precipitate. Where's the need for a maser?

DULK: Electrons are forced to fill the loss cone at the maximum possible rate. Maser empties the loop about as fast as it can be filled. If incidentally the field is asymmetrical, essentially all of the precipitation can be at one end, and I think that this may explain unpaired footpoints.

RAMATY: In a general astrophysical situation, a maser is thought to exist where you have high brightness temperatures. You seem to be using the maser just to produce isotropy. Could you say something about the competing mechanisms for that?

DULK: The electron-cyclotron maser is, in my view, the most satisfactory mechanism for microwave bursts from the Sun and stars, for Jupitar's decametric radiation and the Earth's kilometric radiation. But there are many situations where maser action just won't happen. For example, in a loop of constant cross-section there will be no mirroring and thus no maser effect. In another case, if the field converges only at one end, there would be at most one bounce and hence no maser.

KUNDU: With a maser you will necessarily get very high brightness temperatures. It is an observational fact that we do not see such high temperatures.

DULK: Not for ordinary microwave bursts, but 10^{13} K or more is implied for Slottje's kind of microwave spike bursts at 3 GHz or less.

UCHIDA: This subject is lively enough to reserve for a full workshop treatment this afternoon, but I would now like to invite also discussions that go beyond the so-called masers.

ZIRIN: I agree that this thermal/nonthermal question is an important one, since some energy-release mechanisms will be thermal while others will be inherently nonthermal in terms of the distribution function. But I fail to understand the intensity of the discussion. One has fairly direct X-ray and microwave spectral observations of flares and one can get a fairly good idea about the distributions. It seems to me very unlikely that you can get such tremendous electron energies in the early impulsive phase of a flare. I have never understood why people discuss such things as adiabatic compression, for example. I don't see how such mechanisms can produce 100-keV electrons in the impulsive phase. If that is so, I think that we can rule out a lot of the mechanisms that would just produce thermal distributions.

EMSLIE: Thermal/nonthermal implies to me a distinction between whether the electrons occupy one distribution in equilibrium or not, that is whether the X-radiation comes from a small fraction of the electrons or from the whole ambient distribution.

DULK: I think that depends upon whether you have one population or two - a hot population and a background population. I think you have to address the question about the distribution function separately for each population. But with two populations it isn't really correct to speak of a 'thermal' model. Regarding the precipitating population, certainly the target population is important in that case. But the basic question from theorists to observers must be: What is the primary particle distribution function?

EMSLIE: But a narrow definition based upon the observed X-ray spectrum may not be correct - a power law X-ray spectrum does not require a nonthermal distribution.

DULK: I agree.

EMSLIE: The distinction really lies in whether the energization affects the whole distribution or just a small fraction.

DULK: Right. And for the theorists the distribution function is the important thing.

ENOME: It is important to ask about the mechanism of establishment of equilibrium. Will it be collisions or some other mechanism? In any case, equilibrium can be established only in a certain finite time. If we observe variations on shorter time scales, of course they must be nonthermal in the sense that they are due to non-equilibrium phenomena. Even if the particle distribution happens to be Maxwellian at such a time, I don't think that it is appropriate to use the term 'thermal.'

UCHIDA: Thank you Dr. Enome. I really hate to stop the interesting discussion here, but we must hand the microphone to Dr. Doschek now, since otherwise we squeeze the time for the last topic too much. Dr. Doschek please.

IV. MASS TRANSPORT AND INTERACTION OF LOOPS

DOSCHEK: I think we shouldn't lose sight of the soft X-ray event, the so-called thermal event, which contains as much or more energy as the so-called nonthermal component. We have had a lot of discussion in the last ten years about chromospheric evaporation, the theory being that the soft X-ray source results somehow from the propagation of energetic particles in the chromosphere and the subsequent evaporation of material into loops. I would like to give a more balanced view of this phenomenon.

We do not have yet any direct evidence for this process; we know that material is evaporating - from the X-ray spectrometer observations - but we don't know for certain that it is moving into the loops seen in soft X-rays. The reason for this is that the X-ray spectrometers lack spatial resolution. The Skylab observations gave limited Doppler information. The highest resolution observations are probably from the NRL 'A' instrument on Skylab, but it is hard to interpret these images. Yesterday Brueckner showed the flare of June 15; the Fe XXV emission observed by HINOTORI would probably have come from the region confined by the Fe XXIV emission seen by Skylab. As Brueckner pointed out, there is evidence here for a shift, an erupting filament that appears as a spike in He II radiation, and in very long exposures as Fe XXIV. The first exposure in this flare was taken later than flare maximum. If you will recall the observations with the spectrometers, the shifted component was very weak. The fact that we saw only a slight indication of the evaporation doesn't matter, because we were apparently observing at the wrong time. The point is that this emission is coming not from the footpoints, but from the top of the loop! So we have information that some of the Doppler shifts we observe do not come from the footpoints.

These comments don't mean that evaporation is not happening, but I do think that we should be careful and not accept it as a fact until we get more observational evidence. If it is not happening, what are the alternatives? You have to have some means of getting high densities in loops. One method is to heat a high density transition-region loop, namely one for which the peak temperature is on the order of 10^5 K. Another method is by compression, as with currents and pinches as observed in the laboratory. I would like to ask some of the theorists if such a mechanism could also be operating in the Sun.

This discussion naturally leads to the question of preflare conditions in the flaring loops. When we do a numerical simulation, we must start with something, but it is difficult because we have little knowledge of the actual conditions. What are the temperatures and pressures in the preflare loops? Are they in equilibrium? On the latter point, we find from our numerical simulations of evaporation that if we want to get a high density at the top of the loop, we have to start with a pretty high density. This should be observable, and we should easily see it, but we don't. If on the other hand the temperature is low enough to make the loop unobservable in X-rays, say 10^5 K, then if the pressure is to be high enough, the loop must be too small. Thus if the initial state is a transition region loop, it must be out of equilibrium at the onset.

Finally, I would like to ask about the energy sources for post-flare loops. There is ample evidence for continued energy deposition well after the impulsive phase. Where does the energy come from? Do we have multiple loop interactions and emerging flux? In the June 15 flare, the image was observed to move by about 5 arc sec. Thus we have to ask if we can consider the loop theoretically to be isolated; can we trust the results from the one-dimensional simulations?

ZIRIN: Let me mention two observational facts that I think are important and hard to understand, relating to the thermal event. The first is the observation originally made by Bernard Lyot that what we now call flare loops (not post-flare loops, since we understand these loops to be the real flare) show a peak density at the top of the loop. This is a very difficult thing to imagine a mechanism for. It is an anti-hydrostatic condition, and therefore favors George Doschek's third suggestion of compression. I don't understand how else you can produce the peak density at the top of the loop.

The second crucial piece of evidence is the separation of the two strands, or the growing of the loops. I think that this can only be explained by continuous magnetic reconnection or continuous evolution of a magnetic process that is producing energy. Obviously more and more area is taking part in the energy-release process. I think these two pieces of evidence should be remembered all the time as we discuss these phenomena.

ACTON: It is not necessarily productive at this point to argue about the reality of chromospheric evaporation in some flares. We have many lines of evidence of the deposition of energy at the top of the chromosphere from the flare process and the reaction is an upwelling of the material. We now have high-resolution H-alpha observations during the impulsive phase that cover the line profile, and Dick Canfield says very authoritatively that the observed line profile means that the chromosphere has been made thin. The way you make it thin is by stripping the material away. We have Doppler measurements of upwelling material at coronal temperatures. It may not always happen, and we may not know its exact geometry, but it is clear that the evaporation process actually does happen in some flares.

DOSCHEK: I agree that evaporation is happening somewhere...

BRUECKNER: I would like to make a plea that we use a different word for this phenomenon. We learned in high school that 'evaporation' is a change of state of matter, and it does not change state in the Sun. I think that this is a stupid word.

Everybody agrees that material is moving up, but - and this is one of my standard comments - you can read about it in Waldmeier's book. He called it 'coronal condensation', an equally bad word, but he clearly described these huge coronal enhancements after flare. The material has to come from somewhere, and nobody can doubt that it must come from below. The point of confusion is the mechanism that moves the material. Is it electron heating, or is it a hydromagnetic process occurring deep in the atmosphere and squeezing material upwards? In my opinion, this is the important question.

DOSCHEK: There is no doubt that we see the material leaving the chromosphere, because we see the blue-shifted lines. The important thing that is unknown is the destination of the material. I think the kind of studies that Ester Antonucci reported are very important - the attempt to correlate the observed motion with the rise in emission measure, for example. The HINOTORI people should be encouraged to proceed in this direction, and we also with our P-78 data.

Regarding the name of the phenomenon, there is a very analogous process that occurs in laser fusion, where material is driven off a solid in the form of a very high-temperature gas. The laser-fusion community calls it 'ablation.' This is a reasonable word for it.

DE JAGER: Another simple word for heated gas moving up is 'convection.' I think that this is the physical situation. We see heating to twenty million degrees, and such heated gas will rise convectively. The convection velocities will be on the order of the sound speed, and you will have shock formation. It is a very fierce process.

ZIRIN: What about the non-equilibrium density at the loop top? This doesn't sound like something that could result from a convection process.

DE JAGER: The heating is so strong that it can overcome the density difference that you would have in the preflare situation.

DOSCHEK: Incidentally, the numerical simulations do not show any tendency towards density enhancements at the loop top. They usually show a fairly hydrostatic situation with fairly constant pressure.

UCHIDA: In a dynamical transient situation, you could have higher density at the loop top. For example in the calculation reported by ourselves yesterday, this results from the collision of masses driven by recoiled shocks. If this process continues to take place in successively higher loops with separations below the resolution of the X-ray observations, it may appear to remain at the top of the 'loop' as long as the process continues.

DUIJVEMAN: I would like to go back to our 1980 November 5 flare. The observations show very clearly the growth of the soft X-ray emission in the interconnecting loop; the growth is much more rapid near the foot points than at the top of the loop. This is just what you would expect from the footpoint energy deposition by a beam of energetic electrons producing the evaporation.

Accordingly we did some numerical simulations with Boris Somov. We calculated the beam flux with a thick-target model, and the resulting simulation showed a very large response at the footpoints but very little response at the loop top. This would confirm the point made by Doschek about the hydrostatic nature of the evolution. If you make a temperature and density analysis of the plasma at loop top and footpoint, you find that at all times the pressure is about twice as great at the top as compared with the footpoints. This seems odd if only the evaporation process occurs.

PRIEST: I would just like to make two comments. First, to draw your attention to the existence of two kinds of flares. In the one case, a single loop becomes filled, and you can simulate this with a hydrodynamic code with evaporation taking place. In the second case, the two-ribbon flare, you have a fully magnetohydrodynamic situation, and you can't treat this in terms of a single loop. The loops that you see later on in the flare are different from the ones at the beginning.

You mentioned the pinch effect. I think there are many ideas in laboratory plasma physics that we need to try and understand and take over into solar physics. But we must not take these ideas uncritically. We need to understand the differences between a laboratory plasma and a solar plasma. The pinch effect is one example of this. In the laboratory, you pass a current through a plasma and the magnetic lines of force are perpendicular to this current. You thus get a J x B force, which compresses the plasma. The current and magnetic field are perpendicular. By contrast, in the Sun most of the current is parallel to the field, so you don't get this pinch effect occurring.

UCHIDA: The same question came up in response to Loren Acton's suggestion of interacting loops. A reconnecting simple arcade would be a puzzle without such, isn't that right?

PRIEST: Yes, that's very similar. The current sheets are very special parts of the solar atmosphere, however.

PALLAVICINI: I would like to challenge your conclusion about the scaling law of temperature in the preflare loop. Your conclusion that a preflare loop at transition-region temperature must be too small depends upon this scaling law. There is ample evidence that this scaling law works quite well for active region loops, not only from the point of view of observations, but also from theoretical considerations based upon hydrostatic equilibrium. Why do you have to take a very low temperature for your preflare loop? There is a wide range of loop sizes present in an active region, from 100 km to 100000 km. You can take a very large loop with low pressure, and still have a temperature of three million degrees as a starting point. You could increase the density and pressure as much as you want, but if you keep heating for a long time you still have the same maximum temperature in the flare since the evaporation will continue. It will continue slower and slower because it must work against more material, but you can still increase the pressure in the loop. You can find a pressure at least an order of magnitude greater in the flaring loop than in the preflare state of, say, a few dynes/cm^2. This lets the preflare density be as low as 10^9 if desired.

UCHIDA: Well, I am afraid that the scheduled time is amply over. I am afraid I must close the session of this morning by thanking the discussion leaders and all of you who contributed from the floor, and would like to ask Professor De Jager to chair the closing session.

DE JAGER: As the chairman of the closing session, I invite Dr. Brown to give the summary talk.

CONFERENCE SUMMARY

BROWN: First I would like to comment that this has been a remarkable meeting in several ways. People are always asking if Eric Priest and I work together a lot; the answer of course is almost never. We live about 150 km apart. That distance seldom decreases except when we meet in some remote place. The other evening we discovered that the distance could approach one nanometer if we both rode on a rush-hour commuter train of Tokyo! Eric Priest and I will probably never be so close again.

Turning to matters of science, one always has to find a point of focus at a time like this. A few years ago, around the time of the Ann Arbor meeting, I wrote a couple of reviews (one with Dean Smith) in which we tried to identify reasons why the flare problem has not yet been solved. One rather profound reason we listed was that the plasma physics is complicated - this will always be true - and that the data were inadequate from the point of view of resolution and coordination. In general the theory is not at a predictive level.

Regarding the observations, we can no longer complain that there is a lack of detail or coordination. Compared with the situation ten years ago, we now have an amazing growth in the quality of data. At this meeting we have seen data from about seven different spacecraft. We have seen microwave movies, X-ray movies, stereo observations, direct solar neutrons detected within a few seconds from a flare, hard X-ray footpoints coincident with optical bright points, a hard X-ray source that just sits stationary in the corona for a long time. Many of these things are completely new. They are a huge advance since ten years ago. For a theorist the situation can only be described as 'bamboozling.' To be bamboozled is something like to have the morning-after feeling from a fine reception in which you loaded up simultaneously on Suntory whisky and Sake. The data leave me with something of this feeling.

Speaking as a cynical theorist, I should say that the data I have seen here and elsewhere recently have indeed settled a couple of questions in my mind. We have been looking into the question of signal-to-noise analysis of the footpoint brightenings observed by HXIS. Are the alleged footpoints real, and what is the contrast to the loop itself? We have tried to take into account instrumental effects and to do a deconvolution with the maximum-entropy method. We have been unable to make the footpoints disappear. The maximum-entropy algorithm will not let you draw a picture without the footpoint brightenings and without some hard X-ray emission from between the footpoints. A maximum-entropy image in principle has the least information consistent with the observations. We have also added Poisson statistics.

To sum up, I am convinced that we see bright points in the hard X-ray images. In agreement with Kees De Jager's comment, we don't know for sure whether these are the actual foot points yet.

The steady coronal hard X-ray source seen by HINOTORI impresses me as much as it impressed Alan Kiplinger. I think that this phenomenon deserves a lot of attention from the theorists.

Now to turn to the theoretical side, I do not think that the progress here has been so rapid. I tried to consider a thermal model quantitatively at the IAU meeting in 1973. It was rejected then on the grounds that it would cool much too quickly. The idea was in fact strongly ridiculed. Now, the argument is exactly the contrary - the thermal model is used to prolong the time scale of the emission.

What model do I believe in? I don't 'believe' in any particular model, but I do think that it is difficult to rule out thermal situations, and I will return to this later.

I would like to give some advice to the theorists: Please make the theories more predictive. It is all very well doing mathematical models, but they should result in calculations of diagnostic quantities that can be observed. The results should be given in a predictive fashion, rather than as an explanation by hindsight. A good example of the right way to do it is that of Ramaty's gamma-ray calculations, carried out long before the observations. It is not a prediction to be able to say, yes, that is what I would have expected in my model - this prediction by hindsight is all too common in solar meetings. At the same time that theorists work hard to make their theories close to the observational material, they should keep things as simple as possible.

How is it possible to get 100 keV electrons from a thermal plasma? Hal Zirin asked this question. I agree with him that the heating mechanism is unknown; on the other hand the acceleration mechanism in a nonthermal model is equally unknown. I don't know how to assess the relative impossibility of the two alternatives, but on thermodynamic grounds one might argue that the thermal distribution is better from the point of view of entropy maximization.

If I could resort to some very simple ideas that Don Melrose and I put together earlier, let us consider exceedingly simple physics. If you take a plasma with field B and density n and release the field energy into the plasma, then the temperature you should expect would correspond to about 60 keV, exactly what is observed, for B = 100 gauss and $n = 10^{19}\ cm^{-3}$. Losses would reduce the figure, of course. For emission measure, we get something on the order of $10^{44}\ cm^{-3}$ for a size scale of a couple of arc seconds. Again, this is in the right range. This simple first-law consideration gives you something independent of the details of the mechanism that is near what is observed.

While we may not be at the saturation level in terms of data one can still want a pinhole or a hard X-ray polarimeter - it seems to me that we have a lot of good observations to work with. Therefore, for the next decade, all budgeting should be directed entirely into theory ... And on a more serious note, I would like to thank you all once again for such an interesting meeting.

DE JAGER: Thank you very much, Dr. Brown, for such an interesting summary. I understand that there is a remark from the organizers.

KANE: Now, as one of the organizers, I would like to carry out the pleasant task of thanking those who made this meeting possible: The National Science Foundation and the Japan Society for the Promotion of Science are cordially thanked for their support which enabled us to have this meeting. The real push to get the meeting going in the initial phase was due to Dr. Oda, who unfortunately is not here today. The Organizing Committee would like to thank him warmly. Finally the participants are the most important part of a seminar of this sort, in which a very broad range of fields is covered. I would like to thank you for contributing and sticking with a fairly grueling schedule.

Special thanks must go to our hosts, Yutaka Uchida and Katsuo Tanaka, and also to the members of the local organizing committee, S. Enome, E. Hiei, K. Kai, T. Takakura, and Y. Tanaka, and many others. It has been an excellent organization and I would like to thank you for the excellent job.

DE JAGER: Speaking as a neutral European at this bilateral meeting, I would like to thank the organizers for the way in which they have done it. I thank the Japanese colleagues for their wonderful hospitality, and to the Organizing Committee for bringing us together. This has been a stimulating and useful meeting. We have been overcome by a storm - I should say a typhoon - of new observational details. There is indeed a very rapid progress in this field which may be hard to recognize in the midst of the present activity, but if we later look back at this time from the perspective of a few years, the progress made will be very clear. If we see how the theoretical research is coping with the flood of new observations, we will realize how important these few years of intense observations have been. I also hope that this marvelous momentum will continue for a number of years, and many results will be produced and many papers will be presented. On the other hand I also fear for an eventual decrease of the momentum, and I am convinced of the urgency of new observation in the next solar maximum less than ten years from now. I think it is not too early to think about missions for the next solar maximum and to consider what kinds of observations can be brought about.

This meeting has been quite stimulating and I particularly appreciated the intense discussions this morning. Well, many thanks for your contributions throughout the sessions, and for the efforts of the organizers, Yutaka Uchida, Sharad Kane, Katsuo Tanaka, and Hugh Hudson. I look forward to the printed Proceedings. I now close the meeting.

TABLE OF CONTENTS

ARTICLES

DISCUSSIONS